# THE REUVEN RAMATY HIGH-ENERGY SOLAR SPECTROSCOPIC IMAGER (RHESSI)

*Mission Description and Early Results*

*Edited by*

ROBERT P. LIN
*Space Sciences Laboratory and Physics Department, University of California, Berkeley*

BRIAN R. DENNIS
*NASA Goddard Space Flight Center, Laboratory for Astronomy and Solar Physics, Greenbelt*

and

ARNOLD O. BENZ
*Institute of Astronomy, ETH-Zentrum, Zürich*

Reprinted from *Solar Physics*, Volume 210, Nos. 1–2 (2002)

KLUWER ACADEMIC PUBLISHERS
DORDRECHT / BOSTON / LONDON

A C.I.P. Catalogue record for this book is available from the Library of Congress.

ISBN 1-4020-1107-5

Published by Kluwer Academic Publishers,
P.O. Box 17, 3300 AA Dordrecht, The Netherlands.

Sold and distributed in North, Central and South America
by Kluwer Academic Publishers,
101 Philip Drive, Norwell, MA 02061, U.S.A.

In all other countries, sold and distributed
by Kluwer Academic Publishers,
P.O. Box 322, 3300 AH Dordrecht, The Netherlands

*Printed on acid-free paper*



Printed in the Netherlands

# TABLE OF CONTENTS

Reuven Ramaty
1937 – 2001

Dr. Ramaty was a pioneer in the fields of solar physics, gamma-ray astronomy, nuclear astrophysics, and cosmic rays. He was a HESSI Co-I and one of the founding members of the HESSI team. His active involvement and enthusiastic support were critical for HESSI's selection by NASA as the sixth Small Explorer (SMEX) mission. Following the launch on February 5, 2002, HESSI was renamed in his honor. It became the Reuven Ramaty High-Energy Solar Spectroscopic Imager (RHESSI), the first space mission named after a NASA scientist.

# PREFACE

This volume on the Reuven Ramaty High-Energy Solar Spectroscopic Imager (RHESSI) continues a *Solar Physics* tradition of Special Issues dedicated to major solar space missions. RHESSI is a NASA Small Explorer mission launched on 5 February 2002 to explore the basic physics of particle acceleration and explosive energy release in solar flares. This science goal is accomplished by imaging spectroscopy of X-rays and gamma-rays with unprecedented spectral, spatial and temporal resolution from 3 keV to 17 MeV. Included in this volume are an overview of the science objectives, definitive descriptions of the instrument, spacecraft, operational procedures, and data reduction methods, and early scientific results.

The accompanying CD-ROM contains movies and document files, and color versions of many of the printed black and white figures.

In keeping with the efforts of the RHESSI team to widely and rapidly disseminate information about the mission, its data and results, we did several things to reduce publication time without sacrificing rigor. First, all manuscripts were handled electronically and the journal editors split the work load equally. Second, the technical instrument and system description papers were internally refereed by RHESSI team members and checked by the journal editors without calling upon external referees. Past experience has shown that descriptive papers of this sort are seldom questioned by referees. Third, the early-science papers were reviewed externally by independent, mostly anonymous referees. The authors prepared, in advance, draft title pages and abstracts that were submitted by late July 2002. Using potential referee lists prepared by the Guest Editors, this new step allowed the journal editors to select and contact referees before the final manuscripts were submitted. Most early-science manuscripts were received by the second week of August 2002. Fourth, the Guest Editors reviewed the papers before they were submitted, reducing the effort required by the external referees. Fifth, a deadline for initial reports of ten days after manuscript receipt was a condition accepted and largely adhered to by the referees. Sixth, a deadline of mid-September for acceptance resulted in most contributions being accepted between mid-August and mid-September. These steps compressed the time between submission and acceptance (usually including one revision and a second review) to less than five weeks. For this, we are indebted to the authors for fast response to referee comments and especially to a dedicated group of selfless and constructive referees.

September 2002

R. Lin, B. Dennis, and A. Benz (Guest Editors)
J. Harvey, O. Engvold, and Z. Švestka (Editors)

# THE REUVEN RAMATY HIGH-ENERGY SOLAR SPECTROSCOPIC IMAGER (RHESSI)

R. P. LIN[1,2], B. R. DENNIS[3], G. J. HURFORD[1], D. M. SMITH[1], A. ZEHNDER[4], P. R. HARVEY[1], D. W. CURTIS[1], D. PANKOW[1], P. TURIN[1], M. BESTER[1], A. CSILLAGHY[1,9], M. LEWIS[1], N. MADDEN[5], H. F. VAN BEEK[6], M. APPLEBY[7], T. RAUDORF[8], J. McTIERNAN[1], R. RAMATY[3,10], E. SCHMAHL[3,15], R. SCHWARTZ[3,16], S. KRUCKER[1], R. ABIAD[1], T. QUINN[1], P. BERG[1], M. HASHII[1], R. STERLING[1], R. JACKSON[1], R. PRATT[1], R. D. CAMPBELL[1,10], D. MALONE[1], D. LANDIS[1], C. P. BARRINGTON-LEIGH[1], S. SLASSI-SENNOU[1], C. CORK[5], D. CLARK[3], D. AMATO[3], L. ORWIG[3], R. BOYLE[3], I. S. BANKS[3], K. SHIREY[3], A. K. TOLBERT[3,16], D. ZARRO[3,26], F. SNOW[3], K. THOMSEN[4], R. HENNECK[4], A. MCHEDLISHVILI[4], P. MING[4], M. FIVIAN[1,11], JOHN JORDAN[12], RICHARD WANNER[12], JERRY CRUBB[12], J. PREBLE[12,13], M. MATRANGA[12,14], A. BENZ[17], H. HUDSON[1], R. C. CANFIELD[18], G. D. HOLMAN[3], C. CRANNELL[3], T. KOSUGI[19], A. G. EMSLIE[20], N. VILMER[21], J. C. BROWN[22], C. JOHNS-KRULL[23], M. ASCHWANDEN[24], T. METCALF[24] and A. CONWAY[25]

[1]*Space Sciences Laboratory, University of California Berkeley, Berkeley, CA 94720-7450, U.S.A.*
[2]*Physics Department, University of California, Berkeley CA 94720–7300, U.S.A.*
[3]*NASA/Goddard Space Flight Center, Greenbelt, MD 20771, U.S.A.*
[4]*Paul Scherrer Institut (PSI), CH-5232 Villigen PSI, Switzerland*
[5]*Lawrence Berkeley National Laboratory, Berkeley, CA 94720, U.S.A.*
[6]*H. F. van Beek Consultancy (VBC), 3971 LB Driebergen, The Netherlands*
[7]*Tecomet, Woburn, MA 01801, now at Mikro Systems Inc., Charlottesville, VA, U.S.A.*
[8]*Ortec, Oak Ridge, TN 37831–0895*
[9]*University of Applied Sciences, CH-5210 Windisch, Switzerland*
[10]*Deceased*
[11]*Previously PSI*
[12]*Spectrum Astro, Gilbert, AZ 85233, U.S.A.*
[13]*SpaceWorks Inc., Carefree, AZ 85377–2014, U.S.A.*
[14]*The Charles Stark Draper Laboratory Inc., Cambridge, MA 02139–3563, U.S.A.*
[15]*Universtity of Maryland, College Park, MD 20742, U.S.A.*
[16]*Science Systems & Applications Inc. (SSAI), Lanham, MD 20771, U.S.A.*
[17]*ETHZ, Zürich CH-8092, Switzerland*
[18]*Montana State University, Bozeman, MT 59717, U.S.A.*
[19]*ISAS, Sagamihara City, Kanagawa Prefecture, Japan*
[20]*University of Alabama in Huntsville, Huntsville, AL 35899, U.S.A.*
[21]*Observatoire de Paris-Meudon, France*
[22]*University of Glasgow, Glasgow G128QW, Scotland, U.K.*
[23]*Rice University, Houston, TX 77005, U.S.A.*
[24]*Lockheed-Martin, Palo Alto, CA 94304, U.S.A.*
[25]*The Open University, Milton Keynes MK7 6AA, U.K.*
[26]*L-3 Communications, New York, NY 10016, U.S.A.*

(Received 16 September 2002; accepted 22 September 2002)

*Solar Physics* **210:** 3–32, 2002.


CD ROM

**Abstract.** RHESSI is the sixth in the NASA line of Small Explorer (SMEX) missions and the first managed in the Principal Investigator mode, where the PI is responsible for all aspects of the mission except the launch vehicle. RHESSI is designed to investigate particle acceleration and energy release in solar flares, through imaging and spectroscopy of hard X-ray/gamma-ray continua emitted by energetic electrons, and of gamma-ray lines produced by energetic ions. The single instrument consists of an imager, made up of nine bi-grid rotating modulation collimators (RMCs), in front of a spectrometer with nine cryogenically-cooled germanium detectors (GeDs), one behind each RMC. It provides the first high-resolution hard X-ray imaging spectroscopy, the first high-resolution gamma-ray line spectroscopy, and the first imaging above 100 keV including the first imaging of gamma-ray lines. The spatial resolution is as fine as $\sim 2.3$ arc sec with a full-Sun ($\gtrsim 1°$) field of view, and the spectral resolution is $\sim 1-10$ keV FWHM over the energy range from soft X-rays (3 keV) to gamma-rays (17 MeV). An automated shutter system allows a wide dynamic range ($> 10^7$) of flare intensities to be handled without instrument saturation. Data for every photon is stored in a solid-state memory and telemetered to the ground, thus allowing for versatile data analysis keyed to specific science objectives. The spin-stabilized ($\sim$ 15 rpm) spacecraft is Sun-pointing to within $\sim 0.2°$ and operates autonomously. RHESSI was launched on 5 February 2002, into a nearly circular, 38° inclination, 600-km altitude orbit and began observations a week later. The mission is operated from Berkeley using a dedicated 11-m antenna for telemetry reception and command uplinks. All data and analysis software are made freely and immediately available to the scientific community.

## 1. Introduction

The processes of particle acceleration and impulsive energy release occur in active cosmic plasmas at diverse sites throughout the universe, ranging from planetary magnetospheres to active galactic nuclei. The understanding of these processes is a major goal of space physics and astrophysics, but we are just beginning to perceive the relevant basic physics. The Sun constitutes an unparalleled laboratory for investigating these processes. Its proximity allows measurements over the entire electromagnetic spectrum to be made on physically relevant scales. At the same time, the system as a whole can be studied, and escaping energetic particles and plasma can be sampled directly. Further, the complexity of solar magnetic fields and the solar atmosphere leads to a broad range of acceleration phenomena, mirroring the rich diversity of processes occurring on cosmic scales.

The primary scientific objective of RHESSI is to understand particle acceleration and explosive energy release in the magnetized plasmas at the Sun. The Sun is the most powerful particle accelerator in the solar system, accelerating ions up to tens of GeV and electrons to hundreds of MeV in solar flares and in fast coronal mass ejections (CMEs). Solar flares are the most powerful explosions, releasing up to $10^{32}-10^{33}$ ergs in $10^2-10^3$ s. The flare-accelerated $\sim$ 10–100 keV electrons (and sometimes $\gtrsim$ 1 MeV nucleon ions) appear to contain a significant fraction, $\sim$ 10–50%, of this energy, indicating that the particle acceleration and energy release processes are intimately linked. How the Sun releases this energy, presumably stored in the magnetic fields of the corona, and how it rapidly accelerates electrons and ions with such high efficiency, and to such high energies, is presently unknown.

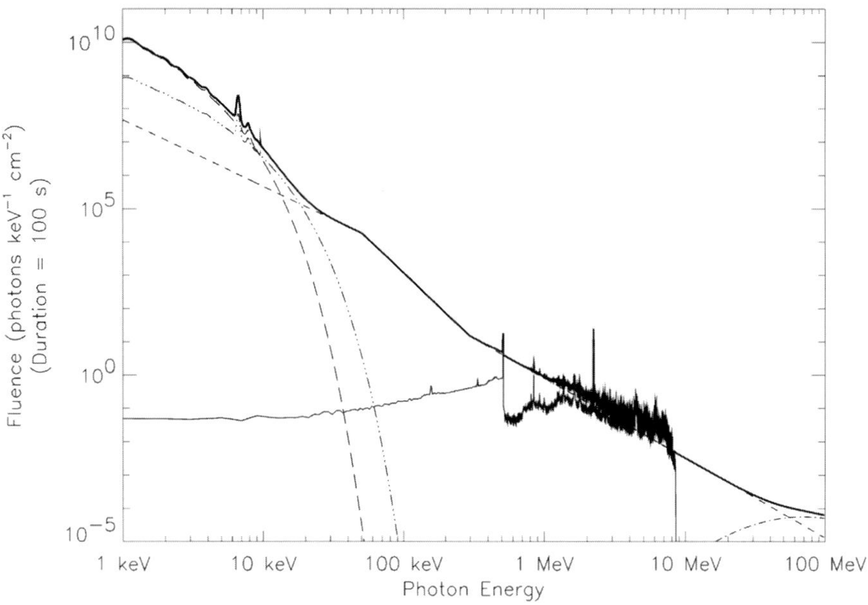


*Figure 1.* Composite X-ray/gamma-ray spectrum from 1 keV to 100 MeV for a large flare. At energies up to $\sim$ 10–30 keV, emission from hot($\sim$ $10^7$ K) and 'superhot' ($\sim$ $3\times10^7$ K) thermal flare plasmas (the *two curves at the left*) dominate. Bremsstrahlung emission from energetic electrons produces the X-ray/gamma-ray continuum (*straight lines*) up to tens of MeV. Broad and narrow gamma-ray lines from nuclear interactions of energetic ions sometimes dominate the spectrum between $\sim$ 1 to 7 MeV. Above a few tens of MeV the photons produced by the decay of pions (*curve at the right*) dominates. RHESSI observations cover almost four orders of magnitude in energy (3 keV to 17 MeV).

High-energy emissions are the most direct signature of the acceleration of electrons, protons and heavier ions in solar flares (Figure 1). Accelerated electrons colliding with the ambient solar atmosphere produce bremsstrahlung hard X-ray and gamma-ray continuum emission, while nuclear collisions of energetic ions result in a complex spectrum of narrow and broad gamma-ray lines. Hot (million-degrees) thermal flare plasmas also emit bremsstrahlung X-rays. RHESSI is designed to provide high resolution imaging and spectroscopy of all these emissions, from soft X-rays (3 keV) to gamma-rays (17 MeV).

These emissions are accompanied by longer wavelength emissions, and sometimes by escaping energetic particles. Their observation by the fleet of spacecraft (Solar and Heliospheric Observatory (SOHO), WIND, Advance Composition Explorer (ACE), *Ulysses*, Transition Region And Coronal Explorer (TRACE), Geostationary Operational Environmental Satellite Project (GOES), Solar Anomalous and Magnetospheric Particle Explorer (SAMPEX), etc.) already in place, and by ground-based instruments, provides the crucial information on the context in which the high energy processes occur.

RHESSI is the first Small Explorer mission carried out in the 'Principal Investigator-mode', where the PI and team are responsible for all aspects of the mission except providing the launch vehicle. This includes developing, integrating, and testing the instrument and spacecraft; providing launch support, telemetry downlink and command uplink, mission and science operations, data processing and distribution; analyzing and archiving the data, and disseminating the results.

The RHESSI mission was selected by NASA in October 1997, with a very aggressive development schedule aimed at a launch in July 2000, near the predicted peak of the $\sim 11$-year solar activity cycle. During environmental testing of the integrated spacecraft at the Jet Propulsion Laboratory in March 2000, however, a malfunction in the shake table subjected RHESSI to a vibration level of $> 25$ G rather than the requested 2 G, resulting in extensive damage to both the instrument and spacecraft. At about this time the failure of a Mars orbiting mission and then a Mars lander led NASA to institute a policy of additional external 'Red Team' reviews. Even with all the extra effort involved in these reviews and the implementation of their recommendations, the instrument and spacecraft were repaired, re-integrated, and re-tested in time for a launch at the end of 2000.

A problem was then discovered on the three-stage solid-propellant Pegasus-XL launch vehicle, and fixing that problem delayed the launch to March 2001, and again to June 2001. In April 2001, the RHESSI spacecraft was integrated with the Pegasus-XL at Orbital Science Corporation's facilities in the Western Test Range, and in early June it was carried underneath an L-1011 aircraft to the Eastern Test Range at Kennedy Space Center. One week before the planned RHESSI launch, NASA attempted a test flight of the prototype X-43 aerospace plane. The modified Pegasus first stage used for launch failed and the X-43 had to be destroyed. This led to a series of further delays while the cause of the failure was investigated. Finally, on 5 February 2002, the RHESSI satellite was launched by the Pegasus-XL, following its release from the L-1011 at $\sim 40\,000$ feet altitude over the Atlantic Ocean. A near perfect, 38° inclination, 600 km altitude circular orbit was achieved.

The mission was renamed from HESSI to RHESSI after launch to honor Dr Reuven Ramaty, a distinguished theoretical high-energy astrophysicist and solar physicist working at Goddard Space Flight Center (GSFC). Dr Ramaty developed much of the theoretical framework for solar gamma-ray line spectroscopy, and he was a Co-I and a strong advocate of RHESSI. He passed away in March 2001, eight months after the planned launch date, but 11 months prior to the actual launch. He is the first NASA scientist to have a space mission named after him.

## 2. Scientific Objectives and Design Considerations

### 2.1. Acceleration of Electrons

Bursts of bremsstrahlung hard X-rays ($\gtrsim 20$ keV), emitted by accelerated electrons colliding with the ambient solar atmosphere, are the most common signature of the impulsive phase of a solar flare (Figure 2). Provided the electron energy $E_e$ is much

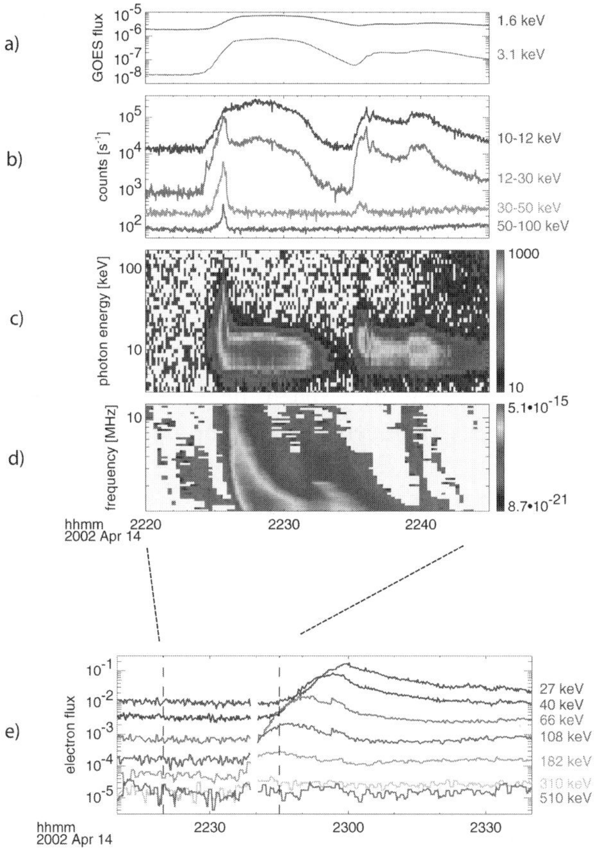


*Figure 2.* [See CD-ROM for color version]. Time profile of X-ray fluxes from GOES (a) and RHESSI (b); counts are multiplied by factors of 8, 50, and 500, for the 30–50 keV, 12–30 keV, and 10–12 keV channels, respectively). The next two panels show spectrograms of (c) the RHESSI counts in 20 logarithmic energy channels, and of (d) 1–14 MHz radio emission detected by the WIND spacecraft (Bougeret *et al.*, 1995). The solar flare with hard X-ray peak at $\sim$22:26 UT is accompanied by a type III solar radio burst, produced by energetic electrons escaping the Sun. The (e) bottom panel (longer time interval) shows the energetic electrons arriving at WIND at 1 AU (Lin *et al.*, 1995), with the faster electrons arriving first, consistent with an impulsive injection of the electrons from the Sun at the time of the X-ray peak.

greater than the average thermal energy, $kT$, of the ambient gas, essentially all of the electron energy will be lost to Coulomb collisions, with only a tiny fraction ($\sim 10^{-5}$) lost to bremsstrahlung. For this non-thermal situation, the hard X-ray fluxes observed in many flares indicate that the energy in accelerated $> 20$ keV electrons must be comparable to the total flare radiative and mechanical output (Lin and Hudson, 1976). Thus, the acceleration of electrons to tens of keV may be the most direct consequence of the basic flare-energy release process.

High-resolution hard X-ray imaging spectroscopy is the key to understanding the electron acceleration and energy release processes in solar flares. High-spectral-resolution measurements of the solar flare hard X-ray spectrum can be directly inverted to obtain the detailed spectrum of the parent X-ray-producing electrons (Johns and Lin, 1992). RHESSI is designed to provide imaging spectroscopy – the photon spectrum is obtained in each spatial element (Figure 3) as a function of time. In principle, this spectrum provides detailed information on $N(E, \mathbf{r}, t)$, the X-ray producing electron number density, as a function of energy $(E)$, position $(\mathbf{r})$, and time $(t)$. With information from context observations on the ambient density, temperature, magnetic field strength and topology, the electron loss processes can be directly evaluated to determine whether the X-ray emission is thermal or non thermal. By using a spatially dependent continuity equation, including loss processes, the spatially and temporally resolved accelerated electron source distribution, $F(E, \mathbf{r}, t)$, can be inferred. Then, detailed quantitative models of the acceleration, energy release, and energy propagation processes can be constructed and tested.

RHESSI is designed to provide spatial (Figure 4(b)) and temporal resolution commensurate with the spatial and temporal scales for the accelerated electrons to lose their energy in the lower corona and upper chromosphere (ambient densities below $\sim 10^{12}$ cm$^{-3}$). To resolve the very steep thermal spectra and determine the lower energy limit of the non-thermal spectrum (critical to determining the energy content in fast electrons), $\sim 1$ keV spectral resolution (Figure 4(a)) is needed. The energy range should extend low enough so the thermal–nonthermal transition can be determined, and as high as practical, but certainly up to relativistic energies where a different acceleration process may be operating. Finally, high sensitivity is required to detect microflares (which may be important for coronal heating), and very wide dynamic range is required to make measurements in the largest flares without saturation.

### 2.2. ACCELERATION OF IONS

Near the Sun, nuclear collisions of accelerated ions with the ambient solar atmosphere result in a rich spectrum (Figure 1) of gamma-ray lines (Ramaty and Murphy, 1987; Chupp, 1990, Share and Murphy, 1995). Energetic protons and alpha-particles colliding with carbon and heavier nuclei produce narrow de-excitation lines (widths of $\sim$ few keV to $\sim 100$ keV), while energetic heavy nuclei colliding with ambient hydrogen and helium produce much broader lines (widths of hundreds of keV to an MeV). Neutron capture on hydrogen and positron annihilation produce delayed narrow lines, at 2.223 MeV and 0.511 MeV, respectively.

The bulk of the gamma-ray line emission is produced by ions with energies of 10–100 MeV nucleon that contain only a small fraction of the energy in the $\gtrsim 20$ keV electrons. However, systematic study of SMM gamma-ray line flares (Share and Murphy, 1995) showed that the 1.634 MeV $^{20}$Ne line is enhanced relative to other lines. Because the cross section for $^{20}$Ne has an unusually low energy threshold ($\sim 2.5$ MeV), this effect appears to be due to large fluxes of low-energy

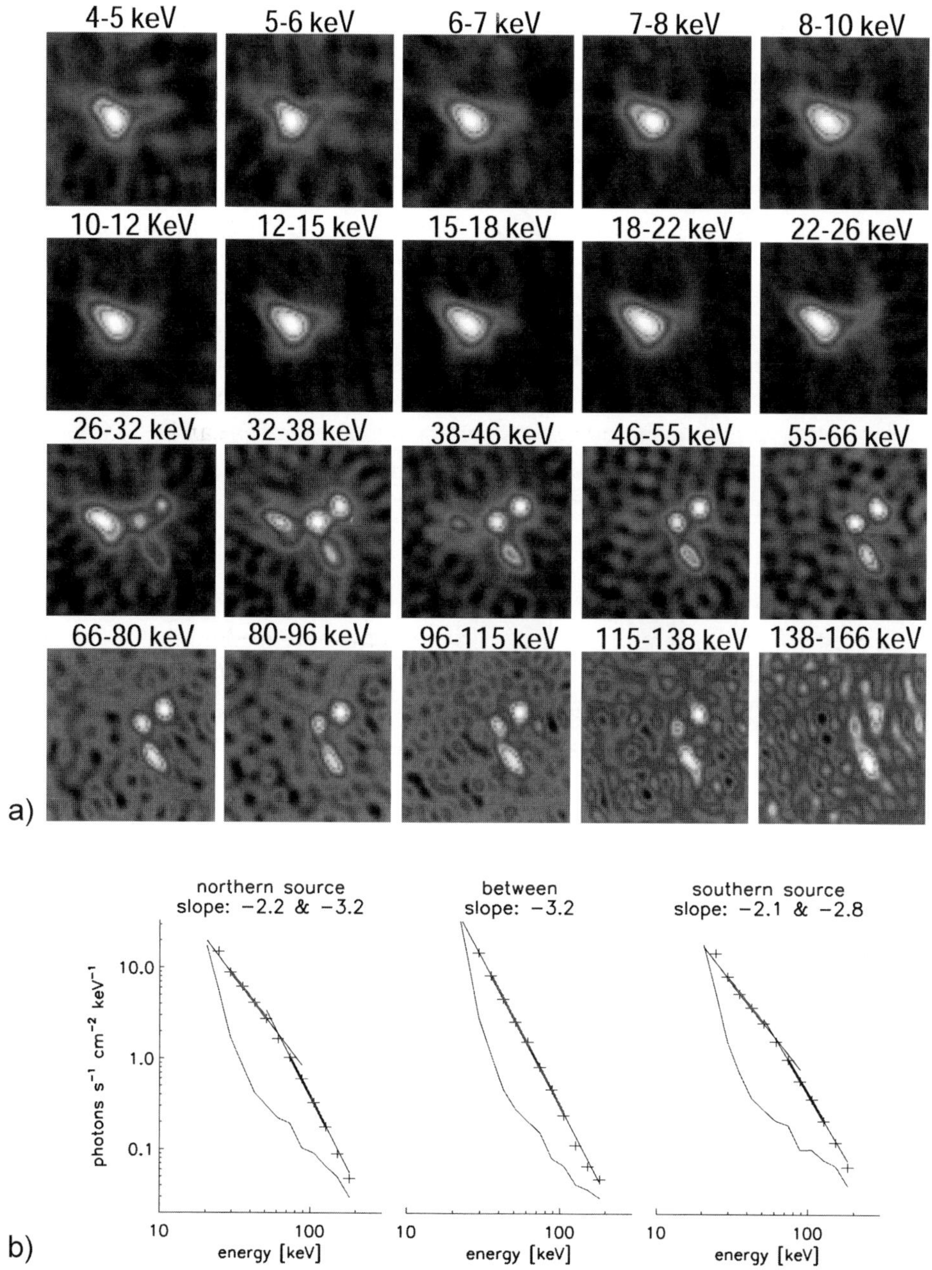


*Figure 3.* [See CD-ROM for color version]. (a) Imaging of the July 23, 2002, X4.8 solar flare in 20 energy bands, from 1 keV wide bins at 4 keV, to 28 keV wide bins (for enough counts to image) at 138 keV, illustrating the changes in sources as a function of energy, from a single dominant elongated source at energies below ~30 keV to three sources above ~40 keV. The images are 64 arc sec on a side; the lower left corner is just at the southeast limb of the Sun. (b) The energy spectra of the three dominant sources at energies above ~40 keV, showing that the spectra are similar for the north and south sources but quite different for the source in between. The *dashed lines* indicate background.

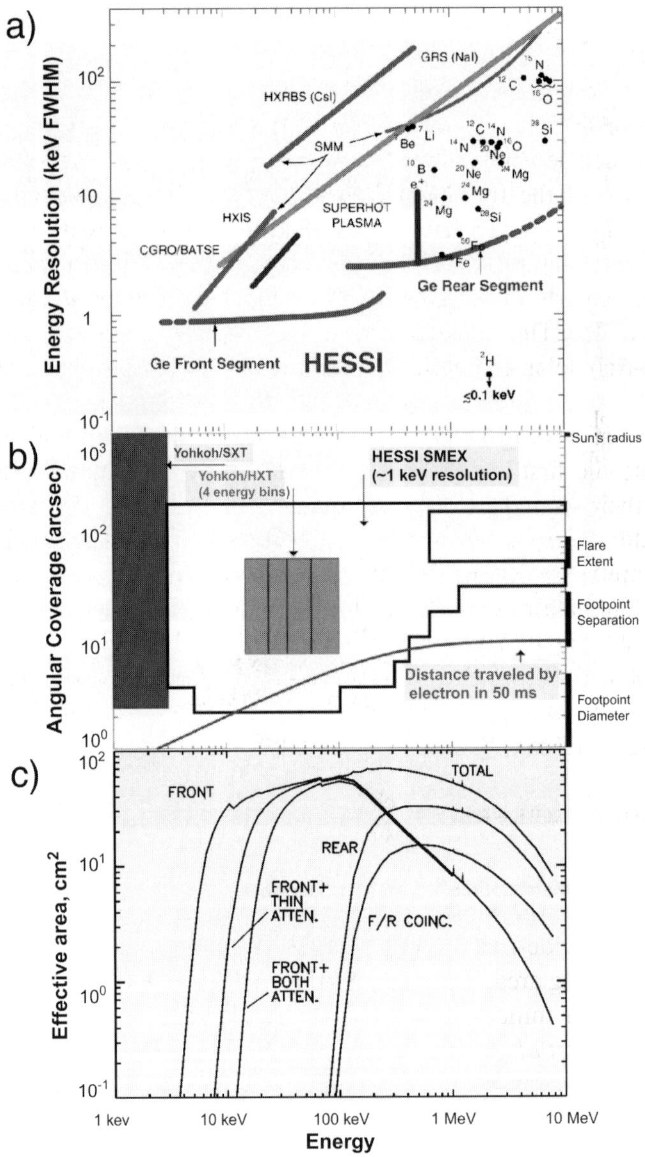


*Figure 4.* [See CD-ROM for color version]. (a) RHESSI's Full Width at Half Maximum (FWHM) energy resolution (*lower curves*), compared with previous instruments (*upper lines*). The *dots* are the expected FWHM widths of the predicted gamma-ray lines, and the *vertical line* the range of widths (temperature and density dependent) for the positron annihilation line. The *short diagonal line* shows the resolution needed to resolve the steep 'superhot thermal' emission. (b) RHESSI's angular resolution versus energy, compared to *Yohkoh* Soft X-ray Telescope (SXT) (*left*) and Hard X-ray Telescope (HXT) (*center*). RHESSI has $\sim 1-10$ keV energy resolution over this energy range and can image up to 17 MeV, while HXT had four broad energy channels from 15 to 100 keV. (c) RHESSI's effective (photopeak) area as a function of energy for solar X-rays/gamma-rays. The *three curves to the left* show the effect on the front segment response (summed over all nine detectors) of no attenuators, thin attenuator, or both thick and thin attenuator inserted over the detectors. The *curves to the right* show the effective area for the rear segment, and for photons that leave energy in both the front and rear segment of the detectors (F/R coincidence).

ions with total energy content often comparable to that of the accelerated electrons (Ramaty *et al.*, 1995; Emslie *et al.*, 1997).

RHESSI is designed to probe ion acceleration with the first high-resolution spectroscopy of solar flare gamma-ray lines (Figure 4(a)), giving the first detailed line shapes, which depend on the angular distribution of the interacting accelerated ions. The shape of the 0.511 MeV positron annihilation line gives information about the density and temperature of the ambient medium since the positrons slow down before annihilating. The high resolution enables closely spaced lines to be separated, particularly important around 1 MeV, where several lines are produced by accelerated $^{3}$He. Their detection would test whether gamma-ray flares and impulsive, $^{3}$He-rich solar energetic particle events have a common origin (Ramaty *et al.*, 1993).

RHESSI is also designed to image solar flare gamma-ray lines (Figure 4(b)), thus providing the first information on the locations of energetic heavy ions and protons and their secondary neutrons and positrons, to compare with the location of the energetic electrons. The instrument's high spectral resolution allows imaging in narrow gamma-ray lines, such as the 2.223 MeV neutron capture line, where line counts dominate over the background. The continuum above $\sim 1$ MeV (especially 4–7 MeV) is often dominated by the broad lines from accelerated heavy ($Z > 2$) ions, and can be imaged to locate them. In a very large gamma-ray line flare with good statistics, the $> 2.5$ MeV protons might be located by imaging the $^{20}$Ne de-excitation line.

## 2.3. Non-solar Science

Although designed as a solar instrument, RHESSI's lack of shielding around the detectors (to minimize weight) makes it an effective high-spectral-resolution ($\sim$ keV FWHM), wide-field-of-view hard X-ray/gamma-ray all-sky monitor, with $\sim 150$ cm$^2$ collecting area. The spacecraft's rotation produces many detector/detector occultations per minute, and two brief Earth occultation/deoccultations occur per orbit (Harmon *et al.*, 1992; Zhang *et al.*, 1993), allowing localization of transients (black-hole X-ray novae, Be/neutron star binary outbursts, etc.) and steady sources. RHESSI is able to resolve cyclotron absorption features from bright transients such as A0535+26, search for line features in gamma-ray bursts, and study the Galactic positron annihilation and $^{26}$Al decay lines from nucleosynthesis in supernovae and massive stellar systems on large angular scales. RHESSI can image the Crab Nebula once a year (when it approaches within 1.6° of the Sun), with unprecedented spatial (2.3 arc sec) and energy ($\sim 1$ keV) resolution in hard X-rays. RHESSI also provides high spectral and temporal resolution measurements of terrestrial X-ray/gamma-ray continuum emissions from electron precipitation, of terrestrial gamma-ray line emission from the impact of cosmic rays and solar energetic particles, and of gamma-ray bursts associated with lightning (Fishman *et al.*, 1994).

## 3. Instrument

The RHESSI scientific objectives are achieved with high-resolution imaging spectroscopy observations from soft X-rays to gamma-rays, utilizing a single instrument consisting of an Imaging System, a Spectrometer, and the Instrument Data Processing Unit (IDPU) containing the instrument electronics. An instrument schematic is shown in Figure 5 and the specifications are given in Table Ia. The Imaging System is made up of nine Rotating Modulation Collimators (RMCs), each consisting of a pair of widely separated grids mounted on a rotating spacecraft. Pointing information is provided by the Solar Aspect System (SAS) and redundant Roll Angle Systems (RASs). The Spectrometer has nine segmented germanium detectors (GeDs), one behind each RMC, to detect photons from 3 keV to 17 MeV. The GeDs are cooled to $\lesssim$75 K by a space-qualified long-life mechanical cryocooler, to achieve the highest spectral resolution (Figure 4(b)) of any presently available gamma-ray detector. As the spacecraft rotates, the RMCs convert the spatial information from the source into temporal modulation of the photon counting rates of the GeDs. The instrument electronics amplify, shape, and digitize the GeD signals, provide low-voltage power and GeD high voltage, format the data, and interface to the spacecraft electronics.

The energy and arrival time of every photon, together with aspect data, are recorded in the spacecraft's on-board 4-Gbyte solid-state memory (sized to hold all the data from the largest flare) and automatically telemetered within 48 hours. With these data, the X-ray/gamma-ray images can be reconstructed on the ground (see Figure 3). The instrument's $\sim 1°$ field of view is much wider than the $\sim 0.5°$ solar diameter, so all flares are detected, and pointing can be automated.

### 3.1. Imaging System

A detailed description of the RHESSI imaging technique is given in Hurford *et al.* (2002). At hard X-ray and gamma-ray energies, unlike soft X-rays, EUV, and longer wavelength emissions, focusing optics are not feasible. The only viable method of obtaining arcsecond-class images in hard X-rays and gamma-rays within the Small Explorer constraints is with Fourier-transform imaging, similar to that used in the pioneering *Hinotori* rotating modulation collimator (Makishima *et al.*, 1977) and *Yohkoh* Hard X-ray Telescope (HXT) (Kosugi *et al.*, 1991). RHESSI uses nine collimators, each made up of a pair of widely separated grids. Each grid is a planar array of equally-spaced, X-ray-opaque slats separated by transparent slits (Figure 6). The slits of each pair of grids are parallel to each other and their pitches (p) are identical, so that the transmission through the grid pair depends on the direction of the incident X-rays. For slits and slats of equal width, the transmission is modulated from zero to 50% and back to zero for a change in source angle to collimator axis (orthogonal to the slits) of $p/L$ where $L$ is the separation between grids. The angular resolution is then defined as $p/(2L)$.

TABLE I
RHESSI characteristics.

| Ia. Instrument characteristics: | |
|---|---|
| Energy range | 3 keV to 17 MeV |
| Energy resolution (FWHM) | $\lesssim$ 1 keV at 3 keV, increasing to $\sim$ 5 keV at 5 MeV |
| Angular resolution | 2.3 arc sec to 100 keV, 7 arc sec to 400 keV, 36 arc sec to 15 MeV |
| Temporal resolution | 2 s for detailed image, tens of ms for basic image |
| Field of view | full Sun ($\sim$ 1°) |
| Effective area (photopeak) | $\sim 10^{-3}$ cm$^2$ at 3 keV, $\sim$ 32 cm$^2$ at 10 keV (with attenuators out), $\sim$ 60 cm$^2$ at 100 keV, $\sim$ 15 cm$^2$ at 5 MeV |
| Detectors | 9 germanium detectors (7.1-cm dia. $\times$ 8.5 cm), cooled to $<$ 75 K with Stirling-cycle mechanical cooler |
| Imager | 9 pairs of grids, with pitches from 34 microns to 2.75 mm, and 1.55-m grid separation |
| Aspect system | Solar Aspect System: Sun center to $<$ 1 arc sec<br>Roll Angle System: roll to $\sim$ 1 arc min |
| Number of flares expected | $\sim$ 1000 imaged to $>$ 100 keV<br>$\sim$ tens with spectroscopy to $\sim$ 10 MeV |
| **Ib. Spacecraft characteristics:** | |
| Mass | Total 291.1 kg, instrument 130.8 kg |
| Power | Total 220.4 W, instrument 142.3 W |
| Size | 1.18 m diameter, 2.06 m height, 5.74 m tip-to-tip with solar panels deployed |
| Telemetry | 4 Mbps, downlink, 2 kbps command uplink |
| On-board storage | 4 Gbyte solid state memory |
| Attitude | 15 rpm spin rate, pointing to 0.2° of Sun center |
| **Ic. Mission characteristics:** | |
| Launch date | 5 February 2002 |
| Launch vehicle | Pegasus XL |
| Orbit | 38° inclination, 600 km altitude apogee, 586 km perigee |
| Nominal mission lifetime | 2 years, 3rd year highly desirable |

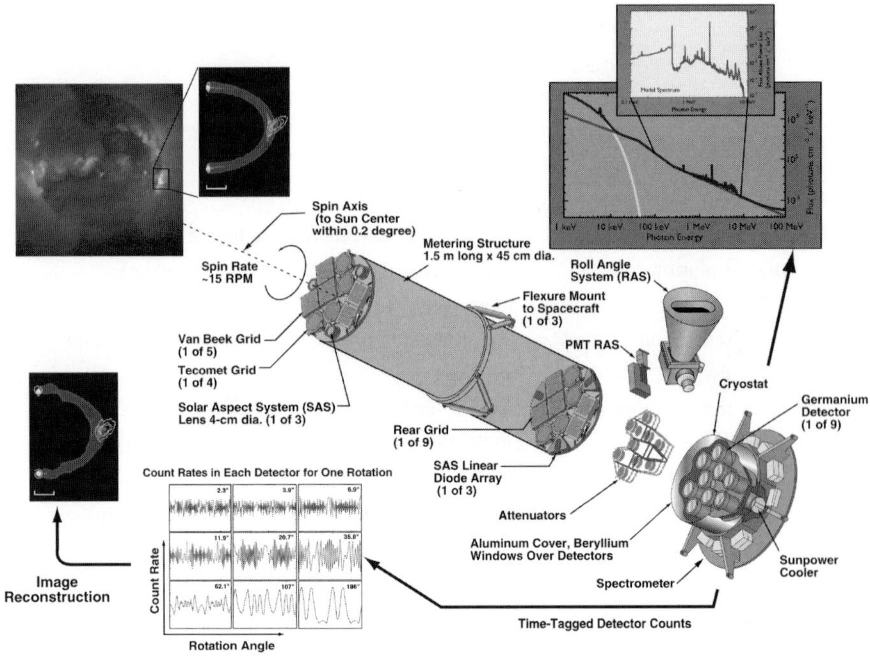


*Figure 5.* [See CD-ROM for color version]. Schematic of the RHESSI instrument illustrating the imaging spectroscopy. X-rays and gamma-rays from the Sun (*upper left*) pass through the slits of the front and rear grids of each of the nine grid pairs to reach the germanium detector. As the spacecraft rotates the detector count rates are temporally modulated (*lower left*). These modulations can be analyzed to reconstruct the image. The germanium detectors are cryogenically cooled to provide high spectral resolution capable of resolving narrow gamma-ray lines and steep solar continuum spectra (*upper right*). The attenuators are inserted automatically when the count rate approaches saturation. The SAS, RAS and PMTRAS provide solar pointing and roll aspect information.

For RHESSI, the transmission of the source photons through the grids is modulated by mounting the instrument on a rotating spacecraft. The X-ray/gamma-ray detector behind the collimator records the arrival time and energy of individual photons, allowing the modulated counting rate to be determined as a function of rotation angle. Note that the detector does not need to have any spatial resolution and hence can be optimized for high sensitivity and energy resolution.

For a parallel incident beam, the modulated waveform generated by a smoothly rotating spacecraft has a distinctive quasi-triangular shape locally. The amplitude is proportional to the intensity of the beam, and the phase and frequency depend on the direction of incidence. For complex sources, and over small rotation angles, the amplitude and phase of the waveform provide a direct measurement of a single Fourier component of the angular distribution of the source (e.g., Prince *et al.*, 1988). Different Fourier components are measured at different rotation angles and with grids of different pitches. For RHESSI, the separation between grids in each RMC is $L = 1.55$ m and the grid pitches range from $p = 34$ $\mu$m to 2.75 mm in steps

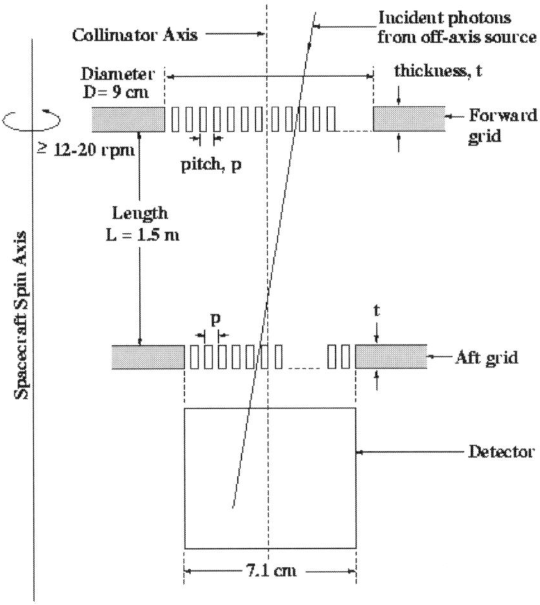


*Figure 6.* Schematic showing the parameters that define the imaging capability.

of $\sqrt{3}$. This provides angular resolutions spaced logarithmically from 2.3 arc sec to $\gtrsim 3$ arc min, allowing sources to be imaged over a wide range of angular scales. Diffuse sources larger than 3 arc min are not imaged but full spectroscopic information is still obtained. Multiple smaller sources are imaged regardless of their separation.

In a half rotation (2 s), the nine RMCs measure amplitudes and phases of $\sim 1100$ Fourier components for a typical source location, compared to 32 Fourier components for the *Yohkoh* HXT, so that far more complex flare images can be resolved. Although one half rotation is required to measure a full set of Fourier components, the measurement of each component takes only a single modulation cycle, which can be as short as 1.3 ms for the finest grids. Thus, when count rates are sufficiently high, crude images (from about ten Fourier components) can be obtained, in principle, on timescales of tens of milliseconds.

Changing the separation ($L$) between grids or displacing the grids parallel to the slits has little effect on imaging performance. A relative displacement perpendicular to the slits affects the phase but not the amplitude of modulation. Any such displacement will be accurately monitored by the SAS, and can be fully compensated for in the image reconstruction process. The critical alignment requirement is associated with the rotation or twist of one grid with respect to the other about the line of sight to the source. A relative twist of $p/D$ ($D$ = diameter of grid) reduces the modulated amplitude almost to zero. Thus, the grid pairs must be well aligned in twist throughout the mission. For the finest grids (2.3 arc sec resolution) a 1-arc min alignment is needed. Thus, HESSI can achieve arc-sec-quality images with an instrument having only arc-min alignment requirements. To minimize twist, the grids are aligned and mounted precisely on grid trays which are attached to opposite ends of a graphite-epoxy support tube.

#### 3.1.1. *Grids*

The main challenges in fabricating the RHESSI grids were: (i) the extremely fine slit and slat widths (20 and 14 microns for 34 micron pitch) required for the finest grid, (ii) the high (50 : 1) aspect ratio of the slit width to grid thickness for maximum absorption consistent with a $\sim 1^\circ$ field of view, and (iii) the fine tolerance on the relative pitches of the two grids in each pair ($< 1$ part in 20 000 for the finest grid pair). Tecomet, Inc., fabricated the four finest grid pairs (plus a third spare grid) using a foil stacking method in which thin sheets of metal were photo-etched and precision stacked with epoxy bonding to produce a solid structure (Figure 7(a), left). Tungsten was used for all of the grids except for the finest; grid pair 1 (Figure 7(b)) was made of molybdenum (sufficiently thin tungsten sheets were not available then), resulting in a maximum energy for modulation of $\sim 100$ keV, rather than $\sim 200$ keV had it been tungsten. The five coarsest grid pairs were made by van Beek Consultancy in The Netherlands using tungsten blades packed together side by side with spacers in between (Figure 7(a), right) to give the required, pitch, slit width, and grid thickness. The thickest grids, No. 6 (Figure 7(c)) and 9 (1.85 and 3 cm thick) were designed to modulate gamma-rays up to 17 MeV (Figure 4(b)).

All the grids were fully characterized at GSFC both optically and using X-rays. Optical images were taken at high magnification of the front and rear of each grid using a customized facility. Each grid was mounted on an XY table with laser readout, and measurements of the pitch, phase, and orientation of the slits in each grid were obtained with micron positional accuracy. A separate X-ray characterization facility determined the average X-ray transmission of each flight grid as a function of photon energy and angle from grid normal. In addition, the variation of this transmission across the full area of each grid was measured with $\sim 1$-cm resolution.

All grids were also fully flight qualified at GSFC by vibration testing and thermal cycling to GEVS (General Environmental Verification Specification) standards. The grids were then transported to Paul Scherrer Institut (PSI) in Switzerland where they were mounted and aligned on the front and rear end trays. Precision coordinate measurement machines were used to establish and verify the correct alignment of each grid pair, so that in the final assembly the slits of the front and rear grids of each pair would be parallel to within the very strict tolerances required for modulation of the incident photon flux.

After assembly of the trays onto the imager tube at PSI, an end-to-end check on the grid alignments was carried out. Since an X-ray beam parallel to $\lesssim 1$ arc sec could not be obtained, a radioactive $^{109}$Cd source was placed behind the spare grid for each pair, to provide a series of diverging fan beams of 22 keV X-rays. By moving this source/grid while maintaining an appropriate distance from the grid

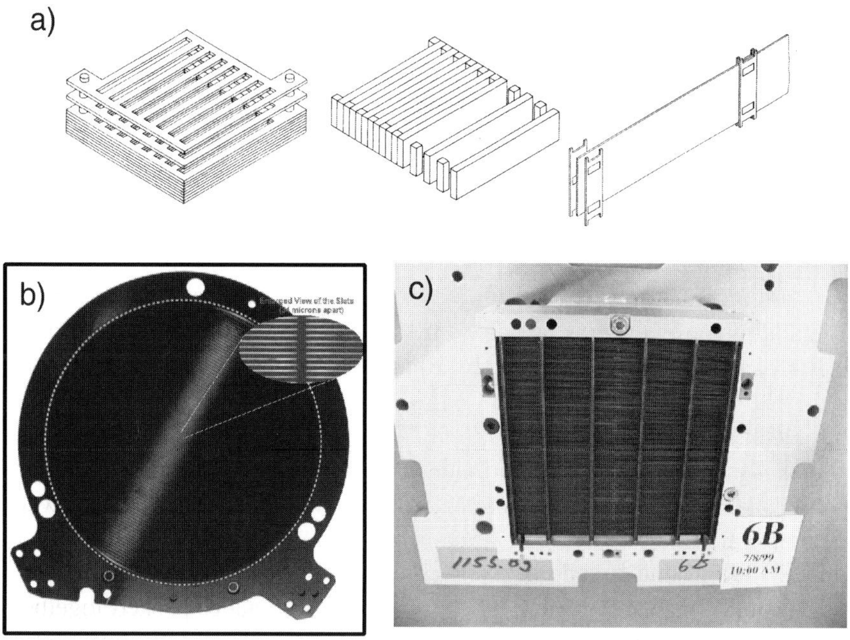


*Figure 7.* [See CD-ROM for color version]. (a) Schematic of the two grid fabrication processes: left, stacking etched foils and epoxying them to obtain the required thickness; middle and right, packing vertical blades with spacers in between. (b) Photograph of a grid No. 1 showing the slits and bridges in the insert. (c) Photograph of flight grid No. 6.

pair, a modulation in transmission would be detected providing that the two grids of that pair were correctly aligned. This test was carried out on all but the two coarsest grid pairs – their alignments could be checked visually.

For long-term monitoring of the grid alignment through all testing and environmental qualification prior to launch, an optical twist monitoring system (TMS) was used repeatedly. This system relied on optical photodiodes mounted behind pinholes in the rims of the rear grids. The light from these photodiodes passed through annuli mounted in the front grids in such a way that the diffracted beams from any one grid pair converged onto a Charge Couple Device (CCD) camera mounted at the appropriate distance in front of the telescope assembly. The positions of the converging beams as determined from the images provided an accurate measure of any change in the relative twist of the front and rear grids of a each pair. In this way, the correct alignment of even the finest grid pair could be verified up to launch. No significant change in alignment was ever measured except after the vibration accident at Jet Propulsion Laboratory (JPL), when the front grid No. 6 was hit and moved on its mount. The complete imaging telescope system was shipped back to PSI, a new mount installed for the front grid 6, and all grid alignments were rechecked and verified, including repeating the TMS check.

#### 3.1.2. *Aspect Systems*

The ability to make arc-sec-quality images and accurately co-align the images with other solar observations, depends on knowing the orientation of the collimators with respect to the direction to the Sun at all times. This essential aspect information is obtained with the Solar Aspect System (SAS), which provides pitch and yaw measurements relative to the solar limbs to arcsecond accuracy on time scales of tens of ms, and two redundant Roll Angle Systems – a CCD-based version (CCD RAS) and a photomultiplier-based version (PMT RAS) – that each provide the roll angle to arc min accuracy several times per rotation with respect to the fixed stars. Full details of these systems are given by Zehnder *et al.* (2002) for the SAS and CCD RAS and by Hurford and Curtis (2002) for the PMTRAS.

#### 3.1.3. *Solar Aspect System (SAS)*

The Solar Aspect System (SAS) provides (1) high-resolution, high-bandwidth aspect information for image reconstruction, (2) monitoring of the relative twist of the two grid trays, and (3) full-Sun white-light images, on occasion, for co-alignment with ground-based images. It consists of three identical lens-filter assemblies mounted on the forward grid tray to form full-Sun images on three $2048 \times 13$-$\mu$m linear diode arrays mounted on the rear grid tray. Simultaneous exposures of three chords of the focused solar images are made every 10 ms by each of the arrays. A digital threshold algorithm is used to select four (or more, commandable) pixels that span each solar limb for inclusion in the telemetry. These digitized pixel outputs allow six precise locations of the solar limb to be obtained on the ground by interpolation, thus providing knowledge of Sun center in pitch and yaw to 1.5 arc sec per readout ($3\,\sigma$).

#### 3.1.4. *Roll Angle System (RAS)*

For image reconstruction on the ground (no impact on real-time spacecraft operations), knowledge of relative roll is required at all times to 3 arc min ($3\,\sigma$). Since all sources of torque on the spacecraft are weak, the required information can be obtained with a star scanner that samples the roll orientation at least once per rotation. Interpolation between measurements allows the roll orientation to be determined at intermediate times with the required accuracy. The CCD RAS consists of a CCD array and electronics behind an $f/1.0$, 50-mm lens. A sunshade limits the field of view so that a 30° band is swept out across the sky at 15° to orthogonal to the spin axis. As the spacecraft rotates, each detected star generates a brief spike in the output of one or two pixels, whose timing defines the roll orientation. For +2 magnitude stars, the detection signal-to-noise is 15 : 1. Allowing for Earth occultation and the recovery time from anticipated earthshine saturation, at least one (and typically seven) such star(s) will be detected each rotation throughout the mission. Measurements of only one star, averaged over a minute, allow the roll angle to be determined to 2.7 arc min ($3\,\sigma$).

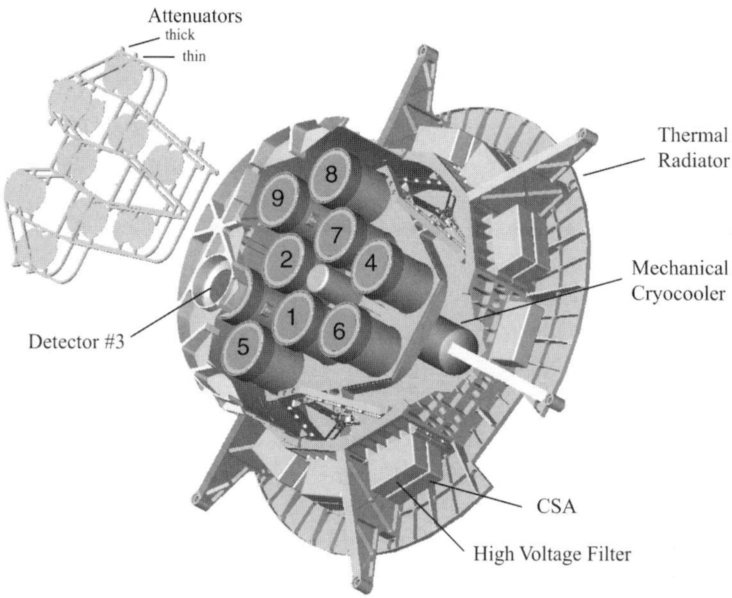


*Figure 8.* [See CD-ROM for color version]. A cutaway of the Spectrometer, showing the location of the germanium detectors under each grid (by number). The Sunpower Stirling-cycle mechanical cooler is below the cold plate holding the detectors. The thermal radiator faces anti-sunward to reject the heat of the cryocooler. The attenuators are automatically moved in when the counting rate exceeds thresholds (commandable from the ground).

The Aspect Data Processor (ADP) receives the data from the SAS and RAS, performs on-line processing. The PMTRAS, consisting of a photomultiplier (PMT) behind a slit to scan for bright stars, was included for redundancy.

## 3.2. SPECTROMETER

Figure 8 shows a cutaway of the Spectrometer, described in detail in Smith *et al.* (2002). The RHESSI germanium detector (GeD) design provides energy coverage from 3 keV to 17 MeV with a single mechanically robust detector. The largest, readily available, hyperpure (n-type) coaxial germanium material (7.1-cm diam × 8.5-cm long) was used. The inner electrode is segmented into two contacts that collect charge from two electrically independent detector segments, to provide the equivalent of a ∼1.5-cm thick planar GeD in front of a ∼7 cm thick coaxial GeD. The top and curved outer surfaces are implanted with a thin (0.3 micron) boron layer to provide a surface transparent down to 3 keV X-rays. With advanced Field Effect Transistors (FETs) and state-of-the-art electronics, the front segments achieve a 3-keV energy threshold.

The front segment thickness is chosen to stop photons up to ∼250 keV (Figure 4(c)), where photoelectric absorption dominates, while minimizing the active volume for background. Front-incident photons that Compton-scatter, and background photons or particles entering from the rear, are rejected by anticoincidence with the rear segment; a passive, graded-Z (Pb, Cu, Sn) ring around the front segment absorbs hard X-rays incident from the side, to provide the low background of a phoswich-type scintillation detector. Photons with energies from $\sim 250$ keV to 17 MeV, including all nuclear gamma-ray lines, stop primarily in the thick rear segment alone, with smaller fractions stopping in the front segment, depositing energy in both the front and rear segments, or in two or more GeDs. All these modes contribute to the total photopeak efficiency (Figure 4(c)).

The intense hard X-ray fluxes that usually accompany large gamma-ray line flares are absorbed by the front segment, so the rear segment will always count at moderate rates. This is essential for gamma-ray line measurements where optimal spectral resolution and low dead time are desired. To accommodate the large dynamic range ($\sim 10^7$) in soft X-ray flux from microflares to very large flares, two sets of aluminum disk attenuators (also called shutters) can be moved in front of the GeDs to absorb low energy (see Figure 4(c)) photons. These attenuators are inserted automatically by heating temperature-sensitive Shape Memory Alloy (SMA) actuators when the count rate increases above pre-set values. The attenuation stays in for a fixed duration currently ($\sim 5$ min), but programmable and then are removed. The cycle repeats as long as the rates stay high.

The GeDs in their modules are mounted on an aluminum cold plate suspended on fiberglass straps to reduce thermal conduction. This is surrounded by multilayer radiation shields and enclosed in an evacuated cryostat. The cryostat's curved sidewall is ribbed thin-wall aluminum near the GeDs to provide $\sim 20$ keV threshold for non-solar X-ray/gamma-rays incident on the side.

The GeDs are cooled on-orbit by a single Sunpower Inc. M77B single-stage, counterbalanced, Stirling-cycle cryocooler. The interface is a flexible aluminum/sapphire cold finger designed to minimize coupling of microphonics into the GeDs and to allow a structured cooldown to avoid condensation of contaminants onto the GeDs. The cryocooler uses a gas bearing/flexure system to prevent contact between moving parts, and a moving-magnet motor to eliminate flexible motor leads. This cooler provides up to 4 W of cooling at 77 K, at an input power of 100 W. The cryocooler is mounted in an external cavity between the radiator and cryostat to minimize losses in the heat rejection path. The anti-Sun facing heat rejection radiator (76 cm diameter) is thermally coupled to the cryostat housing to provide a large thermal mass. The equilibrium radiator temperature is in the range of $-15$ °C to $-30$ °C.

Radiation damage of the GeDs is of concern since RHESSI passes through the South Atlantic Anomaly (SAA) on about 5 orbits a day. Bombardment by high energy SAA protons produces traps in the GeDs which degrade the spectral resolution with time. The Spectrometer is designed with the capability to anneal the GeDs, by heating them up to $\sim 100$ °C, to recover the resolution. This is not expected to be needed within the nominal two-year mission lifetime.

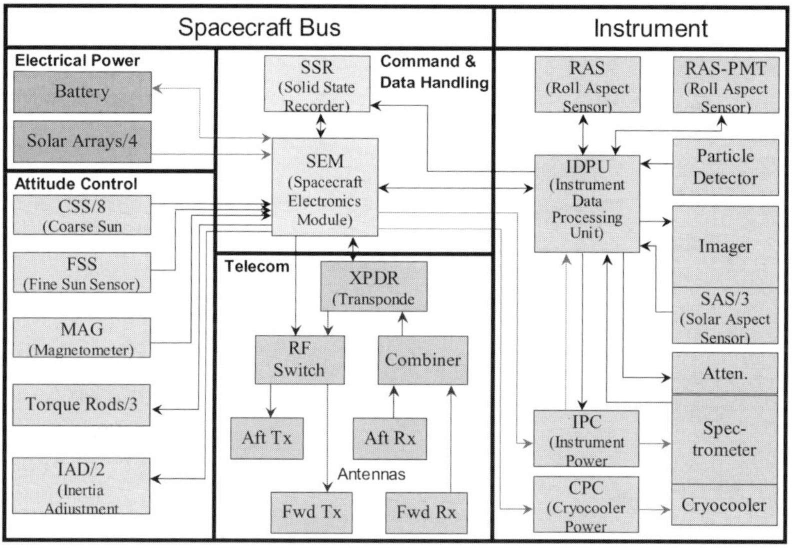


*Figure 9.* Block diagram for the RHESSI spacecraft bus (*left*) and the instrument (*right*).

## 3.3. Instrument Electronics

Each GeD is biased at between 4 and 5 kV by a separate adjustable high-voltage power supply. Photons interacting in a GeD generate charge pulses, which are collected and amplified by a transistor-reset Charge Sensitive Amplifier (CSA) with an advanced 4-terminal type FET to provide the best resolution and high-count rate performance. The CSAs and the high voltage filters are mounted directly to the Spectrometer (Figure 8).

The Instrument Data Processing Unit, described by Curtis *et al.* (2002) (see Figures 9 and 10) contains all the remaining Spectrometer electronics. The signals from the CSAs are shaped, amplified, processed, and digitized by nine identical Detector Interface Boards (DIBs), one for each GeD. The DIBs feature (1) quasi-trapezoidal shaping to compensate for ballistic deficit effects of the charge collection in these large GeDs, (2) dual fast/slow signal processing chains for pulse-pileup rejection, and (3) ultrahigh rate counting in broad energy bands with the fast chain with live time measurements every 0.5 ms to preserve the imaging capability.

The GeD front segment energy range is $\sim 3$ keV up to $\sim 2.7$ MeV in 8192 channels with $\sim 0.33$ keV ch$^{-1}$. The rear segments cover from $\sim 20$ keV to $\sim 2.7$ MeV and have an additional low-gain slow amplifier to cover from $\sim 2.7$ MeV up to $\sim 17$ MeV (2.7 keV ch$^{-1}$). For each detected photon, 14 bits of energy information and the time of arrival to 1 microsecond are encoded together with detector identification and live time into a 24-bit event word. Normally every photon is stored in the spacecraft mass memory and then telemetered to the ground.

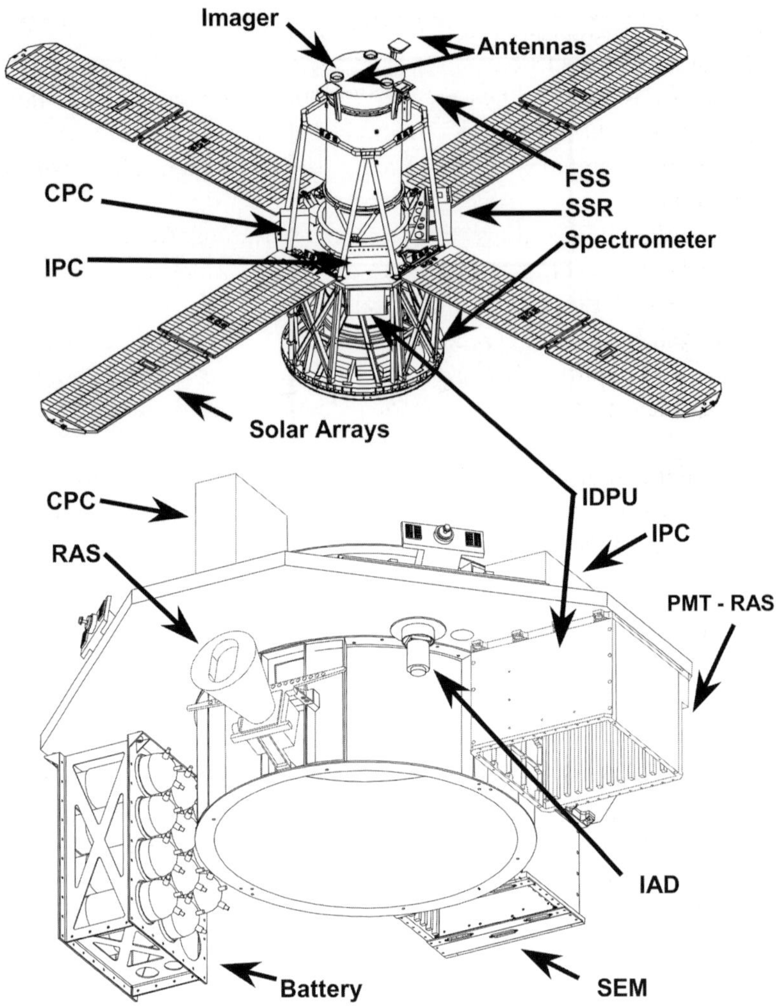


*Figure 10.* [See CD-ROM for color version]. Schematic showing the location of instrument and spacecraft components on the RHESSI spacecraft. The acronyms in the top view are: Fine Sun Sensor (FSS), Solid State Recorder (SSR), Cryocooler Power Converter (CPC), Instrument Power Converter (IPC), Instrument Data Processing Unit (IDPU); in the bottom view, Roll Angle System (RAS), Photomultiplier Roll Angle System (PMT RAS), Inertial Adjustment Device (IAD), Spacecraft Electronics Module (SEM).

A 300-micron thick, 1 cm$^2$ area, passivated-ion-implanted (PIP) silicon detector with a CSA/amp/discriminator electronics, similar to those flown on the Wind 3D Plasma and Energetic Particle instrument (Lin *et al.*, 1995), monitors energetic particle fluxes. The IDPU low voltage power converter provides the 100-V bias supply for the particle detector. Two discriminator levels provide measurements of energetic ($\gtrsim$ 1 MeV) electron and ($\gtrsim$ 10 MeV proton fluxes.

The Controller card collects and formats data from the DIBs and the particle detector, and passes the formatted data to the spacecraft over a high speed (>20 Mbps) bus. A microprocessor on the Controller card controls the IDPU, including the cooler and shutters, and interfaces to the spacecraft over a low speed serial interface for receipt of ground commands and exchange of housekeeping and status information.

A space-qualified power converter (CPC) consisting of two amplifiers was developed for the cryocooler. A pulse-width-modulated amplifier amplifies a temperature dependent IDPU-provided sine wave at 59.6 Hz to drive up to 100 watts peak into the voice-coil-like cryocooler load of $\sim 1.2\ \Omega$. The second amplifier, driven from an IDPU-provided phase-shifted temperature-dependent sine wave at the same frequency, powers the cryocooler counterbalancer to minimize microphonics.

An Instrument Power Controller (IPC) contains the low and high voltage converters for the instrument. The CPC and IPC are housed in physically separate boxes from the rest of the IDPU (Figure 10).

## 4. Spacecraft

### 4.1. Structure and Mechanisms

The RHESSI spacecraft bus (Figure 10) was designed and manufactured by Spectrum Astro, Inc. of Gilbert, AZ. The primary structure supports all spacecraft components during the launch environment and maintains the relative alignment of the Imager and the Spectrometer. Most of the spacecraft components are located on an octagonal aluminum honeycomb equipment deck that is mounted to the spacecraft aft ring. The Imager tube is supported around its center of gravity by three flexure mounts to a machined aluminum imager support ring mounted on the forward side of the equipment deck. The Spectrometer is attached to the spacecraft aft ring with its large thermal radiator flush with the launch vehicle separation plane for an unobstructed field of view.

The solar array consists of four identical wings, each with two panels connected by a hinge, located symmetrically around the equipment deck. A metal tip-mass at the end of each wing increases the deployed spin-axis moment of inertia for spinning stability. To align the spin axis closer to the imager boresight on orbit, the spacecraft spin balance can be fine-tuned with two controllable Inertia Adjustment Devices (IADs) – motorized linear drive screws that move two of the solar array wings. The mid-wing and root hinges of each solar array wing use steel tape-measure material to provide both the deployment force and a rigid latch in the deployed state. Shape Memory Alloy (SMA) actuated release devices preload the array panels against snubbers in the stowed configuration. These actuators stretch to snap the titanium tiedown bolts to release the wings. The solar cells are exposed on the outboard panel so that some power is generated even before the panels are deployed.

### 4.2. Attitude Control

The Attitude Control Subsystem (ACS) enables RHESSI to follow the Sun over time autonomously with a $3\,\sigma$ pointing accuracy of 0.14° (8.4 arc min). The primary attitude sensor is an Adcole Inc. fine Sun sensor (FSS) with a $\pm 32°$ field of view and 0.005° resolution, that is mounted to the front of the imager tube. The pointing error measured by the FSS, together with local magnetic field measurements made by the spacecraft magnetometer, are inputs to the ACS control algorithms in the flight software. This runs on the RAD6000 flight processor in the Command and Data Handling (C&DH) subsystem to drive three orthogonally-mounted Ithaco Inc. 60 Ampère-meter$^2$ Electromagnetic Torque Rods to maintain the spacecraft attitude. Finally, a set of eight coarse sun sensor cells (two mounted on each solar array wing) allow the ACS subsystem to acquire the Sun from any initial attitude after separation from the launch vehicle.

The ACS flight software provides an Acquisition mode which damps rates after launch vehicle separation, a Precession mode which orients the spin axis toward the sun from any starting attitude, a Spin Control mode which adjusts the vehicle spin rate to a commandable value, and a Normal mode which is used during most mission operations to keep the spin axis pointed at the Sun. The ACS also has an Idle mode which does not actively control the vehicle attitude. Because RHESSI is designed to be a stable spinner, the idle mode provides a safe mode that is entered in the event of an anomaly. The ACS flight software is auto-coded using MatrixX software and integrated with the remainder of the flight software.

### 4.3. Command and Data Handling

The Spacecraft Electronics Module (SEM, Figure 10) houses the Charge Control Board (CCB), the Power Control Board (PCB), and the Auxiliary Driver Board (ADB) for the Electrical Power Subsystem (EPS); and Communications Interface Board (CIB), the Payload and Attitude Control Interface (PACI) board, and the flight computer (CPU) board of the Command and Data Handling (C&DH) subsystem. A separate Solid State Recorder (SSR, Figure 10), built by SEAKR Engineering, provides 4 gigabytes of solid-state memory for science data storage.

The Instrument Data Processing Unit provides formatted telemetry packets of science data directly to the SSR recording high-speed parallel interface. Science data are played back from the SSR for downlink via a high-speed parallel interface with the CIB, the command and telemetry interface for the SEM to the RF transponder. The CIB is powered from the essential bus and is operational at all times. It provides command decoding capability for critical functions including the reset and power control of the flight computer, control of the telemetry transmitter, and adjustment of the battery charge control parameters. This hardware command decoding capability of the CIB provides an operational backup for faults which result in the shutdown of CPU or software.

The Payload and Attitude Control Interface (PACI) board is responsible for telemetry encoding and data acquisition. It digitally encodes analog voltage, current and temperature data, and formats telemetry frames for downlink and on-board storage. It provides serial communications interfaces for control and monitoring of the SSR and the IDPU. The PACI board is powered by the essential power bus and is always producing hardware state of health telemetry packets; whenever the transmitter is powered on these packets are transmitted to the ground. This feature along with the CIB hardware command decoding, allows problems to be diagnosed and fixed from the ground, even without the CPU or software running.

The CPU board is a radiation-hardened RAD6000 processor made by BAE Systems. It contains 128 MB of DRAM for data memory storage and cache memory storage, and 3 MB of EEPROM for code memory storage. The CPU board controls the operation of all of the other boards in the SEM. The SEM also houses DC/DC power converters and an oven-controlled crystal oscillator (OCXO). The essential bus +28V power provided by the power subsystem is used to generate secondary +5V, and ±15V services which power the SEM boards. The OCXO provides a stable clock signal at a frequency of $2^{22}$ Hz, which is divided by the CIB to produce clock signals at 1 Hz and $2^{20}$ Hz (approximately 1 MHz). These signals are distributed to the CIB, the PACI, and the IDPU, where they are used to time-stamp data acquisition and frame transmission times.

### 4.4. Flight Software

The Flight Software is hosted on the CPU board. All software tasks execute under a VxWorks® Real Time Operating System, which handles software initialization and scheduling on a priority basis. Most tasks are scheduled to execute in one of three hardware generated cycle rates, the fastest of which is 8 Hz. ACS tasks are generally synchronous, while some C&DH tasks are triggered asynchronously by events. C&DH tasking performs all non-ACS spacecraft and payload functions including clock and schedule management, commanding validation and execution, telemetry collection/formatting, ground communication, power control, payload interfacing as required, and fault management.

### 4.5. Electrical Power Subsystem (EPS)

The Electrical Power Subsystem (EPS) utilizes four triple-junction gallium arsenide (GaAs) solar array wings, each producing 133.5 W for a total of 534 W at 3 years end-of-life. Energy for eclipse operations is stored in a 15 Ampère-hour battery comprised of eleven common pressure vessels, each containing two nickel-hydrogen cells. The battery can operate at 50% depth-of-discharge for the full three year design life, and provide up to 280 W during the nominal 35-min eclipse duration. The Charge Control Board(CCB) uses a direct energy transfer system and is better than 95% efficient. The amount of current produced by the solar array is controlled by pulse-width modulating FET switches between the eight solar cell circuits and the power bus. Unused solar array power is dissipated in the solar array, not in the spacecraft. The CCB uses a temperature-compensated battery voltage algorithm to set the battery charge current. The PCB distributes power to the spacecraft components and provides switched power to those components requiring unregulated power at 28+7/−4 V. It also provides current sensors for telemetry monitoring and over-current protection for the power bus and under-voltage load shedding. The Auxiliary Driver Board (ADB) provides drive signals for the Inertia Adjustment Devices and the electromagnetic torque rods, and controls the solar array wing deployments

### 4.6. Telecommunications

The Telecommunications subsystem provides S-band Radio Frequency (RF) links for telemetry and command. It includes an S-band transponder made by Cincinnati Electronics, an RF combiner, an RF switch, and four patch antennas located at the forward (Figure 10) and aft sections of the spacecraft which are combined and provide nearly $4\pi$ steradian uplink coverage. The system is capable of full duplex operation. Baseband data format is NRZM to eliminate phase ambiguities that can occur in the uplink and downlink receivers. Two kbps command data is BPSK modulated onto a 16 kHz sub-carrier on the main 2040 MHz carrier frequency. The downlink data rate is nominally 4.0 Mbps, and the frequency is 2215 MHz.

The receiver is ‘hardwired’ to the essential power bus and can never be switched off. The RF path to the receiver contains no switches in order to eliminate risk associated with failures. The transmitter is placed into standby mode or powered off between downlink activities through time tagged commands. The downlink signal is switched to either the forward or aft antenna to avoid interference. The ground provides time-tagged commands to select the best antenna based on the known sun-pointed vehicle attitude and the look-angle to the ground station. Link acquisition begins by transmission of a beacon with data transitions. Following downlink signal acquisition, the ground station transmits the uplink signal. The spacecraft is then commanded to transmit scientific data as well as normal state of health telemetry.

### 4.7. Thermal control

The thermal control system for the spacecraft bus is a simple, cold-biased design using flight proven technologies. The thermal radiator surfaces are covered with 10 mil silver-coated teflon, and Multi-Layer Insulation blankets have a 2 mil, Indium Tin Oxide coated/reinforced second surface/kapton outboard layer, 0.25 mil aluminized Mylar inner layers with Dacron mesh separators. Active thermal components include kapton etched foil strip heaters controlled by bimetallic thermostats.

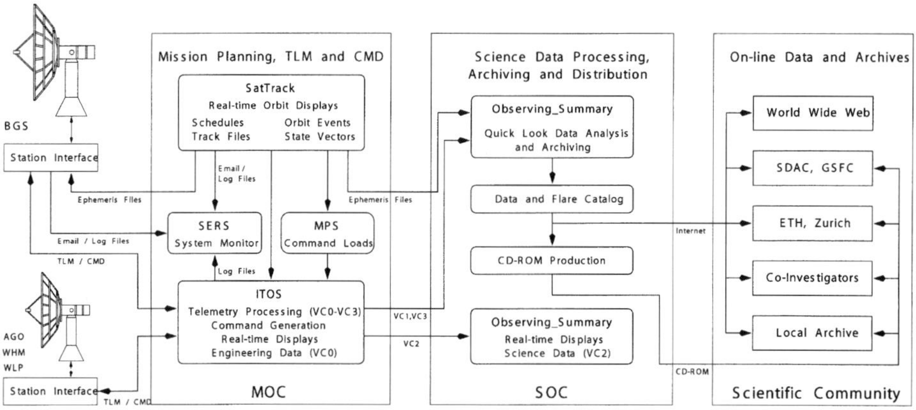


*Figure 11.* RHESSI ground data system.

The thermal performance in orbit matches the analytical predictions very well, with all components well within their design temperature ranges.

## 5. Mission Operations and Ground Data Systems

A schematic of the RHESSI Ground Data System (GDS) is shown in Figure 11. RHESSI is operated from the highly integrated and automated Mission Operations Center (MOC) located at Space Sciences Laboratory of the University of California at Berkeley. The MOC also supports the Fast Auroral Snapshot Explorer (FAST). Co-located with the multi-mission MOC are the RHESSI and FAST Science Operations Center (SOC) and the Berkeley Ground Station (BGS), the primary ground station to support RHESSI on-orbit.

### 5.1. Mission Operations Systems

RHESSI is operated in store-and-dump mode. The spacecraft transmitter is turned on and off by time sequence commands stored on-board. These commands and many others related to configuring instruments for various phases of the orbit are part of an Absolute Time Sequence (ATS) load generated with the Mission Planning System (MPS). Command loads are uploaded to the spacecraft every two days and cover 4–5 days in advance.

The spacecraft command and control system for RHESSI is the Integrated Test and Operations System (ITOS). Since ITOS was also used during mission integration and testing, members of the Berkeley Flight Operations Team were trained early on operating the spacecraft. This approach allowed for a smooth transition from spacecraft integration and testing to normal on-orbit operations.

Flight dynamics and mission planning products are generated by the Berkeley Flight Dynamics System, which is based on the SatTrack Suite V4.4. SatTrack also has heritage with various NASA missions and is used to generate all flight dynamics products such as ground station view periods, link access periods, terminator, high-latitude region, and South Atlantic Anomaly (SAA) crossings, and other orbit events needed as input to MPS. Other tools in the SatTrack Suite are employed to distribute real-time event messages to various ground data system elements such as ITOS and the BGS in an autonomous client/server network environment. SatTrack also provides a multitude of related automation functions as well as 2-D and 3-D real-time orbit displays.

All RHESSI space and ground systems are tied into the Spacecraft Emergency Response System (SERS), which is a data base system that regularly parses through log files and automatically checks for yellow or red limit violations. It also acts on warning and error messages received from various GDS subsystems via electronic mail. In case an anomaly is detected, the on-call operations team member is alerted via 2-way email pager in order to assess and resolve the situation. SERS completes the autonomous ground system and adds a high degree of reliability.

### 5.2. Berkeley Ground Station (BGS)

The Berkeley Ground Station (BGS) is located adjacent to Space Sciences Laboratory. The antenna consists of a pedestal with an 11-m parabolic reflector. A three-axis drive system eliminates the keyhole at the zenith. The antenna is equipped with a full-duplex S-band telemetry and command system. The receiving system has a figure of merit (G/T) of 24.2 dB/K in each of the two receive channels (Left Hand and Right Hand Circular Polarization) for elevations above 5°. The system uses dual receivers with diversity combination. A conical scan feed system provides autotrack capabilities with a typical accuracy of 0.1°. The transmit polarization is selectable, and the nominal RF output power is 100 W (EIRP 63.0 dBW).

### 5.3. Normal Operations

During Normal Operations, communication with the spacecraft is established six times per day via the Berkeley Ground Station to monitor the spacecraft health and safety, and to retrieve science and engineering data. Scheduling and execution of these pass supports is performed fully autonomously. Routine orbit determination functions are carried out by USSPACECOM (formerly NORAD). Updated two-line element sets are automatically downloaded and archived twice a day in order to generate all tracking schedules and mission planning products.

The SatTrack Gateway Server at the Berkeley MOC invokes scripts that regenerate all mission planning products and contact schedules. The updated multi-mission pass support schedule is then loaded into the Gateway Server. Connected clients such as the BGS and various ITOS systems receive support request messages 10 minutes before a pass support. All systems then automatically configure themselves and open network connections for telemetry and command data flows. Once the spacecraft rises above the horizon, the real-time pass support commences by establishing two-way communications with the spacecraft in order to perform health and safety checks and to download stored science and engineering data.

Upon completion of a pass support, the downloaded engineering data are examined for any system anomalies. Yellow or red limit violations trigger immediate notification of operations personnel via the Spacecraft Emergency Response System (SERS). Science data are automatically transferred from the ground station to the level-zero processing (LZP) system. Once LZP is completed, the data are transferred to their on-line archive at Berkeley. In addition, the automated CD-ROM production system will eventually produce multiple copies of the data.

### 5.4. Back-up Telemetry and Command Support

Three additional ground stations are used regularly to provide additional telemetry and/or command support. These ground stations are the Wallops Island ground station in Virginia, operated by NASA, the Weilheim ground station in Germany, operated by DLR, and the Santiago ground station in Chile, operated by the University of Santiago. A dedicated T-1 line from the Berkeley Mission Operations Center to Goddard Space Flight Center, which is shared with the FAST project, is used to establish secure real-time communications with the RHESSI spacecraft through the Wallops ground station. Telemetry data received and stored at the ground stations are transferred to Berkeley post-pass via the open Internet. The averaged link access for the Berkeley and Wallops 11-m ground stations is 55 min day$^{-1}$. Santiago can provide an additional 51 min day$^{-1}$, and Weilheim 16 min day$^{-1}$.

The RHESSI ground system was designed to recover all the data (except for periods of major flare activity) with 6 Berkeley passes daily, but the solar fluxes in the previously unexplored 3–20 keV range, the terrestrial precipitation at $L = 2$–$2.5$, and the background were all higher than anticipated. Thus, typically 4 additional passes, taken at any of the other three stations, are required to recover all the data.

## 6. Science Operations and Data Analysis

### 6.1. Science Operations

RHESSI operations have been designed from the start to be largely autonomous with minimum input in terms of different operating mode or observing plans. All systems are designed to operate automatically with no manual intervention. The main operations task is the management of the on-board solid-state recorder (SSR) during periods of high solar activity or when sufficient ground station dumps are not available to keep the SSR from filling up. When strong flare activity appears likely, we try to keep the SSR below $\sim 20\%$ full at the end of the Berkeley passes, so that there is plenty of capacity remaining for a big X-class flare. An attenuator can be inserted by command to reduce the incoming soft X-ray flux, or some of the data in the SSR can be skipped and not telemetered down.

A team scientist, modeled on the *Yohkoh* 'Tohban' role, monitors solar activity and the instrument operation daily to ensure that the observations are being taken to maximize the scientific return. The Tohban also coordinates with other observatories and notes any special campaign-style observations that may be ongoing to ensure the optimum interchange of information.

The data are generally available for analysis from one to three days after the observation is made. Once the instrument data are recorded in the SSR, it takes up to two days to be telemetered to a ground station depending on how full the SSR becomes. Once on the ground at Berkeley, the packetized data files are converted to Flexible Image Transport System (FITS) format and stored on a Redundant Array of Independent Disks (RAID) system. They are then transmitted over the Internet to Goddard, where they are stored on a similar RAID system. The data are also stored at the HESSI Experimental Data Center (HEDC) at ETH Zurich. At this point, all the data are freely available for downloading by outside users without restriction.

### 6.2. Data Analysis

RHESSI differs from many imagers in that, instead of transmitting a preselected subset of images, the telemetry includes all of the information about each detected photon. Thus, the data analyst can make tradeoffs among time resolution, spectral range and resolution, spatial resolution, image quality, etc., on the ground. These decisions can be made on a case-by-case basis to match the unique characteristics of the event under study and the relevant scientific objective. A key driver of the RHESSI data analysis approach is the preservation of this flexibility to extract the maximum scientific return from the observations. This means that all detailed scientific analysis will use the same primary database with the most current calibration information.

Furthermore, (1) the complete data output of the RHESSI mission is made available promptly to the scientific community, without restriction; and (2) a fully documented analysis package, supported by a range of platforms, is available to the scientific community, with the same toolbox of software used by the PI team. A promptly-generated catalog of summary data products is distributed with the RHESSI data base, to serve as a multi-parameter index and overview of the data base, and to provide data products to users not requiring custom analyses.

The data analysis software is described in the accompanying paper by Schwartz *et al.* (2002). It is also freely available and can be conveniently downloaded as part of the Solar Software (SSW) tree. The extensive RHESSI software package, mostly written in the Interactive Data Language (IDL) programming language, contains all procedures necessary to read the FITS data files, prepare and plot light curves, images, and spectra, and output the results for further customized analysis.

Furthermore, the joint analysis of many different observations of the same events by other observatories is greatly facilitated, since most other solar space missions and many ground-based observatories also have their analysis software in this same SSW tree. A convenient interface is provided to allow easy comparison of RHESSI images and light curves with similar products from SOHO, TRACE, GOES, Big Bear Observatory, etc. The analysis procedures can all be invoked from the IDL command line, or a more user-friendly graphical user interface is also available for basic analysis tasks. All the software is fully compatible with both the Unix and Windows operating systems. The SSW system allows for rapid bug fixes and software upgrades that can be downloaded to each user's own computer at any time from a central server, several of which exist in different countries around the world. For users without IDL, the HEDC provides internet browser software to access and analyze the data.

In an effort to familiarize as many interested scientists as possible with analyzing RHESSI observations, three data analysis workshops have been held. These provided training to about thirty scientists at each workshop in accessing RHESSI data and in the use of the image reconstruction and spectral analysis software. On-line documentation is available for all the software from beginner guides to the detailed manuals required by program developers. This documentation can be accessed through the following web site:

*http://hesperia.gsfc.nasa.gov/rhessidatacenter/.*

## 7. Summary

On 12 February 2002, one week after launch, the germanium detectors were turned on after being cooled down to their operating temperature range by the cryocooler, and RHESSI detected its first flare, a C2 GOES event at 02:14 UT. Since then, RHESSI has been operating continously, and through the end of August 2002, it had detected over 1900 flares above 12 keV and over 600 above 25 keV. It has provided the first imaging spectroscopy of solar flares. It has detected the first 3–10 keV hard X-ray microflares, and found that the Sun is continually emitting hard X-rays above $\sim 3$ keV. On 23 July it obtained the first high-resolution spectrum of solar gamma-ray lines and the first images of a gamma-ray line, from a GOES X4.8 flare.

As indicated by the early results papers in this issue, RHESSI is already providing many exciting new results, particularly gratifying given the many travails the project suffered pre-launch. However, the power of RHESSI lies in its capability for detailed quantitative probing of the particle acceleration and energy release mechanism. That will require careful, comprehensive analysis of the RHESSI data, together with the context measurements from other spacecraft and from the ground.

## Acknowledgements

This work was supported by NASA contract NAS5-98033. The work in Switzerland was supported by a grant from the Swiss National Science Foundation. We wish to acknowledge the efforts of the technical and support staff at the SSL, GSFC, and PSI; of Dennis Lee, Bill Davis, Jim Barrowman,and Tony Comberiate in the Explorer office at GSFC; and of Bill Wagner, Marcus Watkins, Charles Holmes, George Withbroe, George Albright, Lika Guhathakurta, and colleagues in the Sun-Earth Connections Division at NASA Headquarters. The efforts of the various RHESSI review panels are also appreciated, in particular, thanks to the chairs - Tim Gehringer, Joe Wonsever, and Don Miller - and to Gerry Share, who provided science input.

## References

Bougeret, J.-L. *et al.*: 1995, *Space Sci. Rev.* **71**, 231.
Chupp, E. L.: 1990, *Phys. Scripta* **T18**, 15.
Curtis, D. W. *et al.*: 2002, *Solar Phys.*, this volume.
Emslie, G., Brown, J. C. and MacKinnon, A. L.: 1997, *Astrophys. J.* **485**.
Fishman, G. *et al.*: 1994, *Science* **264**, 1313.
Harmon, B. A. *et al.*: 1992, *Proc. of the CGRO Workshop*, p. 69.
Hurford, G. J. and D. W. Curtis: 2002, *Solar Phys.*, this volume.
Hurford, G. J. *et al.*: 2002, *Solar Phys.*, this volume.
Johns, C. and Lin, R. P.: 1992, *Solar Phys.* **137**, 121.
Kosugi, T. *et al.*: 1991, *Solar Phys.* **136**, 17.
Lin, R. P. and Hudson, H. S.: 1976, *Solar Phys.* **50**, 153.
Lin, R. P. *et al.*: 1995, *Space Sci. Rev.* **71**, 125.
Makashima, K. *et al.*: 1977, in K. A. van der Hucht and G. Vaiana (eds.), *New Instrumentation for Space Astronomy*, Pergamon Press, New York.
Prince, T. A. *et al.*: 1998, *Solar Phys.* **118**, 269.
Ramaty, R. and Murphy, R. J.: 1987, *Space Sci. Rev.* **45**, 213.
Ramaty, R. *et al.*: 1993, *Adv. Space Res.* **13**, 275.
Ramaty, R., Mandzhavidze, N., Kozlovsky, B. and Murphy, R. J.: 1995, *Astrophys. J.* **455**, L193.
Schwartz, R. *et al.*: 2002, *Solar Phys.*, this volume.
Share, G. H. and Murphy, R. J.: 1995, *Astrophys. J.* **452**, 933.
Smith, D. M. *et al.*: 2002, *Solar Phys.*, this volume.
Zehnder, A. *et al.*: 2002, *Solar Phys.*, this volume.
Zhang, S. N., Fishman, G. J., Harmon, B. A. and Paciesas, W. S.: 1993, *Nature* **366**, 245.

# THE RHESSI SPECTROMETER

D. M. SMITH[1], R. P. LIN[1], P. TURIN[1], D. W. CURTIS[1], J. H. PRIMBSCH[1], R. D. CAMPBELL[1], R. ABIAD[1], P. SCHROEDER[1], C. P. CORK[2], E. L. HULL[2], D. A. LANDIS[2], N. W. MADDEN[2], D. MALONE[2], R. H. PEHL[2], T. RAUDORF[3], P. SANGSINGKEOW[3], R. BOYLE[4], I. S. BANKS[4], K. SHIREY[4] and RICHARD SCHWARTZ[5]

[1]*Space Sciences Laboratory, University of California, Berkeley, U.S.A.*
[2]*Lawrence Berkeley National Laboratory, U.S.A.*
[3]*ORTEC, U.S.A.*
[4]*NASA Goddard Space Flight Center, U.S.A.*
[5]*NASA Goddard Space Flight Center/SSAI, U.S.A.*

(Received 12 September 2002; accepted 16 September 2002)

**Abstract.** RHESSI observes solar photons over three orders of magnitude in energy (3 keV to 17 MeV) with a single instrument: a set of nine cryogenically cooled coaxial germanium detectors. With their extremely high energy resolution, RHESSI can resolve the line shape of every known solar gamma-ray line except the neutron capture line at 2.223 MeV. High resolution also allows clean separation of thermal and non-thermal hard X-rays and the accurate measurement of even extremely steep power-law spectra. Detector segmentation, fast signal processing, and two sets of movable attenuators allow RHESSI to make high-quality spectra and images of flares across seven orders of magnitude in intensity. Here we describe the configuration and operation of the RHESSI spectrometer, show early results on in-flight performance, and discuss the principles of spectroscopic data analysis used by the RHESSI software.

## 1. Introduction

The Reuven Ramaty High-Energy Solar Spectroscopic Imager (RHESSI) was designed to study high-energy emission from flares over a broad energy range, from thermal X-rays through nuclear gamma-ray lines (Lin *et al.*, 2002). The design of its spectrometer was driven by the need to satisfy many requirements with a single instrument of modest cost and weight. The spectrometer had to be able to process up to millions of hard X-rays per second for imaging, while carefully collecting rare gamma-ray line photons with high efficiency, high energy resolution, and without interference from the X-rays. It was desirable for the energy range to extend down to 3 or 4 keV to clearly image the thermal components of flares and to be highly sensitive to microflares, while extending well above the 4.4 and 6.1 MeV gamma-ray lines to characterize the electron bremsstrahlung spectrum above the nuclear line region. We wanted to be sensitive to events covering at least seven orders of magnitude in intensity, from microflares to the largest X-class events.

Most importantly, RHESSI had to have extremely high energy resolution in order to make advances in spectroscopy comparable to the advances its high-angular-resolution Rotating Modulation Collimator (RMC) system would make in imaging (Hurford *et al.*, 2002). The Solar Maximum Mission Gamma-Ray Spectrometer (SMM/GRS), with its large collecting area and its long and successful mission operations, produced a wealth of data on gamma-ray lines from large flares (Vestrand *et al.*, 1999) and a significant contribution was also made by the instruments on the Compton Gamma-Ray Observatory (Share, Murphy, and Ryan, 1997) and the Gamma-Ray Spectrometer on *Yohkoh* (Yoshimori *et al.*, 1994). But all these instruments were scintillators with energy resolution $\geq 6\%$ full width at half maximum (FWHM) at gamma-ray line energies. In some cases (particularly around 1 MeV) this low resolution can result in the blending of several lines, and, more importantly, it leaves the issue of Doppler shifts and broadening of the nuclear deexcitation lines mostly unaddressed. Energy resolution an order of magnitude finer would allow these Doppler phenomena to be observed unambiguously, revealing information about the angular distribution of accelerated ions in large flares. In the hard X-ray band, high resolution (about 1 keV FWHM from 3–100 keV) would allow a clean separation between thermal and nonthermal emission never before achieved except on balloon payloads flying germanium spectrometers (Lin and Schwartz, 1987).

RHESSI has many new, unique features compared to other gamma-ray spectrometers flown in space, even beyond the obvious advance of using a large array of germanium detectors. These include an efficient, lightweight, and inexpensive mechanical cooler; movable attenuators that respond to the intensity of a flare, so that the detectors can record microflares without saturating even in the biggest flares; and extremely low-noise electronics and segmented detectors to cover from 3 keV to 17 MeV with a single crystal. These and other advances allow RHESSI to meet its goals of resolution, sensitivity, and dynamic range within the cost and weight constraints of a Small Explorer mission.

The sections below will address the composition and operation of the RHESSI spectrometer, the spectrometer data formats, in-flight performance, and the principles of the spectroscopic data analysis.

## 2. Components of the RHESSI Spectrometer

### 2.1. Germanium Detectors

RHESSI's spectrometer is an array of segmented coaxial germanium detectors. When ultrapure germanium is at cryogenic temperatures, no electron-hole pairs are in the conduction band, but a hard X-ray or gamma ray interacting in the crystal will release one or more energetic electrons, which lose energy by creating free pairs. If there is a high electric field (on the order of 1000 V $cm^{-1}$) across the crystal, the electrons and holes will be pulled to each electrode, creating a current

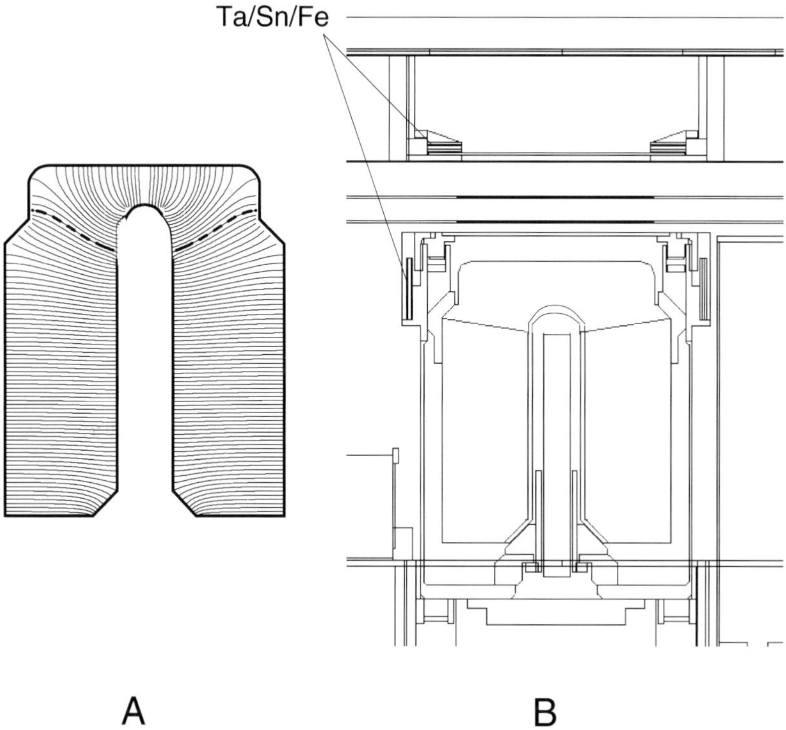


*Figure 1.* Cross-sections of a RHESSI detector. (A) A detector profile with field lines, with the field line marking the segment boundary in *bold dashes*. (B) A detector in the cryostat, showing Ta/Sn/Fe/Al shielding around the side of the front segment and above the shoulder of the rear segment.

pulse that can be amplified and digitized by suitable electronics. The total charge in the current pulse is proportional to the photon energy.

Figure 1 shows two cross-sections of the cylindrically-symmetrical RHESSI detector design. This design was a joint effort of the RHESSI co-investigators at U. C. Berkeley and Lawrence Berkeley National Laboratory and the manufacturer, ORTEC (currently a division of AMETEK). The shape is a variation of a 'closed-end coaxial' detector, the industry standard design for large volumes and high gamma-ray sensitivity. The ultrapure, slightly n-type germanium material is doped in a very thin outer layer with boron on the front and side surfaces, and a thicker, $n$-type layer of diffused lithium ions on the inner bore. The rear surface is left as an insulator. When 2000–4000 V is applied between the inner and outer electrodes, the crystal is depleted of free charge carriers, with enough electric field in the crystal from the space charge and external voltage combined to cause the electron-hole pairs to reach terminal velocity.

For space applications, it is important to fly $n$-type material with this electrode configuration in order to minimize the effect of radiation damage on resolution.

Since radiation damage produces primarily hole traps rather than electron traps, the goal is to minimize the length of germanium traversed by the holes. For a coaxial detector, most of the volume is near the outer radius. Thus most interactions take place near the outside and the holes should therefore travel to the outside electrode. The effects of radiation damage can also be minimized by keeping the detectors very cold (72–76 K in the case of RHESSI) and making sure the detector high voltage is never turned off once there has been significant exposure to protons and neutrons (Hull, 1998; Koenen *et al.*, 1995).

The inner (lithium) contact is discontinuous at a point near the top, and signals are extracted separately from the two halves of this electrode. The line extending from this point to the outside edge of the detector in Figure 1(A) represents a boundary electric field line: photons stopping above this line are detected in the front channel, and those stopping below it in the rear channel. Thus a single crystal becomes a zero-gap stacked pair of detectors. The front segment will absorb all the hard X-rays up to about 100 keV, letting most gamma-ray line photons through. The rear segment will stop many of the latter, so that fine spectroscopy can be done without high deadtime from the X-rays.

The notch on the outer edge of the detector serves two purposes: first, it concentrates the electric field lines at the corner of the notch, so that the field line that originates at the break in the inner contact always hits the proper place on the outside of the detector. In addition, it removes some mass from in front of the rear segment, so that fewer high-energy gamma rays Compton scatter before entering the rear. A ring of 'graded-Z' material keeps the shoulder of the rear segment from being swamped with flare hard X-rays (see below). This shield is just as effective as the front segment in photoelectrically absorbing hard X-rays, but with much less Compton scattering of gamma-rays.

Figure 2 shows a side view of the detector segmentation produced by scanning a 3 mm beam of 81-keV X-rays across the detector in discrete positions corresponding to each pixel. The brightness of each pixel is the sum of the front and rear segment count rates in the photopeak of the line. The segmentation boundary probably looks dark due to a combination of effects which cause charge to be split between the front and rear signal electrodes: Compton scattering and k-shell photon transport, which produce interactions on either side of the boundary, and charge-sharing between the segments for even single-site interactions that happen very close to the boundary.

Note that the segmentation boundary is not independent of azimuth angle, but rather has a wavy appearance, with four peaks and four valleys around the periphery of the detector. This is a natural result of the cubic crystal structure of germanium: electron mobility is highest along the three main crystal axes and lower between them. Thus electrons traveling to the central bore from different spots on the crystal surface at terminal velocity will not travel directly along field lines but instead their paths will bend toward the nearest axis. This distorts the segmentation boundary position as a function of azimuth about the cylindrical axis. The crystals

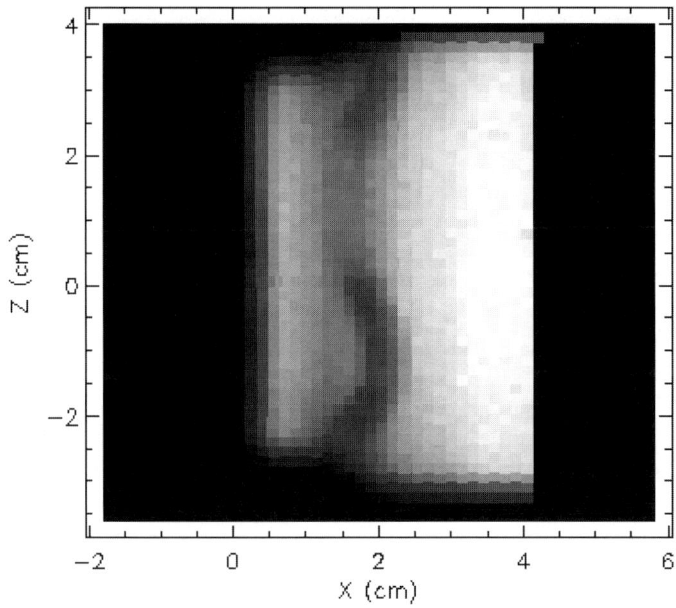


*Figure 2.* Result of a narrow-beam 81 keV X-ray scan of the side of a RHESSI detector showing the ripple in the segmentation boundary (see text). The front of the detector is to the left. The scan does not proceed all the way down the rear segment. Lighter shades represent a higher count rate in the 81 keV line.

are always grown so that one axis is parallel to the axis of the cylinder ('X' in Figure 2).

## 2.2. Detector Electronics

The preamplifiers for the RHESSI detectors feature custom four-terminal, *N* channel junction field effect transisitors (JFETS) with adjustable back-gate voltage for very-low-noise operation, one for each segment. These are immediately behind the detectors, and are thermally connected by wires to the outer thermal shield (see below) to achieve an ideal operating temperature between 130 and 170 K. Their signals are taken out to the rest of the preamplifier circuit, which resides in an individual box for each detector, clustered around the bottom of the spectrometer (see Figure 3), by a harness consisting of thin traces of manganin (for low thermal conductivity) layered in a flexible plastic film.

The preamplifiers use a pulsed reset circuit developed at Lawrence Berkeley National Laboratory (Landis, Cork, and Goulding, 1982). The integrated charge on the preamplifier's feedback capacitor is restored by this circuit when a total charge equivalent to interactions from 40 MeV worth of photons or cosmic rays has been deposited in each segment. The preamplifier's output-level-sensing and charge-restoration logic avoids transient signal reductions associated with high-energy events.

*Figure 3.* The RHESSI spectrometer, before being mounted on the spacecraft. The aluminum attenuator disks are visible on top (one set in the aperture, the other out), and the preamplifiers and high-voltage filter boxes can be seen around the skirt. The detectors are behind the thin, ribbed side wall.

Leakage current in the detectors is so low (a few picoamps) that cosmic rays and background photons dominate the total current through the detectors. Each reset event results in approximately 40 $\mu$s of deadtime in the resetting segment, but as this seldom happens at a rate of more than 100 Hz it represents a small contribution to the overall system deadtime. Leakage current is kept low by avoiding the introduction of any contamination on the back (insulating) surface of the crystals in two ways: (1) careful design of the cryostat (see below) and (2) a proprietary ORTEC passivation process used on the rear germanium surface.

The signals from the preamplifier are taken to a Detector Interface Board (DIB) in the primary Instrument Data Processing Unit (IDPU) (Curtis *et al.*, 2002) box on the spacecraft deck. Each board contains both the analog and digital signal-processing circuitry for both segments of one detector. The signal from each event is split, going to a fast-shaping channel (which produces a triangular pulse 800 ns wide at the base) and a slow-shaping channel (8 $\mu$s peaking time) for spectroscopy.

The fast pulse is used for pileup rejection: two events very close together in time, which would be analyzed as a single higher-energy event by the analog-to-digital converter (ADC), can instead be recognized, and both rejected. If the second event comes shortly after the slow-shaped pulse from the first has already been sampled at its peak and fed to the ADC, then only the second is rejected, since the first was read with its correct energy. The fast lower-level discriminator is set at about 7 keV in the front segments (just above the noise) and about 20 keV in the rear segments. Photons below 7 keV in the front segments may pile up with other counts without triggering the pileup veto, so careful data analysis is necessary for high count rates of soft photons.

The circuit also uses the height of the fast pulse for large events to choose quickly between one of two gain ranges available in the rear segments: a high gain setting that goes up to about 2.7 MeV and a low one that goes to about 17 MeV. The ADC samples the shaped pulse (see below) at its peak to produce a 13-bit (8192 channel) value for either range. Thus the ADC gain can be set before the slow-shaped pulse reaches its peak. Because the energy resolution of the fast pulse is poor and because its short integration time means that each pulse still has a different shape depending on the position of the interactions in the detector, the transition between the two rear gain ranges is gradual rather than abrupt, i.e., events from about 2.6 to 3.2 MeV may appear with either gain. Since the gain range that was used is recorded as part of the event, there is no loss of information. The front segments use only the first (high) gain value.

The bottoms of the usable energy ranges in each segment are set by both electronics and physics. The physical limits are absorption in the materials around the detectors (see below). The electronic thresholds, set by the noise level in the slow-shaping channel, are about 2.7 keV for most of the front segments and 20 keV in the rears. Both the physical and electronic cutoffs are somewhat gradual and have to be modeled precisely for low-energy spectral analysis. The front segment electronics use quasi-triangular shaping to optimize low-energy response, and the electronics for the rear segments use quasi-trapezoidal shaping to minimize ballistic deficit.

The ADC-busy and pileup rejection signals are 'or-ed' together and sampled at 1 MHz by a livetime counter. Because the pileup circuit has the ability to sometimes veto *both* piled-up events, no simple circuit can sample the true system livetime. There is, however, a simple mapping between the recorded and true livetimes as long as all events are above the fast discriminator threshold. When a large fraction of the counts are below this energy, deriving the true livetime is more complicated, but can still be done based on pre-flight and in-flight calibrations.

## 2.3. The RHESSI cryostat

The RHESSI cryostat was designed to provide a lightweight but secure environment for the detectors. This involves maintaining a hard vacuum and very good thermal insulation, so that the detectors can remain at operating temperature (about 75 K) with a very low heat leak. Figures 3–5 depict the interior and exterior of the cryostat and the arrangement of the detectors.

The weight constraints of a Small Explorer did not allow a heavy active shield to veto solar photons that scatter out of the detectors and to keep out background

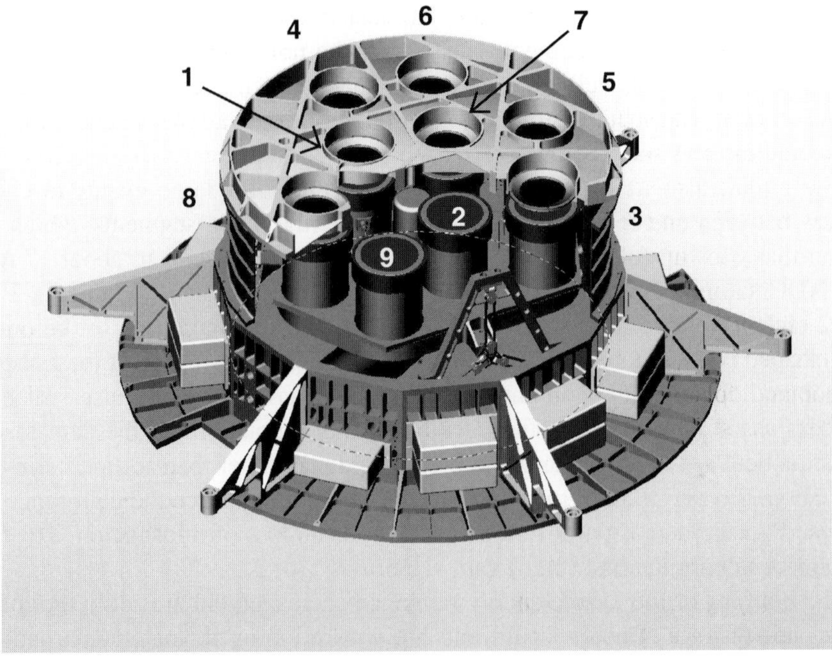


*Figure 4.* Exterior of the spectrometer shown with part of the outer housing cut away, revealing the detector modules inside. The numbering scheme of the detectors matches the grids.

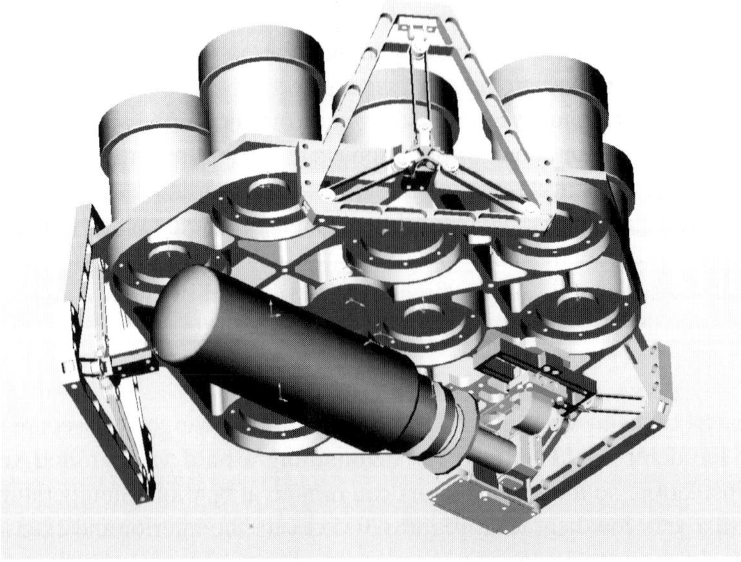

*Figure 5.* The interior of the RHESSI spectrometer, showing the detectors mounted on the coldplate, the Sunpower cryocooler below it, and the fiberglass-strap 'cartwheels' that suspend this assembly in the interior of the cryostat.

photons. We therefore decided instead to design the side walls of the cryostat to be as thin as possible; there is about 4 mm of Al between the rear detector side surfaces and space. This allows hard X-rays (above about 25 keV) and gamma-rays from cosmic sources such as supernova remnants, pulsars, and gamma-ray bursters to enter the spectrometer, providing a wealth of secondary science, including astrophysical topics (Smith *et al.*, 2000, 2002) and the study of emissions from the Earth's atmosphere (Share *et al.*, 2002). Cosmic and terrestrial sources usually appear primarily in the rear segments rather than the front segments, so they can generally be distinguished from solar events.

The detectors are securely housed in independent modules that are not hermetically sealed but have a very small indirect gas conductance path to the outside to minimize contamination during handling. The modules are firmly attached to a common coldplate, and the detector/coldplate assembly is suspended by three sets of fiberglass straps for thermal isolation (see Figure 5). The coldplate assembly is surrounded by two thin aluminum thermal shields at intermediate temperatures, each of which is surrounded by multilayer aluminized-mylar insulation.

The cryocooler is a Sunpower model M77 Stirling-cycle refrigerator, a small, high-efficiency unit modified for long-term use in orbit by the cryogenics group at NASA's Goddard Space Flight Center. The cooler is visible in Figure 5 as the horizontal cylindrical object below the coldplate. It is equipped with an actively-driven balancer mass to cancel vibrations to first order. Its 'gas bearing' system means that there is no rubbing of solid parts during operation. With 55 W of input power, the cooler can provide the approximately 3.5 W of cooling needed to keep the entire coldplate/detector assembly at 75 K. The cooler is coupled to the coldplate by a flexible, S-shaped link made from 700 sheets of 0.025 mm aluminum foil for good heat conduction with minimum transmission of vibration. Also in the thermal path between the cooler's coldfinger and the coldplate is a rod of sapphire. This material has high thermal conductivity at low temperatures and vice-versa, and is important during cooldown and annealing (see below). As part of the thermal isolation, the outer thermal shield is actively cooled by its own separate connection partway down the coldfinger of the cooler, which keeps it at about 155 K. Waste heat from the cooler is radiated to space from the back surface of the spectrometer, which is also the rearmost surface of the spacecraft. It is covered with silver teflon tape. It alternately views deep space and the Earth during each orbit.

### 2.4. The Attenuators or Shutters

Attached to the top of the spectrometer are two lightweight, movable frames, each of which carries nine aluminum disks that can be moved in front of the detectors (see Figure 3). These disks, manufactured to our design by Tecomet, Inc., serve as attenuators, to keep the detectors from being saturated at high counting rates. One set of disks is thicker than the other, but they are not uniform: each has a small, thin spot in the center so that there is always some low-energy response. There is

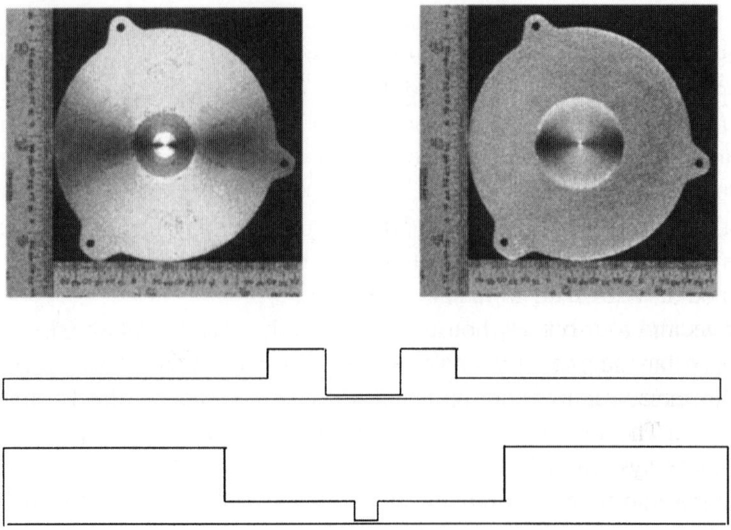

*Figure 6. Top*: photographs of the attenuator disks. *Left*: thin disk. *Right*: thick disk. *Bottom*: cross-section of the thin and thick disks. The vertical scale has been exaggerated by a factor of four to make the features easier to see. The outer diameters are 61.5 mm to cover the detector front segments. Thicknesses vary from 0.05 mm to 1.53 mm.

also a slightly larger region that is thick on the otherwise thin disk and thin on the otherwise thick disk: thus, the full attenuation doesn't occur until both disks are in place (see Figure 6). The attenuator dimensions vary slightly from detector to detector based on the grid-slat spacing in order to avoid introducing biases into the imaging when the attenuators are put in.

The onboard computer monitors the front segment livetimes and puts the disks in (thin first) when deadtime gets too high. The state with the thick shutter in and the thin shutter out is not currently used in spacecraft operations. Because the system is particularly sensitive to pileup at low energies (see above), and because low-energy thermal emission dominates the detected flare photons when both shutters are out, the thin shutter is currently set to come in when deadtime exceeds 8%. The deadtime will then drop sharply. If the flare brightens further, the thick shutter will come in when the deadtime reaches 10%. Early in the mission the shutters were only brought in at higher deadtimes, until it was realized that pileup was too severe in the resulting data. Because bringing in each set of shutters reduces the deadtime so effectively, it is difficult for the onboard software to estimate when to pull them out again. As long as the deadtime is very low, the flight software will try briefly pulling out the last shutter it put in every 4 min until either (1) the deadtime is found to be low enough to leave it out or (2) it exceeds a set maximum number of shutter motions per spacecraft orbit.

Each shutter frame is moved by long, thin shape-memory alloy (SMA) wires that contract when a voltage is applied and current runs through them, producing heat. There are four sets of six wires each: one set for moving each frame in each direction. Each motion takes less than one second, including the delay for heating, and about 20 s is required for cooldown before the same set of wires can be activated again. In addition, each frame is connected to a heavy SMA spring for an emergency backup that can pull it strongly into its default position (thin in, thick out) and lock it there. These are only to be used in the unlikely event of a mechanical jam or other failure of the wire SMA system. A second set of SMA springs was used to uncage each frame after launch.

With neither attenuator in place, the front segments view the Sun through four Be windows and four blankets of multilayer aluminized-mylar insulation (two inside the cryostat and one on each end of the imager tube). The blankets are thinned but not eliminated immediately above each detector to improve the transmission of solar photons. The net transmission is 10% at about 5.5 keV (see Figure 8 below). The attenuator system adds about four orders of magnitude to the dynamic range of microflares and flares detectable by RHESSI. In order to save onboard memory and downlink capacity, the thin shutter is often left in place for one or more days at a time.

## 2.5. Other Spectrometer Subsystems

### 2.5.1. *Charged Particle Detector*

A solid-state silicon charged particle detector (CPD) is mounted on a strut that runs past the side of the spectrometer. The detector is a disk 1.0 mm thick and 5.64 mm in diameter, and is mounted so that it faces outward, its normal axis perpendicular to the spacecraft axis. It is encased in shielding consisting of at least 3.8 mm of Al in all directions except for an outward-facing pinhole of diameter 1.02 mm. This detector is not configured as a spectrometer, but count rates are read out every $\frac{1}{8}$ s above two energy thresholds: 50 keV and 620 keV. The higher threshold is commandable and the lower fixed. The commandable threshold may be set to any energy from about 20 keV (limited by noise) to 2 MeV.

Cosmic rays penetrate the shielding easily and usually trigger both thresholds. Thus the CPD can measure cosmic-ray flux, although the upper level discriminator (ULD) counter in the germanium detectors themselves provides a more accurate measure due to their much larger area. Electron precipitation events generally trigger the lower threshold only. During transit of the SAA, the upper threshold counts mostly trapped protons and the lower mostly trapped electrons, although the upper threshold is partially contaminated by SAA electrons, which are much more numerous, entering the pinhole. Because the detector rotates with the spacecraft it samples the pitch-angle distribution of the particles whenever the spacecraft axis is near perpendicular to the local magnetic field. At these times, the SAA particles are seen to have an extreme pancake distribution (circling at 90° to the field) as expected for mirroring particles.

### 2.5.2. *Beryllium Scatterer*

Since the rear segments see no direct flare photons below 100 keV, we can use them at these low energies as a hard X-ray polarimeter (McConnell *et al.*, 2002). There is a cylinder of beryllium 3 cm in diameter and 3.5 cm long attached to the coldplate, nestled among the rear segments near the center of the spectrometer. Above this cylinder is a thin spot in the spectrometer shell and a hole in the grid trays, so that solar photons $> 20$ keV can reach the cylinder and scatter into the adjacent rear segments. The Compton cross-section, differential in azimuth angle, is a function of the angle from the polarization axis. Thus, by watching the relative counting rates in these rear segments, we will measure the direction and degree of polarization for incoming photons of roughly 20–100 keV. Simulations (McConnell *et al.*, 2002) suggest that we will be able to detect polarization fractions as low as a few percent for the largest flares. The key difficulty in the analysis will be photons scattered from the Earth's atmosphere, which also produce low-energy counts in the rear detectors and which also vary with the spacecraft spin, albeit with one peak per spin instead of two. Since this component will not be a perfect sinusoid, any first-harmonic power it has could be mistaken for a polarization signal. Scattering in the coldplate and other passive materials into the rear detectors does not modulate with the spacecraft spin, and so it only interferes with polarization measurements by increasing the background count rate.

### 2.5.3. *Graded-Z Shielding*

In the absence of a heavy active shield, there were several places were a small amount of passive shielding to block hard X-rays was considered valuable. In each place, the material is a 'graded-Z' laminate of metals: from outside to inside, 0.5 mm of tantalum, 1.0 mm of tin, and 0.5 mm of stainless steel. The tantalum stops most incident photons up to 100 keV. Each successive metal absorbs the K-shell fluorescence photons of the one before it. The iron K-shell photons from the steel are absorbed in the aluminum structures backing the shielding, and the aluminum K-shell photons are both too low in energy to trigger the detectors and mostly stopped in the aluminum itself. The combination is mostly opaque below 100 keV, and fairly transparent above.

A ring of this material surrounds the side surface of the front segments to shield them from background photons from the Earth and space, so that they can be sensitive to the faintest possible microflares. A washer-shaped ring sits above the shoulder of each of the rear segments to protect them from direct solar photons. Both these rings are visible in Figure 1(B). Finally, a larger sheet of the material, with holes for each detector and the Be scatterer, is attached to the top cryostat wall, below the attenuator assemblies. Its purpose is to minimize the number of solar hard X-rays which scatter in the coldplate and other cryostat materials and enter the rear segments at low energies. This serves two functions: to keep the counting rates in the rear segments from saturating during the largest flares, and to reduce the background against which polarization measurements will be made in the rear segments at low energies.

### 2.5.4. *Onboard Radioactive Source*

RHESSI carries a tiny onboard radioactive source (5 nanocuries of $^{137}Cs$) that makes a line at 662 keV, far from any line expected to occur in flares or in RHESSI's variable background. This source is so weak that it only gives a useful number of counts in spectra accumulated over many hours. Having a line of known intensity lets us monitor loss of narrow-line efficiency in the detectors due to radiation damage, which happens on a time scale of months to years. The same function can be served by a line at 1460 keV from naturally occurring $^{40}K$ in the spacecraft, but the absolute flux of this line is not known from before launch since there was always some of the isotope in the laboratory.

### 2.5.5. *Onboard Pulser*

The IDPU can put regular, small pulses onto the detectors' high voltage (HV) line. The electronics see this as equivalent to photon events. The pulse energy can be tuned across the detectors' full range, but the front/rear ratio is fixed at roughly 1:3 because the HV is shared by both segments and their response is proportional to their capacitance. Pulse rates can be commanded separately for each detector at 11 discrete frequencies, spaced by a factor of two, up to 1024 Hz.

## 2.6. Cryostat Thermal Operations

Cooldown of the spectrometer began within a few hours of launch and lasted about six days. Although the cryostat was kept on a vacuum pump until being sealed off shortly before launch, continued outgassing from the interior, particularly water from the mylar thermal blankets, remained a concern. The thermal impedance of the sapphire rod between the coldplate and the cryocooler was included to force the first stage of cooldown of the detectors to proceed extremely slowly. The secondary connection between the cryocooler and the intermediate thermal shield has no such impedance, so the shield cools much more quickly. This allows all the water vapor and other volatiles in the cryostat vacuum to freeze out onto the shield first instead of the detectors, so that we do not need to use detectors that are individually hermetically sealed. At about 200 K the cooling curves cross and the detectors become colder than the shield from then until they reached their operating temperature of around 75 K.

Once the radiation damage that has built up in the detectors is significantly affecting RHESSI's energy resolution (which we predict will occur 2–4 years after launch), we can anneal the detectors at high temperature to remove most of the effects of the damage. Although some damage is due to cosmic rays and the secondary neutrons they generate, most of it will be caused by trapped protons encountered when the spacecraft passes through the SAA. When the anneal cycle begins, the detectors will be powered down and then a one-time-use SMA actuator will be fired to open a valve in the cryostat's back surface to the vacuum of space so that gases released during the warmup can escape. Then, diode heaters on the coldplate will be powered to begin heating it up. The cryocooler will be run at a low power level during the early phases of the warmup to keep the intermediate shield cold, so that it doesn't release any deposited water onto the detector surfaces. The detectors will be held near 100 °C for several days before being cooled down again to operating temperature.

## 3. Data Formats

### 3.1. Event Data

Most of the time, whether a flare is active or not, every photon event in the RHESSI detectors is stored in onboard memory as 4 bytes of data and is telemetered to the Berkeley ground station (or a backup station) within a day or two. Each event contains:

– 5 bits identifying the segment and (if a rear segment) gain setting of the event,

– 13 bits for the energy channel,

– the last 10 bits of the time counter, giving the event time in units of binary microseconds (i.e., $2^{-20}$ seconds), and

– 4 bits that are a partial measurement of the livetime counter (Curtis *et al.*, 2002).

If the onboard memory starts to fill up, a decimation algorithm automatically throws out all but one out of every $N$ events in the front segments below a certain energy $E$, with $N$ from 2–16. $E$ and $N$ are functions of the remaining memory and the position of the attenuators. Decimation in the rear segments can be commanded as a routine way of keeping background (mostly photons from the Earth's atmosphere or the cosmic diffuse background) from filling up the memory. Operating modes that have been used in flight have decimated the rear segments (with $N = 3$ or 4 and $E = 150$ keV to 450 keV) either during spacecraft night or at times when RHESSI is at high magnetic latitudes and electron precipitation is common (see below). For most of the mission, front segment events have been turned off during spacecraft night, and all events are always turned off during passage through the SAA.

### 3.2. Monitor Rates

Monitor rates are counters that are read out every second for each of the 18 segments. They include: number of resets, number of events triggering the slow-channel lower-level discriminator (LLD), number of events triggering the fast-channel LLD, number of high-energy (ULD) events beyond the scale of the ADC (usually cosmic rays), and livetime. This information is normally used to check the health of the detectors and is not necessary for spectroscopy.

### 3.3. Fast Rates

Fast rate data are produced only when the count rates are very high – they are the last line of defense against detector saturation for flares big enough to overwhelm the normal electronics chain with both attenuators in place. The data are count rates in four broad energy bands, roughly 7–15 keV, 15–25 keV, 25–70 keV, and 70 keV and higher, with significant variation from detector to detector. The pulses are sampled from the fast electronics chain. The rates for the three detectors under the finest grids (and therefore the fastest imaging modulations) are sampled at 16 kHz; the next three at 4 kHz; and the three coarsest grids at 1 kHz. Events are not shut off when the fast rate data turn on; however, at these very high count rates the event data will naturally taper off due to very high deadtime.

## 4. RHESSI's Performance in Flight

### 4.1. Resolution

At low energies (below about 200 keV in the RHESSI front segments), the width of a spectral line in a germanium detector is dominated by noise in the electronics, and is roughly constant. At higher energies, one of two noise components will dominate: the counting statistics of the electron-hole pairs, which increases as the square root of energy, and broadening due to trapping of electrons or holes in the crystal, which goes linearly with energy. Electron trapping is intrinsic to the detectors, due to impurities when they are grown. Hole trapping is the result of radiation damage in space.

Eight of the nine RHESSI detectors have been performing well since the start of the mission, with energy resolution of about 1 keV FWHM and a low-energy cutoff of about 2.7 keV in the front segments and with resolution of better than 3 keV FWHM in the rear segments to over 1 MeV (see Table I). Figure 7 shows the 2.223 MeV line from neutron capture on hydrogen seen by RHESSI from the X4.8 flare of 23 July 2002. This is a sum of gain-corrected spectra from the rear segments of all the detectors except the one below grid No. 2 ('G2'; see below). The FWHM is 4.4 keV. The line is expected to be intrinsically $\ll$ 1 keV because the neutrons at the Sun thermalize before being captured, so this is a good measurement of the current resolution capability of the instrument at this energy. The slight asymmetry of the line is probably due to a small amount of trapping. If interpreted as due to radiation damage, it is similar to our prediction for this phase of the mission (six months after launch). The resolution of the summed rear segments at low energies (i.e., the 93.3 keV background line) is 2.5 keV FWHM, which is the contribution of the electronics plus any blurring due to imperfect gain calibration.

TABLE I

RHESSI in-flight resolution by detector segment. Resolution is measured at 93.9 keV in the front segments and 1117 keV in the rears.

| Detector | Front FWHM (keV) | Rear FWHM (keV) |
|---|---|---|
| G1 | 1.13 | 2.90 |
| G2 | 7.94 | ... |
| G3 | 0.98 | 2.77 |
| G4 | 0.98 | 2.82 |
| G5 | 1.47 | 2.73 |
| G6 | 1.01 | 3.05 |
| G7 | 3.15 | 2.98 |
| G8 | 1.26 | 3.36 |
| G9 | 1.19 | 2.27 |

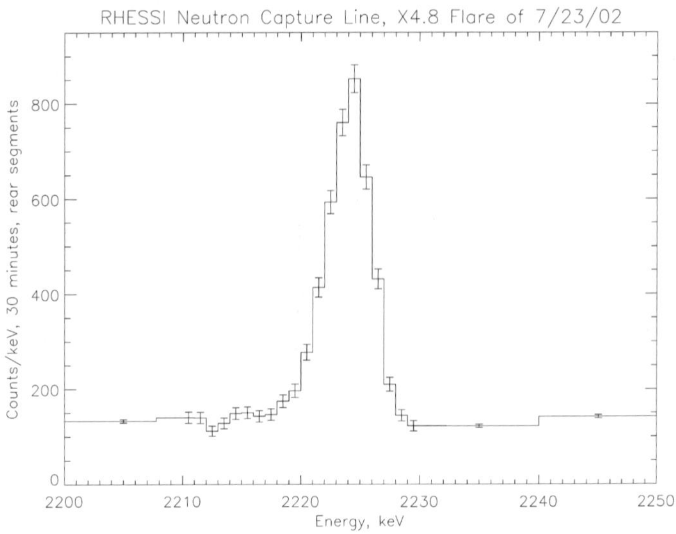


*Figure 7.* The 2.223 MeV neutron-capture line from the X4.8 solar flare of 23 July 2002, as observed by the RHESSI rear segments.

The gain calibration of each segment is currently re-calculated about once a week, using a full orbit's worth of background data. The entire channel-to-energy conversion is represented as a purely linear fit over each energy scale (0.0–2.7 MeV for the front segments and 0.0–2.7 and 2.7–17 MeV in the rear segments). The gain and offset of the linear relation are currently determined by finding the centroids of the two brightest background lines, a germanium activation line at 93.3 keV and the positron annihilation line at 511 keV. The high degree of linearity of the electronics

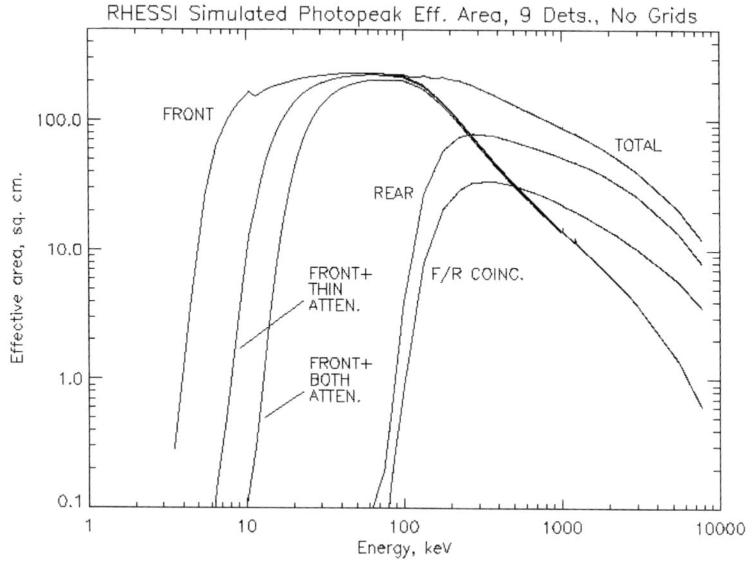


*Figure 8.* RHESSI front and rear segment effective areas for photopeak absorption, summed over all 9 detectors. The front-segment traces peak at lower energy, and from top to bottom represent the case with no attenuator in, with the thin attenuator only in, and with both in. Below the rear-segment trace is the trace representing events which are split between the front and rear segment of one detector. The total of the front (no attenuator), rear, and coincidence modes is also shown.

allows this fit to produce results as good as Figure 7 even though the energy is extrapolated by a factor of four from the higher line used in the gain solution. Gain variations within one orbit and within one day are negligible thanks to the thermal isolation of the detectors and the careful design of the electronics to be temperature-independent. Gain solutions using more background lines and longer integrations are currently being produced and may yield a slight improvement on the already excellent results.

### 4.2. Dynamic Range and Effective Area

With both sets of attenuators out, RHESSI can see microflares in the 3–10 keV range that are too small to be seen by GOES (Krucker *et al.*, 2002). With both sets in, the X4.8 flare of 23 July 2002, at its peak, caused no more than 30% deadtime in the front segments. This still requires a pileup correction for good hard X-ray spectroscopy (see below), but not a particularly difficult or nonlinear one. The count rate at the peak of the X4.8 flare was between 20 000 and 25 000 counts $s^{-1}$ in each front segment, close to the maximum throughput. The maximum throughput is about 20% higher when both shutters are out and the flare is dominated by soft photons that don't trigger the pileup suppression. A count rate several times higher can be processed by the fast rate counters.

Figure 8 shows the effective area of the RHESSI array for the front segments alone and the rear segments alone, excluding the effects of the grids. The effect of the attenuators on the front segment effective area is also shown; their effect on the rear segment is negligible. The curve labeled 'F/R COINC.' represents events that leave part of their energy in each segment, adding up to the full energy. These events are not yet used by the analysis software, but the adaptation, which requires new ways of dealing with background and gain, is in progress. Such 'coincidence' events between segments of *different* detectors present even more complications and add less added effective area, so they will be addressed only after front/rear coincidences have been incorporated.

The grids reduce the effective areas shown by factors strongly dependent on energy and the offset of the flare from the imaging axis, ranging from roughly 4 to 6 below 100 keV. The response of the spectrometer is generated by GEANT3, a high-energy photon and particle transport code that is used to generate the response matrix of the instrument (see below). Before flight, the GEANT3 performance was calibrated by taking spectrometer data in the laboratory using radioisotopes with lines from 3.7 keV to 6.1 MeV, placed in many positions above and around the spectrometer. The laboratory data were compared with GEANT3 simulations of the same configurations, and the detectors' internal segmentation boundaries in GEANT3 were adjusted until the GEANT3 results matched the data.

Although the effective areas of the front and rear segments in Figure 8 cross at about 250 keV, this is not where the *sensitivity* of the two segments crosses. Because the background in the rear segments is a factor of 10 higher (see below), the front segments are more sensitive in a background-dominated situation up to about 800 keV.

### 4.3. Background

RHESSI is not designed to be a low-background instrument. Heavy shielding was not possible within the Small Explorer weight constraints. A launch into an equatorial orbit was not available, which would have reduced every background component except cosmic diffuse radiation by minimizing cosmic-ray exposure and eliminating exposure to trapped protons in the South Atlantic Anomaly (SAA). Fortunately, solar flares are bright, and most RHESSI flare science will still be done using data that are source-count limited rather than limited by statistical fluctuations in the background.

Figure 9 shows the background in the RHESSI front and rear segments. The front spectrum is shown both with the segment operated on its own and in anticoincidence with the rear segment. The latter mode is similar to 'phoswich' detectors made of two layers of scintillator, i.e. the rear segment serves as a closely-placed anticoincidence shield to reject events that Compton scatter into or out of the front segment and charged-particle events. At most energies continuum is dominant, which is typical for an unshielded instrument. The continuum is comparable to

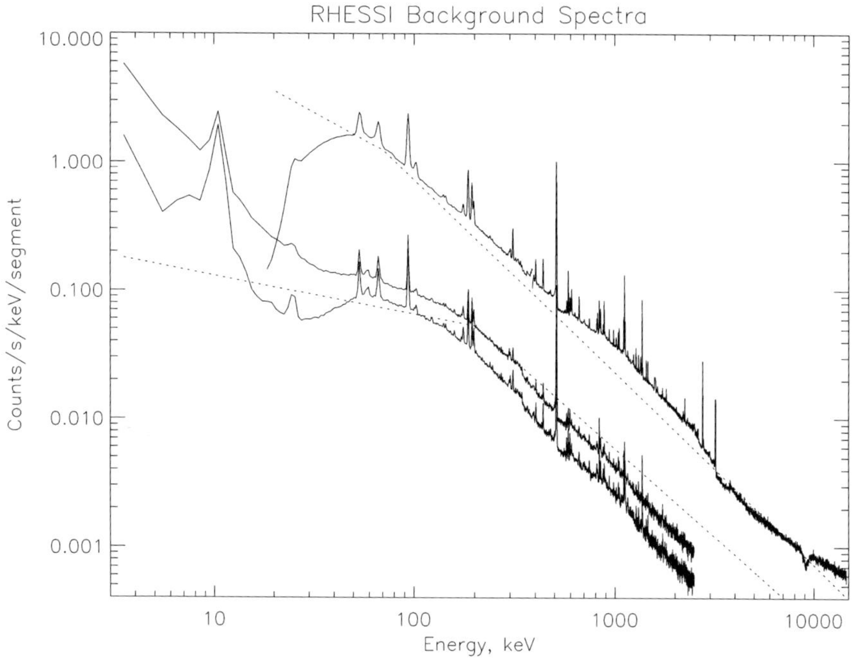


*Figure 9.* Typical RHESSI background spectra, accumulated over 2.5 hr. The three *solid traces*, top to bottom, are the backgrounds in the rear segments, the front segments, and the front segments using front/rear anticoincidence ('phoswich mode'). The *dotted lines* are pre-flight predictions of the background that included continuum from the cosmic diffuse background and from cosmic rays interacting in the Earth's atmosphere and in the spacecraft.

preflight estimates (dotted lines in Figure 9). In the simulations that produced the estimates, the dominant background component is continuum from the Earth's atmosphere. Below about 100 keV in the rear segments, cosmic diffuse emission dominates. Prompt cosmic-ray interactions with the spacecraft, induced radioactivity of spacecraft materials and $\beta$-decays in the detectors also contribute.

Most of the prominent lines in the background spectra are due to the activation of the germanium detectors themselves (e.g., 10, 54, 67, 93, and 198 keV). The brightest higher-energy line, however, is the positron-annihilation line at 511 keV. Much of this comes from the atmosphere, but there is a significant contribution due to the activation of spacecraft materials: immediately after RHESSI exits the South Atlantic Anomaly (SAA), the 511 keV line flux is several times its average value for orbits that don't include the SAA, and the excess decays over the next hour or so.

The three primary sources of variation in the RHESSI background are passes through the SAA (several a day on successive orbits), smooth modulations due to changes in geomagnetic latitude (and therefore cosmic-ray flux) over each orbit, and occasional periods of electron precipitation from the outer radiation belt when the spacecraft is at its highest geomagnetic latitudes (about 40–50°). Bremsstrahlung produced by these electrons in the spacecraft and the Earth's atmosphere can temporarily increase the background rate by more than an order of magnitude at times. These events can be distinguished from solar flares because they appear more strongly in the larger rear segments, whereas flares always produce more counts in the front segments. In addition, the CPD counts electrons at a significant rate in all but the smallest precipitation events.

### 4.4. INSTRUMENTAL ANOMALIES

#### 4.4.1. *Detector Dropouts*

On orbit, we find that each detector segment can spontaneously shut off for a brief period ranging from tens to hundreds of milliseconds. These dropouts occur up to once every few seconds in the front segments and generally much less often in the rear segments, although this is reversed in detector G5. The cause is still under investigation, but we expect it is a response of the electronics to heavy cosmic-ray nuclei, which leave more energy in the detector than anything available on the ground during pre-flight testing. This phenomenon does not have a significant effect on spectroscopy, since it is energy independent, nor on imaging, since the dropouts do not correlate with the count-rate modulations used to reconstruct images. It does have an effect on photometry and lightcurves, and this is addressed by the data-analysis software, which can recognize the dropouts and compensate for them by adjusting the livetime accordingly. During periods when the count rate is relatively low, dropouts are distinguished from merely quiet moments because they always begin with a preamplifier-reset event.

#### 4.4.2. *Individual Detector Anomalies*

Shortly after its high voltage was turned on, detector G2 suffered from a breakdown either on the rear surface of the crystal or somewhere along its high voltage circuit, such that it can no longer be operated above about 2400 V. At this voltage, the innermost part of the crystal is not depleted, and therefore the contact break on the inner bore (see Figure 1(A)) is not seen: G2 operates as a monolithic detector, with all events anywhere in the crystal going through the front segment electronics. In addition to losing the various advantages of segmentation, this also results in extremely noisy operation: energy resolution of about 10 keV FWHM and a low-energy cutoff closer to 20 keV than the 2.7 keV of most of the front segments. Therefore this detector should not be used for spectroscopy, although it can still be used for imaging from about 20 keV to the energy at which its grids, which are the second-finest of the nine, become transparent.

Detector G8, which is located just a few centimeters from the aft spacecraft antenna, becomes noisy when that antenna is switched on. This can occur several times a day for a few minutes at a time during contacts with the spacecraft. A database of times when this antenna is active is on line, but has not yet been integrated with the data analysis software.

Finally, detector G7 has a slight problem with its front segment signal contact, which causes its energy resolution to be degraded from 1 keV to 3 keV FWHM and its energy threshold to be raised from about the normal 2.7 keV to 7 keV. This behavior has been understood and unchanged since before launch.

#### 4.4.3. *Spectral Artifacts*

A very small fraction of events near or above 3 MeV that should be analyzed with the high-energy (low-gain) range of the rear segments do not trigger the gain shift in the fast channel. These pile up in the top 64 channels (8128–8191) of the high-gain scale, and produce a spectral artifact there. A similar artifact appears in the 64 channels preceding the halfway point of the spectrum (i.e., channels 4032–4095) (Share *et al.*, 2002). Routines to automatically eliminate these channel ranges for each detector will be implemented but are not yet in place.

#### 4.4.4. *Image Events*

When the clouds of electrons and holes liberated by a gamma-ray event move through the detector, they create induced charges on the electrodes of the segment in which they are moving. The change in time of this induced charge is the current pulse that is amplified and integrated by the electronics to become the detected event. However, they also induce charges on the electrodes of the empty segment. The difference is that this image charge reverses sign in the empty segment as the clouds approach the electrode of the segment in which they are actually moving. The result is that the image signals (current versus time) in the empty segment are bipolar in shape and integrate to zero charge.

The RHESSI electronics, however, do not integrate the signals for an infinite time. Therefore a small amount at the very end of each pulse is not counted. Since it is the negative part of the bipolar signal in an empty segment that comes last, the result is that there is a very small, positive residual from the bipolar signal: in other words, a 1 MeV event detected in one segment will create a simultaneous (but false) event in the other segment of a few keV. Fortunately, the timing of an image event is a couple of microseconds different from that of a front/rear coincidence due to Compton scattering, so the image events can be weeded out of the data on the ground without eliminating Compton scattered events, if that is desired.

## 5. Spectral Data Analysis

Spectral analysis is an inverse problem that begins with a spectrum of counts per spectrometer channel and seeks to recover the spectrum of photons per energy interval initially incident on the spacecraft. Figure 10 is a block diagram of the data analysis process for RHESSI spectra. The RHESSI data analysis software is written to be part of the SolarSoft system (Freeland and Handy, 1996), and, like all the data, is publicly available without a proprietary period. Spectral analysis in par-

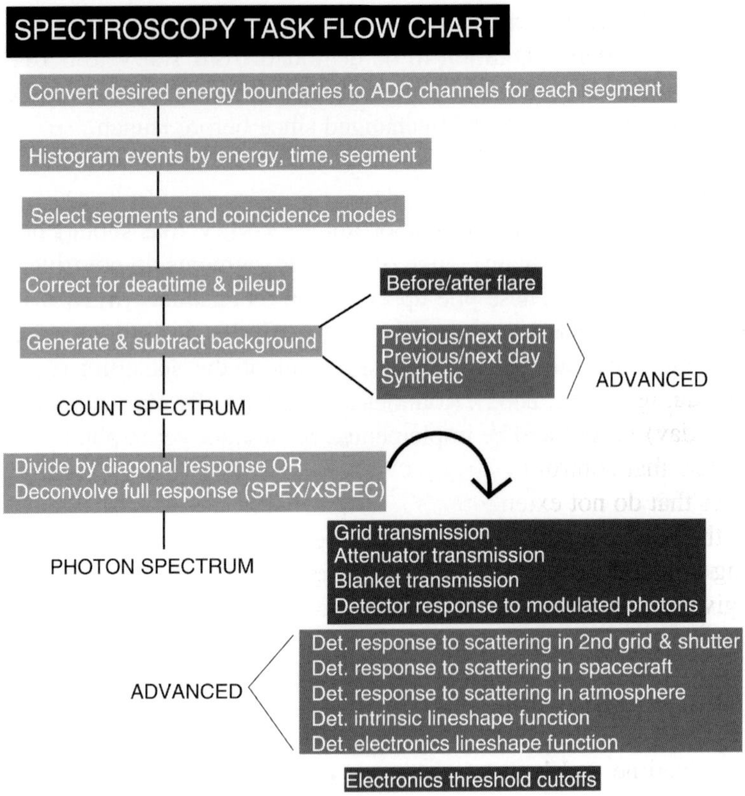


*Figure 10.* Flow chart of a spectroscopy analysis. 'Advanced' features are needed only for a few flares, mostly the highest-energy ones.

ticular is accomplished with the SPEX spectral-inversion code (Schwartz, 1996), which has heritage from the Burst and Transient Source Experiment (BATSE) on the Compton Gamma-Ray Observatory. It would be beyond the scope of this paper to begin a tutorial on the use of the software, but it is worthwhile to summarize the steps that are needed for an accurate spectral result. As the software evolves, more and more of this will become automated and transparent to the user.

### 5.1. Gain, Livetime, and Background

The first stage of analysis is to correct for gain drift and deadtime. Even our preliminary gain analysis is yielding excellent results (see Figure 7) due to the high linearity of the electronics and the narrow, easily identified background lines. The deadtime correction routines start with the livetime counter in the electronics, correct for the double rejection of piled-up events, and compensate for data dropouts.

The next task is to identify and subtract background. At present this is most easily done by selecting data intervals just before and just after the flare and subtracting the spectra in those intervals from the spectrum at the time of interest (whether this is a small part of the flare or its entire duration). When the flare is long, this can be inaccurate, since the background varies significantly over tens of minutes. The results are improved by SPEX's ability to take several such intervals before and after the flare and fit a low-order polynomial to the background variation in each energy channel. At low energies, most flares are bright enough that the background is small compared to the flare flux, and this method of subtraction is suitable.

For a very few flares, more precise methods of background estimation are desirable. These include very long flares and flares with faint hard emission or faint gamma-ray lines. We are developing several additional background-subtraction options, including: (1) using as background the periods exactly 15 orbits (approximately one day) before and after the flare interval, since at these times most of the parameters that control the background had the same value as during the flare; (2) for flares that do not extend much above 50 keV, using the simultaneous background in the rear segments to predict the background in the front segments; and (3) training a neural network to produce a predictive model of the background at any time given a few parameters as input (such as geomagnetic latitude, time since last passage through the SAA, etc.).

### 5.2. Response Matrix

Once gain, livetime and background are accounted for, a series of instrumental effects that together constitute the 'response matrix' of the instrument to flare photons must be removed. There are many effects that modify the input spectrum, including:

- absorption in the mylar blankets, cryostat windows, and grids;
- Compton scattering into and out of the detectors;
- Compton scattering off the Earth's atmosphere, which can dominate the flare count rate in the rear segments below 100 keV;
- noise in the electronics;
- resolution degradation due to radiation damage; and
- the low-energy cutoff imposed by the electronics.

All these effects are accounted for when the analysis software creates a response matrix. Each of these effects contributes either to the diagonal elements of the matrix (i.e. the efficiency of the instrument for detecting photons at their proper energy) or to the off-diagonal elements (moving photons from their true energy to another energy, usually lower), or to both.

When the user is only interested in isolated gamma-ray lines, the response is just the efficiency for photopeak detection, and the conversion from counts to photons is done immediately by dividing the count spectrum by the diagonal terms of the response matrix. This is also adequate for hard X ray flares with no significant component above 100 keV, since the response of the front segments below this

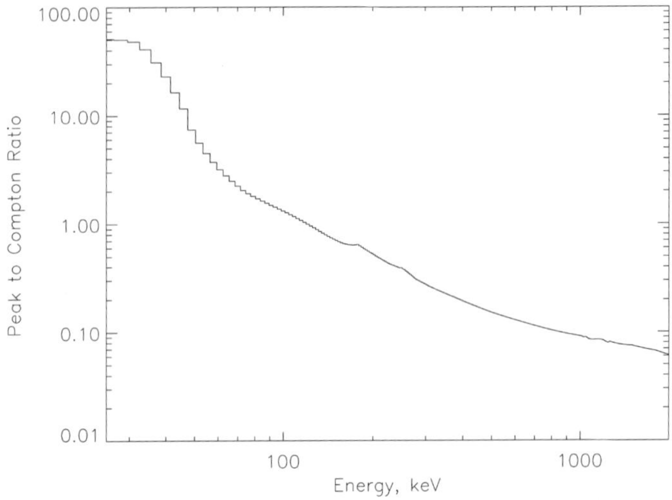


*Figure 11.* RHESSI response matrices: photopeak to Compton-continuum ratio versus energy. The continuum contains every simulated photon which leaves less than its full energy but more than 3 keV in the detectors.

point is dominated by complete absorption, not scattering. Figure 11 shows the ratio of photopeak to partial-energy (Compton-scattered) detections in RHESSI versus energy. Since most flare spectra are steeply falling, the measured count rate in any given energy band is usually dominated by photopeak counts up to an energy well beyond 100 keV. One exception is at the lowest energies (below 15 keV) when one or both of the shutters are in. In this case, the true low-energy counts are strongly reduced and the spectrum can be dominated by non-photopeak counts created when a 15–25 keV photon is photoelectrically absorbed in the germanium but the K-shell fluorescence photon (about 10 keV) escapes.

For studying low energies with the shutters in, or for studying continuum and broad-band emission at high energies, the non-diagonal response must be accounted for. This can be done automatically by SPEX using forward-folding. The user specifies a model form for the flare spectrum, which can be a combination of simple functions (power laws, Gaussians, etc.) and physics-based spectral forms (e.g., a set of known nuclear lines from a particular element bombarded by energetic protons, or a thin-target bremsstrahlung spectrum from a monoenergetic electron beam). The software will then multiply this spectrum by the full response matrix, check the goodness of fit to the observed count spectrum, and repeat the process, varying the parameters of the input model until the best fit is found. The output of this process is either the best-fit parameters themselves, or else a spectrum created by multiplying the observed count spectrum by the ratio of the model photon spectrum to the model count spectrum.

Figure 12 shows the modeled RHESSI front-segment response to input photons at three energies: 50, 350, and 2500 keV. At 50 keV, the photopeak dominates but

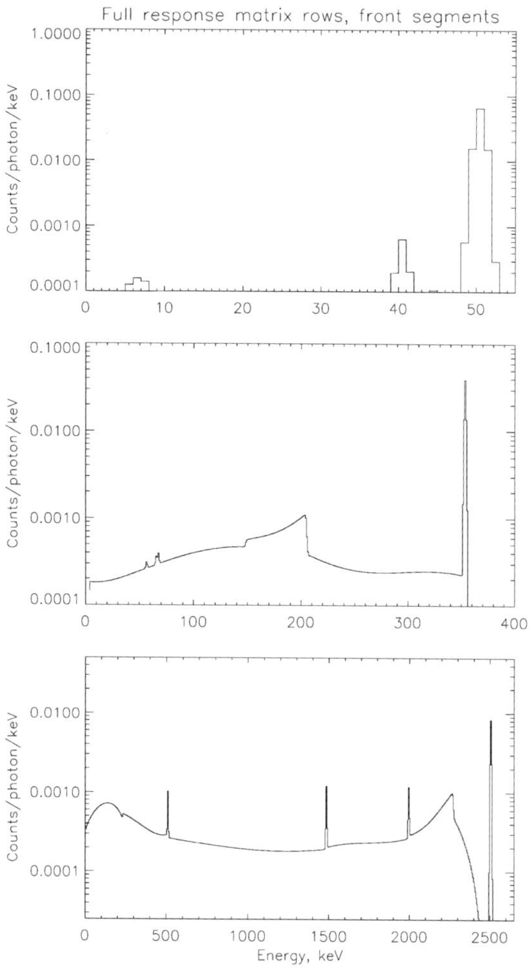


*Figure 12.* RHESSI response matrices: sample responses at 50, 350, and 2500 keV (*top to bottom*).

the K-shell escape peak is visible. At 350 keV, a strong Compton continuum is seen, along with small K-shell fluorescence peaks from nearby passive material, including the tungsten RMC grids. At 2500 keV, pair-production can occur and there are three narrow lines due to positron annihilation (511 keV) and to the escape of one or two positron-annihilation photons from the detector.

In addition to SPEX, we are developing tools to export the RHESSI count spectra and response matrices to the XSPEC package (Arnaud, 1996). XSPEC has a wider variety of built-in spectral models than SPEX and is often used to combine simultaneous spectra from different instruments, which may be useful in the future if RHESSI and INTEGRAL or another instrument observe the same flare. We also plan to develop an algorithm for model-independent inversion of the spectra (Johns and Lin, 1992; Smith *et al.*, 1995). This algorithm will probably be most useful for

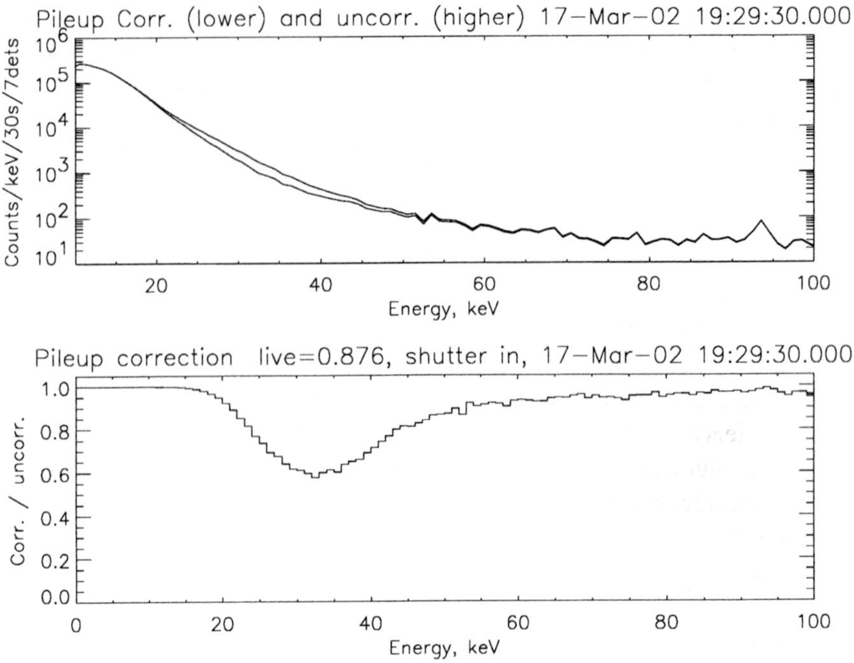


*Figure 13*. Pileup correction at modest deadtime with the thin shutter in place. *Top*: count spectra before and after pileup correction. *Bottom*: ratio of corrected to uncorrected spectrum as a function of energy.

spectra that are dominated by the continuum, not lines, but that extend to high enough energies that simply dividing by the efficiency is insufficiently accurate.

## 5.3. PILEUP

The pileup correction is the most difficult to implement. It is an off-diagonal effect in that it redistributes counts to different energies, but, unlike Compton scattering, its effect is also non-linear: the contribution of 100 keV photons to the piled-up spectrum depends not only on the number of 100 keV photons coming in, but also on the shape of the rest of the spectrum as well. For this reason it cannot be made part of the response matrix. A full, correct treatment of pileup requires it to be treated as a *separate* stage of forward folding, occurring after the model spectrum has been folded through the response matrix. The (real) background must be added to the forward-folded model flux before pileup is simulated. Furthermore, since the deadtime and degree of pileup will vary rapidly as the grids modulate a bright flare, working with the *average* count rate to estimate pileup is only an approximation. For imaging in particular, future livetime and pileup corrections will have to be fully time-dependent down to the millisecond scale.

These refinements are still far in the future for the RHESSI analysis software, but in the meantime we are implementing an approximate pileup correction that can be applied early in the analysis flow (see Figure 10) without the need of a spectral model. Figure 13 shows the effect of that pileup correction on a spectrum taken with the thin shutter in and an average of 12.4% deadtime. As we noted above, under the current flight software version the thick shutter would normally be brought in to cut down on pileup at 10% deadtime.

## Acknowledgements

The authors would like to thank Gerald Share of the Naval Research Laboratory for providing a valuable external perspective and advice throughout the development process and since launch. We would like to recognize our colleagues at the Space Sciences Laboratory who participated in calibration and testing of the spectrometer: engineers Peter Harvey (also the RHESSI Project Manager) and Jane Hoberman, technician Michelle Galloway, and students Daniel Main, Albert Shih, Lewis Hyatt, Eunyong Cho, Scott Pollack, Mark Chu, and Carmel Levitan. Finally, we would like to thank the technicians, machinists, and others who helped build the RHESSI spectrometer.

The instrument could not have been what it is without the contributions of our colleagues who passed away before RHESSI was launched: Reuven Ramaty and Natalie Mandzhavidze, whose work on gamma-ray line emission in flares inspired its high-resolution design, and Robert D. Campbell, who designed its particle detector and worked tirelessly to characterize and calibrate its germanium detectors. We remember them with gratitude and respect.

## References

Arnaud, K.: 1996, in G. Jacoby and J. Barnes (eds.), *Astronomical Data Analysis Software and Systems V*, Astronomical Society of the Pacific, San Francisco, U.S.A., p. 17.

Curtis, D. W. *et al.*: 2002, *Solar Phys.*, this volume.

Freeland, S. L. and Handy, B. N.: 1998, *Solar Phys.* **182**, 497.

Hull, E. L.: 1998, Ph. D. dissertation, Indiana University, U.S.A.

Hurford, G. *et al.*: 2002, *Solar Phys.*, this volume.

Johns, C. M. and Lin, R. P.: 1992, *Solar Phys.* **137**, 121 (Erratum: *Solar Phys.* **142**, 219).

Koenen, M., Bruckner, J., Korfer, M., Taylor, I. and Wanke, H.: 1995, *IEEE Trans. Nucl. Sci.* **42**, 653.

Krucker, S., Christe, S., Lin, R.P., Hurford, G.J., and Schwartz, R.A.: 2002, *Solar Phys.*, this volume.

Landis, D. A., Cork, C. P., and Goulding, F. S.: 1982, *IEEE Trans. Nucl. Sci.* **29**, 1125.

Lin, R. P. and Schwartz, R.: 1987, *Astrophys. J.* **312**, 462.

Lin, R. P. *et al.*: 2002, *Solar Phys.*, this volume.

McConnell, M.L., Ryan, J.M., Smith, D.M., Lin, R.P., and Emslie, A.G.: 2002, *Solar Phys.*, this volume.

Schwartz, R. A. 1996, 'Compton Gamma Ray Observatory Phase 4 Guest Investigator Program: Solar Flare Hard X-ray Spectroscopy,' Technical Report, NASA Goddard Space Flight Center.

Share, G. H., Murphy, R. J., and Ryan, J.: 1997, in C. D. Dermer, M. S. Strickman, and J. D. Kurfess (eds.), *Proceedings of the Fourth Compton Symposium*, American Institute of Physics, Woodbury, New York, U.S.A., p. 17.

Share, G. H. *et al.*: 2002, *Solar Phys.*, this volume.

Smith, D. M. *et al.*: 1995, *J. Geophys. Res.* **100**, 19675.

Smith, D. M. *et al.*: 2000, 'Extra-Solar Astrophysics with the High-Energy Solar Spectroscopic Imager (HESSI),' *AIP Conf. Proc., The 5th Compton Symposium*, AIP New York, AIP), pp. 510, 671.

Smith, D. M. *et al.*: 2002, 'Non-solar Astronomy with the Reuven Ramaty High-Energy Solar Spectroscopic Imager (RHESSI),' *SPIE Conf. Proc.* **4851**, in press.

Vestrand, W. T. *et al.*: 1999, *Astrophys. J. Suppl.* **120**, 409.

Yoshimori, M. *et al.*: 1994, *Astrophys. J. Suppl.* **90**, 639.

# THE RHESSI IMAGING CONCEPT

G. J. HURFORD[1], E. J. SCHMAHL[2,3], R. A. SCHWARTZ[3,4], A. J. CONWAY[5], M. J. ASCHWANDEN[6], A. CSILLAGHY[1,7], B. R. DENNIS[3], C. JOHNS-KRULL[8], S. KRUCKER[1], R. P. LIN[1,9], J. MCTIERNAN[1], T. R. METCALF[6], J. SATO[10] and D. M. SMITH[1]

[1]*Space Sciences Laboratory, University of California, Berkeley, CA 94720, U.S.A.*
[2]*Astronomy Department, University of Maryland, College Park, MD 20742, U.S.A.*
[3]*Lab for Astronomy and Solar Physics, NASA Goddard Space Flight Center, Greenbelt, MD 20771, U.S.A.*
[4]*NASA Goddard Space Flight Center / SSAI, Greenbelt, MD 20771, U.S.A.*
[5]*Department of Physics and Astronomy, The Open University, Milton Keynes MK7 6AA, U.K.*
[6]*Lockheed Martin Advanced Technology Center, Solar & Astrophysics Laboratory, Dept. L9-41, Palo Alto, CA 94304, U.S.A.*
[7]*University of Applied Sciences, CH-5210 Windisch, Switzerland*
[8]*Department of Physics and Astronomy, Rice University, Houston, TX 77005, U.S.A.*
[9]*Department of Physics, University of California, Berkeley, CA 94720, U.S.A.*
[10]*Department of Physics, Montana State University, Bozeman, MT 59717, U.S.A.*

(Received 17 September 2002; accepted 17 September 2002)

**Abstract.** The Reuven Ramaty High-Energy Solar Spectroscopic Imager (RHESSI) observes solar hard X-rays and gamma-rays from 3 keV to 17 MeV with spatial resolution as high as 2.3 arc sec. Instead of focusing optics, imaging is based on nine rotating modulation collimators that time-modulate the incident flux as the spacecraft rotates. Starting from the arrival time of individual photons, ground-based software then uses the modulated signals to reconstruct images of the source. The purpose of this paper is to convey both an intuitive feel and the mathematical basis for this imaging process. Following a review of the relevant hardware, the imaging principles and the basic back-projection method are described, along with their relation to Fourier transforms. Several specific algorithms (Clean, MEM, Pixons and Forward-Fitting) applicable to RHESSI imaging are briefly described. The characteristic strengths and weaknesses of this type of imaging are summarized.

## 1. Introduction

The primary scientific objective of the Reuven Ramaty High-Energy Solar Spectroscopic Imager (RHESSI) is the study of energy release and particle acceleration in solar flares. This is accomplished by imaging-spectroscopy of solar hard X-rays and gamma-rays over a 3-keV to 17-MeV energy range with energy resolution of $\sim 1$ keV, time resolution of $\sim 2$ s or better, and spatial resolution as high as 2.3 arc sec.

The only practical method of combining such angular resolution with high sensitivity in this energy range within the cost, mass and launch constraints of a small satellite is to use collimator-based Fourier-transform imaging. Coded-aperture imag-

*Solar Physics* **210:** 61–86, 2002.

ing, often used in astrophysics, would not be practical in this situation, since it would require a hard X-ray detector with $\sim 20$ micron spatial resolution to achieve this angular resolution. (See Prince *et al.* (1988) for a review of imaging techniques.) One of the most powerful of the Fourier family of techniques is rotational modulation synthesis, first proposed by Mertz (1967) and implemented by Schnopper, Thompson and Watt (1968) for non-solar observations.

In a solar context, previous related instrumentation included the Hard X-ray Imaging Spectrometer (HXIS) (Van Beek *et al.*, 1980), a direct-imaging hard X-ray telescope flown on the Solar Maximum Mission. Although its multi-grid collimator (divided into $\sim 10^3$ subcollimators) achieved 8 arc sec resolution, its one-to-one association of imaging pixels to detector elements seriously limited its sensitivity. A Rotating Modulation Collimator (RMC) was used for solar flare X-ray imaging with angular resolution of $28''$ in the 20–40 keV energy range on the *Hinotori* mission (Makishima *et al.*, 1977; Ohki *et al.*, 1982; Enome, 1982). Subsequently, the Hard X-ray Telescope (HXT) on the *Yohkoh* satellite used non-rotating Fourier synthesis with angular resolution of $\sim 8''$ in the 20–100 keV energy range (Kosugi *et al.*, 1991). A balloon-borne solar telescope with two RMCs, the High Energy Imaging Device (HEIDI) (Crannell, 1994) was used as a vehicle for making several relevant engineering advances including the demonstration of a high-bandwidth, high-resolution solar aspect system and the development of error-analysis techniques for the RMC optical design.

Among the new features of RHESSI imaging are its high angular resolution, its use of aspect knowledge in place of precision pointing, its fine energy resolution and wide energy range viewed with a common set of grid 'optics', its relative immunity to alignment errors, its ability to self-calibrate its own instrumental response and its measurement of a large number ($\sim 10^3$) of Fourier components for improved image quality.

An overview of the RHESSI mission is provided by Lin *et al.* (2002). Further information can be found at the following web sites: *http://ssl .berkeley.edu/hessi* and *http://hesperia.gsfc.nasa.gov/rhessidatacenter*. The purpose of this paper is to describe the concepts and techniques of RMC imaging as implemented on RHESSI. An appreciation of the strengths and weaknesses of these techniques may prove useful both for making such images and for evaluating them.

## 2. The RHESSI Imager – a Brief Hardware Description

### 2.1. The Subcollimators

The RHESSI imaging hardware is described in detail by Zehnder *et al.* (2002). A schematic view (Figure 1) shows a set of nine bi-grid subcollimators, each consisting of a pair of widely separated grids in front of a corresponding non-imaging X-ray/gamma-ray detector. Each grid consists of a planar array of equally-spaced,

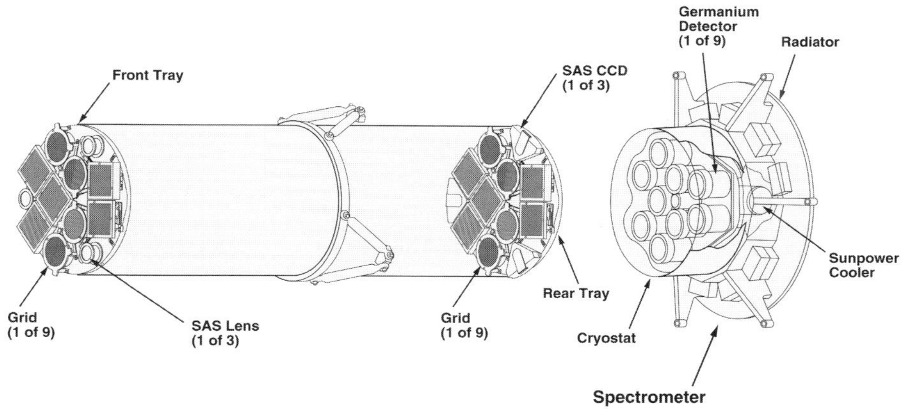


*Figure 1.* Perspective of the RHESSI imager. The key imager components are two identical sets of nine grids mounted on front and rear grid trays. A corresponding set of nine cooled germanium detectors is mounted behind the rear grids. The solar aspect system (SAS) consists of three lenses mounted on the front grid tray which focus optical images onto SAS CCDs on the rear grid tray.

TABLE I

RHESSI – nominal grid parameters.

| Subcollimator number | 1 | 2 | 3 | 4 | 5 | 6 | 7 | 8 | 9 |
|---|---|---|---|---|---|---|---|---|---|
| Pitch (mm) | 0.034 | 0.059 | 0.102 | 0.177 | 0.306 | 0.530 | 0.918 | 1.590 | 2.754 |
| slit width (mm) | 0.020 | 0.035 | 0.061 | 0.106 | 0.184 | 0.318 | 0.477 | 0.811 | 1.487 |
| FWHM resolution (arc sec) | 2.26 | 3.92 | 6.79 | 11.76 | 20.36 | 35.27 | 61.08 | 105.8 | 183.2 |
| Max. transmission | 0.60 | 0.60 | 0.60 | 0.60 | 0.60 | 0.60 | 0.52 | 0.51 | 0.54 |
| Grid thickness (mm) | 1.2 | 2.1 | 3.6 | 6.2 | 10.7 | 18.6 | 6.2 | 6.2 | 30.0 |
| Slat material | Mo | W | W | W | W | W | W | W | W |
| Field of view (deg) | 1.0 | 1.0 | 1.0 | 1.0 | 1.0 | 1.0 | 4.4 | 7.5 | 2.8 |

X-ray-opaque slats separated by transparent slits. Within each subcollimator, the slits of the two grids are parallel and their pitches are identical. The nominal parameters of the RHESSI grids are listed in Table I. Details of the grid geometry, calibration and response as a function of energy will be given in a forthcoming paper by Hurford *et al.*

As illustrated in Figure 2, the transmission through the grid pair depends on the direction of the incident X-rays. If the direction of incidence is changed as a function of time, the transmission of the grid pair is modulated in time as the shadow of the slats in the top grid alternately falls on the slits or slats in the rear grid. For slits and slats of equal width, the transmission is time-modulated from zero to 50% and back to zero as the direction to the source changes. One cycle

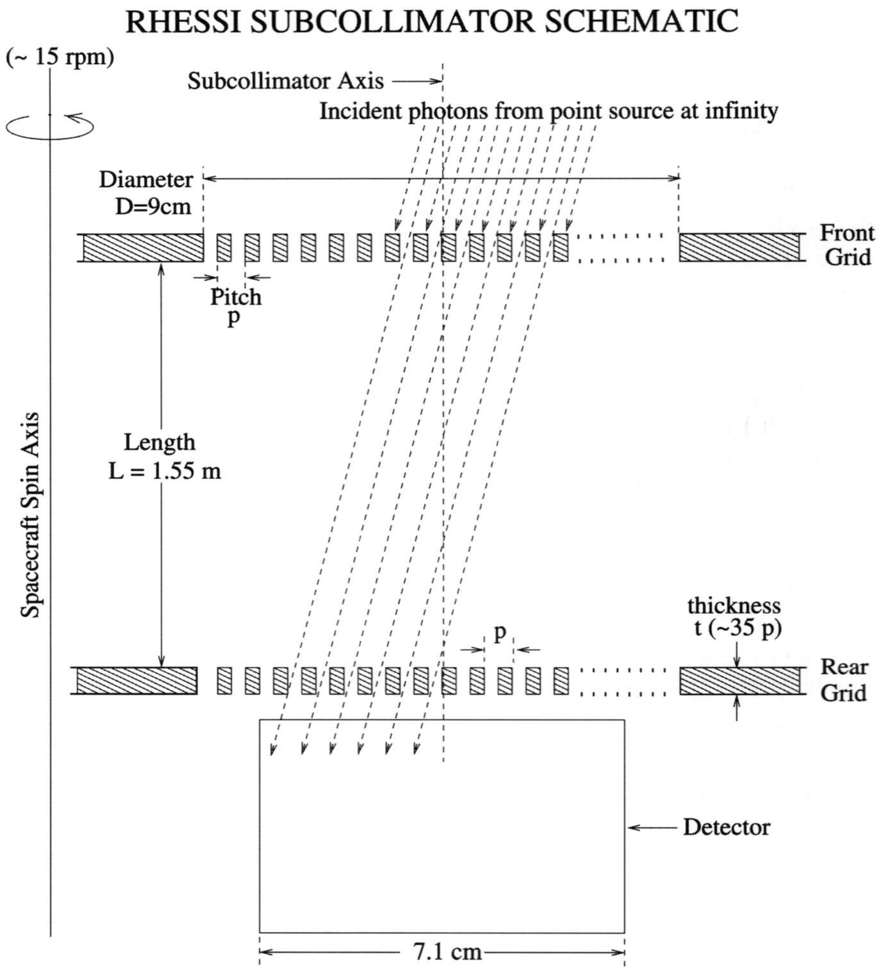


*Figure 2.* Schematic geometry of the RHESSI subcollimators, showing representative incident photons with respect to the collimator axis.

of this periodic transmission occurs for a change in source angle (in the plane orthogonal to the slits) of $p/L$ where $L$ is the separation between grids (1550 mm). The angular resolution is defined as $p/(2L)$. For off-axis sources, changing the angle between the source and collimator is achieved by rotating the spacecraft at $\sim 15$ revolutions per minute.

### 2.2. OTHER SUBSYSTEMS

The role of the detector and data system is to record the arrival time and energy of each photon detected, allowing the modulated count rate to be determined as a function of rotation angle.

The detectors, described by Smith *et al.* (2002), are high-purity germanium crystals, each electrically divided into front and rear segments. The $\sim$ 1-cm thick front segment is sensitive from 3 keV to $\sim$ 200 keV while the $\sim$ 7-cm thick rear segment extends the response to 17 MeV. Detector segmentation shields the rear segment from the intense flux of low-energy X-rays during moderate flares. Since the detectors themselves have no spatial resolution, they are optimized for high sensitivity and energy resolution. The detectors are mounted in a cryostat and are mechanically cooled to $\sim$ 75 K.

The electrical output pulse from each detected photon is amplified, shaped, digitized and passed to the Instrument Data Processing Unit (IDPU) (Curtis *et al.*, 2002). The IDPU stores information about each photon as a 32-bit event word that includes the detector ID, a 14-bit energy tag and the arrival time. This time is recorded with 1 microsecond resolution to enable all combinations of coincidence events between segments and detectors to be inferred during data analysis. A 4-Gbyte solid-state recorder, corresponding to almost $10^9$ events, stores the data for periodic downloading. An average of 1.8 Gbytes per day is transmitted to the ground.

A high-bandwidth solar aspect system (SAS) (Zehnder *et al.*, 2002) permits substitution of accurate pointing knowledge for pointing stability, which need only be controlled to arcminutes. The SAS consists of 3 lenses in the front grid plane that focus solar images onto a set of 3 linear diode arrays in the rear grid plane. The solar limb is determined at 6 positions (2 per array) at rates up to 128 Hz (16 Hz typical), giving pitch and yaw to $\sim 0.4''$ r.m.s.

Roll aspect is provided by one of two redundant star scanners. Imaging to date has used a Photo-Multiplier-Tube Roll Aspect System (PMTRAS) (Hurford and Curtis, 2002). This system views the star field perpendicular to the Earth–Sun line and determines absolute roll to $\sim$ 1 arcminute by noting the times at which bright stars are detected as the spacecraft rotates. Data are also available from the other roll aspect system (RAS) (Zehnder *et al.*, 2002), which is based on a similar concept but uses a linear photodiode in place of the photomultiplier.

The data handling concept outlined above must accommodate potential count rates in excess of $10^6$ s$^{-1}$ detector$^{-1}$ where the finite response time of the detector and data handling system become relevant (Smith *et al.*, 2002). This accommodation employs four techniques, the goals of which are to preserve the sensitivity to rare high-energy gamma-rays while maintaining the ability to image lower energy X-rays, whose incident flux can be $\sim$ 12 orders of magnitude higher.

The first technique is to apply corrections for the estimated dead time of the detector/electronics. Spare bits in the event words are used to encode this dead time with 512 microsecond time resolution. This is sufficient to follow the effects of modulation for all the grids.

Second, as the dead time becomes larger, on-board software commands either or both of two sets of nine aluminum attenuators (Smith *et al.*, 2002) to be mechanically inserted between the rear grids and the detectors. The attenuators reduce the flux of low-energy photons that reach the detectors while having no effect above $\sim 100$ keV.

Third, when count rates become sufficiently high that the modulation is suppressed by dead time, an additional technique, capable of handling higher count rates, is automatically activated. In this 'fast rate mode', front detector counts are sorted into just 4 energy channels with time bins that are sufficiently short to preserve the modulation. This permits imaging at higher count rates than is possible by fully digitizing and time-tagging individual photons.

A fourth technique is used to deal with the finite size of the solid-state recorder. As the recorder becomes filled, a decimation scheme is automatically enabled to digitally discard a fixed fraction of front-segment events below an energy threshold. Both the fraction and the energy thresholds can be preset by ground commands. Discarding events in this way does not introduce any bias that would affect the modulation.

Two features of the detector response, both discussed by Smith *et al.* (2002), are particularly relevant in an imaging context. The first is that substantial data gaps (up to several hundred milliseconds) are observed. Since these gaps have a characteristic signature, the time and duration of their occurrence can be determined independently and are treated as periods of zero live time.

The second feature is pulse pileup (Datlowe, 1975) whereby two or more low-energy photons arrive 'simultaneously' and are indistinguishable from a single higher-energy photon. This has long been a complication for solar X-ray spectroscopy and at very high count rates it can also introduce artifacts in the imaging. The pileup counts are generated at a rate that goes roughly as the square of the detected low-energy count rates. Therefore, they are also modulated but appear at higher energies. This can result in 'ghost' low-energy sources appearing in images nominally formed from high-energy photons.

## 3. Modulation Principles

### 3.1. The Modulation Profile

As described above, the RHESSI imaging hardware uses a set of rotating collimators to time-modulate the detected photon flux. To understand how this encodes imaging information, it is conceptually useful to adopt the perspective of a rotating coordinate system fixed on the collimator. From this perspective, in the typical case of a distant source that is slightly offset from the collimator axis (illustrated in Figure 2), the source moves in a circle about the rotation axis. The component of source motion parallel to the slits does not cause modulation. It is the simple harmonic motion perpendicular to the slits that modulates the count rates. (A characteristic of the resulting modulation is that while its frequency varies, it is locally periodic in time over a limited range of rotation angles.)

For a single source, the various panels in Figure 3 show how the resulting modulated count rates (modulation profiles) depend on the source intensity, location and size. The first panel, with which the others may be compared, shows the modulation profile of a single point source. The second panel assumes a source with the same location, but one half the intensity. Since the response of the the collimator/detector system is linear, changing the intensity of the source just decreases the amplitude of the modulation without changing its shape. The third panel shows the effect of moving the source in azimuth about the rotation axis. This shifts the modulation profile in time. Moving the source further off-axis increases the number of modulation cycles per rotation as shown in panel 4. Increasing the diameter of the source while keeping its total intensity the same (panel 5) reduces the amplitude of the modulation while leaving the time-averaged transmission unaffected. Further increases in source size (compared to the resolution of the collimator) reduce the modulation still further (panel 6).

In practice, of course, real sources may be more complex, and the sum of their multiple components yields a modulation profile such as illustrated in the last panel in Figure 3. Thus, the central data analysis task for RHESSI imaging is the inverse problem of deducing the source geometry, given a set of observed modulation profiles from the different subcollimators. Much of the remainder of this paper will be devoted to the different approaches taken to the solution of this problem.

We can set the stage for solving this inverse problem by digitizing the modulation profile into a series of time bins, and by describing the brightness distribution in the source plane as a pixelized image $F_m$, where $F_m$ is the photon flux (photons cm$^{-2}$ s$^{-1}$) from pixel $m$ incident on RHESSI's front grids.

Neglecting background and for a specific energy interval, the following formal equation shows that the expected counts in the $i$th time bin is given by

$$\mathcal{C}_i = A \sum_m P_{im} F_m \Delta t_i . \tag{1}$$

In this notation, $P_{im}$ is the probability that a photon originating in pixel $m$ and incident on the front grid will be counted in the $i$th time bin during interval $\Delta t_i$ by a detector with area, $A$. (Note that since $m$ is the index of a two-dimensional map, $P_{im}$ is really a numerical 'cube'.) Since $P_{im}$ may be calculated from the grid properties and collimator aspect, the inverse problem can be summarized as follows: find the source map, $F_m$, given a measurement of count rates $\mathcal{C}_i$ in each time bin.

### 3.2. Describing the Grid Response

The solution of the inverse problem posed by Equation (1) requires a knowledge of $P_{im}$, the probability that a photon from map pixel $m$ will be detected in the $i$th time bin. This requires calculation of the transmission probability of a grid pair as a function of energy and direction of incidence. In this section, we describe the approach taken to accomplish this.

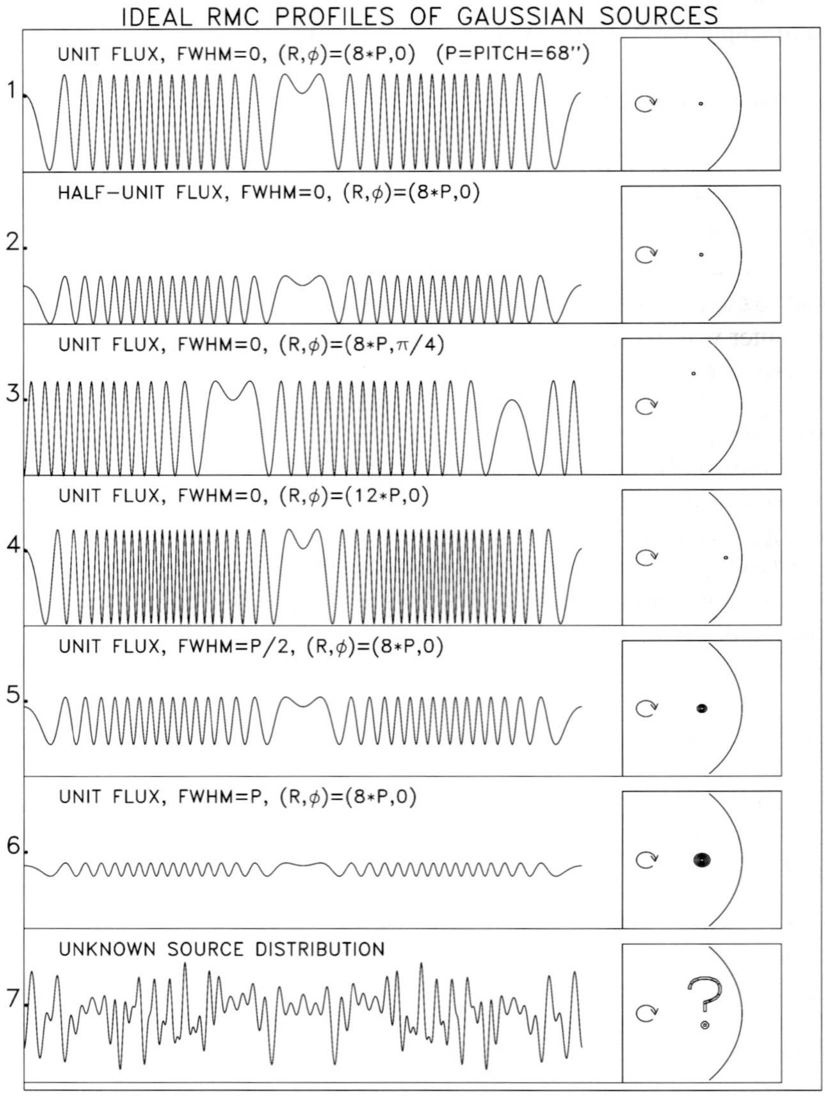


*Figure 3.* Modulation profiles plotted for one complete rotation for various configurations of an off-axis source, assuming ideal grids of pitch P with equal slits and slats mounted on a collimator that is rotating uniformly about a fixed axis. As discussed in the text, successive panels show the effect on the modulation profile of changing the source characteristics. R and $\phi$ are the radial offset and azimuth of the source position relative to the axis of rotation.

To introduce this issue, we return briefly to the idealized case of thin, opaque grids with equal slit and slat widths, for which the rapid modulation takes on a triangular form ranging between 0 and 50%. Since this time profile is periodic as a function of angle of incidence in the plane orthogonal to the slits, the transmission probability can be described in terms of the first few harmonics of its expansion as a Fourier cosine series containing only odd harmonics:

$$\wp_{\text{triangle}}(\Phi) = \frac{1}{4}\left(1 + \frac{8}{\pi^2}\cos(\Phi) + \frac{8}{9\pi^2}\cos(3\Phi) + \frac{8}{25\pi^2}\cos(5\Phi) + \cdots\right), \quad (2)$$

where $\Phi = 2\pi\theta L/p$. ($p$ is the grid pitch; $L$ is the distance between front and rear grids; $\theta$ is the angle of incidence in the plane orthogonal to the slits.) This is equivalent to the form used by Schnopper *et al.* (1970) for their observations of the Galactic center with a rocket-borne RMC.

There are several reasons why the triangular functions described by Equation (2) cannot be applied directly to RHESSI. As photon energies increase, the grids become transparent. The slits and slats are, in general, not equal. The peak transmissions are decreased by internal shadowing within each grid due to the large ratio of grid thickness to slit width (Table I). The grids have imperfections, due to fabrication tolerances and to structural design features. Diffraction effects become relevant below $\sim$4 keV for subcollimator 1. Furthermore, these effects are not independent since, for example, some moderate energy photons, incapable of penetrating the grids directly, can penetrate the corner of a slat when they are incident from a few arc min off-axis.

In previous missions such as *Yohkoh* HXT, such effects were not as severe and so could be accommodated by modifying the triangular functional form and introducing more parameters (Sato, Kosugi, and and Makishima, 1999). However, for RHESSI, a different approach is necessary. All the factors identified above can be quantitatively described by generalizing Equation (2) to express the transmission of the grid pair in the form

$$\wp(\Phi) = T(1 + a_1\cos[\Phi - \Psi_1] + a_2\cos(2[\Phi - \Psi_2]) + a_3\cos(3[\Phi - \Psi_3]) + \cdots), \quad (3)$$

where $T$ is the average collimator transmission, $a_i$ and $\Psi_i$ are relative modulation amplitudes and collimator phases for the $i$th harmonic. (In this terminology, harmonic 1 is the fundamental.) For ideal grids, these parameters would have fixed values, but for RHESSI, $T$ and the other coefficients become slowly varying functions of angular offset and energy. Note that the expansion includes terms with both even and odd harmonics. Use of the higher-order terms is in fact desirable since the 2nd and 3rd harmonics effectively increase the angular resolution of each grid by factors of 2 and 3, respectively, albeit with lower sensitivity. For RHESSI imaging to date, and for much of the remainder of this paper, we will assume that only the fundamental term is used. Because of the 'orthogonality' of the harmonics, this simplification introduces no 'bias' and only slightly degrades the signal to noise.

In summary then, the characteristics of real grids can be accommodated by treating the idealized triangles as sinusoids. As we shall see, this approximation also enables considerable computational efficiency in the imaging algorithms.

### 3.3. The Imaging Geometry

Equation (3) shows how the grid transmission probability, $\wp(\Phi)$, can be expressed in terms of the angular offset $\theta$ orthogonal to the slits. In this section we establish how this is related to the $P_{im}$ 'cube', in Equation (1), in the context of a mapping geometry for RHESSI imaging.

To begin the mapping process, the user chooses a map center, field of view and pixel size. Although the selected map center is arbitrary, it should be close to the centroid of the emission sources since the grid response parameters are calculated for the map center and their variation across the map is subsequently neglected. In practice, the map center can be found by coarse mapping or context observations, or by using previous knowledge of the flare position.

Figure 4 illustrates a typical map geometry with the map center defined in a heliographic coordinate system. The coordinates $(x_m, y_m)$ of individual map pixels are defined in turn with respect to the map center.

An 'imaging coordinate system', fixed on the spacecraft, has its $Z$-axis defined by a vector between two fixed points on the front and rear grid trays. The projection of this vector onto the Sun is shown as the 'imaging axis' in Figure 4. In general, this axis is not coincident with the spacecraft spin axis, so the imaging axis will move with time. At any instant, however, its location relative to Sun center is provided by the aspect solution (Fivian and Zehnder, 2002). The orientations of the $X$ and $Y$ axes of the imaging coordinate system are fixed with respect to the spacecraft. As a result, their projection onto the Sun rotates (clockwise looking toward the Sun), with the $Y_{\text{imaging}}$ axis making an angle $\alpha(\text{t})$ with the heliographic $Y$ axis.

A pair of grids is fixed in the imaging coordinate system with the slits oriented at a constant angle $\beta$, the 'grid-orientation', measured counter-clockwise with respect to the imaging $Y$ axis. Typical lines of maximum transmission are indicated by the parallel dashed lines labeled 1–12 in Figure 4. Note that a line of maximum transmission does not necessarily pass through the imaging axis and so for each subcollimator there is a phase offset (labeled $Q$). The quantity $Q$ is the same as $-\Psi_1$ in Equation (3).

The map center lies at a radial distance $r_0$ from the imaging axis. Its phase with respect to that of the imaging axis depends on the component of $\mathbf{r_0}$ orthogonal to the grids. This is shown by the dashed line labeled $\theta$ in Figure 4. For a source at map center, this is equivalent to the $\theta$ mentioned earlier as the angle of incidence perpendicular to the slits.

To simplify the calculation of the phase, we use a wavevector $\mathbf{K}$ with magnitude $K = 2\pi/p$ and orientation directed orthogonal to the slits (parallel to the line

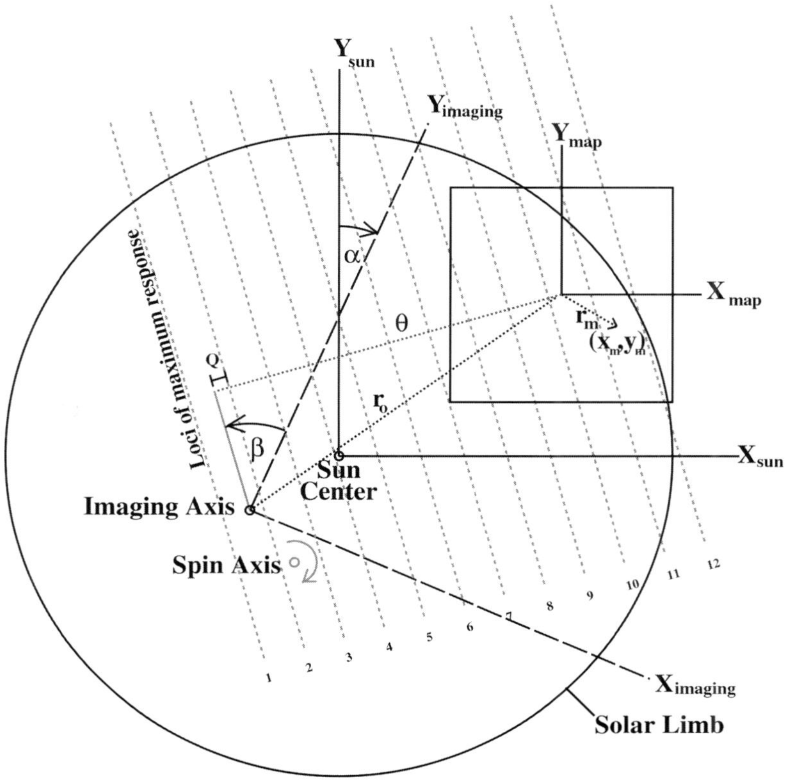


*Figure 4.* The imaging geometry. *Solid lines* and axes are fixed in heliocentric coordinates. *Dashed lines* and axes rotate clockwise with the spacecraft. Details are discussed in the text.

labeled, $\theta$, in Figure 4). Then at time bin $i$ the map-center phase with respect to the imaging axis is given by

$$\Phi_{0i} = 2\pi\theta/p = \mathbf{K_i} \cdot \mathbf{r_{0i}}. \tag{4}$$

Next we generalize this to sources anywhere in the map field of view. For a given subcollimator, the phase $\Phi$ (relative to the imaging axis) of a pixel $m$ displaced from map center by $\mathbf{r}_\mathrm{m}$, is given by

$$\Phi_{im} = \Phi_{0i} + \mathbf{K}_i \cdot \mathbf{r_m}. \tag{5}$$

In principle, with this knowledge of $\Phi_{im}$ and grid parameters, $T_i$, $a_i$ and $\Psi_i$ appropriate to the energy and offset $\mathbf{r_0}$, we can calculate the response $P_{im}$ as used in Equation (1). In practice, this calculation is aided by the introduction of an additional concept, the 'modulation pattern', discussed in the next section.

### 3.4. The Modulation Pattern

The two-dimensional *modulation pattern* characterizes the instantanteous response of the collimator. (It is not to be confused with the one-dimensional *modulation profile* which describes the response of a collimator vs time.) The modulation pattern can be thought of as a probability map on the Sun of the possible origin of a photon that was detected at a specific time. It is the ($i$ = constant) plane in the 'cube' $P_{im}$ whose role is described by Equation (1).

The modulation pattern is found, in one guise or another, in other domains of Fourier imaging. In radio astronomy, the modulation pattern is a projection of the complex *fringe pattern*. In the case of *Yohkoh*/HXT, there are 64 fixed modulation patterns, one for each of its subcollimators.

For RHESSI the modulation pattern for each subcollimator is fixed with respect to the rotating spacecraft. Relative to the grids, Equation (3) shows that by neglecting higher harmonics, the modulation pattern has a sinusoidal profile. The lines of maximum response shown in Figure 4 represent the peak contours of these profiles. Over the mapped area, we neglect any variation in the grid-dependent parameters in Equation (3).

Although the modulation pattern is fixed in the *rotating* frame of the spacecraft, a computational challenge arises because in the course of imaging, the 'cube' $P_{im}$ must be evaluated over the *non-rotating* solar map which itself is drifting with respect to the imaging axis. This would require rotating and shifting the modulation pattern for each time bin before evaluating it at each map pixel. Computation and storage requirements render this impractical if done in a straightforward manner.

Two simplifications are used to deal with this. First, the mapping is done in polar coordinates, with an arbitrary origin well outside the mapping field of view. (The final map is converted back to Cartesian coordinates before display so that the use of polar coordinates is not seen by the user.)

Second, for each subcollimator the instrument response is calculated in terms of '*universal modulation patterns*'. One such pattern represents the modulation pattern (in polar coordinates) in a coordinate system fixed with respect to the imager. The second pattern is the same, except for a 90-deg phase shift. (This mimics the use of visibilities, described in the Appendix.) Consequently, the modulation pattern for any combination of rotation and shift can be calculated as the weighted sum of a subset of the elements of these pre-calculated *universal modulation patterns*. This approach saves an order of magnitude in computation time and memory, while remaining transparent to the user. Using these techniques, the $P_{im}$ 'cube' can be evaluated efficiently so that a map can be reconstructed by inverting the observed modulation profile as described by Equation (1). In the next section, we will outline the algorithms by which this is accomplished.

Before doing this, however, it is necessary to correct the observed modulation profile for the detector livetime. Specifically, in order to interpret the number of photons $\mathcal{C}_i$ incident on the detector, during the $i$th time bin, one must incorporate the live time $\tau_i$ (live time = observation time – dead time) (Smith *et al.*, 2002). When this is done, the expected number of counts, $\mathcal{C}_{im}$ in the $i$th time bin from a source with photon flux, $F_m$, at map pixel $m$ is given by

$$\mathcal{C}_{im} = AF_m T_i \tau_i \{1 + \sum_n a_n^i \cos[n(\Phi_{im} - \Psi_n^i)]\}, \tag{6}$$

where the subscript n refers to the $n$th harmonic and the energy-dependent subcollimator transmission $T_i$, phase offsets, $\Psi_n^i$, and amplitudes $a_n^i$ are evaluated at map center.

Exploiting the linearity of the detector and subcollimator response, the predicted modulation profile, $\mathcal{C}_i$ for any source, is just the sum over $\mathcal{C}_{im}$ for all nonzero pixels. As will be discussed in later sections, this predictive principle is used by several of the RHESSI reconstruction algorithms to assess the consistency between the reconstructed image and the observed modulation profiles.

## 4. Image Reconstruction

In this section we review the various image reconstruction algorithms that are used for RHESSI imaging. The general approach is to use a 'back projection' algorithm to generate an initial estimate of the image. This estimate represents a convolution of the source with the instrumental response and so has sidelobes. To improve the image quality (*viz.*, to reduce the sidelobes), one can then use a variety of techniques to be discussed below (Clean, MEM, MEMVIS, Pixons, Forward-Fitting). Each of these makes broad assumptions about the character of the source, and most proceed in a cycle of predicting the modulation profile for a test image, comparing the predicted and observed modulation profiles, modifying the test image and iterating until an 'acceptable' agreement is obtained.

### 4.1. Back projection and Fourier transforms

Back projection (Mertz, Nakano, and Kilner, 1986) is the most straightforward and basic method of image reconstruction. It is equivalent to a 2D inverse Fourier transform (Kilner and Nakano, 1989). A map constructed by this method is the analog of the radio astronomer's initial Fourier transform of the observed visibilities (the so-called 'dirty map'). Back projection is a linear process: maps for arbitrary time intervals may be added together, and maps for different pitches and harmonics may be summed, generally leading to improvement of the image. Further improvements to the image by Clean or MEM (for example) do not share this property of linearity.

As discussed above, at a given instant in time, the response of a detector is characterized by the modulation pattern of the corresponding subcollimator oriented according to the roll angle. Detected photons are most likely to have come from regions where the modulation pattern has its highest values.

Back projection builds up a map by summing over time bins. For each time bin the map is incremented by the product of the counts in that time bin and the corresponding modulation pattern. This process preferentially populates map pixels corresponding to the real source(s). Other pixels will be built up as well, but to a lesser extent, as determined by statistics and by the symmetries in the set of modulation patterns used. (It can be shown, for example, that for a single subcollimator with a sinusoidal modulation pattern, rotationally averaging such a pattern results in a map which is proportional to a zero order Bessel function $J_0(2\pi r L/p)$ whose first positive sidelobe is 30% of the peak.) For adequate statistics, many time bins, several detectors and a mapped area that does not include the axis of rotation, the highest points in the back projection map will be at the location of the point source(s).

Mathematically, back projection can be described as follows: the modulation patterns $P_{im}$ are computed and appropriately summed for all time bins $i$ and map pixels $m$. Flat-fielding is necessary since the sensitivity is proportional to the variance of the modulation profile for each map point. This correction and normalization is implemented using a modified version of a recipe described by Durouchoux, *et al.* (1983). Subtract out the mean over all roll angles at each pixel: $\widetilde{P}_{im} = P_{im} - \langle P_m \rangle$. Then divide each value by its variance over all roll angles: $\widehat{P}_{im} = \widetilde{P}_{im} / \langle \widetilde{P}_m^2 \rangle$.

The intensity $I_m$ of each pixel (m) in the back-projection map is defined by the following linear combination of the count rates $\mathfrak{C}_i$:

$$I_m = \frac{1}{A} \sum_{i=1}^{N} [\mathfrak{C}_i / \Delta t_i] \widehat{P}_{im}, \tag{7}$$

where $A$ is the effective area (cm$^2$) of the detector and $\Delta t$ is the time-bin duration (s). The normalization and the division by $A\Delta t$ ensures that the expectation value at the peak of the dirty map equals the strength of a dominant source. The map then has units of counts cm$^{-2}$ s$^{-1}$. In practice, a slightly more complex formulation is used to distinguish between the variance component due to modulation and the variance component due to the twice-per-rotation changes in transmission caused by internal shadowing in individual grids.

Figure 5 shows an example of a back-projection image for a compact flare source whose spatial size is smaller than the subcollimator's angular resolution. The full-disk image, which uses the three coarsest subcollimators, shows the characteristic ringed patterns caused by incomplete sampling of the Fourier plane. Noteworthy is the 'mirror' source seen in the lower left quadrant of the full-disk map. This is a consequence of not having both 'sine' and 'cosine' subcollimators, but is not a problem in practice since the mapped area usually includes only the real source. In this case, consistency among maps made by individual subcollimators shows that the true source is on the limb in the upper right. Exactly halfway between the true and mirror sources is the spin axis, where no modulation occurs. The small-field back-projection map, made with subcollimators 3–8, shows the

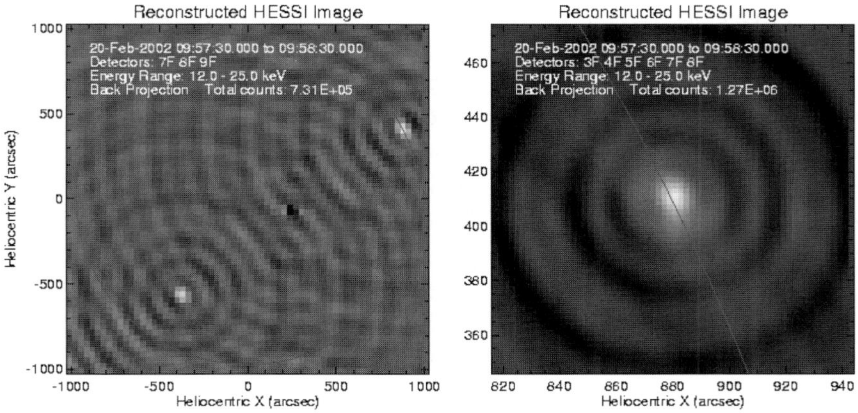


*Figure 5.* RHESSI full-Sun and small-field back-projection images for a compact flare that occurred on 23 February 2002. The pixel sizes are 32″ and 2″, respectively. Insets indicate the subcollimators used and other details.

flare surrounded by rings whose amplitude is smaller than in the full-disk map due to the use of a larger number of subcollimators.

All of the practical image reconstruction algorithms for improving a back-projection ('dirty') map are nonlinear. In the following section, we describe some of the options currently implemented.

### 4.2. CLEAN

Clean is an iterative algorithm, originally developed for radio astronomy, which is based on the assumption that the image can be well represented by a superposition of point sources. Nevertheless, it is often satisfactory for extended sources as well. Adapted to RHESSI, the basic Clean method, developed by Högbom (1974), postulates that the observed 'dirty' map is a convolution of a set of point sources with the instrument Point Spread Function (PSF) (*viz.*, the imager's response to a delta function source). That is,

$$D = \mathcal{P} \otimes I_{\mathrm{source}}, \tag{8}$$

where $\mathcal{P}$ is the PSF for one or many subcollimators and/or harmonics, $I_{\mathrm{source}}$ is the source distribution, $D$ is the back projection ('dirty') map and $\otimes$ denotes a convolution.

The algorithm proceeds as follows: an image called the 'residual map' is initialized with the back-projection map, to a value $I_0$. Then the position $(x_j, y_j)$ of the pixel with the largest flux $F_j$ in the residual map is saved in a 'Clean component' table. The PSF $\mathcal{P}$ at $(x_j, y_j)$, normalized to $\mu F_j$ (where $\mu$ ($\leq 1$) is the so-called 'loop gain'), is subtracted from the current residual map, $I_n$ to yield a new residual map $I_{n+1}$. This process is continued iteratively until a specified number of iterations is reached, or until the residual map has a negative peak which exceeds the largest positive peak, or until the observed modulation profile agrees well with that predicted from the Clean components as indicated by a chi-squared test.

Since the final residual map, ($I_{\text{final}}$) presumably consists mainly of noise, the information content of the Clean map is contained in the table of the amplitudes and locations of the Clean components. For the purposes of display, these Clean components are convolved with a Clean PSF (or 'Clean Beam'), $\mathcal{P}_{\text{Clean}}$, which is simply a gaussian whose FWHM reflects the effective resolution of the subcollimators used for the dirty map. The final Clean map

$$I_{\text{clean}} = \sum_j \mathcal{P}(x_j, y_j)\mu F_j + I_{\text{final}} \tag{9}$$

represents the sum of the Clean components, convolved with the Clean PSF, plus the final residual map. With Clean (unlike the other algorithms), the latter is added to provide a visual estimate of noise in the final map.

A final detail is that, for computational efficiency, the Clean processing is done in polar coordinates, although results are transformed back to rectangular coordinates for display.

Figure 6 shows examples of back projection, Clean, and other techniques discussed below.

### 4.3. Maximum Entropy Methods

The goal of Maximum Entropy Methods (MEM) is to find the map that is both consistent with the data and that contains the least information about the source. The latter condition generally makes MEM maps appear quite 'smooth' compared to other techniques.

MEM has a rich history and has been used as the main image reconstruction algorithm for *Yohkoh*/HXT. Although the details of the 'MEM-Sato' implementation are somewhat different for RHESSI, the basic theory is still relevant (Sato, 1998; Sato, Kosugi, and Makishima, 1999).

To implement MEM-Sato for RHESSI, the list of time-tagged photons for each subcollimator and energy band are converted to a set of counts in time bins, each of which corresponds to a particular roll angle. Traditionally, consistency of a map with the data is measured using a $\chi^2$ measure on the counts:

$$\chi^2 = \sum_i \frac{(\mathcal{C}_i - \mathcal{E}_i)^2}{\sigma_i^2}, \tag{10}$$

where $\mathcal{C}_i$ is the observed count for time bin $i$; $\mathcal{E}_i$ is the 'expected' count from the reconstructed image; and $\sigma_i$ is the estimated uncertainty of count $\mathcal{C}_i$. If only photon counting noise is important, then $\sigma_i^2 = \mathcal{E}_i$, but in practice systematic errors and the background may also be significant and so should be included in $\sigma_i$. Since it is quite common to have fewer than 10 counts per time bin (in which case $\chi^2$ is inaccurate),

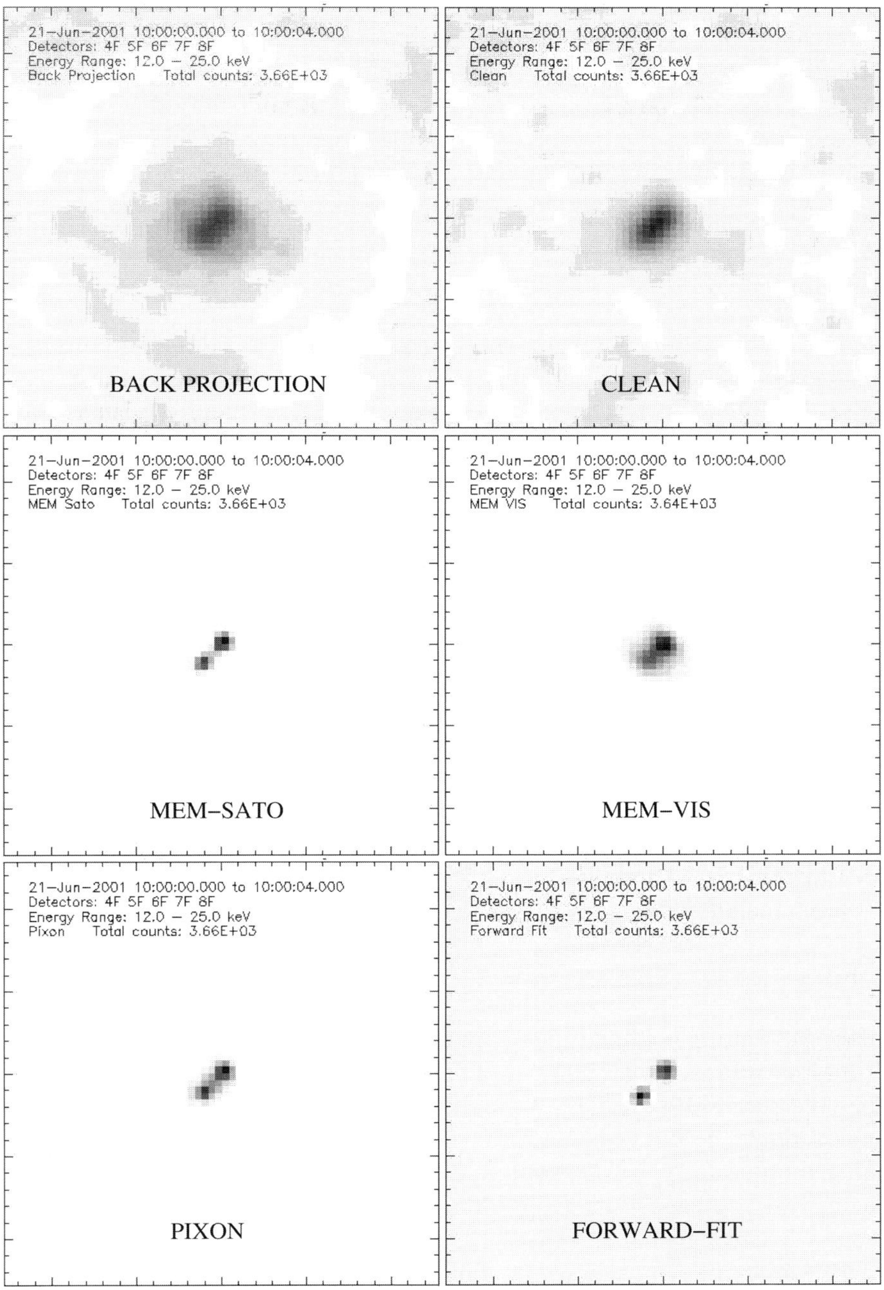


*Figure 6.* Examples of back-projection, Clean, MEM-Sato, MEM-vis, Pixons, and pixelized Forward-Fit maps of a simulated double source. Two identical point sources were located 17″ apart. The pixels are 4″ and the FOV is $256 \times 256''$. 5 subcollimators with resolutions ranging from 12″ to 106″ were used.

the MEM-Sato algorithm uses the more rigorous C-statistic (Cash, 1979) instead. (For simplicity, however, we will continue to use the term $\chi^2$ in this paper.)

The condition of minimizing the information about the source is quantified using an entropy measure. In one view, the entropy is a quantity that, when maximized, produces a positive image with a compressed range in pixel values (Cornwell, 1984), but according to Gull and Skilling (1984) and Sivia (1996) the only function which guarantees that no unwanted correlation be imposed is

$$\mathcal{H} = -\sum_m F_m \log F_m, \tag{11}$$

where $F_m$ is the flux in pixel $m$. The balance between maximizing entropy and consistency with the data is implemented by maximizing $\mathcal{Q} = \mathcal{H} - \frac{1}{2}\lambda\chi^2$ for the smallest possible $\lambda$. In practice $\lambda$ is set to a small value initially and an image is sought iteratively that has $\chi^2 = 1$ (viz., consistent with observations). If such an image cannot be found, the entropy constraint is eased by increasing $\lambda$. Additional constraints (e.g., total flux) can be added to $\mathcal{Q}$ if desired.

Another approach to the Maximum Entropy Method involves the conversion of the list of time-tagged photons to visibilities. As explained in the Appendix, this effectively converts the observed modulation profile to an equivalent representation of amplitudes and phases as a function of grid orientation. Another RHESSI image reconstruction algorithm, MEMVIS, then applies MEM to the visibilities instead of to the binned counts. This algorithm will be discussed in detail in a forthcoming paper by Conway, Schmahl, and Hurford (in preparation).

The main advantages of MEMVIS are its efficiency (lower memory requirements, use of fast Fourier transforms) and the ease by which the visibilities can be integrated over time by *literally* just adding them together. Simulations have shown that MEMVIS appears to be quite robust at low count rates.

Figure 6 includes examples of MEMVIS and MEM-SATO maps.

### 4.4. Forward-Fitting

The Forward-Fitting method is based on the assumption that the spatial map can be well represented by a small number of elemental source structures, each of which can be characterized by just a few parameters. In a solar hard X-ray context, such structures might be circular, elliptical, or curved elliptical Gaussians (which resemble a loop). Such Gaussian sources can be represented by 4, 6, or 7 parameters respectively. Examples of fits to hard X-ray sources observed by *Yohkoh*/HXT may be seen in Aschwanden *et al.* (1999).

In the Forward-Fitting algorithm implemented for RHESSI (Aschwanden *et al.*, 2002), 'pixelized' maps of multiple Gaussian components are created and used to calculate a model modulation profile which is then compared to the observed modulation profile. The map iteratively evolves into a best-fit image which yields well-quantified parameters. Alternative forms of Forward-Fitting without discrete pixelization have also been used for RHESSI source parametrization (Schmahl and Hurford, 2002).

### 4.5. The Pixon Method

The Pixon method is another technique which removes the sidelobe pattern of a telescope while mitigating the problems of correlated residuals and spurious sources which are commonly seen in Fourier deconvolution and maximum entropy approaches.

The goal of the Pixon method is to construct the simplest model for the image that is consistent with the data (i.e., having an acceptable $\chi^2$ fit). Being the simplest model, the derived image would be artifact free with no spurious sources. The model is necessarily the most tightly constrained by the data, and consequently has the most accurately determined parameters (Puetter, 1995; Metcalf *et al.*, 1996; Alexander and Metcalf, 1997).

MEM imposes a global condition that minimizes the difference between the image and a grey map in a manner that is consistent with the data. The Pixon method is related to MEM, but applies a local constraint such that structure is allowed only where required by the data. (From an information science point of view, one selects a model from the family of multi-resolution basis functions (Pixons) that both has the minimum information content and statistically fits the data.)

Since the model has minimum complexity, spurious sources are unlikely to arise. Each parameter is determined using a larger fraction of the data and so is determined more accurately. This usually results in superior photometric and positional accuracy and, since the minimum number of parameters are used, the data cannot be over-fitted. The Pixon method, however, pays a price for its photometry: the method is one to two orders of magnitude slower than the other reconstruction methods. Therefore, as with *Yohkoh*/HXT, it will probably be used only after the faster reconstruction techniques have been used to optimize the time and energy binning. Pixons can then be used to check the strength, shape or presence of weak sources as required.

## 5. Imaging Performance

### 5.1. Imaging Strengths and Limitations

Although it is too early in the mission to provide a definitive summary on imaging performance, the general properties of RHESSI imaging can be broadly outlined. This is important since RMC imaging has very distinctive strengths and weaknesses compared to conventional imaging instruments, and these properties are directly relevant to the kind of science that can be done with RHESSI imaging.

RHESSI was not designed to provide images with the kind of morphological richness and detail that we have come to appreciate from TRACE, *Yohkoh*/SXT,

SOHO/EIT and other direct-imaging instruments. Instead, RHESSI imaging determines the flux, location, size and shape of the dominant source components. Its measure of image quality is 'dynamic range', defined as the ratio of the surface brightness of the strongest source to the weakest credible source in the field of view. The design goal (and current expectation) is that, in favorable circumstances, a dynamic range of $\sim 100:1$ can be achieved.

There are four broad factors that limit the quality of RHESSI imaging. The first is the limited number of spatial frequencies that are measured. For typical maps, sampling the Fourier components measured by the 9 rotating subcollimators provides $\sim 10^3$ independent measurements. Although this represents a substantial gain over *Yohkoh*/HXT, it still represents a fundamental limit to the complexity of the images that can be generated. This limitation becomes more severe if imaging is done on timescales of less than one half of the rotation period (i.e., $<2$ s).

The second factor is photon statistics, which can be an important constraint. In very favorable circumstances, a point source can be detected and located with as few as $\sim 10^2$ counts. Usually, however, $\sim 10^3$ are required and for more complex images it is preferable to have $10^4$ or $10^5$ counts. (For comparison, a large flare returns over $10^8$ counts.)

Third, the imaging algorithms ultimately rely upon comparisons of observed count rates to 'predictions' based on convolving test maps with the instrument response. Therefore, in circumstances with good statistics and simple sources, imperfect knowledge of the grid response, detector response and/or aspect solution can provide a limitation. At present the grid response is believed to be known at the $\sim 2\%$ level. Good progress continues to be made in understanding the detector response (Smith *et al.*, 2002), which can be relevant at high count rates.

The fourth factor is the validity of assumptions that are implicit in the imaging algorithms. Such assumptions include the short duration of time bins compared to the modulation period, the uniformity of instrument response over the selected energy range, the uniformity of instrument response over the imaging field of view, the absence of temporal structure in the flux (on time scales that are commensurate with the modulation), the neglect of background in current implementation of some algorithms, the neglect of off-diagonal elements in the spectral response matrix (Smith *et al.*, 2002), and the accuracy of corrections for live time and pulse pileup.

In general, RHESSI imaging has proven to be remarkably robust, a feature which permitted the generation of viable images early in the mission and in situations where the foregoing assumptions were not strictly valid.

The strengths of this imaging technique include the ability to accurately and absolutely locate the source components on the Sun to arc sec. This supports both reliable comparisons of images as a function of time and energy and accurate co-location with images in other wavelength regimes.

Photometric reliability of RHESSI images depends on the circumstances and is expected to improve. At present, it is typically at the $\sim 10\%$ level, which is sufficient to support some feature-based imaging spectroscopy. Source sizes can also be determined (e.g., Schmahl and Hurford, 2002), although at present this requires careful interpretation of the images.

Finally, one of the underlying strengths of RHESSI imaging, derived from the photon-based nature of the data, is the analyst's ability to select the time resolution and range, energy resolution and range, imaging field of view, resolution and reconstruction technique (Schwartz *et al.*, 2002). These choices can be made and iterated during data analysis on the ground rather than in the mission design or operations phase. This enables these tradeoffs to be made in a manner that takes into account the characteristics of the flare and the specific science objectives of the analyst. In practice, this is perhaps one of the most valuable and novel features of RHESSI imaging.

### 5.2. Self-calibration

As discussed above, one of the potential limitations to image quality is the knowledge of the grid response. However, one of the features of RHESSI imaging is the ability to self-calibrate this response. While a discussion of this feature is beyond the scope of this paper, it is perhaps worth mentioning the progress made to date. In particular, the relative 'phases' of the grids are vital parameters. For each grid this represents the average position of the center of the slats with respect to the imaging coordinate system defined by the solar aspect system. The current determination of these phases is based on maximizing the response of the system to unresolved sources. The accuracy of these determinations appears to be $\sim 10^{-2}$ of the grid pitch. A more sensitive technique, based on the stability of imaging using even and odd half-rotations, will be used to improve this further.

Other grid response parameters, such as the average transmission, can be verified by comparison of spatially integrated spectra obtained from each of the nine detectors. Interpretation of these ratios as a function of energy and offset angle is a sensitive diagnostic that can be used to fine-tune our knowledge of the grid parameters.

### 5.3. Expected evolution of imaging

During the course of the RHESSI mission, lessons are being learned about unanticipated features of the hardware and software. An example of such a lesson is the existence of data gaps (Smith *et al.*, 2002). As in other missions, the imaging software is expected to improve as bugs are eradicated and better algorithms are developed. In particular, we anticipate introducing an alternate approach to combining counts from multiple time bins that will greatly facilitate integration over long time periods. Also, the thus far neglected 2nd and 3rd harmonics of the grid response will be taken into account. For strong flares, this has the potential to improve the angular resolution of the grids by factors of 2 and 3, respectively. An important feature of these improvements is that in all cases they will be applicable retroactively to data acquired from the start of the mission.

### 5.4. Conclusion

In choosing an imaging concept for RHESSI, the selection of rotating modulation collimators was driven by the goals of high angular resolution, high sensitivity, good image quality and a 4-decade range of energies, all to be implemented within the practical constraints of a SMEX-class mission. As an additional feature, the photon-tagged data stream has afforded an unprecedented level of analysis flexibility to the user. Initial indications are that the RHESSI imaging technique has succeeded in providing imaging spectroscopy to meet these goals.

Nevertheless, the imaging concept does carry a cost, one imposed by its unfamiliarity and by the need for users to reconstruct images before they can proceed toward their real goal of addressing science issues. Effective image reconstruction requires integrating the hardware design, the analysis software and the user. This paper has described the imaging concept that unifies these three elements.

## Acknowledgements

We thank the PSI team led by Alex Zehnder for the care, technical expertise, and professionalism demonstrated in fabricating the imager, and M. Appleby for his skill and leadership in providing the fine grids. We acknowledge the role of H. F. van Beek in providing the coarse grids and many years of invaluable advice and insights. We are also grateful to Dave Clark at GSFC for numerous contributions to the imager alignment and calibration. Support for EJS came from NASA grant NAG 5-10180 to the University of Maryland. This work is supported by NASA grant NAS5-98033-05/03.

## Appendix: Visibilities

The concept of visibilities is borrowed from radio astronomy. In the context of RHESSI, 'counts' and 'visibilities' can be regarded as two time series that present the same information: the response of RHESSI to a spatial flux distribution. The visibilities are complex and effectively contain two count profiles. The real (cosine) part can be regarded as (mean-subtracted) counts from a cosine subcollimator, and the imaginary (sine) part corresponds to an identical subcollimator shifted by one quarter pitch perpendicular to the slats (a shift of 90° in phase) i.e., a sine subcollimator. (In Equations (2) and (3), $\Phi$ would be replaced by $\Phi + \pi/2$. ) Roughly speaking, the two parts of each visibility together contain the phase information provided by the aspect system and the amplitude provided by the detectors. The visibilities include the corrections for pointing excursions, transmission variations, and detector sensitivity effects, and so the inverse Fourier transform need only be applied to obtain a 'dirty' map of the source, with most instrumental artifacts removed.

It is worth mentioning that in early versions of RMC design, there were both sine and cosine subcollimators (e.g., Mertz, Nakano, and Kilner, 1986; Murphy, 1990), which would give the visibilities more directly. But our experience has shown this to be unnecessary in hardware, since with sufficient sampling of the modulation profile, relative amplitudes of the sine and cosine components can be determined from the modulation profile itself.

Clearly RHESSI imaging has much in common with radio astronomy interferometry and could benefit from the enormous efforts invested in that subject. Also, we have found that the use of visibilities with RHESSI has several intrinsic and unexpected advantages. Here we outline how a visibility can be related to the count modulation profiles for a point source. A more complete account of visibilities and Fourier space in the context of RHESSI will be given in a forthcoming paper by Conway, Schmahl and Hurford.

A visibility is the value of the 2D Fourier transform of a spatial flux distribution at a point in Fourier space. It can either be expressed in terms of a discrete pixel image, as for the counts in Equation (1) with $P_{jm} = \exp(i\Phi_{jm})$, $j$ and $m$ labeling time bin and pixel respectively, or for a continuous distribution $F(x, y)$ at time $j$:

$$\mathcal{V}(u, v) = \int\int F(x, y)e^{i\Phi(x,y;u,v)}\,\mathrm{d}x\,\mathrm{d}y, \tag{12}$$

where $(u, v)$ are the coordinates in Fourier space determined by the subcollimator's pitch $p$ and the current angle $\alpha_j$ of the grid slats. In fact the polar coordinates in Fourier space are just $(1/p, \alpha)$ where $u = (\cos\alpha)/p$ and $v = (\sin\alpha)/p$. The pitch determines the angular spatial frequency of a subcollimator $K = 2\pi/p$.

The visibility of a point source of flux $F_0$ at coordinate $(x, y)$ or $(r, \theta)$ is therefore

$$\mathcal{V}(u, v) = F_0 e^{i\Phi}, \quad \text{where} \quad \Phi = 2\pi(ux + vy) = Kr\,\cos(\theta - \alpha). \tag{13}$$

For the same source, a RHESSI count modulation profile can be expressed as follows by combining Equations (2) and (3):

$$\mathcal{C} = F_0 T\tau\,(1 + a_1\,\cos(\Phi - \Psi_1) + \text{harmonics})\,. \tag{14}$$

Comparison of Equations (13) and (14) reveals that the real part of the visibility, $F_0\,\cos\Phi$, is similar to the second term in (14), aside from a known amplitude factor of $T\tau a_1$ and a known phase offset of $\Psi_1$. From this it is clear that a RHESSI subcollimator of a particular pitch at a particular roll angle is providing a measure closely related to the Fourier transform, i.e., visibility $\mathcal{V}(u, v)$, at a particular point in Fourier space. Over the course of a rotation, the $(u, v)$ points for a particular subcollimator sweep out a complete circle in Fourier space. Subcollimators of smaller pitch sweep out larger circles on the Fourier plane. Note that while coverage along the circles can be excellent with RHESSI (given enough counts to fill the time bins), the Fourier transform in regions between the circles remains unknown.

The details of how counts are converted to visibilities are beyond the scope of this paper, though we will outline the process here. We assume, as do the other imaging algorithms, that the X-ray flux is restricted to a localized area of the Sun. This allows us to calculate the phase at map center for each time bin. The mean-subtracted counts are fitted with cosine and sine time series. This is done so that only the first harmonic is fitted in the time series, which is all we need for the visibilities. The coefficients from these fits can then be used to yield the real and imaginary parts of visibilities. However these visibilities are based on an origin on the spin axis, which moves with respect to the Sun. It is then necessary to apply a correction, which simply involves multiplying each visibility by a complex number, to bring the visibilities to some fixed center (either Sun center or map center).

Figure 7 illustrates the close correspondence between count rate profiles and visibility profiles. The first panel shows a simulated modulation profile for two point sources about $1\frac{1}{2}$ angular pitches apart. This shows the beating of the sources in the form of slow modulation of the amplitude. The second panel shows the visibilities computed for the same sources. The cosine component (real part) of the visibility precisely mimics the amplitude and phase of the count modulation. The dotted curve shows the sine component (imaginary part), 90° out of phase with the cosine component. The amplitude and phase are shown in the third panel. The amplitude, which equals the root quadratic sum of the sine and cosine components, is the envelope of the modulated curves. The phase wraps from $-180°$ to 180° and back again as the lines of constant phase (see Figure 4) move across the source.

The units used in visibilities warrant some further discussion. According to Equations (13) and (14), the minima-to-maxima variation of the visibilities is several times larger than that of the counts. (Compare the ordinates of the first two panels of Figure 7.) In fact, when the subcollimator pattern is ideal and perfectly triangular, ($a_1 T = 2/\pi^2$), the minima-to-maxima variation of $Re[\mathcal{V}_j]$ is $\pi^2/2 = 4.9$ times that of the count profile. Note also that our definition of visibility is such that the amplitude of the visibility for a point source is equal to its flux $F_0$. The amplitude for an extended source of the same flux will be less that $F_0$ if the source size exceeds the angular resolution. This means that the individual visibilities do not necessarily contain information on the total flux of the source, $F_0$ – an important fact to remember in creating images using visibilities. In practice, this is not a problem because $F_0$ can be estimated by comparing visibilities at different spatial frequencies.

In working with RHESSI data, visibilities offer a convenient intermediary between the observed counts and images that have been corrected for all of the spin-axis excursions, slit-shadowing, transmission and dead time effects. Since visibilities are essentially counts from an ideal subcollimator, it is easy to perform a graphical analysis on the visibilities, as illustrated in Figure 3. This kind of analysis can yield estimates of the position and size of a source that are free from the assumptions (usually smoothness) that must be made to construct images.

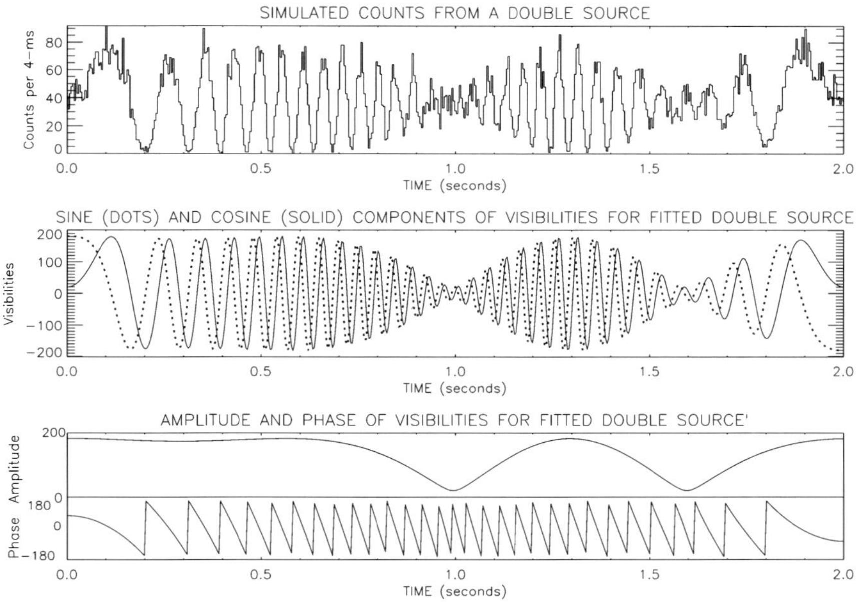


*Figure 7. Top*: simulated modulation profile for one subcollimator for a double source totaling 40 000 photons. Note the beating of the sources, which are about 1.5 angular pitches apart. *Middle*: the *solid curve* is the real part (cosine component) of the visibility, and the *dotted curve* is the imaginary part (sine component). The cosine component reproduces the count rate profile precisely in amplitude and phase. *Bottom*: Amplitude and phase of the visibilities.

Potential disadvantages of using visibilities are the introduction of a systematic error and a correlation of noise on visibilities in neighboring time bins. There is also the problem that since visibilities are constructed from two or more adjacent time bins, successive visibilities are not truly independent quantities. However, from our experience with many simulated sources, we can summarize the advantages of using visibilities as follows:

(1) The aspect correction can easily be applied while constructing visibilities from the observed counts.

(2) Due to 1, visibility modulation patterns do not need to be corrected for the aspect solution for each time bin.

(3) Due to 2, operations involving modulation patterns can be performed with a convolution based on Fast Fourier Transform methods rather than a matrix multiplication. This allows a speed up of order $M^2$ as compared to $M \log M$, where $M$ is the number of time bins.

(4) Due to 2, integration of visibility values over several rotations simply involves adding together visibility series from those rotations.

(5) In constructing visibilities from observed counts, there is effectively a smoothing, which relaxes problems arising from zero counts in some time bins.

(6) Visibilities provide a platform for using widely available imaging programs from radio astronomy. Although this capability is not presently available in the RHESSI software, only a module that writes visibility FITS files is needed to export RHESSI data to AIPS or other imaging systems.

## References

Alexander, D. and Metcalf, T. R.: 1997, *Astrophys. J.* **489**, 422.

Aschwanden, M., Schmahl, E. J. and the RHESSI Team: 2002, *Solar Phys.*, this volume.

Aschwanden, M., Fletcher, Sakao, Kosugi, T., and Hudson, H. S.: 1999, *Astrophys. J.* **517**, 977.

Cash, W.: 1979, *Astrophys. J.* **228**, 939.

Cornwell, T. J.: 1984, in J. A. Roberts (ed.), *Indirect Imaging, Proc. IAU/URSI Symp.*, Cambridge University Press, Cambridge, p. 291.

Crannell, C. J.: 1994, American Institute of Aeronautics and Astronautics: Washington DC, AIAA-94-0299.

Crannell, C. J., Hurford, G. J., Orwig, L. E., and Prince, T. A.: 1986, *SPIE Proc.* **571**, 142.

Curtis, D. W. *et al.*: 2002, *Solar Phys.*, this volume.

Datlowe, D. W.: 1975, *Space Sci. Instrument.* **1**, 389.

Durouchoux, P., Hudson, H., Hurford, G., Hurley, K., Matteson, J., and Orsal, E.: 1983, *Astron. Astrophys.* **120**, 150.

Enome, S.: 1982, *Adv. Space Res.* **2/11**, 201.

Fivian, M., Hemmeck, R., Mchedlishvili and Zehnder, A.: 2002, *Solar Phys.*, this volume.

Gull, S. F. and Skilling, J.: 1984, *IEE Proc.* **131**, 646.

Högbom, J. A., 1974: *Astron. Astrophys.* **15**, 417.

Hurford, G. J. and Curtis, D.: 2002, *Solar Phys.*, this volume.

Kilner, J. R. and Nakano, G. H.: 1989, *SPIE Proc.* **1159**, 27.

Kosugi, T. *et al.*: 1991, *Solar Phys.* **136**, 17.

Lin, R. *et al.*, 2002: *Solar Phys.*, this volume.

Makishima, K. *et al.*: 1977, in K. A. van der Hucht and G. Vaiana (eds.), *New Instrumentation for Space Astronomy*, Pergamon Press, New York.

Mertz, L. N., 1967: *Microwave Research Institute Symposia Series, Proc. Symp. on Modern Optics* **17** Polytechnic Institute of Brooklyn, New York, p. 277.

Mertz, L. N., Nakano, G. H., and Kilner, J. R.: 1986, *J. Opt. Soc. Am.* **466**, 585.

Metcalf, T. R. *et al.*, 1996: *Astrophys. J.* **466**, 585.

Murphy, M. J.: 1990, *Nucl. Instr. Methods Phys. Res.* **A290**, 551.

Ohki, K. *et al.*: 1982, *Proc. Hinotori Symp. Solar Flares*, ISAS, Tokyo, Japan, p. 102.

Prince, T. A., Hurford, G. J., Hudson, H. S., and Crannell, C. J.: 1988, *Solar Phys.* **118**, 269.

Puetter, R. C., 1995: *Int. J. Image Systems Technol.* **6**, 314.

Sato, J.: 1998, Ph.D. Thesis, NAO.

Sato, J., Kosugi, T., and Makishima, K.: 1999, *Publ. Astron. Soc. Japan* **51**, 127.

Schmahl, E. J. and Hurford, G. J.: 2002, *Solar Phys.*, this volume.

Schwartz, R. *et al.*: 2002, *Solar Phys.*, this volume.

Sivia, D. S.: 1996, *Data Analysis, a Bayesian Tutorial*, Clarendon Press, Oxford.

Schnopper, H. W., Thompson, R. I., and Watt, S.: 1968, *Space Sci. Rev.* **8**, 534.

Schnopper, H. W. *et al.*: 1970, *Astrophys. J.* **161L**, 161.

Smith, D.M. *et al.*., 2002, *Solar Phys.*, this volume.

Van Beek, H. F., Hoyng, P., Lafleur, B., and Simnett, G. M.: 1980, *Solar Phys.* **65**, 39.

Zehnder, A. *et al.*: 2002, *SPIE Proc.* **4853**, in press.

# RHESSI ASPECT RECONSTRUCTION

M. FIVIAN, R. HEMMECK, A. MCHEDLISHVILI and A. ZEHNDER
*Paul Scherrer Institute, CH-5232 Villigen PSI, Switzerland*

(Received and accepted 20 September 2002)

**Abstract.** The Reuven Ramaty High-Energy Solar Spectroscopic Imager RHESSI spacecraft spins at about 15 rpm around an axis close to Sun center. Precise knowledge of the pointing and the roll angle of the rotating spacecraft is needed in order to reconstruct images with 2 arc sec resolution using the modulation patterns seen on each of the detectors behind the bi-grid rotating collimators. Therefore, the aspect system consists of two subsystems of sensors, the Solar Aspect System (SAS) and Roll Angle System (RAS). The measured data are sent to the Aspect Data Processor (ADP), where a data compression of about a factor of 1000 and a formatting into telemetry packets takes place. The transmitted data consist of 'solar limb' data from the SAS and 'star event' data from the RAS. Knowing the geometry of all features of the SAS, the position of the Sun center, with respect to a spacecraft fixed coordinate system, can be reconstructed. Similarly, a list of position angle marks can be generated by fitting of the star events and subsequent comparison with a star catalog. Integrating this information allows correcting and interpolating the roll angles that provides a precision of $< 0.4$ arc sec (pitch and yaw) and $< 1$ arc min (roll).

## 1. Introduction

The Reuven Ramaty High-Energy Solar Spectroscopic Imager RHESSI is a Small Explorer mission that was launched by NASA on 5 February 2002 into a circular orbit (600 km altitude, 38 deg inclination). The primary scientific aim of the mission is to investigate the physics of particle acceleration and energy release in solar flares. RHESSI observes the full Sun with an unprecedented combination of spatial resolution (as fine as 2.3 arc sec) and energy resolution (1 keV to 5 keV) in the energy range from 3 keV to 17 MeV. The single instrument on a spin stabilized spacecraft utilizes Fourier-transform imaging with 9 bi-grid rotating modulation collimators spaced at 1.55 m and 9 cooled germanium detectors (Lin *et al.*, 2002).

Since the transmission probability for a photon through the collimators depends on the incident angle with respect to the telescope axis (Hurford *et al.*, 2002), and because the spacecraft can only be controlled on an arc-min level, the spatial resolution depends critically on a precise knowledge of the instantaneous pointing direction and roll angle. In order to have a spatial resolution of 2.3 arc sec (for the finest grid) and to precisely correlate with observations at other wavelengths, two precise aspect systems are implemented, the Solar Aspect System (SAS) and the Roll Angle System (RAS). The SAS yields sub-arc sec knowledge of the radial pointing with respect to Sun center and the RAS provides precise knowledge on the roll angle of the rotating spacecraft (about 15 rpm). The combined SAS/RAS

*Solar Physics* **210:** 87–99, 2002.

aspect system provides a knowledge of the absolute pointing with an accuracy of 1 arc sec.

There are two totally independent instruments on RHESSI measuring the roll angle. At one stage of the development of the RHESSI hardware, the schedule and some performance uncertainties of the roll angle system (RAS) motivated the team to add a redundant roll sensor as a backup system. The latter, the Photo-Multiplier Tube Roll Angle System (PMTRAS), is based on a photo-multiplier tube and is described by Hurford and Curtis (2002). In many aspects, the two instruments are complementary (Hurford and Curtis, 2002). Nevertheless, the RAS is measuring the polar angle of every star event and therefore the RAS data can also be used to reconstruct solar off-pointing for non-solar science. As it turned out, both RAS and PMTRAS are operational, but initial RHESSI imaging has been based on the PMTRAS.

The design and the alignment of the RHESSI imager, i.e., the telescope tube with the grids and the SAS/RAS aspect optics and sensors are discussed by Zehnder *et al.* (2002) and only a short overview is given in the next section. The current status of the calibration of the aspect system is reported by Fivian, Henneck, and Zehnder (2002). In this paper, the aspect data processing and the reconstruction of the aspect solution are presented.

## 2. The Design of the Aspect Systems

A pair of widely separated grids in front of a cooled Ge detector (Smith *et al.*, 2002) produces the characteristic intensity modulation. Each grid consists of a planar array of equally spaced, X-ray-opaque slats separated by transparent slits. If the slits of each pair of grids are parallel to each other and if their pitches ($p$) are identical, then the transmission is modulated depending on the direction of the incident X-rays. For slits and slats of equal width, the transmission is modulated from zero to 50% and back to zero for a change $a = p/L$, where $L$ is the separation between the grids. The angular resolution is then defined as $p/(2L)$.

The efficiency of the modulation for each of the 9 collimators depends critically on the alignment of the corresponding grid pair. In particular, the finest grids with a pitch of 34 $\mu$m must always be aligned within 20 arc sec in twist ($r_{\mathrm{Grid}} \times \Delta\Phi \ll p/2$, with twist $\Delta\Phi$ and $r_{\mathrm{Grid}}$ radius of grid). Another necessary ingredient for image reconstruction with a resolution as high as 2.3 arc sec is the accuracy of the measurement of the pointing with respect to the collimators. Therefore, a structure had to be found which guaranteed the stability of the collimators and the aspect systems throughout the integration procedure, the launch and the thermal stress over the orbit (Zehnder *et al.*, 2002). In addition, the definition of the coordinates for measuring the pointing have to be chosen in such a way that the axis of the collimators can always be measured with respect to Sun center fulfilling the requirements in accuracy.

### 2.1. Solar Aspect System (SAS)

The SAS is a set of 3 identical lens/sensor subsystems. Each subsystem consists of a 4 cm diameter lens located on the front grid tray of the imager tube. Each lens focuses a solar optical image onto a 2048-element $\times$ (13 $\mu$m)$^2$ linear CCD located on the rear grid tray. The orientation of the three linear CCDs are equally spaced at 120 deg. The focal length, given by the length of the imager tube, is 1.55 m resulting in a plate scale of 1.73 arc sec pixel$^{-1}$. In order to reduce chromatic aberration, a bandpass filter at 670 nm with 12 nm bandwidth (FWHM) has been used.

In principle, the measured Sun pointing is related to the SAS features only. In order to have the measurements immune to bending of the imager tube, a design has been chosen to have the lenses coplanar with the front grids and the CCDs coplanar with the rear grids.

While each of the three CCDs are sampling solar profiles at 128 Hz using a programmable integration time between 0.02 and 2 ms, the digitized signal is sent to the Aspect Data Processor (ADP) where a hardware-programmed algorithm triggers one or several pixels around the solar limb using an adjustable threshold. The amplitudes of typically 4 pixels for each of the 6 limbs are written to the telemetry packets using 8 or 10 bits precision.

Each SAS sensor is running with a constant cycle timing at 128 Hz. Within every cycle of $\approx$ 8 ms, the sunlight is integrated on the pixels of the CCD. The accumulated charge is shifted out with a frequency of 1 MHz and sampled twice, once at the offset level and once at the signal level. The subsequent digitization at the 12-bit ADC (analog digital converter) provides a resolution of 0.5 mV channel$^{-1}$ (2 V/$2^{12}$channels $\approx$ 0.5 mV channel$^{-1}$). To remove the charge of the continuously incident sunlight the readout of the CCD follows a pre-readout that removes most of the charge accumulated during off-integration time. The front-end electronics incorporates a Field Programmable Gate Array (FPGA) providing basic control of the sensor. In the normal operational mode, the difference of the two samples is calculated and bits 1–10 are sent to the ADP via a synchronous serial link with a speed of 0.1 $\mu$s bit$^{-1}$ or 1 $\mu$s pixel$^{-1}$. By using the difference of the two samples, the reset noise of the CCD is essentially eliminated. In order to have the full resolution on the 10-bit range of the digital system, the integration time is currently programmed to be $\approx$ 700 $\mu$s as described by Fivian, Henneck, and Zehnder (2002). Beside the normal operational mode, several test and diagnostic modes (first-level modes) are implemented in the front-end FPGA.

On the ADP, the SAS data are processed first by a hardware program in the second-level FPGA (SAS/RAS FPGA) and then by the software running on the Data Signal Processor (DSP). For every SAS cycle (i.e., every 8 ms), the three measured solar profiles are stored in the SAS/RAS memory on the ADP. At the same time, while the data are flowing through the FPGA logic, the addresses of all trigger level transitions are stored. The trigger levels are programmable and are stored in a FPGA register. In order to minimize the data rate consistent with the required pointing accuracy, a cadency divider is implemented which produces SAS interrupts to the DSP program with 128, 64, 32, 16, or 8 Hz. The nominal operational cadency of 32 Hz results in an additional pointing error of the order of 0.05 arc sec for linear interpolation between the measured points. However, the limited telemetry rate and a photon count rate that is higher than expected prior to launch required a reduction of the cadency to 16 Hz for a substantial fraction of operation time adding an error of $\approx$0.2 arc sec. An overview of the ADP software is given in Section 2.3.

### 2.2. Roll Angle System (RAS)

The RAS is a star scanner pointing radially outwards with respect to the rotating spacecraft and observes stars within a field of view of 60 to 90 deg in polar angle. Stars within the field of view are focused by a $f/1.4$ lens system with a focal length of 50 mm. While the spacecraft is rotating, the images of the stars travel over a linear CCD array and trigger one or several pixels. Even though the sensor is a TDI CCD (Time Delay and Integration CCD) with 2048 $\times$ 96 pixels with a pixel size of 13 $\times$ 13 $\mu$m$^2$, it is operated in a mode which makes it equivalent to a linear CCD with 2048 pixels with a size of 13 $\times$ 1248 $\mu$m$^2$ each. Thus, the field of view in roll angle is $\approx$ 1.4 deg. Assuming a rotational speed of 15 rpm, the passage time of a star event is between 15.3 ms and 17.6 ms depending on the polar angle with respect to the rotation axis. The size of a pixel corresponds to 0.9 arc min in polar angle.

The combined system with the optics and the CCD shows a spectral responsivity between 4000 Å and 10 000 Å with a maximum around 7000 Å. Thus, the sensitivity for a star detection depends strongly on its spectral type. The system is about twice as sensitive to M-class stars than to B-class stars. Although the integration time is programmable, the currently configured integration time is $\approx$8.8 ms corresponding to $\approx$47 arc min in roll angle. For a star with visual magnitude $m_V = 3$, this results in a response of 3.3 mV for a B0 star and 7.2 mV for a M0 star.

The front-end electronics with the first-level FPGA is very similar to that of the SAS with the difference that the cycle time is the same as the integration time which means that the CCD is always integrating light. The integrated charge is shifted out through the CCD shifter and digitized on a 12-bit ADC providing a resolution of 0.5 mV channel$^{-1}$. A double sampling and subsequent subtraction essentially eliminates the reset noise as described in Section 2.1. The timing for shifting the signals is the same as for the SAS. Besides the normal operational mode, several test and diagnostic modes (first-level modes) are implemented in the front-end FPGA.

As with the SAS subsystem, the RAS data are processed first by a hardware program in the second-level FPGA on the ADP and then by the software running on the DSP processor. In Section 2.3 an overview of the ADP hardware and software is given. Nevertheless, the RAS relevant part of the second-level FPGA is described in the following two paragraphs.

At the same time as the data are flowing through the second-level FPGA logic, the addresses for all pixel values exceeding their individually programmable threshold values are stored. It has been shown by Fivian, Henneck, and Zehnder (2002), that a threshold of about 5 channels above the dark signal is marginal to trigger at least one star for every revolution of the spacecraft. Therefore, a pixel summation algorithm has been implemented in order to increase the signal to noise ratio. The FWHM of the point spread function is $\approx 2.6$ pixels as pointed out by Zehnder *et al.* (2002) and Fivian, Henneck and Zehnder (2002). Thus a summation of two adjacent pixels combined with an adjusted threshold level provides an increase of a signal to noise of almost $\sqrt{2}$. Another factor of almost $\sqrt{2}$ can be gained by summing two temporal consecutive values of the same pixel number since the passage time of an event is almost double the integration time.

The low altitude orbit of RHESSI implies that almost $2\pi$ of the celestial sphere may be covered by the Earth. The signal induced by the Earth albedo can reach a level of 1000 times the saturation level of the RAS. Many CCD chips provide an anti-blooming feature, i.e. an internal chip capability to let the extra charge flow through an additional gate. However, the selected CCD in the RAS does not have that feature and therefore an extra mechanism had to be implemented. On the first-level FPGA, a programmable earth albedo threshold is set. If 64 pixels of a CCD readout are above this threshold value, the sensor is switched to maximum cadence of $\approx 2.2$ ms per cycle and once the readout level is below the threshold again, the sensor is kept in this fast readout mode for a programmable additional time before being switched back to normal mode. This results in an additional blind time after rotating out of the Earth albedo of up to 0.8 s. Nevertheless, the flight data shows a change of the dark signal of several channels for many configurations of the Earth albedo with respect to the spacecraft. This causes many false trigger events which can be removed with the analysis software.

### 2.3. Aspect Data Processor (ADP)

The ADP controls the SAS and RAS front-ends, does on-board processing of CCD images and manages the data transfer between the front-ends and the Instrument Data Processing Unit (IDPU) (Curtis *et al.*, 2002). The processing is necessary in order to reduce the data rate since the aspect sensors generate $\approx 2$ Mb s$^{-1}$ of raw data and the data has to be formatted into telemetry packets to be read by the IDPU. The achieved data reduction is in the order of a factor of 1000. The ADP can be separated into three different hardware parts. In Figure 1 the three columns of units represent the SAS/RAS interface, the central processor and the communication control. The different parts including the software can be controlled by parameters which have to be loaded into the memory prior to running the software.

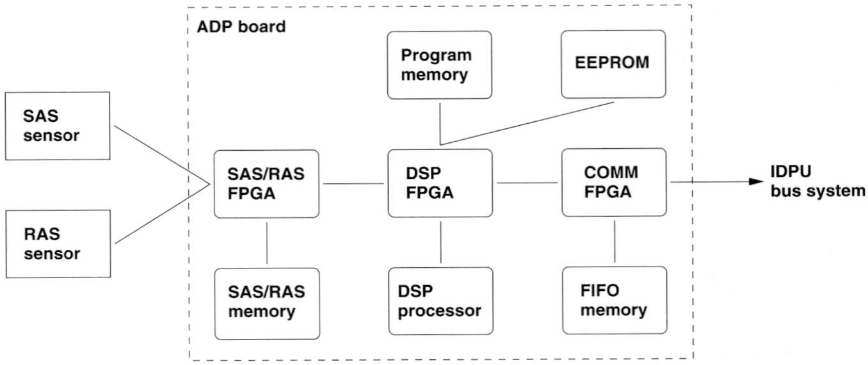


*Figure 1.* Simplified logical scheme of the Aspect Data Processor (ADP). For description see text in Section 2.3.

After initialization, the SAS/RAS interface acts as an independent unit and becomes responsible for the functioning of the SAS/RAS front-ends. The FPGA program which is driven by an internal sequencer is essentially described in Sections 2.1 and 2.2. It generates an interrupt to the central processing part for every SAS and RAS image which has been processed.

On the central processing unit, a DSP program processes the SAS and RAS data and generates basically four different classes of data to be stored in the FIFO. RAS star events, SAS limbs and SAS/RAS images are handled as different data streams. Whereas the images have to be formatted into telemetry packets only, the event-driven data from the SAS and the RAS need more processing. Images are stored frequently with a programmable duty cycle for later control of the gain and the dark signal.

In order to store the information for the solar limbs, the address and a programmable number of pixels for each limb are written to the telemetry packet. For every SAS cycle, all triggers from the first subsystem are stored first followed by the second and the third. The number of stored pixels above and below the threshold level can be programmed separately. It turned out that $2 + 2$ pixels is a optimum number to reach the required accuracy. For later analysis of the pointing, the cycles have to be counted in order to reconstruct the timing. The full resolution of 10 bits and a reduced resolution of the 8 most significant bits can be commanded.

Each RAS event consists of a time stamp, the address of the triggering pixel and a number of pixel values. For later reconstruction of the dark signal, a programmable number of pixels on either side of the pixels above threshold are stored. In general, a star event triggers several adjacent pixels which are treated as one event. Also with temporal and spatial summation enabled, the measured pixel values are stored for later analysis.

The third part of the ADP, the communication control, provides an interface between the DSP processor, the FIFO memory and the IDPU bus system. In addition, it stores the spacecraft time and incorporates power control.

A full list of all programmable parameters and a full list of the used APP-IDs (application identification numbers) identifying the different data streams are presented by Fivian (2002).

## 3. Reconstruction of the Aspect Solution

The telescope-fixed or spacecraft-fixed imaging coordinate system has been defined in such a way that any geometrical error has the least effect to the image reconstruction. Therefore, it has to be closely related to the features of the collimators and the SAS. The imaging $\tilde{z}$-axis or 'Imaging Axis' is defined to be parallel to the line from the center of gravity of the CCD centers to the center of gravity of the optical centers of the lenses. The origin is the center of gravity of the CCD centers and the sense of the $\tilde{z}$-axis is positive towards the Sun. Furthermore, the $\tilde{x}$- and $\tilde{y}$-axis are chosen to build a right-handed Cartesian coordinate system where the orientation of the $\tilde{y}$-axis is basically arbitrary. However, the $\tilde{y}$-axis is pointing approximately towards SAS10, the first SAS subsystem and is well defined with respect to the RAS defining the roll angle. All features from the front and rear grids, the SAS, the RAS and the PMTRAS have to be given in that coordinate system.

As indicated in Section 1 and described in detail by Hurford *et al.* (2002), the pointing of the rotating spacecraft and the roll angle have to be known at any given time in order to reconstruct images. A set of a well-defined pointing and roll angles is defined as the aspect solution. Figure 2 shows a instantaneous relation between the solar limb with solar-fixed $xy$-coordinates and the projection of the imaging $\tilde{x}\tilde{y}\tilde{z}$-coordinate system onto the Sun. The direction of the imaging $\tilde{z}$-axis is marked as 'Imaging Axis'. The pointing $\mathbf{P}_{\mathrm{ASPECT}}$ is defined to be the angle between the direction from the spacecraft towards the Sun center and the imaging axis. Furthermore, the roll angle $\varphi$ is defined as the angle between the solar fixed $y$-axis and the projection of the spacecraft fixed imaging $\tilde{y}$-axis. The sign of the roll angle is defined to increase with time. From the spacecraft looking towards the Sun, the spacecraft is rotating clockwise, and therefore, a positive roll angle is a rotation from the solar $y$-axis towards the rotating $\tilde{y}$-axis as indicated in Figure 2.

Since the SAS is a part of the telescope mounted as close as possible to the grids, the solar pointing origins from the measurement of the position of the Sun center in the body frame. This vector is the same as the vector $\mathbf{P}_{\mathrm{ASPECT}}$ in Figure 2 with the opposite sign, but measured in the rotating $\tilde{x} - \tilde{y}$ imaging coordinate system. Let $\mathbf{P}_{\mathrm{SAS}}$ be the direction of the Sun center measured by the SAS and $\mathbf{P}_{\mathrm{ASPECT}}$ the solar pointing of the telescope as defined earlier in this section, then the transformation is as follows:

$$\mathbf{P}_{\mathrm{ASPECT}} = \begin{pmatrix} \cos\varphi & -\sin\varphi \\ -\sin\varphi & -\cos\varphi \end{pmatrix} \mathbf{P}_{\mathrm{SAS}}, \tag{1}$$

where $\varphi$ is the roll angle. The matrix of the inverse transformation is identical.

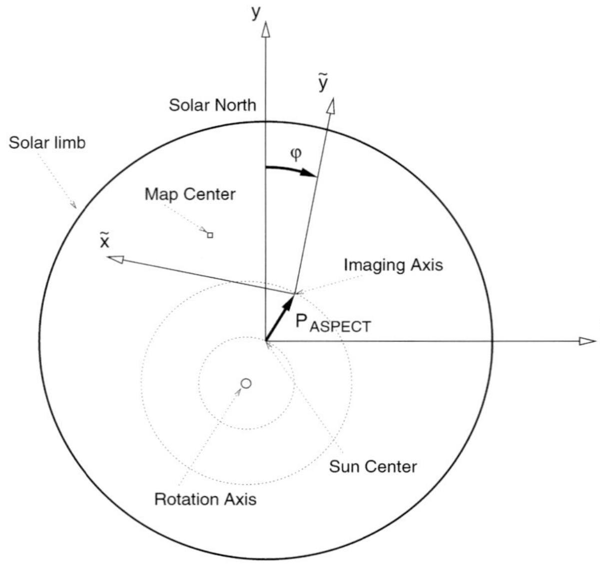


*Figure 2.* The *solid circle* represents the solar limb and the $x - y$ coordinate system refers to solar coordinates with the $y$-axis pointing toward solar north. At a given time, the telescope axis may point at any point on this projection plane. That point is marked with 'imaging axis' and the $\tilde{x} - \tilde{y}$ axis represents the spacecraft fixed coordinate system. The pointing of the spacecraft is defined as the angle between the Sun center and the imaging axis and is represented by the vector $\mathbf{P}_{\text{ASPECT}}$. The roll angle $\varphi$ is defined as the angle between the solar fixed $y$-axis and the projection of the spacecraft fixed $\tilde{y}$-axis.

### 3.1. The Solar Pointing

The transmitted data of each of the six solar limbs are processed off-line by the aspect reconstruction software which is integrated in the ground based RHESSI data analysis software (Schwartz *et al.*, 2002). First the position on the CCDs for each limb is determined. For every given SAS cycle of transmitted solar limbs, the mid-perpendiculars to the straight lines between the two limb positions are constructed as shown schematically on the left part of Figure 3. In order to correlate the results from the three independent subsystems, conceptually the three systems are linearly shifted to share the point of the Sun center as shown on the right side of Figure 3. For ideal circumstances the mid-perpendiculars would then intersect in one point, the instantaneous position of the Sun center.

In general, the mid-perpendiculars form a residual triangle as indicated in Figure 4 and the center of gravity of the intersections is the best guess for the position of the Sun center. Hence, the solar pointing for every SAS cycle is determined and the high cadence of measurements allows linear interpolation to determine the solar pointing for any given time.

This reconstruction suffers from two main error sources. One is resulting from errors in the fit of the limb positions and the other is due to uncertainties in the

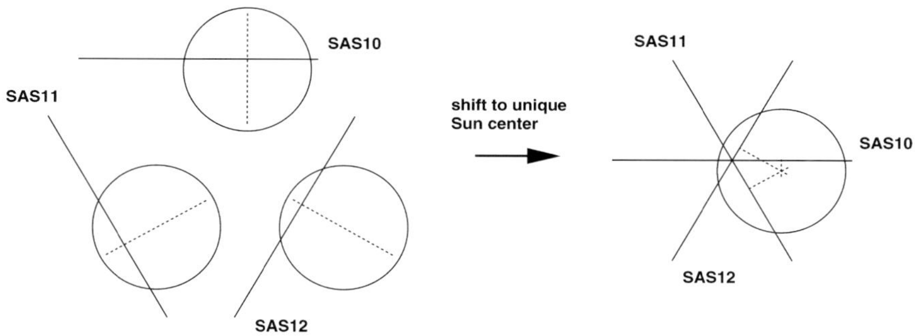


*Figure 3.* The CCD with the focused solar limb is shown for each subsystem *on the left part* of the schematic figure. Shifting the coordinates to share the point of the Sun center leads to the reconstruction of the instantaneous Sun center.

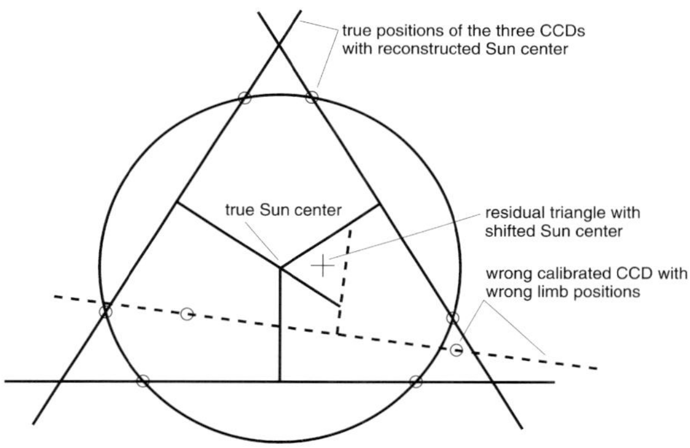


*Figure 4.* The *circle* indicates the solar limb and the *solid lines* are at the true positions of the CCDs with the corresponding reconstruction of the Sun center. A wrong calibration of the position of a CCD leads to a residual triangle in the reconstruction and therefore a shifted, incorrectly reconstructed Sun center as shown with the *dashed lines*.

knowledge of the geometry of the imager, therefore implying errors in correlating the three subsystems. If the fit of the limb positions had a systematic error, they would have the opposite sign for the positions of the two limbs on each of the CCDs. Thus, by calculating the mid-point of the two limb positions this error cancels essentially. The analysis of the flight data show that a linear fit of the pixel values and a subsequent fixed thresholding is accurate enough as pointed out later in this section.

The second source of pointing errors is a consequence of dimensional uncertainties. It can be well controlled by applying a self-calibration algorithm as described in detail by Fivian, Henneck and Zehnder (2002). In Figure 4, the case with an error for only one subsystem is represented. For the general case with a remaining residual triangle, the statistics of $1/\sqrt{3}$ of the length of the residual triangle or 2/3 of its height (Fivian, 2002), is a measure for the relative pointing error. Figure 5 shows the result for a typical orbit. Apparently, the spacecraft shows an off-pointing

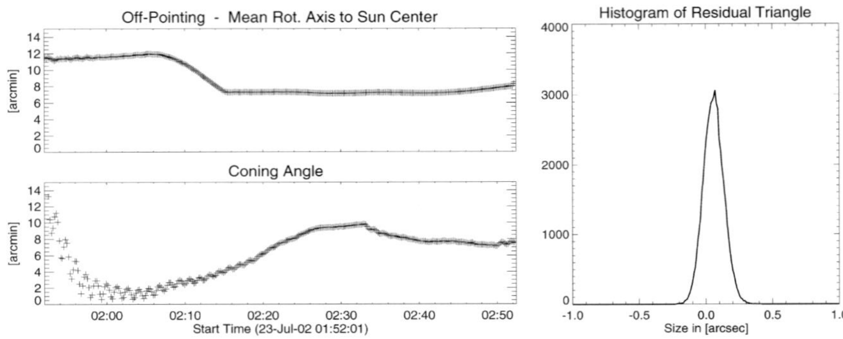


*Figure 5.* The plots show data for a typical orbit with the spacecraft being in direct sunlight. After sunrise, the spacecraft is pointed off the Sun center by $\approx 11$ arc min and has a coning angle of $\approx 14$ arc min, i.e., the rotation axis differs from the imaging axis. The angular momentum applied by the torque rods induces a precession which can be seen as wiggles in the coning angle for about 20 min. The complicated dynamics of changing the off-pointing and coning angle by several arc min is perfectly measured by the SAS, which is illustrated by plotting the histogram of the residual triangle for the same time range. The mean of the characteristic size of the residual triangle is $\approx 0.06$ arc sec and the width of the distribution corresponds to $\sigma \approx 0.08$ arc sec. An additional error induced by interpolating measured data points adds to the given value (see Section 2.1).

from the Sun center of up to 12 arc min and a coning angle of up to 10 arc min. The latter is the angle between the imaging axis and the instantaneous rotation axis. With the current calibration the systematic error of the relative pointing is $\approx 0.06$ arc sec and the statistical error is in the order of $\sigma \approx 0.08$ arc sec.

An overall shift or rotation of the imaging coordinates is called the absolute error of the solar pointing. Even though this can be seen as a cross-calibration between the SAS and the collimators, the calibration of the grid parameters corrects for initial mis-alignment. This calibration is based on 'focusing' the X-ray image by modifying the relative grid phases and keeping the imaging coordinates unchanged. In principle one can say, that any calibration error contributes to a blur of the X-ray image. The relation between the SAS and the grids is not fully understood yet. Nevertheless, images with a resolution of the order of 2–3 arc sec have been achieved.

### 3.2. The Roll Angle

The basic RAS principle is to determine the momentary roll orientation of the spacecraft with respect to known stars. In the rotating plane the roll angle determination can be reduced to a time measurement. Every star event contains several pixel values from the same CCD readout. In a first step, a Gaussian with an arbitrary baseline is fitted determining the polar angle, the dark signal and the response of that particular event. The response depends on the brightness of the star, its spectral type and the fraction of the integration time the star illuminated the CCD. Since the integration time is about two times shorter than the star passage time, a star event

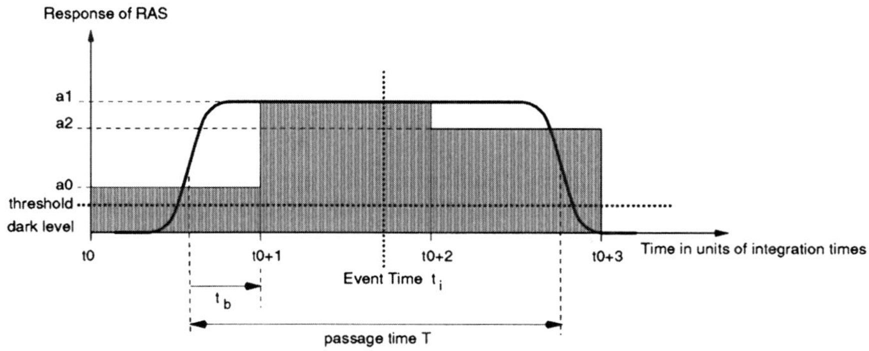


*Figure 6.* The *solid line* shows the response of the RAS CCD as a continuous function of time. Integration over a fixed time interval gives a histogram like function (*shadowed rectangles*) with values $a_0$, $a_1$, and $a_2$ for the integration cycles starting at $t_0$. After every integration time, the response level is compared to a threshold level for triggering the readout of the CCD pixels. The amplitudes $a_i$ and the triggering time $t_0$ defines the temporal centroid $t_i$ of the event.

generates a response on more than one consecutive time frame as represented in Figure 6. Thus, the problem has been reduced to a precise determination of the timing of a star event and the identification of an event with a specific star.

Assuming that the time dependency of the incident light for a star event can be treated as a function which is comparable to a rectangular shape. The width (FWHM) of the point spread function of the RAS optics is $\approx 2$ arc min and this angle corresponds to $\approx 0.4$ msec given the rotational speed of 15 rpm. Therefore, the convolution of the point spread parameterized in time with a rectangular function of the width of $\approx$ 15 msec passage time gives a small enough rise time.

Hence, the timing of a star event can be described by the equation system

$$a_0 = rt_b, \tag{2}$$

$$a_1 = rt_{\text{int}}, \tag{3}$$

$$a_2 = r(T - t_{\text{int}} - t_b), \tag{4}$$

which is made evident with Figure 6. For $i = 0, 1, 2$, $a_i$ are the fitted responses for the particular time frames, $r$ is the responsivity of the system, $t_b$ is the time since the star entered the field of view to the next integration time, $T$ is the passage time of a star event and $t_{\text{int}}$ is the integration time. With a fully calibrated system and an identified star event, the equation system over-defines the only unknown $t_b$. The event time for that particular event is then $t_i = t_0 + t_{\text{int}} - t_b + T/2$.

The measured responses $a_0$, $a_1$, and $a_2$ have noise. Assuming that $a_0$ and $a_2$ have the same accuracy, $t_b$ can be calculated in a symmetric way according to Equation (5) and the responsivity $r$ can be calibrated by the statistics from many stars using $r = a_1/t_{\text{int}}$:

$$t_b = \frac{1}{2}\left(\frac{a_0 - a_2}{r} + T - t_{\text{int}}\right). \tag{5}$$

In order to achieve an accurate enough timing using Equation (5), the responsivity $r$ has to be well calibrated and the passage time $T$ has to be known. Since the responsivity depends on the star, it is basically unknown until the star has been identified but the sum of Equations (2)–(4) gives an initial estimate using $r = (a_0 + a_1 + a_2)/T$ leading to a timing according to Equation (6):

$$t_b = \frac{1}{2}\left(\left(\frac{a_0 - a_2}{a_0 + a_1 + a_2} + 1\right)T - t_{\text{int}}\right). \tag{6}$$

On the other hand, the passage time $T$ is initially unknown and it has to be estimated using the statistics from a bigger number of star events. Once the responsivity and the passage time are estimated, the stars can be identified by comparing the timing, polar angle and the response with a star catalog. The RAS specific responsivity can be calibrated and stored in the star catalog too. Thus, after identifying the star events the responsivity is known and the estimate of the passage time can be improved with

$$T = \frac{a_0 + a_1 + a_2}{r}. \tag{7}$$

Hence, the calculation of the timing has to be done twice, first to achieve good estimates in order to identify the star events and second to result in an accurate timing.

For fainter stars, the response of the first and the last time frame might be too low to trigger RAS pixels. If only one of those is available, Equation (5) can be used in an asymmetrical way with only $a_0$ or $a_2$. If a star event results in pixels for only one time frame starting at $t_0$, the best estimate for the timing is simply $t_i = t_0 + t_{\text{int}}/2$.

The analysis and its software for the RAS aren't completed yet. The current status of the calibration is presented by Fivian, Henneck, and Zehnder (2002). However, the aspect reconstruction software currently uses the PMTRAS system to determine the roll angle. This system was designed as a backup roll angle system and is described by Hurford and Curtis (2002).

## Acknowledgements

The authors thank Gordon Hurford, Peter Ming and the PSI workshop and design office for their substantial work on the design, building and integration of the hardware. This work has partially been funded by the Swiss National Science Foundation.

## References

Curtis, D. W. *et al.*: 2002, *Solar Phys.*, this volume.
Fivian, M.: 2002, Thesis ETH.
Fivian, M., Henneck, R., and Zehnder, A.: 2002, *SPIE Proc.* **4853**, Waikoloa, Hawaii.
Hurford, G. J. and Curtis, D.: 2002, *Solar Phys.*, this volume.
Hurford, G. J. *et al.*: 2002, *Solar Phys.*, this volume.
Lin, R. P. *et al.*: 2002, *Solar Phys.*, this volume.
Schwartz, R. A. *et al.*: 2002, *Solar Phys.*, this volume.
Smith, D. M. *et al.*: 2002, *Solar Phys.*, this volume.
Zehnder, A. *et al.*: 2002, *SPIE Proc.* **4853**, Waikoloa, Hawaii.

# THE PMTRAS ROLL ASPECT SYSTEM ON RHESSI

G. J. HURFORD and D. W. CURTIS
*Space Sciences Laboratory, University of California, Berkeley, CA 94720-7450, U.S.A.*
*(e-mail: ghurford@ssl.berkeley.edu)*

(Received 17 August 2002; accepted 17 August 2002)

**Abstract.** High-resolution solar hard X-ray imaging on the Reuven Ramaty High-Energy Solar Spectroscopic Imager (RHESSI) spacecraft is achieved by a set of rotating modulation collimators. The interpretation of the observed time-modulated X-ray flux in terms of high-resolution, accurately located images requires continuous, arc-minute roll aspect, which at present is provided by the 'Photo-Multiplier Tube Roll Aspect System' (PMTRAS). This paper describes the PMTRAS operating principles, hardware implementation, calibration, performance and data analysis approach, with emphasis on its effect on RHESSI imaging.

## 1. Introduction

The Reuven Ramaty High-Energy Solar Spectroscopic Imager (RHESSI) is a NASA Small Explorer (SMEX) mission to study the acceleration and transport of high-energy particles in solar flares (Lin *et al.*, 2002). Observationally this is accomplished by high-resolution imaging spectroscopy of X-rays and gamma-rays from 3 keV through 17 MeV with angular resolution as high as 2.3 arc sec. Achieving high angular resolution in this energy range requires indirect imaging techniques. In this case a set of 9 modulation collimators are mounted on a rotating spacecraft. Each collimator consists of a pair of widely spaced grids that alternately transmits and blocks part of the incident X-ray flux as the spacecraft rotates. The imaging information is encoded in the time-modulated lightcurves (Hurford *et al.*, 2002). The data consist primarily of time- and energy-tagged photons from which the image is subsequently reconstructed by ground-based analysis software (Schwartz *et al.*, 2002).

The RHESSI spacecraft rotates at a nominal 15 rpm about an axis directed at the Sun. One of the enabling features of this imaging approach is that spacecraft pointing requirements (0.2 deg) are based on the 1-degree imager field of view, instead of the 2.3-arc sec angular resolution. A high-bandwidth subarc-second pitch and yaw aspect system (Zehnder *et al.*, 2002; Fivian *et al.*, 2002) can be used to compensate for time-varying pointing. In addition, accurate absolute knowledge (1 arc min, $1\sigma$) of the roll orientation is required both to coherently interpret the time-modulated light-curves and to accurately place the reconstructed image on the Sun.

*Solar Physics* **210:** 101–113, 2002.

A number of options were considered for obtaining this roll knowledge. Available non-solar aspect sensors included horizon crossing indicators, star cameras, magnetometers, and differential GPS. Horizon crossing indicators generally have insufficient angular resolution to meet the requirements, as does differential GPS on a spacecraft the size of RHESSI. Model magnetic fields would be required to use a magnetometer as an aspect sensor, and these are neither sufficiently accurate nor stable to meet RHESSI's requirements. In addition having significant cost, power, mass and telemetry implications, most off-the-shelf star cameras are designed for 3-axis stabilized spacecraft and are not well suited to deal with the RHESSI spin rate. Another significant challenge for star sensors is handling the Earth albedo. The sensor will look directly at the sunlit Earth for almost half of rotation, and must come out of saturation in time to measure relatively dim stars.

To meet these requirements, RHESSI has two redundant roll aspect systems: a CCD-based Roll Aspect System (RAS) (Zehnder *et al.*, 2002; Fivian and Zehnder, 2002) and a Photo-Multiplier Tube Roll Aspect System (PMTRAS). The latter was added as a backup system at a late stage of the development of the RHESSI hardware. Both systems are star scanners that view a star field perpendicular to the spin axis. As the spacecraft rotates, the instruments accurately note the times at which bright stars pass through their field of view. With knowledge of the stars' location and confidence in the inertial rotation of the spacecraft, the roll orientation of the spacecraft at any intermediate time can then be interpolated. At the nominal spin rate of 15 rpm, the 1-arc min angular resolution requirement is equivalent to a timing requirement of 0.2 ms.

The PMTRAS was implemented by adapting a low-cost flight-proven star scanner, developed by M. Lampton for rocket payloads, to meet RHESSI's requirements. Both RAS and PMTRAS are operational, but RHESSI imaging to date has been based on the PMTRAS.

The purpose of this paper is to describe the PMTRAS principles of operation, hardware implementation, calibration, performance and software. In Section 2, the hardware is summarized. Pre- and post-launch calibrations are described in Section 3. Performance issues are discussed in Section 4, followed by an outline of the data analysis approach in Section 5.

## 2. Hardware Description

### 2.1. Optics

A schematic view of the PMTRAS optics is shown in Figure 1. It consists of a primary 75 mm focal length $f/3$ plano-convex singlet lens, stopped down to 22 mm, that focuses an image of the star field onto a $6.5 \times 0.02$ mm slit. Optical distortions due to the use of a singlet, primary lens are not important since only the integrated light distributed along the slit is relevant. A secondary re-imaging

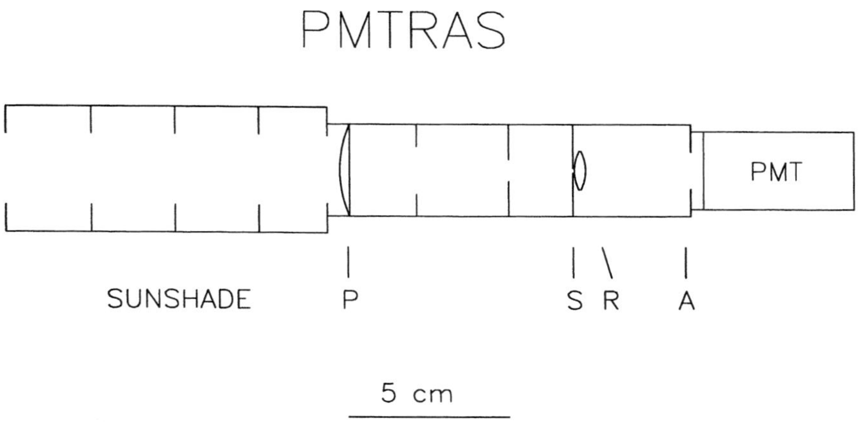


*Figure 1.* Profile of the PMTRAS optics in the plane orthogonal to the slit, S, showing the primary lens, P, the reimaging lens, R, and PMT aperture, A.

singlet lens with a 25 mm focal length distributes the light passed by the slit over a 12 mm diameter aperture over the face of a 1-inch R1924-01 Hamamatsu photomultiplier tube (PMT) whose bialkali photocathode is nominally sensitive from 300 to 650 nm. The peak system response is at 420 nm. Fused silica lenses were used to provide good ultraviolet transmission and to minimize radiation damage.

The use of a singlet lens results in a focal length that is wavelength dependent but for most star classes provides a nominal 0.07 mm r.m.s. focal spot. Convolving this with the slit width implies a FWHM response of 10 arc min on the sky. This corresponds to a 2 ms FWHM transient signal (*blip*) at 15 rpm. The field of view parallel to the spin axis is 5 deg, centered at a polar angle of 90 deg. A sunshade suppresses illumination by a factor of $> 10^6$ outside an unobstructed 15-deg field of view.

### 2.2. Electronics

The PMTRAS electronics includes the PMT, along with its bias supply and bleeder chain, a charge sensitive preamp/discriminator, and logic circuitry that accumulates counts and interfaces to the RHESSI Instrument Data Processing Unit (IDPU) (Curtis *et al.*, 2002). Since the PMTRAS was added late in the RHESSI development process, it was constrained to fit with the existing spacecraft and instrument hardware design, while using as much off-the-shelf design and hardware as possible. An existing flight-spare PMT and HV bias supply from the WIND 3D Plasma instrument were found to meet the sensitivity requirements. An Amptek A111 charge-sensitive preamp / discriminator was used to convert the anode charge pulses into logic pulses. These pulses were then passed to a Field Programmable Gate Array (FPGA). This FPGA contained counters and logic to interface the

PMTRAS data to the RHESSI IDPU. The FPGA was mounted on a small daughter board attached directly to the IDPU backplane.

One issue that had to be addressed was that the PMTRAS would face the bright Earth albedo once per rotation during the sunlit portion of the orbit. To speed recovery from this condition, modifications were made to the bleeder chain that distributes high voltage among the subsequent dynodes in the PMT. The photocathode current is sensed, and when it exceeds a set level, the bias to the first dynode of the PMT is shorted out. This greatly reduces the gain of the PMT, avoiding excessive rates at the later dynodes that can cause aging. This also speeds the PMT recovery during transitions from Earth- to space-viewing each spacecraft rotation.

The PMT and its bleeder chain and bias supply were packaged with the PMTRAS optics and attached directly to the side of the IDPU. This provided the desired FOV without affecting spacecraft interfaces.

### 2.3. Telemetry

The basic data acquired by the PMTRAS consist of digitized output pulses from the PMT. These pulses are summed in counters with *accumulation intervals* of a few milliseconds (programmable by ground command to $\frac{1}{256}$, $\frac{1}{512}$ or $\frac{1}{1024}$ s).

To accommodate this 8 kbps data rate within the instrument diagnostic telemetry allocation of 128 bps, the FPGA looks for the maximum count in a $\frac{1}{4}$ s *readout interval* (also programmable in powers of 2). The telemetry then includes the time of the sample containing the maximum number of counts in the readout interval, the maximum counts in that interval and the counts in the preceding and following intervals. Nominally, the maximum corresponds to the brightest object (star) seen in the readout interval. Since counts from a typical star are divided among 2 or 3 of the recorded accumulation intervals, interpolation to accuracy much better than the 2 ms accumulation time is possible. After compressing the counts to 8 bits, each readout requires 4 bytes. This meets the 128 bps telemetry requirement while providing information on up to 16 stars per rotation.

## 3. Calibration

### 3.1. Prelaunch Calibration

The key calibration requirement is to determine the direction of maximum response of the PMTRAS optics with respect to the orientation of the imager grids to 1 arc min. It is important to note that errors in this calibration have no effect whatsoever on the morphology of the reconstructed image. Rather, this knowledge serves only to place the image at the correct position angle with respect to Sun center. For a spin axis at Sun center and a limb flare, a 1-arc min roll error corresponds to a 0.3 arc sec image location error.

This calibration was conducted after the PMTRAS was mounted on the spacecraft by noting the amplitude of the PMTRAS response to a blinking diode that was slowly moved in azimuth at two successive different radii ($\sim 3$ and $\sim 6$ m). Theodolites were used to relate the diode positions at maximum response to tooling balls on the imager hardware, whose positions, in turn, were known with respect to grid orientations to $\ll 1$ arc min. Redundant measurements provided an internal consistency of $\pm 0.01$ deg that met the requirement.

Knowledge of the dead time of the PMT response is necessary for correcting count rates observed for the brightest stars. This was determined by noting the relative count rates as a fixed blinking source was moved radially from the PMTRAS. The measured dead time was 0.9 $\mu$s, which enables input count rates up to $\sim 2 \times 10^6$ counts $s^{-1}$ to be corrected.

### 3.2. Post-launch Calibration

Post-launch calibrations permit improvements in both the relative and absolute accuracy of the roll data. The most important of these calibrations is based on observations of the Crab nebula during its annual passage within 1.5 deg of the Sun on 15 June. This is sufficiently close that while nominally observing the Sun, the Crab passes within the imaging field of view. When analysis of these data is complete, the reconstructed location of the Crab pulsar will confirm (or correct) our knowledge of the orientation of the roll aspect system with respect to the grids to a level of accuracy of $\sim 0.1$ arc min, well in excess of the solar requirement.

Statistical analyses of the observed star field can be used to calibrate some instrument parameters. For example, an initial analysis of which stars were detected (or not) was used to confirm the polar alignment and field of view of the slit to $\pm 0.2$ deg. In the future, the precision of the roll solution will benefit from self-calibration of the orientation of the slit, which can be inferred from systematic phase shifts from stars observed at different polar angles.

Statistical analysis of the distribution of counts for each star can be used to infer the pulse shape and width, since the instant of maximum count rate should be randomly distributed within integration intervals. Preliminary analysis has yielded an average pulse width of about 2.6 ms, slightly higher than expected for a rotation rate of 14.5 rpm. The observed distributions also suggest that some star classes generate an asymmetrical pulse shape.

## 4. Performance

One of the advantages of using a PMT in this application is that signal-to-noise is dominated by the integrated background star field rather than electronic noise. Since reaching its nominal spin period of 15 rpm on 27 Februray 2002, normal operation of the PMTRAS has used the 2 ms integration time option. The number of counts due to the integrated background star field varies widely depending on proximity to the galactic plane and on orbital phase. In a 2 ms interval, typical values range from a few tens to a few hundreds of counts. This dominates the PMT noise, which is estimated to be about 2 counts. Prelaunch expectations for counts from magnitude 4 stars ranged from $\sim 30$ to $\sim 150$ counts in a 2 ms interval, depending on spectral class (M0 to B0). In-flight sensitivity has proved to be consistent with this. No significant degradation in PMTRAS sensitivity has been noted in the first few months of operation.

For any given orbit, the visible star field represents a 5-deg band that intersects the ecliptic poles. During the year, this band rotates about the poles so that a substantially different star field is sampled every few days. As a result, the number of observed stars varies significantly both with time of year as well as with orbital phase. Typically several tens of valid blips are detected per minute, representing from 1 to $\sim 10$ different stars.

PMTRAS measurements of spin period for typical orbits are shown in Figure 2. Although ideally RHESSI would spin with a constant angular velocity, the spin period is in fact not uniform. Within each orbit there is an asymmetrical cyclic variation in the spin period of 2 ms peak-to-peak. This is associated with thermal expansion and contraction of the solar panels due to the day/night cycle and with power generation that further heats the panels during the first part of the sunlit portion of each orbit. (The solar panels expand at higher temperatures, increasing their moment of inertia and so lengthening the spin period.) The temporal profile is generally smooth, except for sharp ($\sim 30$ s) ‘snaps’ that occur as the spacecraft enters and leaves sunlight. In addition, there is a systematic $\sim 0.25$ ms increase in rotation period each orbit, which is probably due to secondary effects of the torque-rod-based aspect control system.

## 5. Data Analysis

### 5.1. Data Analysis Steps

As previously discussed, the raw PMTRAS data consist of 4 potential star ‘blips’ per second, each of which defines the time and the values of 3 successive 2-ms accumulations of counts. The data analysis objective is to convert these data into knowledge of absolute roll angle as a function of time.

Since many of the blips are due to causes other than stars, the first step is to filter the 4-Hz dataset. Albedo response is suppressed by rejecting blips if the *minimum* of the 3 counts exceeds 1000 counts. A diagnostic check of PMT operation flags blips if the *maximum* count is less than 2. To suppress cases where the remaining albedo or background star field still dominates, a requirement is imposed that the difference between the minimum and maximum counts among the 3 samples be statistically significant at the $4.5\sigma$ level. After such filtering, most of the remaining

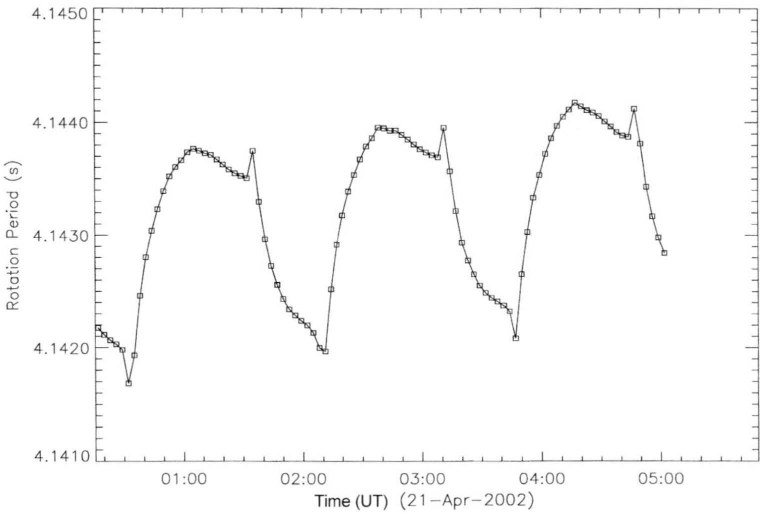


*Figure 2.* Preliminary estimates of rotation period for a typical 3-orbit interval that in this case includes RHESSI's first X-class flare on 21 April 2002. The six sharp changes occur at successive entries into sunlight and into eclipse.

valid blips represent well-observed stars although a few have a terrestrial origin ('city lights').

The next step is to determine a more precise occurrence time of each valid blip within its identified 2-ms integration interval. To that end, a timing parameter, $R$, is defined as

$$R = \frac{(\mathrm{C2} - \mathrm{C0})}{(\mathrm{C1} - \min(\mathrm{C2}, \mathrm{C0}))},$$

where C0, C1, C2 are the counts observed in the 3 successive integration intervals (C1 > C0, C2). $R$ has the property that it is independent of background and source intensity, and for any given pulse profile will increase monotonically as the occurrence time of the blip advances through the 2-ms accumulation interval. The complete set of valid blips are time-ordered on the basis of $R$ and assigned equally spaced time slots within the 2-ms interval slots on the basis of this ordering. (The rationale is that blip times should occur randomly within a 2-ms interval.) For a dataset containing a large number of blips, the interpolated time has an accuracy that is dominated by counting statistics and does not depend on knowledge of the blip time profile. For typical analyses, interpolated blip times are good to $\sim 0.1$ ms.

A set of preliminary estimates of rotation period are obtained every 3 min by identifying recurring time differences (and integral multiples thereof) among the blips. This technique exploits the high timing accuracy of the blips and is robust in the face of missing blips, irregular occurrence of multiple blips per rotation, and the presence of blips of non-stellar origin. Figure 2 shows a typical sequence of such periods.

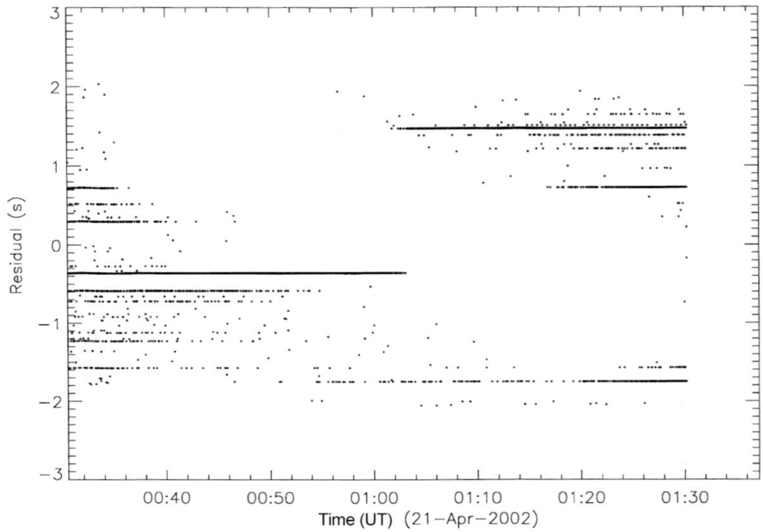


*Figure 3.* Preliminary timing residuals for the first orbit shown in Figure 2. Each point represents an observed 'blip'. The timing residual corresponds to the difference between the observed time of the blip and the time at which the spacecraft had a fixed (but arbitrary) roll orientation as determined from the periods in Figure 2. Each star has the same timing residual on successive rotations and so appears as a horizontal line. The diagonal blank area corresponds to intervals in each rotation when the PMTRAS was viewing the Earth.

The sequence of average preliminary rotational periods determines a set of reference times at which the spacecraft has the same position angle. The observed blip times are then expressed as a set of a time residuals, relative to the closest reference time (Figure 3).

The timing residuals for each blip are converted to rotational phases (still with an arbitrary zero) and are associated with their intensity as shown in Figure 4. In principle, at this stage all blips associated with a given star should have the same intensity and relative phase. In practice, as Figure 4 suggests, the phase scatter is much better than that of the intensity, which typically varies by $\sim \times 2$ for a given star. (The poor photometry arises because inaccurate background subtraction results in a dependence of apparent intensity on the occurrence time of a blip within the accumulation interval.) The relative phases (and to a lesser extent their intensities) are used to identify up to $\sim 10$ groups of blips, each associated with individual stars whose identities are unknown at this stage. Averaging within each 'blip group' then yields a best estimate of the relative phase for each blip group.

To associate the blip groups with specific stars, a database, derived from the Yale bright star catalog, is used to convert precessed star positions to a set of potentially detectable stars for a given date and time. Such a subset typically contains $\sim 100$ stars whose positions are expressed in terms of position angle. To minimize potential identification errors, stars with closely spaced position angles are discarded

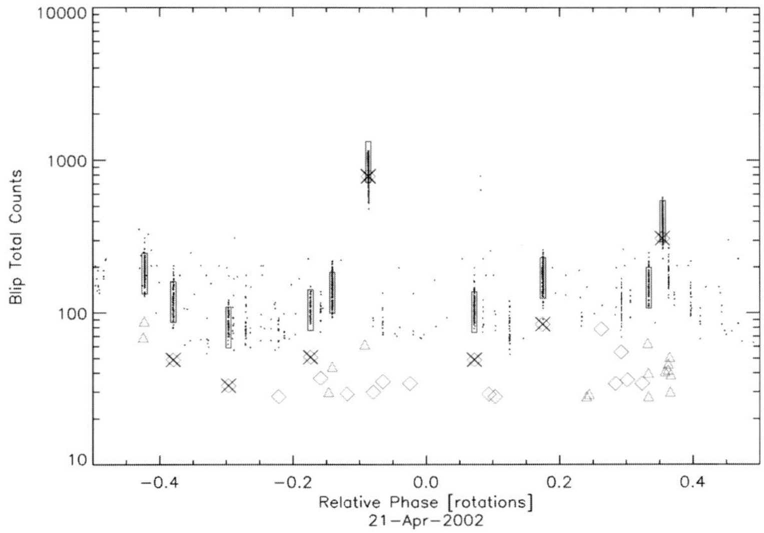


*Figure 4.* Intensity/phase correlation for the same observed 'blips' shown in Figure 3. The relative phase is the ratio of the timing residual to the rotation period. The tall *rectangles* identify groups of blips that were associated with a specific star. The *diamonds* represent stars that could potentially be observed, plotted as predicted star intensity/phase in the same units and same relative phase as the observed blips. The 7 '×'s indicate stars that were identified with the blip group that appears above them. *Triangles* represent potential stars that were not considered because of proximity in position angle to other stars.

from this list unless one is much brighter than the other, in which case the stronger one is retained.

At this stage, we have $M$ ($\sim 10$) observed 'blip groups' with accurate *relative* rotational phases but whose absolute position angles are unknown. The next objective is to associate these with specific stars (with known position angles) selected from the $N$ ($\sim 100$) element subset of bright stars. Given the $\sim N^M$ possibilities, this cannot be done on a trial-and-error basis. Instead, the more tractable $M \times N$ possible cases of association of a single star are considered. In each case the position angles of all 'blip groups' are shifted on the basis of the trial association and position angle differences between the remaining 'blip groups' and nearest star are evaluated. The best association is based on the number and proximity of matches between the two groups. This works very quickly and reliably, provided accurate relative observed blip-group phases are available. Typically, star associations for about 2 to 6 blip groups are obtained.

By this point, all blips within 'associated blip groups' have been associated with specific stars. All other blips and all preliminary estimates of rotation period are discarded. The associated blips now define a sequence of monotonically increasing position angles as a function of time. After rejection of a few outliers and converting to a table of spacecraft $Y$-axis orientations (clockwise relative to solar north)

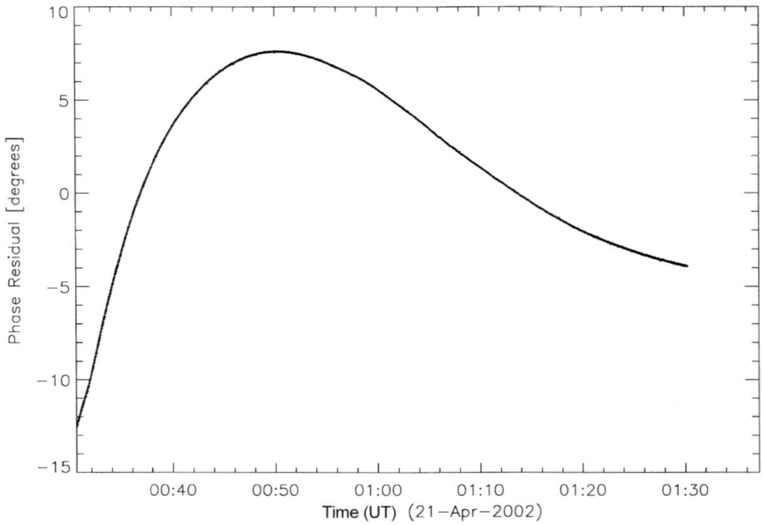


*Figure 5.* The final roll solution, plotted as a phase residual relative to a best-fit fixed period of 4.143114 s. The data consist of all star-associated blips in Figure 4.

as a function of time, this becomes the output of the PMTRAS. Figure 5 shows a typical plot of the results, expressed as a phase relative to a best-fit fixed period. The magnitude of the phase variations demonstrates the need to accommodate the non-uniform rotation period for integrations of a few minutes or more. The single-valued nature of the curve, which is based on multiple stars, provides strong confirmation of the correct star association. Figure 6 shows a subset of this data over a more limited time to illustrate typical scatter for individual blips. At any time, the scatter is $\sim 0.03$ deg peak-to-peak. Even without averaging, these data can be linearly interpolated to yield position angle with an accuracy of $< 1$ arc min r.m.s.

Note that for this level of accuracy, and for solar pointing accurate to 0.2 deg, the PMTRAS solution does not require any correction based on the pitch and yaw pointing of the spacecraft to meet the accuracy requirements for solar imaging.

### 5.2. Data analysis issues

There are a number of assumptions that must be fulfilled to make the foregoing sequence of analysis steps work in practice. First, several different stars must be observed during the analysis interval. Second, appropriate thresholds must be set so that there are not too many potential stars with which to evaluate trial associations. Third, during the analysis interval there must be no significant PMTRAS data gaps or extended periods of no viable star observations. Satisfaction of this condition is necessary so that coherent initial phases can be derived to support star identification. These conditions can be satisfied by appropriate selection of PMTRAS

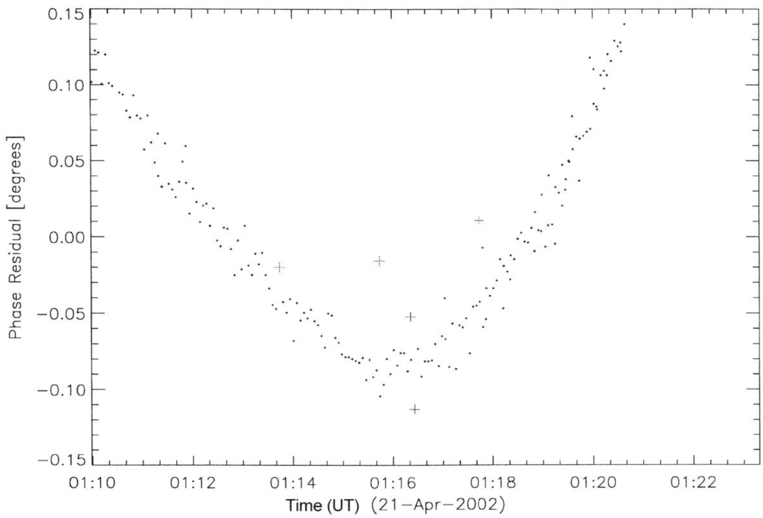


*Figure 6.* An expanded view over a 10-min interval in Figure 5, shown relative to a locally-fit fixed period of 4.143686 s. The peak-to-peak scatter of individual measurements at any specific time is $\sim 0.03$ deg.

analysis parameters. Although this can almost always be done manually, automated selection of such parameters sometimes results in one or more of these conditions not being satisfied.

The fundamental reason for the failure of automated selection of PMTRAS analysis parameters is rooted in the character of the RHESSI analysis software. In order to retain the flexibility that RHESSI provides for users to select analysis parameters (such as image-, time- and energy- resolution and range), most users analyzing RHESSI data start from level-0 data and perform calibration tasks on-the-fly. This approach works well for X-ray analyses over a wide range of time intervals. However, the optimum PMTRAS analysis intervals are based on other considerations. For example, sometimes it is desirable to span a substantial portion of an orbit to support star identification, but at other times this is undesirable, since an extended interval may incorporate an unbridgeable data gap. Failure to select appropriate parameters can result in unnecessary gaps in the PMTRAS solution or, more usually, in false star associations. A roll solution with false star association will still provide images with the correct morphology, but their placement in position angle about Sun center is arbitrary.

While manual PMTRAS analysis with appropriate parameters is certainly a viable solution to this dilemma, this approach would be incompatible with our goal of making image reconstruction as transparent as possible to the user.

A second set of problems is associated with the presence of data gaps in the PMTRAS dataset. These arise because the PMTRAS shares a small (1000 packet) data buffer with other diagnostic information. In circumstances where (intentionally or inadvertently) the data buffer becomes full, no PMTRAS data is stored until the next ground contact. In the first six months of the mission, such gaps average about 4% of the total mission time and typically last many minutes. In some cases, interpolation can be used to achieve a useful solution.

### 5.3. Future developments

The highest priority future development associated with PMTRAS roll aspect solution is the elimination of problems associated with automated selection of analysis parameters. Although it means establishment of a secondary database, a mission-long roll solution will be generated that will provide a roll solution applicable to any user-specified time interval. Such an approach is also well suited to the expected eventual integration of RAS and PMTRAS analysis outputs and to non-solar analyses over extended time intervals. This database will also incorporate the obvious improvement of averaging the PMTRAS output on a $\sim 1$ min timescale. This should reduce the current $\sim 1$ arc min scatter by an order of magnitude.

Additional improvements in the roll calibration will result from analysis of the Crab pulsar observations described above and from further analysis of the in-flight self-calibrations.

## 6. Discussion

The RAS and PMTRAS systems provide redundant information on RHESSI roll aspect. They are complementary in several respects: the PMTRAS sensitivity is limited by the background star field whereas the RAS sensitivity is limited by electronic noise; the RAS and PMTRAS sensitivities peak in the red and blue respectively; the RAS has lower sensitivity but a wider (30 deg) field of view and views a different field; they employ different techniques for albedo suppression and recovery. Thus they are well suited to providing redundant roll information.

Although the PMTRAS was a late and inexpensive addition to the RHESSI mission, it has proved successful in terms of providing roll aspect with the required accuracy and reliability. The primary difficulty has been due to the application of flexible, user-specified analysis parameters to a subsystem whose data analysis is better matched to the traditional approach that provides users with calibrated output. This will be addressed in the near future with the generation and distribution of a roll-solution database.

## Acknowledgements

We thank Paul Turin of SSL who was responsible for the mechanical design and assembly of the PMTRAS. We are also grateful to Dave Clark of GSFC for his exemplary work (as always) in the optical calibration. This work was supported by NASA grant NAS5-98033-05/03.

## References

Curtis, D. W. *et al.*: 2002, *Solar Phys.*, this volume.
Fivian, M., Hemeck, R., Mchedlishvili, and Zehnder, A.: 2002, *Solar Phys.*, this volume.
Hurford, G. J. *et al.*: 2002, *Solar Phys.*, this volume.
Lin, R. P. *et al.*: 2002, *Solar Phys.*, this volume.
Schwartz, R. A., Csillaghy, A., Tolbert, A. K., Hurford, G. J., McTiernan, J., and Zarro, D.: 2002, *Solar Phys.*, this volume.
Zehnder, A. *et al.*: 2002, *SPIE Proc.* **4853**, in press.

# THE RHESSI SPACECRAFT INSTRUMENT DATA PROCESSING UNIT

D. W. CURTIS[1], P. BERG[1], D. GORDON[3], P. R. HARVEY[1], D. M. SMITH[1] and A. ZEHNDER[2]

[1]*University of California Space Sciences Laboratory, Berkeley, CA 94720-7450 U.S.A.*

[2]*Laboratory for Astrophysics, Paul Scherrer Institut, CH 5232 - Villigen PSI*

[3]*ELF Electronics, Brookline MA 02446 USA*

(Received 18 September 2002; accepted 19 September 2002)

**Abstract.** The Ramaty High-Energy Spectroscopic Imager (RHESSI) spacecraft is a NASA Small Explorer (SMEX) class mission. RHESSI is designed to image solar X-rays and gamma rays with high-energy resolution. The Instrument Data Processing Unit (IDPU) serves as the central RHESSI instrument on-board data-processing element. It controls and monitors the instrument operations, and provides a flexible telemetry collection and formatting system. The system responds autonomously to optimize science data collection over a wide dynamic range of conditions, handling up to 40 Mbps of telemetry during solar flares. This paper presents an overview of the IDPU hardware and software design.

## 1. Introduction

The Ramaty High-Energy Spectroscopic Imager (RHESSI) spacecraft is a NASA Small Explorer (SMEX) class mission. RHESSI is designed to image solar X-rays and gamma rays with high energy resolution (Lin *et al.*, 2002). The RHESSI Instrument Data Processing Unit (IDPU) serves as the central instrument data processing element for all RHESSI science package. It controls the instruments and collects and formats the instrument telemetry, and provides a single point interface between the spacecraft and instruments for commands, telemetry, and power. All instrument electronics not closely coupled to the instrument detectors reside in the IDPU.

The IDPU has significant performance challenges, while being significantly constrained by mass and power. State-of-the-art signal processing electronics is required for the germanium detectors to provide the throughput and resolution, but at a fraction of the power of previous systems. A variable data throughput rate up to 40 Mbits per second must be accommodated from the instruments to the spacecraft Solid State Recorder (SSR). The system must manage this throughput in an intelligent fashion to optimize the data return while not over-filling the SSR. The large quantity of aspect data collected must be reduced to a manageable volume by intelligent selection of critical subsets of the data to send. Active instrument control is required to maintain optimal instrument performance, such as spectrometer and imager thermal control, automatic detection of the excessive background generated when the spacecraft passes over the South Atlantic Anomaly (SAA) and

*Solar Physics* **210:** 115–124, 2002.

appropriate response to limit data collection during that time, and sunlight/darkness detection and instrument re-configuration.

A key element to achieving high data throughput at low power is having the bulk of the data collection and formatting handled in field programmable gate arrays (FPGA) rather than in a microprocessor. This is similar to the FAST IDPU (Harvey *et al.*, 2001), allowing a modest microprocessor to provide the required intelligence while the routine tasks are carried out more efficiently in the FPGAs.

The IDPU is implemented in three physical boxes: the Instrument Power Controller (IPC), the Cryo Power Controller (CPC), and the VME chassis. The IPC contains the low and high voltage power converters for the instrument. The CPC is a power amplifier that generates the AC power required to run the cryocooler. The VME chassis contains the analog and digital signal processing elements. The VME chassis contains 12 6-U VME-sized cards, a backplane, and a small daughter card attached to the backplane. These cards include:

– A Data Controller Board (DCB) containing the central processor, spacecraft signal interfaces, backplane controller, and data formatter. The DCB also contains the analog and digital processing electronics for the particle detector.

– Nine Detector Interface Boards (DIB), which contain the analog and digital signal processing for each of the nine germanium detectors in the RHESSI spectrometer.

– An Aspect Data Processor (ADP) that processes the imager aspect sensor data from the Roll Aspect Sensor (RAS) and Sun Aspect Sensors (SAS).

– A Power Controller Board (PCB) which contains the various instrument power switching and controls, plus signal conditioning for the bulk of the instrument analog monitors such as currents, voltages, and temperatures.

The small daughter card attached to the backplane contains the interface to the second Roll Aspect Sensor, the RAS-PMT.

Figure 1 is a block diagram of the IDPU and how it connects to the rest of the instrument and spacecraft.

## 2. Spectrometer Data Flow

Signals from the germanium detectors in the spectrometer are processed by charge sensitive amplifiers attached to the spectrometer before being passed to the IDPU (Smith *et al.*, 2002). A total of 18 signals are generated, one for each segment of each detector (2 segments per detector). These signals are connected to the DIBs (2 segments are processed on each DIB board) where they are shaped, thresholded, and converted. The measured pulse height, corresponding to the energy deposited in the segment by the incident photon, is passed on to the detector interface field programmable gate array (DIF) on the DIB. The DIF combines the energy information with a time tag, segment identifier, and live time information to make a 32-bit event word per measured event. Event time provides microsecond accuracy,

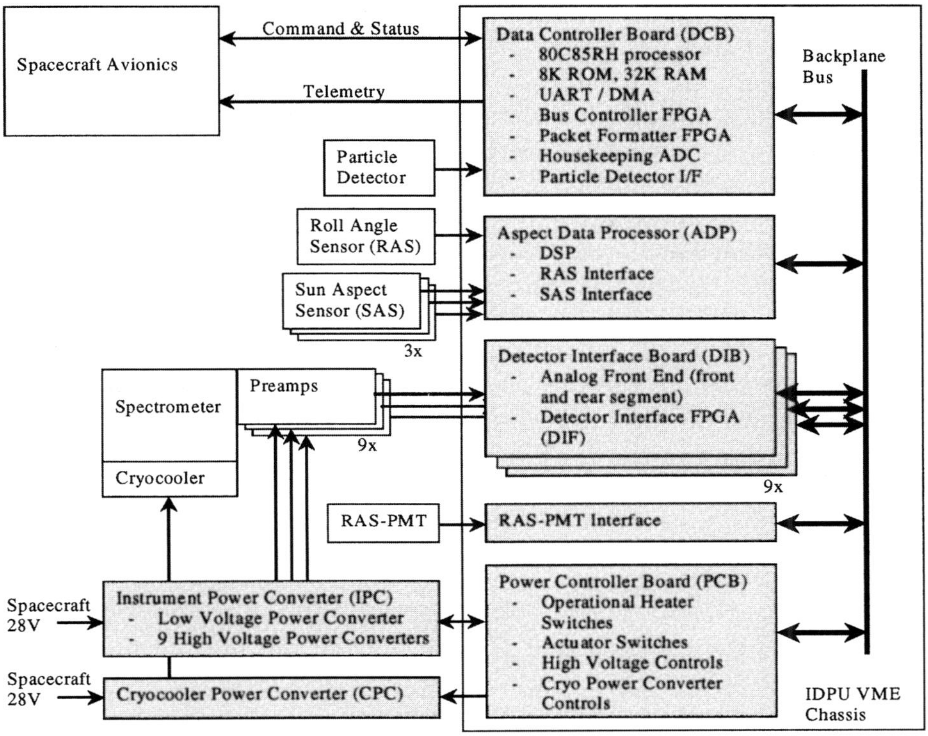


*Figure 1.* IDPU block diagram.

with millisecond dynamic range. These timing bits must be taken together with time tags events and the packet header time included in the event telemetry stream by the DCB to reconstruct the full event time.

Live time bits in each event provide information about the DIB throughput necessary to normalize the measured event rates. Live time is measured for each detector segment by a counter running at 1 MHz that is disabled when the DIB is not ready to process a new event because it is busy with the previous event. This live time counter is sampled at 2 KHz and read out in the live time field of the following 3 events (a few bits of the counter in each event). If events are happening so slowly that the DIB processes fewer than 3 events in the time between live time measurements, some of the live time bits are lost. This is acceptable, because at such low rates live time is relatively unimportant.

The DIF also incorporates monitor and fast rates counters, threshold DAC control, test pulser control, and DIB power control. Threshold DACs provide adjustable low-energy thresholding for events. The test pulser circuit generates a selectable rate of selectable amplitude pulses that are capacitively coupled to each detector. This provides a simple ground test for the electronics in the absence of photons, and also a calibration point for the processing electronics. The power control provides the ability to power off the DIBs individually, and also provides over-current protection.

The DIF talks to the DCB over the backplane. The backplane consists of a 16-bit wide parallel data bus with addressing that allows the DCB to collect data or write control registers on each of the DIBs (or the ADP or PCB) on request. The backplane also accommodates one Event Transfer Request (ETR) signal from each DIB. This signal informs the DCB that the DIB has an event to be collected.

The DCB incorporates a microprocessor system plus a pair of field programmable gate arrays, the Bus Controller FPGA (BCF), and Packet Formatter FPGA (PFF). The FPGAs deal with the high throughput data under the direction of the processor, but without the processor having to deal with the data directly. This allows a very modest 8-bit processor to be used. The BCF arbitrates the backplane bus, collecting event data in response to ETR signals and rate data periodically. This data is passed on to the PFF, which then formats this data into CCSDS packets. Three types of packets are generated by the PFF:

– Event packets, consisting of a series of 32-bit events plus embedded time stamps.

– Monitor Rates packets, consisting of sets of monitor rate counter read-outs from the DIBs, each read out once a second.

– Fast Rates packets, consisting of readouts of the fast rates counters, read at rates of up to 16 kHz.

All packets include an automatically generated CCSDS packet header plus a science header containing key status information.

The PFF then passes completed packets on to the solid state recorder (SSR). The SSR is used in a FIFO mode, with data being continuously fed in by the IDPU, and read out during contacts to the transponder.

The BCF also allows the processor access to the backplane so that the processor can set instrument control registers and read back status registers from the DIBs and other boards.

## 3. Imager Aspect Sensor Data Flow

The imager includes a Roll Aspect Sensor (RAS) and three Sun Aspect Sensors (SAS) (Zehnder *et al.*, 2002). These sensors are optical instruments used to accurately determine the orientation of RHESSI with respect to the Sun. The data from these sensors is passed to the Aspect Data Processor (ADP) in the IDPU. The data is there processed using a DSP to greatly reduce the data rate to just that information that is required. The ADP then formats the data into CCSDS packets.

The BCF collects Imager Aspect data packets from the ADP on request (using another ETR signal). As ADP data is pre-formatted into packets, the BCF and PFF just pass them on without further processing. The processor can access ADP registers to collect status information and load control information and software updates.

## 4. RAS-PMT Interface

A second Roll Aspect Sensor, the RAS-PMT, was added late in the development cycle to add redundancy to the original RAS (Hurford *et al.*, 2002). This new RAS requires a simple low-speed interface to the IDPU backplane. The circuit consists of a single RAS-PMT FPGA mounted to a daughter board attached to the backplane. The processor collects the data at a low rate (4 Hz), formats it into packets, and passes it to the spacecraft using the low speed state-of-heath interface rather than to the SSR. In addition to collecting data, the processor controls a static collection mode register and a DAC that sets the high voltage supply output voltage via the RAS-PMT FPGA.

## 5. Particle Detector

The particle detector is another relatively low-speed instrument used to monitor particle background and identify when the spacecraft is in the South Atlantic Anomaly (SAA). A charge sensitive amplifier is mounted together with the particle detector. The output of the amplifier is passed to the DCB, which contains shaping and two discriminators (one with a programmable threshold), followed by two counters included in the BCF. The BCF includes these particle detector rate counters in the Monitor Rates packets, sampled at 8 Hz. In addition, the processor reads these rates counters for use in the SAA detection software.

## 6. Spacecraft Data Interfaces

Spacecraft data interfaces to the IDPU include a high-speed (40 Mbps) interface to the SSR, a low-speed (9600 baud) RS232 interface to the spacecraft processor, and a stable clock interface. The SSR interface consists of an 8-bit data bus, a ready signal from the SSR, and a write strobe from the IDPU. Once ready, the SSR can ingest data continuously at 5 Mbytes per second with no handshaking required. The signals are conditioned with differential RS422 drivers/receivers.

The RS232 serial interface is used to exchange low-speed data between the IDPU and spacecraft, including:

– A housekeeping packet, containing IDPU state of health data.

– A diagnostic packet from the IDPU.

– Data from the Imager Aspect system, to be used as a possible backup to the Fine Sun Sensor by the Attitude Control System.

– Command packets from the spacecraft to the IDPU.

– Spacecraft status information, including instrument power, transponder, SSR, and ACS data, plus spacecraft time.

Data is exchanged once a second, following the 1Hz time tick. Diagnostic data is provided as required by the IDPU, with a fill packet sent if no diagnostic data is required. The spacecraft has a stable, oven-controlled oscillator that provides timing for all the spacecraft systems. This clock is provided to the IDPU in the form of a 1 MHz clock signal, a 1 Hz clock pulse (synchronous with the 1 MHz clock), and a time code provided over the RS232 interface once a second, corresponding to the time at the 1 Hz clock pulse. This time interface is used to synchronize and time tag all instrument and spacecraft data, providing relative timing accuracy to 1 $\mu$s, and absolute time accuracy to about 1 ms (after correlation of the spacecraft clock to UTC on the ground).

## 7. Processor

The processor provides control and monitoring of the instruments and data flow, collection of instrument housekeeping, and interaction with the spacecraft over the serial interface. The processing resources required for the IDPU are modest because the bulk of the high-speed data handling is done by FPGAs. The processor selected is an 80C85RH, an 8-bit radiation hardened processor with very low power requirements ($< 100$ mW). The processor runs at 3.3 MHz, and has an 8 kbyte PROM plus a 32 kbyte RAM. An 82C37RH DMA controller is used to facilitate data transfer between the IDPU and spacecraft over an RS232 serial interface contained in the BCF. The processor can access the backplane via the BCF as processor I/O registers. The BCF arbitrates processor access requests along with other use of the backplane.

An Analog to Digital Converter (ADC) on the DCB is used by the processor to collect analog housekeeping. The inputs to the ADC come via the backplane. Each board has an analog housekeeping multiplexer attached to a common analog housekeeping signal on the backplane, which is routed to the ADC. The IDPU controls this distributed multiplexer tree via registers on the boards controlled over the backplane.

## 8. Power Control Board

The Power Control Board (PCB) contains a number of ancillary functions for the IDPU, including:

– Operational heater power switching for the instruments (Grid tray heaters, RAS CCD heater, cold plate (anneal) heaters, imager grid tray heaters, spectrometer heater).

– Actuator power switches for the instrument actuators (spectrometer attenuators, vacuum valve, and RAS shutter).

– Cryocooler power and balancer waveform generators (58.6 Hz sine waves with separately programmable amplitudes and programmable relative phase).

– Detector high-voltage supply digital to analog converters, providing a control voltage to the detector high-voltage supplies on the IPC.

– Instrument voltage, current, temperature, and accelerometer monitors (via the DCB analog housekeeping system).

## 9. Software

The RHESSI IDPU flight software coordinates all instrument activities, as described in the processor section above. The software consists of a variety of tasks with moderate timing requirements; the high bandwidth tasks are performed by the FPGAs. The software is organized into 14 modules, totaling about 8000 lines of assembly code. The code is contained in a bipolar PROM, which is copied into RAM on boot-up. Code in RAM can easily be patched on-orbit as required by command load.

The IDPU software includes a very simple operating system for distributing processor time amongst tasks, handling interrupts, and handling hardware interfaces. Tasks are allocated time either as part of a foreground ‘round-robin’ polling loop, or at a fixed time slice as part of the 256 Hz timer-tick interrupt handler. The fixed time slice scheme allows deterministic timing and load leveling, while the foreground polling loop provides for non time-critical tasks which take too long to execute in an interrupt service routine. One additional interrupt associated with the DMA system allows that system to respond quickly to completed DMA transfers.

The software handles a number of routine tasks such as housekeeping data collection and formatting, command decoding and routing, instrument mode and power control, and thermal control. In addition, the software has a few tasks where it is required to take action on its own in response to data. A few of the more important tasks, at least in terms of understanding the RHESSI data set, are described below.

### 9.1. Attenuator Control

The attenuators are used to extend the detector photon flux dynamic range. When the detector count rate is low, the attenuators are left open to maximize the detector sensitivity. As the count rate rises towards detector saturation first the thin and then, if count rates continue to increase, the thick attenuators are put in to cut down on the number of low energy photons that reach the detectors. The detector live times (from the monitor rates) are used to identify when the detectors are approaching saturation. Determining when to remove the attenuators is a little more difficult.

The amount of attenuation is a function of the photon spectra, which is information not available to the processor. This means that the processor cannot predict in advance what the rates will be when the attenuators are removed. Currently the software uses a fixed live time threshold to decide when to remove an attenuator, but this often results in the attenuator being put right back in. A time delay between motions is enforced to keep this from happening too often.

### 9.2. Decimation

The detector interface FPGAs have a decimation system for reducing electronically the number of low energy photons that are processed. This scheme causes a selectable fraction of the photons below a selectable energy level to be eliminated from the telemetry stream (they are still counted by the rate counters). This system does not keep the detector from becoming saturated, but does reduce the telemetry volume. The detector front segment decimators are used to reduce the data volume when the SSR starts to get full. Eight levels of increasing decimation are applied as the SSR fill level increases. When the SSR reaches 99% capacity, all events (front and rear segment) are eliminated as a last ditch effort to avoid over-filling the SSR while maintaining at least detector diagnostic data (rate counters).

Rear segment decimation is controlled from the ground. While rear segment flare count rates are rarely very high, they have a background rate that is a significant contributor to filling the SSR. The ground can select decimation of rear segment events at times of less interest, such as when the sun is not visible (behind the Earth), at the expense of non-solar science. Rear decimation can also be used in high latitude regions (as predicted on the ground) to limit background rates from penetrating particles.

### 9.3. South Atlantic Anomaly detection

The South Atlantic Anomaly (SAA) generates a high background level in the detectors from penetrating particles. To avoid filling the SSR with this background, it is desired to disable all detector events during these parts of the orbit. This can be done from the ground using a predicted envelope for this region, but the region moves around some. A conservative prediction results in good data being lost, while a less conservative prediction results in filling the SSR with background events. To improve the ability to determine when the spacecraft is in one of these regions, the software looks at the particle detector count rate. When the particle count rate is high, detector event rates are disabled. A conservative prediction of the SAA location is used on the ground to enable this feature when approaching the SAA.

## 10. Cryo Power Converter

The Cryo Power Converter (CPC) is a power amplifier that converts unregulated spacecraft 28 V to the AC power waveform required by the spectrometer cryocooler. A separately switched and current limited spacecraft 28 V service is provided to the CPC. The CPC actually has two amplifiers with independent waveforms: one for the cryocooler drive, and one for the active balancer used to minimize cryocooler vibrations. Waveforms for both amplifiers are provided by the Power Control Board. The CPC provides up to 100 W of power to the cryocooler at about 89% efficiency.

## 11. Instrument Power Converter

The Instrument Power Converter (IPC) takes unregulated spacecraft 28V power and generates the secondary voltages used by the IDPU and instruments. Three separately switched and current limited spacecraft power services are provided; one for the normal instrument operating power, and one for switched loads such as operational heaters and actuators, and a third for survival heaters. Survival heater power is routed to the instrument survival heater/thermostat circuits without conditioning. Switched power goes to the Power Control Board where FET switches control its distribution. In addition a 10 V supply powered by the switched power service provides power for the actuators. Actuator power is also routed to the Power Control Board where FET switches control its distribution.

The normal power bus is regulated and converted into a number of secondary voltages, including +5 V digital, ± 5 V analog, ± 12 V analog, +15 V, +28 V, and +100 V, with about 75% efficiency. Most of the secondaries are provided with about 0.5 V over-voltage so that they can run low-drop-out regulator/current limiters downstream on the boards. This provides clean power on the boards to reduce cross-talk, and also provides power isolation to isolate a failure on a board.

The regulated +28 V is used to power the detector high voltage supplies, also part of the IPC. These supplies provide a programmable output (controlled by voltages provided by the Power Control Board), up to 5000 V to bias the detectors in the spectrometer. Another 100 V programmable power supply is used to bias the particle detector (again using a control voltage provided by the Power Control Board).

The 100 V output of the IPC is used to supply the special Zener diodes used to heat the spectrometer cold plate for anneal cycles (and also to perform controlled detector warm-ups on the ground). The 100 V supply is routed via the Power Control Board, where it is switched and limited to provide a fixed current to the heater diodes.

## 12. On-Orbit Results

The RHESSI mission was launched in February 2002, and has had no problem meeting all its requirements to date. The IDPU continues to function flawlessly, providing data reliably with sufficient flexibility to accommodate all conditions encountered.

## References

Fivian, M., Hemmeck, R., Mchedlishvili, A., and Zehnder, A.: 2002, *Solar Phys.*, this volume.

Harvey, P. R., Curtis, D. W., Heetderks, H. D., Pankow, D., Rauch-Leiba, J. M., Wittenbrock, S. K., and McFadden, J. P.: 2001, *Space Sci. Rev.* **98**, 1.

Hurford, G. J. and D. W. Curtis: 2002, *Solar Phys.*, this volume.

Lin, R. P. *et al.*: 2002, *Solar Phys.*, this volume.

Smith, D. M. *et al.*: 2002, *Solar Phys.*, this volume.

# RHESSI AS A HARD X-RAY POLARIMETER

MARK L. MCCONNELL[1], JAMES M. RYAN[1], DAVID M. SMITH[2], ROBERT P. LIN[3] and A. GORDON EMSLIE[4]

[1]*Space Science Center, University of New Hampshire, Durham, NH 03824, U.S.A.*

[2]*Space Sciences Laboratory, University of California, Berkeley, CA 94720, U.S.A.*

[3]*Physics Department and Space Sciences Laboratory, University of California, Berkeley, CA 94720, U.S.A.*

[4]*Physics Department, University of Alabama, Huntsvile, AL 35899, U.S.A.*

(Received 8 August 2002; accepted 10 September 2002)

**Abstract.** Although designed primarily as a hard X-ray imager and spectrometer, the Ramaty High-Energy Solar Spectroscopic Imager (RHESSI) is also capable of measuring the polarization of hard X-rays (20–100 keV) from solar flares. This capability arises from the inclusion of a small unobstructed Be scattering element that is strategically located within the cryostat that houses the array of nine germanium detectors. The Ge detectors are segmented, with both a front and rear active volume. Low-energy photons (below about 100 keV) can reach a rear segment of a Ge detector only indirectly, by scattering. Low-energy photons from the Sun have a direct path to the Be and have a high probability of Compton scattering into a rear segment of a Ge detector. The azimuthal distribution of these scattered photons carries with it a signature of the linear polarization of the incident flux. Sensitivity estimates, based on Monte Carlo simulations and in-flight background measurements, indicate that a 20–100 keV polarization sensitivity of less than a few percent can be achieved for X-class flares.

## 1. Introduction

Hard X-ray emission from solar flares, like any other form of electromagnetic radiation, has four, and only four properties. Each photon can be completely characterized by its time of arrival, its energy, its direction of arrival, and its polarization state. RHESSI is capable, to varying degrees, of determining all four of these quantities. The first two of these properties (time of arrival and energy) are measured directly as photon interactions in the Ge detectors. The third and fourth properties (arrival direction and polarization state) are determined through an analysis of the grouping of photons in time at each detector. RHESSI is capable of providing information on the linear polarization of photons between roughly 20 and 100 keV. The study of polarization at hard X-ray energies is especially appealing in that the hard X-ray emission from any bremsstrahlung source (such as a solar flare) will be polarized if the phase-space distribution of the emitting electrons is anisotropic. Polarization measurements therefore provide a direct handle on the extent to which the accelerated electrons are beamed, which, in turn, has important implications for particle acceleration models.

*Solar Physics* **210:** 125–142, 2002.

Studies of $\gamma$-ray line data from the SMM Gamma Ray Spectrometer (GRS) suggest that protons and $\alpha$-particles are likely being accelerated in a rather broad angular distribution (Share and Murphy, 1997; Share *et al.*, 2002). There is no reason to expect, however, that electrons are being accelerated in a similar fashion. The ability of RHESSI to make polarization measurements, coupled with its ability to (independently) image the hard X-ray emission, provides a powerful tool for studying electron acceleration in solar flares. In this section, we first review the current status of efforts to measure the electron beaming (via photon directivity) using studies of the distribution of observed events across the solar disk and stereoscopic studies of individual flares. We then review the potential value of polarization measurements and the checkered history of polarization studies of solar flares.

### 1.1. Hard X-ray Directivity Measurements

Related to polarization is the directionality of the hard X-ray radiation yield. An anisotropic ensemble of bremsstrahlung-producing electrons will produce a radiation field that is not only polarized, but anisotropic. Measurements of the hard X-ray photon directivity can therefore provide a probe of the extent to which the accelerated electrons are beamed.

One technique for studying hard X-ray directivity on a statistical basis is to look for center-to-limb variations. Correlations between flare longitude and flare intensity or spectrum reflect the anisotropy of the X-ray emission and hence any directivity of the energetic electrons. For example, if the radiation is preferably emitted in a direction parallel to the surface of the Sun, then a flare located near the limb will look brighter than the same flare near the disk center. Various statistical studies of X-ray flares at energies below 300 keV reported no significant center-to-limb variation of the observed intensity (e.g., Datlowe, Elcan, and Hudson, 1974; Pizzichini, Spizzichino, and Vespignani, 1974; Datlowe *et al.*, 1977). The conclusion from these early studies was that either the structure of the X-ray emitting region is variable from event to event or that the average motion of the energetic electrons is random. However, a statistically significant center-to-limb variation in the spectral shape of the spectra of these events was found by Roy and Datlowe (1975), suggesting that perhaps some directivity may be present.

The large sample of flares detected at energies greater than 300 keV by SMM GRS allowed, for the first time, a statistical search for directivity at higher energies. Photons at these energies are well above the thermal emission from the flare. In addition, the greatly reduced Compton backscatter from the photosphere and the increased directionality of the bremsstrahlung cross-section make relatively strong radiation anisotropies a possibility at these energies. Analysis of high-energy data (>300 keV) from SMM GRS collected during cycle 21 provided the first clear evidence for directed emission, with a tendency for the high-energy events to be located near the limb (Vestrand *et al.*, 1987; Bai, 1988). Using an independent set of flares detected by instruments on the *Venera* 13 and 14 spacecraft, Bogovlov *et al.* (1985) also reported evidence for anisotropies at hard X-ray energies. They also attributed this disk center-to-limb spectral variation to bremsstrahlung directivity. Observations from SMM GRS during cycle 22 provided further support for directivity (Vestrand, Forrest, and Rieger, 1991). However, several high-energy events were also observed near the disk center by a number of different experiments during cycle 22 (e.g., on GRANAT and CGRO; see Vilmer, 1994, for a summary), perhaps suggesting a more complex pattern. Li (1995) used data from SMM to confirm the general results of Vestrand *et al.* (1987) and concluded that there was evidence for increasing directivity with increasing energy. Quantifying the magnitude of the directivity from these statistical measurements is difficult. For example, one needs to know the size-frequency distribution for flares as well as the form of the electron distribution to derive the predicted limb fraction (e.g., Petrosian, 1985). Furthermore, the results only represent an average for the flare sample. Different flaring regions are not likely to have identical geometry. Nor are individual flares likely to have time-independent electron distributions.

Another method for studying the directivity in individual flares is the stereoscopic technique first employed by Catalano and Van Allen (1973) at soft X-ray energies. This method compares simultaneous observations made on two spacecraft that view the flare from different directions. Kane *et al.* (1988) combined observations from spectrometers aboard the PVO (Pioneer Venus Orbiter) satellite and the ISEE-3 (Third International Sun–Earth Explorer) satellite to produce stereoscopic flare observations of 39 flares in the energy range from 100 keV to 1 MeV. While the range of flux ratios measured by Kane *et al.* (1988) is consistent with the results of statistical studies (Vestrand and Ghosh, 1987), the deviations of the ratio from unity show no clear correlation with increasing difference in viewing angles. Li *et al.* (1994) used sterescopic observations of 28 flares from the HXRBS and GRS experiments on SMM and the SIGNE experiments on *Venera* 13 and *Venera* 14 and concluded that directivity could only be marginally present in the 100–500 keV energy range. Most recently, Kane *et al.* (1998) reported on a study of 13 flares in the 20–125 keV enery range using stereoscopic observations with X-ray instruments on both *Ulysses* and *Yohkoh*. They concluded that there was no clear evidence for directivity in the 20–125 keV energy range, although there were large calibration uncertainties present in the analysis. Stereoscopic observations require good calibration of individual instruments so that differences in the absolute fluxes observed by different instruments can be correctly interpreted.

### 1.2. Hard X-ray Polarization Measurements

The difficulties of statistical and stereoscopic observations for measuring hard X-ray directivity suggest the need for a technique that can measure time-dependent anisotropies for individual flares using a single instrument. Polarization is a diagnostic that can meet these requirements.

Many models of nonthermal (e.g., thick-target) hard X-ray production predict a clear and significant polarization signal, with polarization levels > 10% (Brown, 1972; Langer and Petrosian, 1977; Bai and Ramaty, 1978; Emslie and Vlahos, 1980; Leach and Petrosian, 1983; Zharkova, Brown, and Syniavskii, 1995; Charikov, Guzman, and Kudryavtsev, 1996). The precise level of polarization depends on both energy and viewing angle. Some fraction of the observed 20–100 keV hard X-ray flux will be flux that is backscattered from the solar photosphere, the precise magnitude of which will depend, in part, on the polarization of the primary flux. The reflected component will, in turn, influence the degree of polarization of the observed flux, since even if the electrons are accelerated isotropically, backscattering will introduce polarization fractions of a few percent at energies below 100 keV (e.g., Langer and Petrosian, 1977; Bai and Ramaty, 1978). Even thermal models of the hard X-ray source predict a finite polarization of order a few percent, due to the anisotropy in the electron distribution function caused by a thermal conductive flux out of the emitting region into the cooler surroundings (Emslie and Brown, 1980). The thermal component, with its rather low polarization, tends to dominate the emission from all flares at energies below about 25 keV. At these energies, it therefore becomes difficult to distinguish the non-thermal component, with its intrinsic directivity signature, from the thermal component. This has led to the argument that polarization measurements can best be performed at higher energies (Chanan, Emslie, and Novick, 1988).

These predictions, while clearly testable, could be criticized on the grounds that the modeling assumptions they contain may be oversimplistic. For example, each model to date assumes a single, simple magnetic field structure about which the emitting electrons spiral. It could be argued that any real flare, particularly one sufficiently large to produce a signal of sufficient strength to enable a polarization measurement, will in all probability contain a mix of structures that would average out any polarization signal present. However, hard X-ray imaging observations in the impulsive phase generally show a fairly simple geometry, consisting of two footpoint sources and perhaps a loop-top source (e.g., Sakao *et al.*, 1992; Masuda *et al.*, 1995). These observations suggest that simple magnetic structures are responsible for the energetic emissions and give support to the possibility that a statistically significant polarization signal could be produced in a large event.

The history of observations of hard X-ray polarization from solar flares is a fascinating subject in its own. The first measurements of X-ray polarization from solar flares (at energies of ~ 15 keV) were made by Soviet experimenters using polarimeters aboard the Intercosmos satellites. In their initial study, Tindo *et al.* (1970) reported an average polarization for three small X-ray flares of $P = 40\%$ ($\pm 20\%$). This study was followed by an analysis of three flares in October and November of 1970 (Tindo *et al.*, 1972a, b) that showed polarizations of approximately 20% during the hard impulsive phase. These reports were met with considerable skepticism, on the grounds that they did not adequately allow for detector cross-calibration issues and limited photon statistics (Brown, McClymont, and McLean,

1974). Subsequent observations with an instrument on the OSO-7 satellite seemed to confirm the existence and magnitudes of the polarizations ($\sim 10\%$), but these data were compromised by in-flight gain shifts (Nakada, Neupert, and Thomas, 1974). In a later study using a polarimeter on *Intercosmos* 11, Tindo, Shuryghin, and Steffen (1976) measured polarizations of only a few percent at $\sim 15$ keV for two flares in July 1974. This small but finite polarization is consistent with the predictions for purely thermal emission that contains an admixture of polarized backscattered radiation (Bai and Ramaty, 1978). A small polarimeter was flown on an early shuttle flight (STS-3) and made measurements of eight C- and M-class flares in the 5–20 keV energy range. Upper limits in the range of 2.5% to 12.7% were measured, although contamination of the Li scattering material invalidated the pre-flight calibration (Tramiel, Chanan, and Novick, 1984).

The polarimetry capability of RHESSI will permit high-quality observations of large flares. The RHESSI energy range for measuring polarization (extending to $\sim 100$ keV) may be sufficiently high to reduce some of the thermal contamination that has plagued many previous measurements. In addition, some fraction of the hard X-ray flux that will be observed by RHESSI will not be primary emission from the flare source, but rather will be flux that is backscattered from the solar photosphere. These photospherically backscattered photons should form an 'albedo patch' (Brown and Hoyng, 1975), which may be directly imaged by RHESSI as a constraint on the contribution of such backscattered photons to the primary signal. Clearly, the capability of RHESSI to simultaneously image the hard X-ray emission represents a major advantage over previous efforts to measure hard X-ray polarization.

## 2. Compton Polarimetry

The basic physical process used to measure polarization in the 20–100 keV energy range is Compton scattering, the cross-section for which is given by the Klein–Nishina formula,

$$d\sigma = \frac{r_0^2}{2} d\Omega \left(\frac{\nu'}{\nu_o}\right)^2 \left(\frac{\nu_o}{\nu'} + \frac{\nu'}{\nu_o} - 2\sin^2\theta\cos^2\eta\right), \tag{1}$$

where

$$\frac{\nu'}{\nu_o} = \frac{1}{1 + \left(\frac{h\nu_o}{mc^2}\right)(1 - \cos\theta)}. \tag{2}$$

Here $\nu_o$ is the frequency of the incident photon, $\nu'$ is the frequency of the scattered photon, $\theta$ is the scattering angle of the scattered photon measured from the direction of the incident photon, and $\eta$ is the azimuthal angle of the scattered photon measured from the plane containing the electric vector of the incident photon. For a given value of $\theta$, the scattering cross-section for polarized radiation reaches a minimum at $\eta = 0°$ and a maximum at $\eta = 90°$. In other words, photons tend to be scattered at a right angle with respect to the incident electric field vector. In the case of an unpolarized beam of incident photons, there will be no net positive electric field vector and therefore no preferred azimuthal scattering angle ($\eta$); the distribution of scattered photons with respect to $\eta$ will be uniform. However, in the polarized case, the incident photons will exhibit a net positive electric field vector and the distribution in $\eta$ will be asymmetric. The asymmetry will be most pronounced for scatter angles near $\theta = 90°$.

In general, a Compton scatter polarimeter consists of two detectors to determine the energies of both the scattered photon and the scattered electron (e.g., Lei, Dean, and Hills, 1997; McConnell *et al.*, 2002). One detector, the scattering detector, provides the medium for the Compton interaction to take place. This detector must be designed to maximize the probability of there being a single Compton interaction with a subsequent escape of the scattered photon. This requires a low-$Z$ material in order to minimize photoelectric interactions. The primary purpose of the second detector, the calorimeter, is to absorb the full energy of the scattered photon.

The relative placement of the two detectors defines the scattering geometry. For incident photon energies below 100 keV, the azimuthal modulation of the scattered photons is maximized if the two detectors are placed at a right angle relative to the incident photon beam ($\theta = 90°$). The two detectors must also be arranged relatively close to each other so that there is a finite solid angle for scattering to achieve the required detection efficiency. At the same time, a larger separation between the detectors provides more precise scattering-geometry information. The accuracy with which the scattering geometry can be measured determines the ability to define the modulation pattern (Figure 1) and therefore has a direct impact on the polarization sensitivity (see below).

In principle, the scattering detector need not be active. At soft X-ray energies (<10 keV), for example, the scattering process results in a very small energy loss that can be difficult to detect. For polarimeters designed to measure soft X-rays, the scattering element is usually passive, its only requirement being that it be designed to maximize the probability of a single photon scatter. At higher energies, an active element is certainly preferred as a means to minimize background, but it is certainly not required.

The ultimate goal of a Compton scatter polarimeter is to measure the azimuthal modulation pattern of the scattered photons. From Equation (1), we see that the azimuthal modulation follows a $\cos 2\eta$ distribution. More specifically, we can write the integrated azimuthal distribution of the scattered photons as

$$C(\eta) = A \cos\left(2\left(\eta - \phi + \frac{\pi}{2}\right)\right) + B, \tag{3}$$

where $\phi$ is the polarization angle of the incident photons; $A$ and $B$ are constants used to fit the modulation pattern (cf., Figure 1). In practice, a measured distribution must also be corrected for geometrical effects based on the corresponding

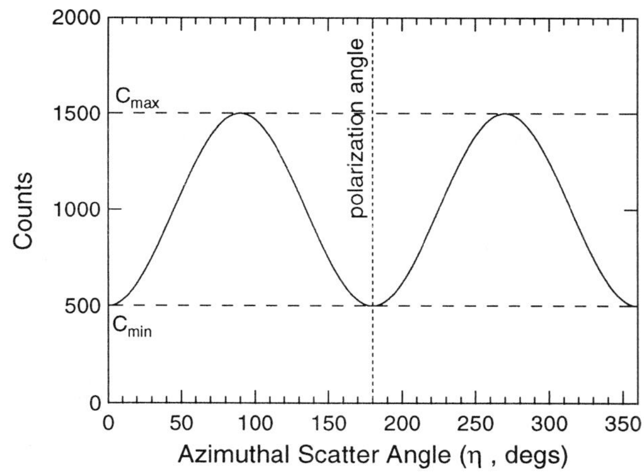


*Figure 1.* An idealized modulation pattern produced by Compton scattering of polarized radiation. The minimum of the modulation pattern defines the plane of polarization of the incident flux.

distribution for an unpolarized beam (Lei, Dean, and Hills, 1997). The quality of the polarization signal is quantified by the polarization modulation factor (Novick, 1975; Lei, Dean, and Hills, 1997). For a given energy and incidence angle for an incoming photon beam, this can be expressed as

$$\mu_p = \frac{C_{p,\max} - C_{p,\min}}{C_{p,\max} + C_{p,\min}} = \frac{A_p}{B_p}, \tag{4}$$

where $C_{p,\max}$ and $C_{p,\min}$ refer to the maximum and minimum number of counts registered in the polarimeter, respectively, with respect to $\eta$; $A_p$ and $B_p$ refer to the corresponding parameters in Equation (3). In this case the $p$ subscript denotes that this refers to the measurement of a beam with unknown polarization. In order to determine the polarization of the measured beam, we need first to know how the polarimeter would respond to a similar beam, but with 100% polarization. This can be done using Monte Carlo simulations. We then define a corresponding modulation factor for an incident beam that is 100% polarized,

$$\mu_{100} = \frac{C_{100,\max} - C_{100,\min}}{C_{100,\max} + C_{100,\min}} = \frac{A_{100}}{B_{100}}. \tag{5}$$

Then, following Novick (1975), we can then use this result, in conjunction with the observed modulation factor ($\mu_p$), to determine the level of polarization in a measured beam,

$$P = \frac{\mu_p}{\mu_{100}}, \tag{6}$$

where $P$ is the measured polarization. The minimum detectable polarization (MDP) can be expressed as (Novick, 1975)

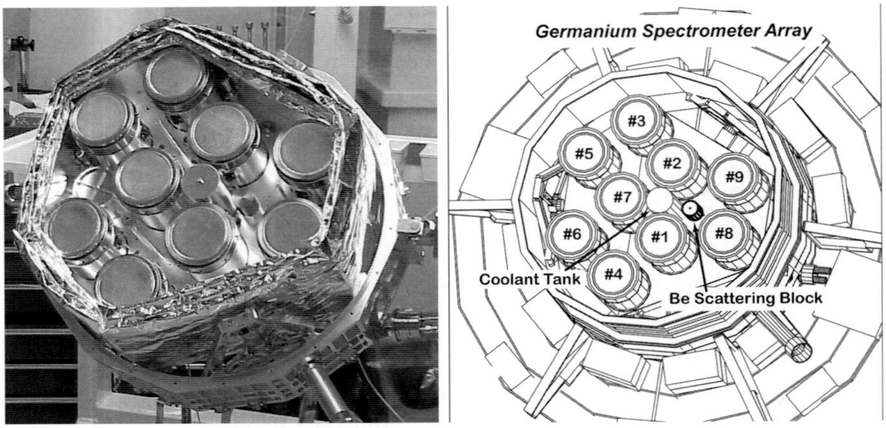


*Figure 2.* The RHESSI spectrometer array, as seen in a pre-flight photo (*left*) and as depicted by the GEANT3 mass model (*right*).

$$\mathrm{MDP}(\%) = \frac{n_\sigma}{\mu_{100} R_{src}} \sqrt{\frac{2(R_{src} + R_{bgd})}{T}}, \tag{7}$$

where $n_\sigma$ is the significance level (number of sigma), $R_{src}$ is the total source counting rate, $R_{bgd}$ is the total background counting rate and $T$ is the total observation time.

## 3. The RHESSI Spectrometer

The principal scientific objectives of RHESSI are being achieved with high-resolution imaging spectroscopy observations from soft X-rays to gamma-rays (Lin *et al.*, 2002). The imaging system is made up of nine Rotating Modulation Collimators (RMCs), each consisting of a pair of widely separated grids mounted on the rotating spacecraft. The spectrometer (Figure 2) consists of nine segmented germanium detectors, one behind each RMC, to detect photons from 3 keV to 17 MeV (Smith *et al.*, 2002). These detectors represent the largest, readily available, hyperpure (n-type) coaxial Ge material ($\sim$ 7.1 cm diam. $\times$8.5 cm long). The detectors are cooled to $\sim$ 75° K by a space-qualified long-life mechanical cryocooler. Each detector (Figure 3) is made from a single germanium crystal, which is electrically divided into independent front and rear segments to provide an optimum response for low and high energy photons, respectively. The inner electrode is segmented into two contacts that collect charge from two electrically independent detector segments, defined by the electric field pattern. This provides the equivalent of a $\sim$ 1 cm thick planar Ge detector (the front segment) in front of a thick $\sim$ 7 cm coaxial Ge detector (the rear segment).

The front segment thickness is chosen to stop photons incident from the front (solar-facing side) of the instrument up to $\sim$ 100 keV, where photoelectric absorp-

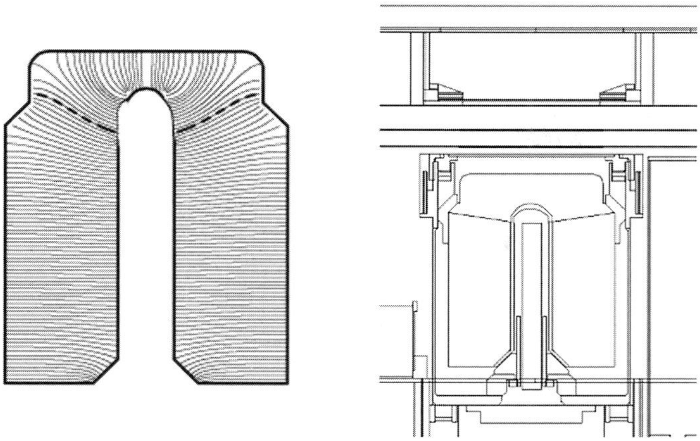

*Figure 3.* Cross-section of a RHESSI detector. *Left*: a detector profile with electric field lines, the *dashed field line* indicating the separation between the front and back segments. *Right*: a detector in the cryostat, showing Ta/Sn/Fe/Al shielding around the side of the front segment and above the shoulder of the rear segment.

tion dominates, while minimizing the active volume for background. Front-incident photons that Compton-scatter, and background photons or particles entering from the rear, are rejected by anticoincidence with the rear segment. A passive, graded-$Z$ ring around the front segment (a laminate of Ta/Sn/Fe) absorbs hard X-rays incident from the side, to provide the low background of a phoswich-type scintillation detector.

Solar photons with energies from $\sim 100$ keV to $\sim 17$ MeV, including most nuclear gamma-ray lines, stop primarily in the thick rear segment alone. A smaller fraction of these high energy photons will stop in the front segment after first Compton scattering in the rear segment, thus depositing energy in both the front and rear segments. Alternatively, some of these photons may deposit energy in two or more Ge detectors. The intense 3–100 keV X-ray fluxes that usually accompany large $\gamma$-ray line flares are absorbed by the front segment, so the rear segment will always count at moderate rates. Photons with energy above 20 keV from non-solar sources can penetrate the thin aluminum cryostat wall from the side and also be detected by the Ge detector rear segments.

RHESSI is a spinning spacecraft, with a spin rate of $\sim 15$ rpm. The high angular resolution imaging capability of RHESSI imposes severe requirements on the knowledge of the instrument orientation direction at any given time. Two spacecraft systems provide the necessary aspect solution. A solar aspect system provides knowledge of Sun center in pitch and yaw to 1.5 arc sec. A star scanner is used to sample the roll orientation at least once per rotation. Interpolation between star measurements allows the roll orientation to be determined with an accuracy of $\sim 2.7$ arc min. The energy and arrival time of every photon, together with spacecraft orientation data, are recorded in the spacecraft's on-board 4-Gbyte solid-state memory (sized to hold all the data from the largest flare) and automatically telemetered to the ground within 48 hours. These data can then be used to generate X-ray/$\gamma$-ray images with an angular resolution of $\sim 2$ arc sec and a field of view of $\sim 1^\circ$.

## 4. RHESSI as a Polarimeter

It was realized during the RHESSI Phase-A study that the essential ingredients for measuring the polarization of solar flare hard X-rays, namely, an array of detectors in a rotating spacecraft, were already present in the RHESSI spectrometer. All that was needed was the addition of a strategically placed cylinder of beryllium in the cryostat to Compton scatter the hard X-rays (20–100 keV) into the rear segments of the adjacent Ge detectors. For the purpose of making polarization measurements, the important feature of the spectrometer design is that photons below 100 keV that are detected in the rear segments are either non-solar background or solar photons scattered into the rear segments (e.g., from the Be scattering block). The distribution of events scattered into the adjacent rear segments (with respect to some fixed reference frame) provides a polarization signature, since the direction of the scattering depends on the orientation of the electric field vectors, or plane of polarization, of the incoming photons. The spacecraft rotation reduces the impact of systematic differences in detection efficiency amongst the different Ge detectors and increases the sampling frequency with respect to the azimuthal scatter angle ($\eta$ in Equation (1)).

The beryllium scattering block (3 cm in diameter by 3.5 cm long) was placed within the spectrometer cryostat, near the center of the Ge detector array (Figure 2). Directly in front of the cryostat is a graded-$Z$ shield (a laminate of Ta/Sn/Fe) that is designed to absorb a large fraction of the flux below 100 keV, flux which tends to dominate the flare event. Openings in this shield provide an unattenuated path for these low energy photons from the Sun to reach the front surface of the cryostat directly in front of each Ge detector and directly in front of the Be scattering block. The cryostat aluminum is thinned in front of the Be scatterer (to get 20 keV threshold), but in front of the Ge detectors, it is cut out entirely and replaced with a thin Be window (to get 3 keV threshold). In addition, the trays which hold the RMC grids each have a hole directly in front of the Be scattering block. The total mass directly in front of the Be block amounts to $\sim 1$ mm of aluminum in the cryostat plus some aluminized mylar thermal blankets, for a total mass of $\lesssim 2$ mm aluminum equivalent.

The sensitive energy range for polarimetry with RHESSI is about 20–100 keV. The graded-$Z$ shield in front of the cryostat insures that solar flare photons of this energy which enter the cryostat will interact either in the front surface of a Ge detector or in the Be block. The segmented design of each Ge detector further insures that there are no direct solar flare photons in the rear segments within the 20–100 keV energy range. Flare photons within this energy range can only reach the rear segments by first scattering within the Be block. There is, however, a significant background from higher energy flare photons which scatter into the rear segments from other parts of the spacecraft and from flare photons which reflect off the Earth's atmosphere (the Earth albedo flux). Photons that scatter into the Ge detectors from other parts of the spacecraft will generally not vary with spin angle. This unmodulated component will interfere with the polarization measurement by effectively reducing the modulation factor.

The albedo flux poses a potentially significant problem in that the level of albedo flux reaching a given Ge detector can vary significantly with spin phase (depending on the orientation of the spacecraft with respect to the local zenith). The magnitude of the albedo flux can be quite large, as much as 40–50% of the direct solar flare flux at energies below 100 keV. Fortunately, the albedo flux has only 1 maximum per spin period instead of the two maxima exhibited by the polarization signal and can therefore be recognized using an appropriate analysis.

The impact of the albedo flux may be somewhat further complicated by the fact that the precise distribution of the albedo flux across the visible disk of the Earth can also depend upon the polarization parameters of the direct flux from the flare. McConnell *et al.* (1996a, b) described this effect and how it might be used to measure the (energy-averaged) polarization parameters of hard X-rays from solar flares or $\gamma$-ray bursts using the BATSE detectors on CGRO. While we expect that this effect will be quite difficult to observe with RHESSI, and hence unimportant in the polarization analysis, it is nontheless desirable to check this assumption with appropriate simulations. If, by some chance, this effect could be measured by RHESSI, it would provide a useful complement to the direct polarization measurement.

Finally, the variations in count rate within the detectors which measure the Be-scattered flux must be separated from intrinsic variations in the flare, but this can be done by examining the summed rate of the Ge detectors. Any variations in count rate that may be related to the source modulation by the grids happen much faster than a spin period, and should average out for the polarization measurements.

## 5. Simulated Polarization Response

To simulate the polarimetric response of RHESSI, we have used a modified version of the GEANT3 code that includes the effects of polarization in Compton scattering and tracks the polarization of the incident photon. We make use of a detailed mass model developed by the RHESSI team. This is a comprehensive mass model that includes not only the spectrometer, but also the complete telescope assembly, supporting electronics, spacecraft support structure and solar panels.

In principle, each of the nine Ge detectors provides a polarization signature. However, the quality of the signature varies, depending on the distance of the detector from the Be and the extent to which that detector is blocked from view with respect to the Be. A particular detector may be blocked from view by another detector and/or by a cryogenic coolant tank (used only for ground operations and empty in flight). Furthermore, the Ge detectors that are furthest from the Be do not provide a polarization signature with sufficient signal-to-noise to be useful in polarization studies. The Ge detectors that are most useful in polarization studies are detectors 1, 2, 8, and 9 (cf., Figure 2). Unfortunately, detector 2 is currently not operating in segmentation mode and is therefore not usable for polarization studies. The simulations presented below assume that only detectors 1, 8, and 9 are used.

The initial simulations involved directing a narrow mono-energetic beam directly at the Be block. In this case, the beam area corresponded to the area of the Be block (3 cm diameter). These simulations allowed us to study the intrinsic polarimetric capabilities of RHESSI. A second set of simulations used a much broader beam (60 cm diameter) that covered the full area of the telescope tube assembly. These more realistic simulations allowed us to study the effects of photon scattering within the telescope tube and the surrounding spacecraft materials. In this case, photon scattering refers to the solar photons which interact in rear Ge segments after scattering from some spacecraft component other than the Be. These photons carry no polarization signature and serve only to increase the level of background.

Figure 4 shows the simulated modulation patterns (number of counts versus azimuthal scatter angle) as derived from both the narrow- and broad-beam simulations. Only those events showing energy deposits above the 20 keV threshold in a single rear Ge segment (in detector 1, 8, or 9) are included. At lower energies (50 keV), the modulation patterns are very similar. But at higher energies (80 keV), the broad beam simulation shows that many more photons are detected. Unfortunately, most of these photons are randomly scattered into the Ge detectors from material other than the Be. These additional events therefore carry no useful polarization signature. Figure 5 shows the modulation patterns for three different energies (30, 50, and 80 keV), as derived from the broad beam simulations.

We characterize the polarization response of RHESSI using two parameters: (1) the effective area ($A_{\mathrm{eff}}$), which represents the effective area for events satisfying the necessary criteria (single energy deposit in rear segments of the selected Ge detectors); and (2) the polarization modulation factor ($\mu_{100}$), which can be thought of as a measure of the quality of the polarization signature (Equation (6)). These two quantities can be used to define a third parameter, the figure-of-merit, that is useful in characterizing the polarization response,

$$\mathrm{FoM} = \mu_{100}\sqrt{A_{\mathrm{eff}}}. \tag{8}$$

As defined here, this figure-of-merit does not incorporate the effects of detector background (cf., Figure 9). The characterization of the simulated response is shown in Figures 6–8. As before, we include only those events that deposit energy above the threshold (> 20 keV) in a single rear segment of detector 1, 8, or 9. Of particular

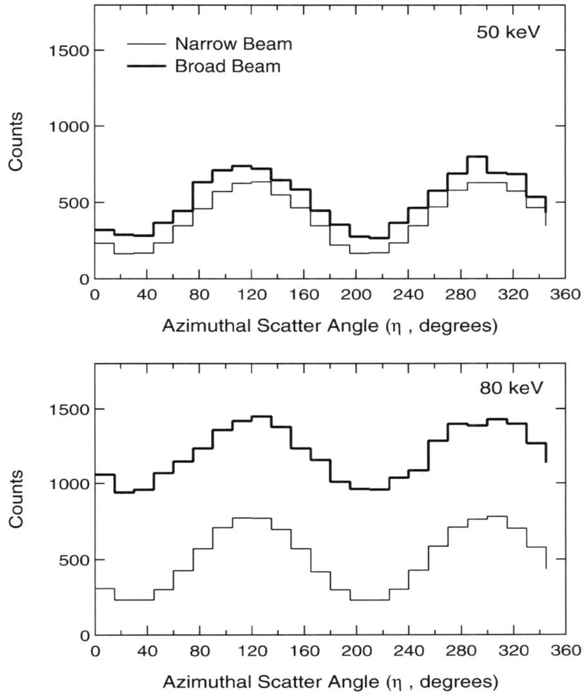


*Figure 4.* The modulation patterns produced from both narrow- and broad-beam simulations for 50 keV (*top*) and 80 keV (*bottom*). Although the total number of events increases dramatically at high energies using the more realistic broad-beam simulation, most of these are unmodulated.

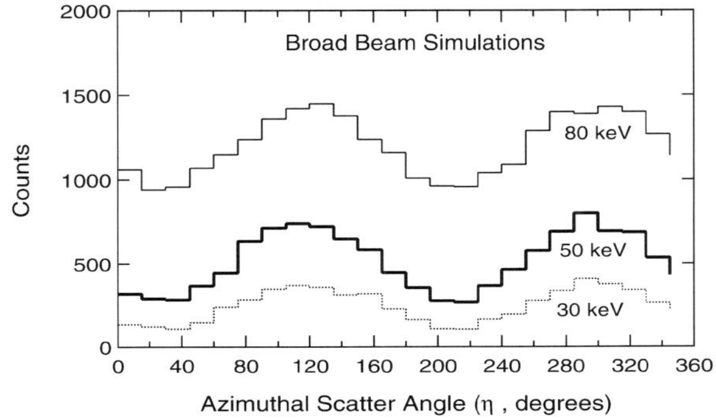


*Figure 5.* The modulation patterns derived from the more realistic broad-beam simulations for three different energies (30, 50, and 80 keV). As energy increases, the total number of detected events increases, but the modulation factor decreases as a result of scattered (unmodulated) events.

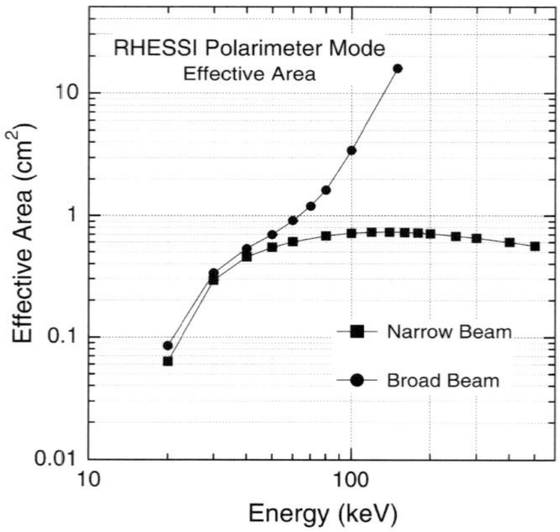


*Figure 6.* Effective area of the polarimetry mode as a function of energy for both the narrow-beam and broad-beam simulations. For the broad-beam simulations, the effective area increases dramatically at higher energies, as a result of photons scattering into the Ge.

importance here is the significant impact of scattered photons, which is most easily seen in terms of the effective area (Figure 6). The effective area is larger with the broad beam simulations as a result of the larger number of valid events due to photon scattering. Above about 50 keV, scattered photons become increasingly important. At energies near 100 keV and above, scattering completely dominates the response. The effects of scattering are also seen in the plot of modulation factor versus energy (Figure 7). Since the scattered component carries with it no polarization signature, the modulation factor decreases at energies above $\sim 30$ keV, as scattering becomes more and more important. Figure 8 shows that, in a source-dominated measurement, the efficacy of the RHESSI polarimeter mode peaks near 60 keV.

These results indicate that RHESSI is most effective as a polarimeter at energies below $\sim 100$ keV. One can also see that having an active scattering element, which would have allowed for rejection of scattered photons by selecting on only those events scattered in the Be, would have provided a very significant improvement in the polarimetric capability of RHESSI. Although multiple scatter events between Ge detectors could be used for polarization measurements, the high-Z nature of the Ge will limit the sensitivity of this approach. Nonetheless, we plan to investigate this mode with future simulations.

## 6. Polarization Sensitivity

The measured RHESSI background (Figure 9), along with the simulated polarimetric characteristics, can be used to estimate the polarization sensitivity for a typical solar flare. Unfortunately, it is difficult to define a 'typical' solar flare to use as a

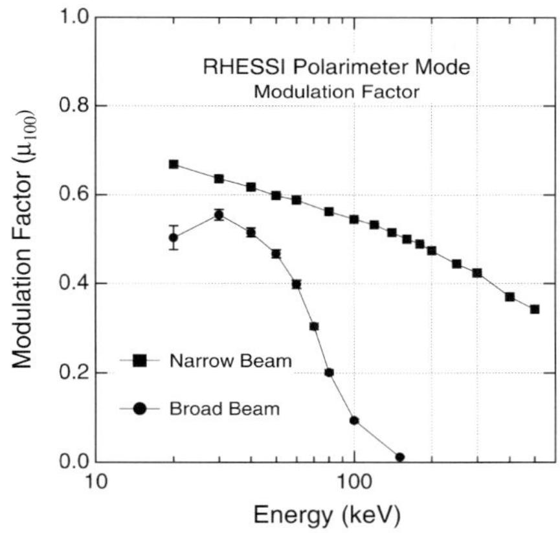


*Figure 7.* Modulation factor of the polarimetry mode as a function of energy for both the narrow-beam and broad-beam simulations. For the broad-beam simulations, the modulation factor decreases dramatically at higher energies, as a result of unmodulated photons scattering into the Ge.

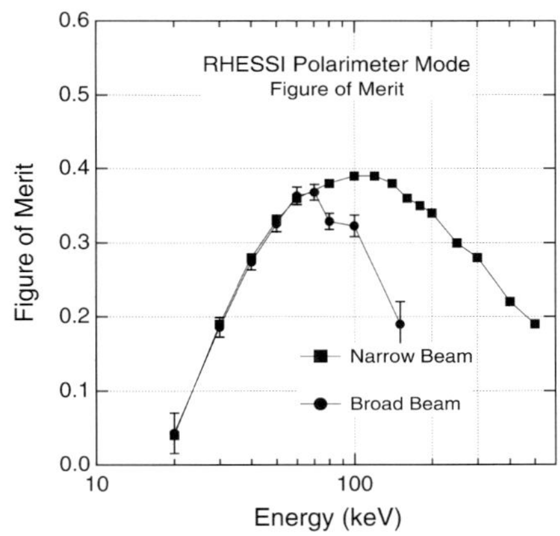


*Figure 8.* Figure of merit of the polarimetry mode as a function of energy for both the narrow-beam and broad-beam simulations.

baseline for estimating polarization sensitivities. Here we have used an X2-class flare having a spectrum of the form (Chanan, Emslie, and Novick, 1988)

$$\frac{\mathrm{d}F}{\mathrm{d}E} = 3.6 \times 10^7 E_{\mathrm{keV}}^{-4.3} \ \mathrm{cm}^{-2}\ \mathrm{s}^{-1}\ \mathrm{keV}^{-1}. \tag{9}$$

The X-ray classification depends only on the *peak* X-ray flux. Therefore, the polarization sensitivity for a given class flare will depend on the duration of the event. We have used durations ranging from 20 seconds up to 500 seconds, with an average spectrum corresponding to that given above. The results of these estimates are shown in Table I. Results for both an X2 class flare and an X10 class flare are shown, with the X10 spectrum assumed to have the same form as that above, but

TABLE I
Minimum detectable polarization.

| | Event duration | | |
|---|---|---|---|
| | 20 s | 100 s | 500 s |
| X2 class flare | | | |
| 20–40 keV | 11% | 5% | 2% |
| 40–60 keV | 53% | 24% | 11% |
| 60–80 keV | – | – | 46% |
| X10 class flare | | | |
| 20–40 keV | 5% | 2% | 1% |
| 40–60 keV | 17% | 7% | 3% |
| 60–80 keV | 61% | 27% | 12% |

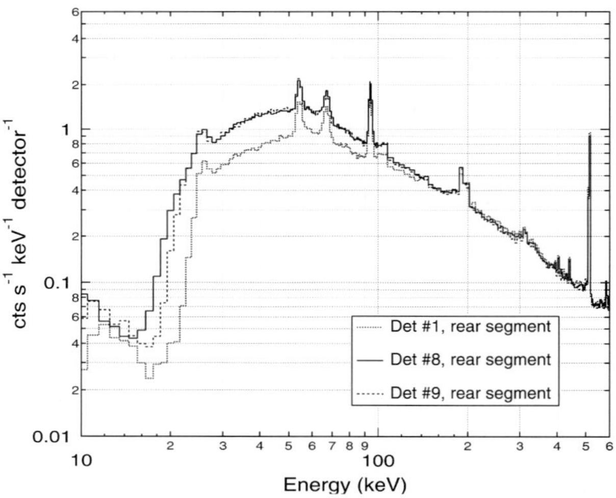


*Figure 9.* Measured background spectra for RHESSI detectors 1, 8, and 9. This spectrum represents an average over about one orbit, excluding periods near the SAA and particle precipitation events.

with a factor of 5 increase in intensity. The polarization sensitivity is given as the minimum detectable polarization (MDP; cf., Equation (7)). The values represent a $3\sigma$ sensitivity in terms of percent polarization.

These estimates indicate that RHESSI has sufficient sensitivity to measure the polarization of X-class flares down to a level below 10% and, in some cases, below 1%. This level of sensitivity will be useful in constraining various models that have been published in the literature.

Our sensitivity estimates have not yet considered the effects of the atmospheric albedo flux. As much as 30–40% of the solar flux gets Compton scattered off the atmosphere at these energies. If the level of this component reaches a level comparable to the intrinsic detector background (as might be expected), the quoted sensitivities would be about 50% worse than shown. Even such a degraded sensitivity will still provide important scientific results.

## 7. Prospects for Solar Flare Polarization Studies

The RHESSI spacecraft was successfully launched on 5 February 2002. To date, there have been several M- and X-class flares that are good candidates for polarization studies. We are currently working to develop the detailed procedures for analyzing these data. These efforts are concentrating first on data from the the X4.8 event of 23 July 2002, which remained at the X-class level of emission for at least 900 s, and the M4.0 event of 17 March 2002, which occurred near local noon where modulation of the albedo flux should be minimal. Given the quality of these data, along with the predicted sensitivity levels, we anticipate being able to achieve sensitivity levels below 10% for some of these events.

## Acknowledgements

This work has been supported by NASA grant NAG5-10203 at UNH and UAH, and by NASA contract NAS5-98033 at UC Berkeley.

## References

Bai, T.: 1988, *Astrophys. J.* **334**, 1049.
Bai, T. and Ramaty, R.: 1978, *Astrophys. J.* **219**, 705.
Bogovlov, S. V. *et al.*: 1985, *Soviet Astr. Lett.* **11**, 322.
Brown, J. C.: 1972, *Solar Phys.* **26**, 441.
Brown, J. C. and Hoyng, P. 1975, *Astrophys. J.* **200**, 734.
Brown, J. C., McClymont, A. N., and McLean, I. S.: 1974, *Nature* **247**, 448.
Catalano, C. P. and Van Allen, J. A.: 1973, *Astrophys. J.* **185**, 335.
Chanan, G., Emslie, A. G., and Novick, R.: 1988, *Solar Phys.* **118**, 309.
Charikov, Ju. E. , Guzman, A. B., and Kudryavtsev, I. V.: 1996, *Astron. Astrophys.* **308**, 924.
Datlowe, D.W., Elcan, M. J., and Hudson, H. S.: 1974, *Solar Phys.* **39**, 155.
Datlowe, D. W., O'Dell, S. L., Peterson, L. E., and Elcan, M. J.: 1977, *Astrophys. J.* **212**, 561.
Emslie, A. G. and Brown, J. C.: 1980, *Astrophys. J.* **237**, 1015.
Emslie, A. G. and Vlahos, L.: 1980, *Astrophys. J.* **242**, 359.
Kane, S. R. *et al.*: 1988, *Astrophys. J.* **326**, 1017.
Kane, S. R. *et al.*: 1998, *Astrophys. J.* **500**, 1003.
Langer, S. H. and Petrosian, V.: 1977, *Astrophys. J.* **215**, 666.
Leach, J. and Petrosian, V.: 1983, *Astrophys. J.* **269**, 715.
Lei, F., Dean A. J., and Hills, G. L.: 1997, *Space Sci. Rev.* **82**, 309.
Li, P.: 1995, *Astrophys. J.* **443**, 855.

Li, P., Hurley, K., Barat, C., Niel, M., Talon, R., and Kurt, V.: 1994, *Astrophys. J.* **426**, 758.
Lin, R. P. *et al.*: 2002, *Solar Phys.*, this volume.
Masuda, S., Kosugi, T., Hara, H., Sakao, T., Shibata, K., and Tsuneta, S.: 1995, *Publ. Astron. Soc. Japan* **47**, 677.
McConnell, M. *et al.*: 1996a, in R. Ramaty, N. Mandzhavidze and X.-M. Hua (eds), *High Energy Solar Physics AIP Conf. Proc.* **374**, 368.
McConnell, M., *et al.*: 1996b, in C. Kouveliotou, M. F. Briggs and G. J. Fishman (eds), *Gamma Ray Bursts, 3rd Huntsville Symposium AIP Conf. Proc.* **384**, 851.
McConnell, M., Ledoux, J. R., Macri, J. R., and Ryan, J. M.: 2002, *SPIE Proc.* **4851**, in press.
Nakada, M. P., Neupert, W. M., and Thomas, R. J.: 1974, *Solar Phys.* **37**, 429.
Novick, R.: 1975, *Space Sci. Rev.* **18**, 389.
Petrosian, V.: 1985, *Astrophys. J.* **299**, 987.
Pizzichini, G., Spizzichino, A., and Vespignani, G. R.: 1974, *Solar Phys.* **35**, 431.
Roy, J. R. and Datlowe, D. W.: 1975, *Solar Phys.* **40**, 165.
Sakao, T. *et al.*: 1992, *Publ. Astron. Soc. Japan* **44**, L83.
Share, G. H. and Murphy, R. J.: 1997, *Astrophys. J.* **485**, 409.
Share, G. H., Murphy, R. J., Kiener, J., and de Séréville, N.: 2002, *Astrophys. J.* **573**, 464.
Smith, D. M. *et al.*: 2002, *Solar Phys.*, this volume.
Tindo, I. P., Shuryghin, A. I., and Steffen, W.: 1976, *Solar Phys.* **46**, 219.
Tindo, I. P., Ivanov, V. D., Mandel'stam, S. L., and Shuryghin, A. I.: 1970, *Solar Phys.* **14**, 204.
Tindo, I. P., Ivanov, V. D., Mandel'stam, S. L., and Shuryghin, A. I.: 1972a, *Solar Phys.* **24**, 429.
Tindo, I. P., Ivanov, V. D., Valníćek, B., and Livshits, M. A.: 1972b, *Solar Phys.* **27**, 426.
Tramiel, L. J., Chanan, G. A., and Novick, R.: 1984, *Astrophys. J.* **280**, 440.
Vestrand, W. T. and Ghosh, A.: 1987, *Proc. 20th Int. Cosmic Ray Conf.* **3**, 57.
Vestrand, W. T., Forrest, D. J., and Rieger, E.: 1991, *Proc. 22nd Int. Cosmic Ray Conf.* **3**, 69.
Vestrand, W. T., Forrest, D. J., Chupp, E. L., Rieger, E., and Share, G.: 1987, *Astrophys. J.* **322**, 1010.
Vilmer, N.: 1994, *Astrophys. J. Supp.* **90**, 611.
Zharkova, V. V., Brown, J. C., and Syniavskii, D. V.: 1995, *Astron. Astrophys.* **304**, 284.

# THE RHESSI EXPERIMENTAL DATA CENTER

PASCAL SAINT-HILAIRE[1,2], CHRISTOPH VON PRAUN[4], ETZARD STOLTE[3], GUSTAVO ALONSO[3], ARNOLD O. BENZ[1] and THOMAS GROSS[4]

[1]*Institute of Astronomy, ETH Zürich, CH-8092 Zurich, Switzerland*
*(e-mail: shilaire@astro.phys.ethz.ch)*
[2]*Paul Scherrer Institute, CH-5232 Villigen, Switzerland*
[3]*Institute of Information Systems, ETH Zürich, CH-8092 Zürich, Switzerland*
[4]*Laboratory for Software Technology, ETH Zürich, CH-8092 Zürich, Switzerland*

(Received 18 September 2002; accepted 19 September 2002)

**Abstract.** The RHESSI Experimental Data Center (HEDC) at ETH Zürich aims to facilitate the use of RHESSI data. It explores new ways to speed up browsing and selecting events such as solar flares. HEDC provides pre-processed data for on-line use and allows basic data processing remotely over the Internet. In this article, we describe the functionality and contents of HEDC, as well as first experiences by users. HEDC can be accessed at *http://www.hedc.ethz.ch.* Additional graphical material and color versions of most figures are available on the CD-ROM accompanying this volume.

## 1. Introduction

The Reuven Ramaty High-Energy Solar Spectroscopic Imager (RHESSI) images the Sun in the 3–17 000 keV range (soft and hard X-rays, as well as gamma rays) with unprecedented spatial, temporal and spectral resolutions. RHESSI's main objective is to deepen our understanding of energy release and particle acceleration taking place in solar flares. Furthermore, it provides spectral information on cosmic gamma-ray bursts, and even images such extra-solar objects as the Crab Nebula (Lin *et al.*, 2002).

RHESSI's nine rotating modulation collimators (RMC) modulate the incoming X-rays on the detectors at the end of each RMC. As each RMC's grid pair has different pitches, a set of nine Fourier-like components are obtained, and are used to reconstruct the original image (Hurford *et al.*, 2002). As each incoming photon is tagged in time, energy and RMC, these photons can be binned in a specified energy band for image reconstructions, lightcurves, and spectra.

A good proportion of existing astrophysical databases concentrate on archiving raw data from one or a few instruments such as *Yohkoh*, TRACE, RHESSI, or ground-based observatories such as *Phoenix-2*. Examples are the Solar UK Research Facility (SURF[1]) and the TRACE Data Center[2]. Other databases offer a selection of data (e.g., synoptic maps) from several observatories and provide

[1]http://surfwww.mssl.ucl.ac.uk/surf/
[2]http://vestige.lmsal.com/TRACE/DataCenter




CD ROM

usually easier or faster access to these data than the primary archives. Examples are the Base de données Solaire Sol 2000 (BASS2000[3]) (Mendiboure, 1998) and the SOHO Summary[4] and SOHO Synoptic[5] search engines.

Most other observatories' data are stored in an immediately usable form, albeit not (fully) calibrated, but suitable for quick perusal. RHESSI data poses the problem that it must be reconstructed to be of any use, much like Hard X-ray Telescope (HXT) data.

Image reconstruction can take from less than a minute (for a basic *back projection*) to several hours (for a tedious *pixon* reconstruction) on a state-of-the-art workstation. Image reconstruction must be done by each data analyst and requires significant hardware and software resources, including the commercial product Interactive Data Language (IDL).

Each flare may require many images, at different times, energy intervals, accumulation time intervals, image sizes and resolution, etc. If we consider the importance of spectra, lightcurves and other ancillary data, we realize that large amounts of computing time are needed to create the derived (i.e., Level-1) data an observer will sift through, to home in on data sets of interest.

These preparatory, but necessary, computational activities partially overlap each observer and project. This realization provides the starting point for the RHESSI Experimental Data Center (HEDC). Its purpose is to automatically generate an exhaustive amount of 'quicklook' data products and assemble them in an on-line data warehouse that will allow fast browsing and other services.

The 'E' in the HEDC acronym stands for 'Experimental': HEDC introduces several innovations but also serves as a platform for deepening our understanding of scientific data warehouses. HEDC is not simply another data repository, but also a database of scientifically useful derived data. Furthermore, every derived data item is accessible on-line, e.g., through the use of any Web browser.

On-line data processing (on the HEDC's servers) by users supplements the available data products. Users can add their own data products to the database, and the derived data on events of special interest thus increase in a self-organizing way. User participation will increase the scientific return of HEDC, and ultimately of RHESSI.

HEDC is a joint project of several groups at ETH Zürich: the Institute of Astronomy, the Institute of Information Systems, and the Laboratory for Software Technology (the last two are in the Department of Computer Science).

Section 2 describes the HEDC system. Section 3 explains how to use HEDC. Section 4 summarizes first experiences by users and concludes this paper. Appendix A describes the contents of the HEDC Extended Catalog (events and data products). Appendix B lists all user-relevant query attributes on HEDC.

[3]http://bass2000.bagn.obs-mip.fr/New2001/Pages/page_acceuil.php3

[4]http://sohowww.nascom.nasa.gov/cgi-bin/summary_query_form

[5]http://sohowww.nascom.nasa.gov/cgi-bin/synop_query_form

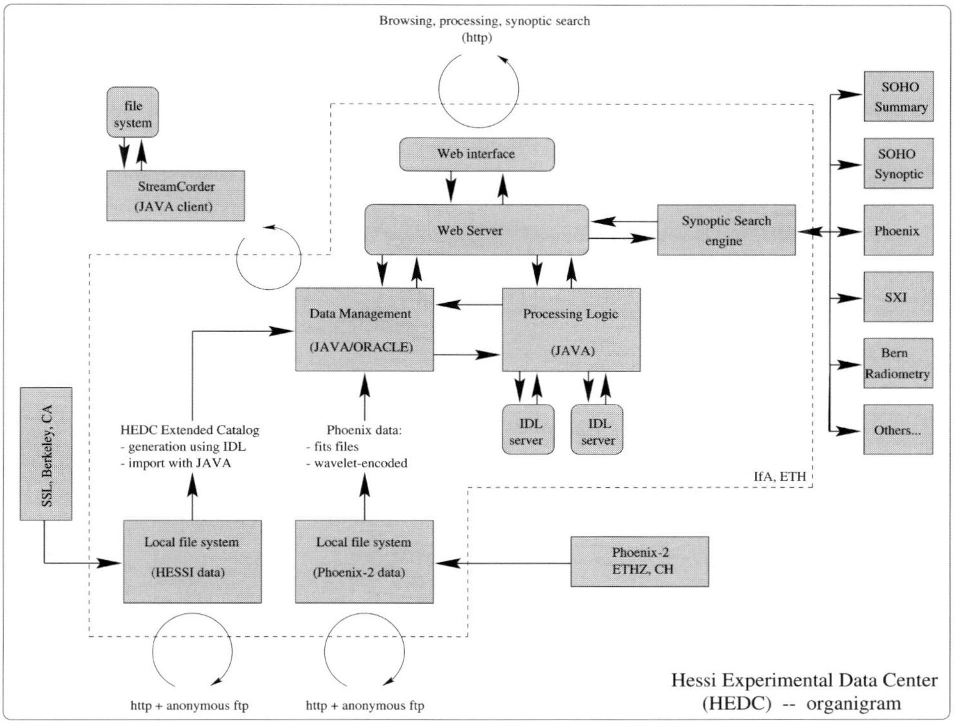


*Figure 1.* The HESSI Experimental Data Center at ETH Zürich.

## 2. Description of HEDC

HEDC classifies its data as *events* and *data products*. An event (sometimes more specifically referred to as an *HEDC event*) can be a solar flare, a gamma-ray burst, or a terrestrial electron precipitation. The term *other event* is reserved for future extensions or events that are not yet determined. An event consists of a list of attributes (such as start and end times, total counts, peak count rates in certain energy bands, etc.) and a list of associated data products. Data product is the generic term that refers to all derived data: images, lightcurves, spectra, spectrograms, etc. A data product consists of a list of attributes (in this case, the more important parameters that were used to generate the derived data, such as accumulation times, energy bands, etc.) and a picture (PNG or JPEG format). The list of attributes for both events and data products are tables for database querying.

The main services provided by HEDC are:

– On-line database for events and data products.

– On-line RHESSI data processing.

– On-line RHESSI Level-0 data repository that contains the raw data.

– Other on-line services such as a Synoptic Search engine that quickly retrieves other solar quicklook data of relevance.

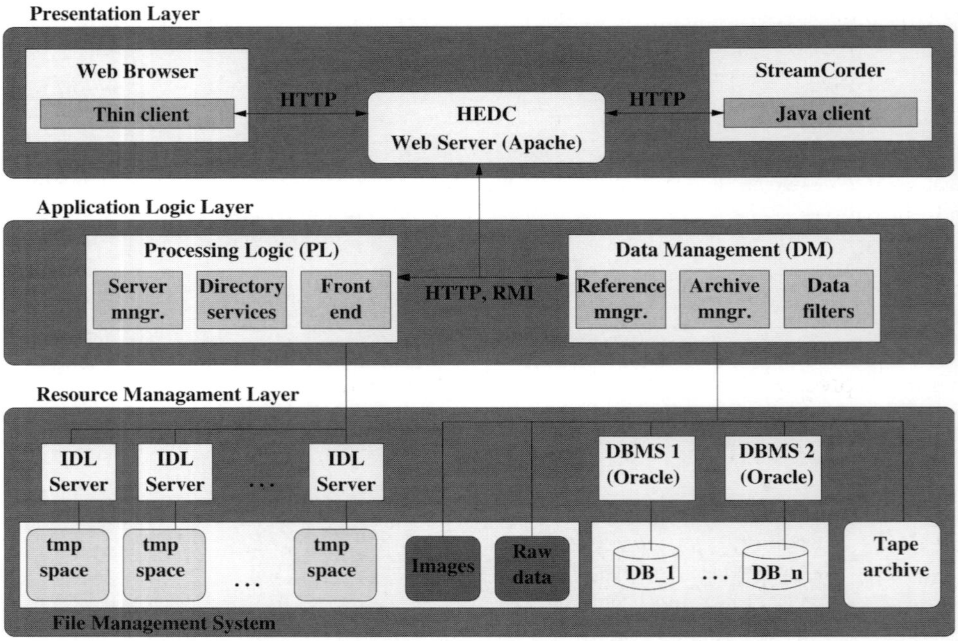


*Figure 2.* Architecture of HEDC.

Figure 1 shows an organigram of the different parts of HEDC. Its principal components are:

– A file system for all Level-0 RHESSI and Phoenix-2 (Messmer, Benz, and Monstein, 1999) calibrated data files.

– The HEDC Extended Catalog generation (using IDL) subunit.

– The Data Management (DM) subunit.

– The Processing Logic (PL) subunit and the IDL servers.

– The Synoptic Search engine.

– A Web server providing the main user interface.

– The StreamCorder, an alternative to the Web (browser) interface that provides more flexibility.

## 2.1. Architecture

HEDC is implemented as a 3-layer architecture (Figure 2), where the intermediate application logic layer relays requests for data and/or processing by *clients* (programs that run on local workstations) to an Oracle Relational Database Management System (DBMS) and a number of processing *servers* (remote computers at ETH) at the resource management layer. The application logic layer consists of two components: (1) The Data Management (DM) component takes care of all data storage issues. (2) The Processing Logic (PL) component acts as an intermediary between the DBMS and the IDL servers. Both are implemented in Java and run as stand-alone programs. HEDC can be accessed through either a Web-based client (e.g., a conventional browser)[6] or a Java-based client, the StreamCorder. With any of the two interfaces, users can query and download raw data, view data products and perform new processing steps. The Java-based client offers some functionality not available in the Web interface, e.g., tools for data visualization and system administration.

HEDC currently runs on a SUN Enterprise Server with 2 GBytes RAM, two 450 MHz processors, two hard-disk drives of 36 GBytes each, and two RAIDs of 654 and 1795 GBytes. Critical data, such as the database configuration information, is stored on RAID with tape backup, as are the derived data and the raw data files. The IDL processing servers execute on the SUN server and on Linux PCs.

### 2.2. On-line RHESSI data repository

RHESSI raw data (the Level-0 data) are stored in the form of FITS files, from which all RHESSI derived data are produced, using the RHESSI data analysis software[7] (Schwartz *et al.*, 2002).

The raw science data are mirrored daily from the Space Science Laboratory in Berkeley (CA, USA) in their entirety. In keeping with RHESSI's open data policy, all RHESSI data files are publicly available via anonymous FTP[8] or by HTTP access from the HEDC home page. The raw data come in at a rate of about 1.8 GBytes per day, taking about three hours to download. The data are usually available one or two days after observation, slightly more in case of a big flare (because of the limited downlink time between spacecraft and ground stations).

### 2.3. HEDC Extended Catalog generation

Along with the Level-0 data files usually comes 'quicklook' derived data (included are images and a flare list, amongst other items, see the online software documentation for further details). HEDC includes those quicklook data products (called here the *standard* catalog) and adds a large amount of other derived data, giving rise to the HEDC *extended* catalog.

The raw data are scanned for events of interest: this is presently done mostly via the flare list incorporated with the quicklook data. Parameters of interest for each event (total counts, peak count rates, etc.) are then extracted and stored in the DBMS as attributes for database queries. A set of data products is then generated for each such event: spectra at different times, images at different times using different energy bands, lightcurves, etc. Whenever possible, additional data relevant

[6] http://www.hedc.ethz.ch/
[7] http://www.RHESSI.ethz.ch/software/
[8] ftp://www.hedc.ethz.ch/pub/hessi/data/

to the event are added, e.g., *Phoenix-2* radio spectrograms, or the quicklook images or spectra.

Appendix A gives the current list of the data products computed automatically for each event. All new events and data products are inserted in the HEDC database, and may be recovered by anyone.

The whole process of creating the extended catalog, using IDL and the *SolarSoft*[9] (SSW) libraries, and inserting its elements into the database is fully automated, producing a standarized set of data products for each event.

The automatic generation of the extended catalog is still complicated by the changing SSW environment and the (as yet) lack of reliable flare positions. It will be constantly improved and completed over the next months, and derived data generated for the extended catalog will be reprocessed regularly until a final satisfactory stage is reached.

### 2.4. The Data Management (DM) Component

The Data Management (DM) component handles data requests by external clients and by the processing logic (PL). It offers HTTP and Remote Method Invocation (RMI) interfaces that hide the complexity of the lowest layer and provide an abstraction from the database schema (i.e., the list of database attributes).

Web clients access the DM through the HEDC Web server. Requests are first analyzed to determine the sub-systems needed to create the response. Then the relevant data and data references are retrieved from the database and data repositories, and a dynamic HTML response page is generated. StreamCorder clients access the DM directly through RMI, so that no HTML pages need to be created.

In astrophysics, system architectures are typically kept simple to cope with the high amount of data provided. Often, access systems are based on FTP combined with simple query scripts. Within the DM component, HEDC uses a commercial object-relational DBMS to manage the meta-data describing the derived data created by users and automated routines. Using a database simplifies the design of the data center as complex tasks can be left to the database rather than implemented anew. Furthermore, it offers greater flexibility for adding new features as need dictates.

Advantages of using a DBMS rather than a file system include consistent data updates with concurrent users, automatic as well as dynamic creation and maintenance of indexes, flexible query capabilities, efficient in-memory data caching for faster access and query processing, view materialization to avoid repetitive work in answering complex queries, and a flexible framework as the number of users increases and HEDC widens its scope. These are all important features that help HEDC to provide much more sophisticated capabilities than file-based systems at a lower development and maintenance cost. The database also takes care of efficient disk utilization and reorganization in an automatic manner and without requiring

[9] http://www.lmsal.com/solarsoft/

manual intervention. This is an important feature given the amount of data involved and the high update frequency. These and many other advantages have also been observed by other projects that have followed a similar approach (see, for instance, the work related to the Sloan Digital Sky Survey (Szalay *et al.*, 2000, 2002)).

Moreover, the fact that HEDC is built upon a database has allowed us to extend the architecture in several interesting directions that would have been cumbersome to pursue if HEDC would have been based on a file system. For instance, one of the HEDC interfaces, the StreamCorder, is meant to reside in the user's computer. Under the covers, the StreamCorder contains another object-relational DBMS that mirrors the database schema at the server. As the users work and process HEDC data using the StreamCorder, raw data and derived data are cached locally, thereby greatly speeding up overall processing. Users of the StreamCorder can work disconnected from the HEDC server using the data stored locally. They can also update this local data with new data products which will be automatically uploaded into the HEDC server once the user is on-line again. The StreamCorder has been designed so that users can gradually create their own HEDC data center provided they have the storage space. Everytime a user performs an operation with the StreamCorder (executes a processing, formulates a query), the StreamCorder checks whether the data are available locally. If that is not the case, it retrieves the data from the HEDC server and caches them locally. Over time and without any effort on their part, users will find that most of the data relevant for their work are stored locally in their small version of the HEDC server. This mechanism helps both HEDC, since it reduces the load at the server, and the final users, who get a much faster access to the data relevant to them.

### 2.5. The Processing Logic (PL) Component

The Processing Logic (PL) allows each user to compute data products beyond those that are part of the existing extended catalog provided by HEDC. Currently, three types of processing are supported, corresponding to the three main objects of the RHESSI software: images, lightcurves and spectra. Each processing activity can be configured with a set of basic parameters.

Users access the PL through the Web-based interface. Each user interacts with the system in a personalized session. Processing steps are specified as tasks that are handled by the system in accordance with the availability of computing resources. Processing is done in the background so that users can submit several tasks at once. Each task is an individual batch operation. This design choice has been taken to avoid user-specific resource reservations on the server. This leaves the scheduling of individual tasks unconstrained, leading to improved resource utilization.

On-line processing is integrated with the DM in the following two ways: (1) At the user interface, the PL can be easily accessed while browsing so that the standard attributes of a data product from the extended catalog are copied into the *processing submission form* of the PL. (2) Processing results (PNG and FITS format) can be submitted to the DM and stored permanently into the database.

The implementation of the PL is based on an object-oriented software framework. The system has been designed to easily accommodate changes and additions in the supported types of processing activities; the task control and scheduling system is strictly separated from application-specific issues, such that more than 95% of the code is independent of the three currently implemented processing activity types. The structure of the system is based on service modules that execute independently and allow distributed task execution in a network of workstations. The session and task management overhead is marginal compared to the cost of data transfers and computation.

Duration of individual tasks can vary significantly. We limited the total Central Processing Unit (CPU) time per task to 20 min and constrained the set of admissible input parameters (i.e., no *pixon* image reconstructions). Those limitations may be alleviated in the future, depending on usage of available computing resources.

### 2.6. Other Services

#### 2.6.1. *Synoptic Search*

The synoptic search subsystem serves to quickly browse through ancillary data related to a particular event in remote astro-archives. The data obtained are usually daily GIF or JPEG images.

The query mechanism resembles a Web-crawler: first, online requests are issued to several remote archives in parallel; then, the results are collected, grouped and displayed to the user.

This service operates largely independent from other subsystems of HEDC. The service is best effort (if a query to a remote archive times out, no results are available). This light-weight approach of rendering synoptic data accessible through HEDC has proved to be practical and robust. In its current configuration, six remote archive sites are searched, including the SOHO summary/synoptic data archive and the *Phoenix-2* archive at ETH Zürich. Due to its flexible software architecture, additional Internet-archives can be easily integrated.

Data in FITS format are better suited for data analysis, but are not displayed by the usual Web browsers, hence slowing down the search for datasets of interest. The RHESSI Synoptic Data Archive[10] provides an ample amount of such FITS data from other observatories, concentrated around RHESSI flare times.

#### 2.6.2. *The StreamCorder*

HEDC is not only accessible through a Web browser, but also through the StreamCorder (see Figure 3), which must be installed locally[11]. Except for some performance-sensitive hardware-dependent routines, it has been completely implemented

[10] http://orpheus.nascom.nasa.gov/~zarro/synop/

[11] http://www.hedc.ethz.ch/release/

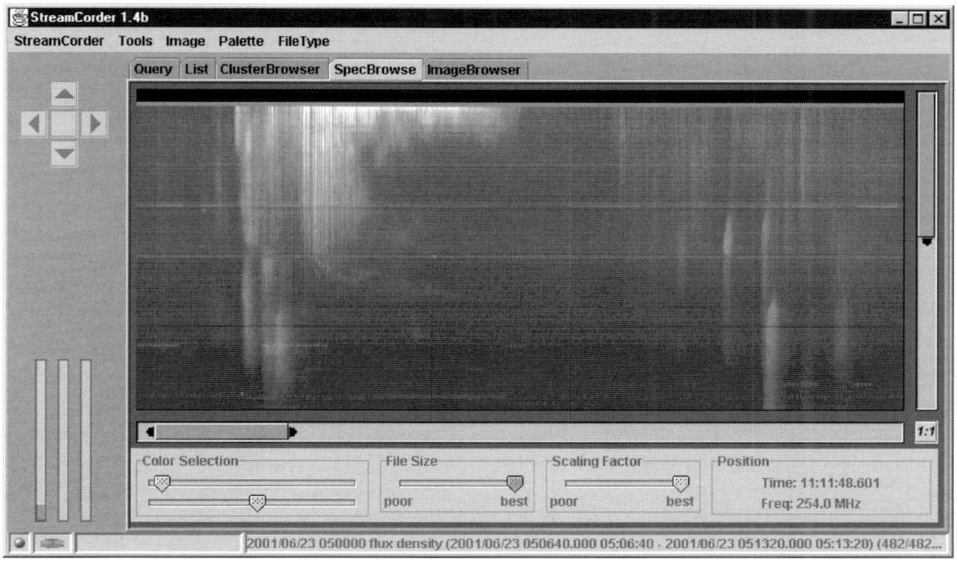


*Figure 3.* The StreamCorder is an alternative to the Web interface. It can search and display all HEDC data products. Here, it is displaying a wavelet-encoded radio spectrogram from *Phoenix-2*. (See the color reproduction on the accompanying CD-ROM.)

in Java. The architecture is extensible and modules are loaded according to the current data context. Modules may access core services such as stream management, request queues and local analysis programs. Currently available modules support browsing and download of all data types stored in HEDC, allow local and remote processing and offer administrative tools. During RHESSI data processing, the StreamCorder coordinates the asynchronous download, caching, decoding and processing of the data. A local database transparently caches query results and manages downloaded files. The local DBMS schema and the structure of the local file-system archive are identical to the ones on the server. Thus, offline work is possible.

#### 2.6.3. *Phoenix-2 Archive*

HEDC also holds the *Phoenix-2* radio spectrometer archive, both in FITS format and in a wavelet-encoded format, the latter for speedy spectrogram viewing with the StreamCorder.

## 3. Using HEDC

The term *browsing* is used in this paper to refer to one of these activities: making database queries (either by event or by data products), exploring the result set by

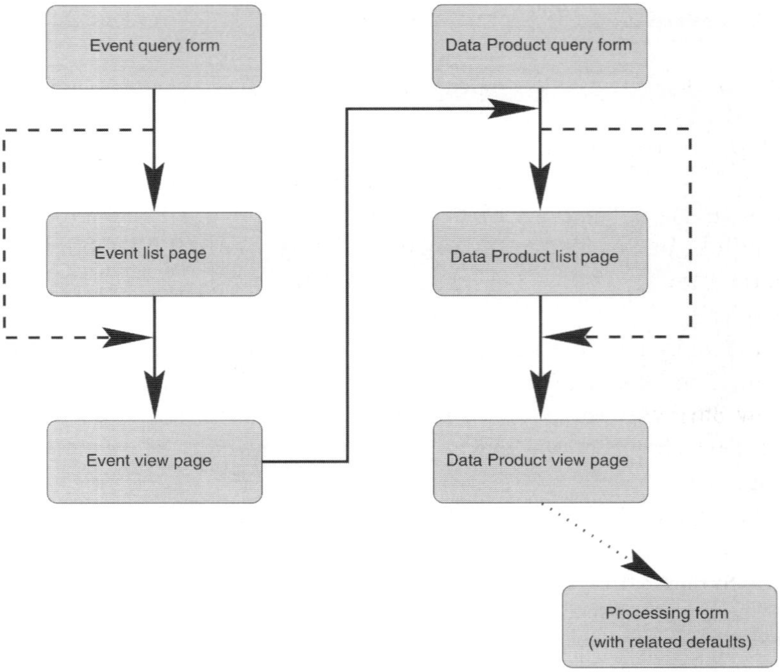


*Figure 4.* Browsing the HEDC using the Web interface. There are two possible entry points: the event query form or the data product query form. After submitting the appropriate query form, a list page containing the result set appears. Clicking on one of its elements leads to the view page, where the full set of database attributes can be examined, as well as a picture in the case of data products. If a query results in a single set, then the list page is bypassed. To the event view page is appended a data product list page, containing the data products associated with that event. The data product view page also has icons to access the processing form: using those instead of directly going to the processing form via the main link on the left of the Web page has the advantage that most of the processing form's attributes are already defaulted to those of the data product just examined.

going through the links, making another query (perhaps a finer one, or an entirely new one), etc.

The standard workflow model shown in Figure 4 for users of HEDC is to browse back and forth for events and/or data products, eventually to make new data products online, and to add them to the HEDC database. Once a user has zeroed in on a dataset of interest, he or she will have to make a thorough scientific analysis on his or her workstation, perhaps downloading some of the images previously made on HEDC. Of course, a user may decide to use only parts of HEDC: browsing, processing, or synoptic searching.

### 3.1. Browsing with the Web Interface

The HEDC Web interface currently offers two types of querying: either by event or by data products (Figure 4). Both possibilities can be done using either a standard Web form, or a more advanced one.

The *standard* Web form is intended for use by casual users. It is simple and intuitive: there should be no need to consult the on-line documentation. However, only a handful of query fields are available. The *advanced* form should be used by people more familiar with the system, and offers the full range of user-accessible database queries.

Once the query has been submitted, a *list page* appears. There is a current limit of 100 entries on this list. Each entry represents a different event or data product, with a few attributes (time intervals, energy bands, imaging algorithm, etc.) to guide the next choice of the user. Each entry is actually a link to an event (or data product).

Clicking on an event will lead the user to an *event view page*, an HTML page that displays all the event's database attributes (e.g., count rates), as well as a list of all associated data products. A data product view form also lists that data product's attributes (imaging algorithm, etc.), as well as a picture (i.e., 2-D plots or 3-D intensity maps).

A comprehensive set of examples are available in the online documentation.

All RHESSI images on HEDC may be downloaded in FITS format (as produced by the RHESSI software *fitswrite* method), by clicking on the appropriate icon in the *data product view page*.

### 3.2. Processing with the Web Interface

Whereas browsing is open to the public, an account is needed to perform processings on HEDC (account requests can be made online).

Once a filled processing form is submitted, a *job list* appears. It is a listing of all job requests that were sent with their states (pending, running, finished, failed). Submitted jobs do not share CPU time: rather, each job is queued or executed as fast as possible using one of the available CPUs.

If a job successfully ends, an icon that displays the resulting picture appears in the job list. Clicking on this icon allows the user to view the full picture, as well as to obtain relevant database attributes and other items pertaining to it. One such item is the IDL output, particularly useful in understanding the cause of a failed job. The current setup allows a maximum of 10 jobs at any time per user.

Jobs stay on the job list until the user logs out, up to a maximum of one week. Clicking on the 'update' button will update the job list to its latest status.

#### 3.2.1. *User Events: Folders for Users' Data Products*

Once a new data product has been generated and is being viewed by a user, it is possible to store it permanently in the HEDC database. Each user-made data product must be saved in a *user event*, which is just another event on HEDC, and serves as folder (or directory) for users' data products. In this manner, individual users can create several different folders, one for each of their projects, and put in them whatever data products they process on HEDC. User events do not have any attributes, except for a *code* starting with the username. This means that a query for *events* using time intervals does not reveal user-made events, even if the user-made data products stored inside are within that time interval.

If one is interested in all data products ever generated for a given time interval, one should browse using the *data product query form*. Both user-made and HEDC-made data products are shown (HEDC-made data products have a code similar to their parent event, always starting with 'HX').

### 3.3. Synoptic Engine

Using the HEDC's Synoptic Engine is straightforward: a user enters an approximative date and time of interest and submits the request. A list of available links to pictures appears. Choosing a longer time-out than the default value may result in more links found.

### 3.4. The StreamCorder

The StreamCorder provides the user with most of the previously described functionalities but in a more flexible manner than the browser-based interface. As it is a client side application (it runs on a user's local workstation), it uses local resources for processing, in contrast to the browser-based interface that employs the HEDC server. This offloads work from the server and allows for faster interactions with the system in case of repetitive queries and processings.

The StreamCorder also implements additional features that would be too expensive (in terms of CPU requirements) to provide in a centralized manner. Some examples are (1) a 'MovieCordlet' extension allows users to rapidly view a sequence of images made on HEDC, (2) the 'Spectrum Browser' enables users to look at wavelet-encoded Phoenix-2 radio spectrograms, allowing for fast exploration of the raw data (Stolte and Alonso, 2002a, b), (3) IDL sessions can be run locally or remotely, (4) a 'Cluster Browser' allows users to visualize the density of *event* or *data product* population in a phase space, where a 'phase' corresponds to any numerical database attribute (Stolte and Alonzo, 2002a, b). The StreamCorder is fully operational, although slight improvements are still being applied to increase ease-of-use.

## 4. First Experiences and Conclusions

From a user's standpoint, HEDC addresses the major software constraints that create a barrier to starting the analysis of RHESSI data: (1) the purchase, maintenance, and installation of IDL, (2) the installation, configuration, regular update of software unique to RHESSI, (3) the need to learn the detailed use of RHESSI software. Users can easily and quickly examine a huge variety of RHESSI data products, or create their own, with only a Web browser. As a side effect, HEDC is appreciated by those working at home, and who do not have available the necessary software or who lack sufficient transmission bandwidth to download raw data files.

As HEDC uses the SSW/RHESSI software to produce data products, the results are exactly the same as if they were produced from a standard IDL session. Relying on the SSW/RHESSI software makes it easy to compare results obtained at HEDC with other results. Furthermore, the substantial effort in writing (and evolving) this software is not duplicated. Of course, HEDC is therefore tightly coupled to the overall SSW/RHESSI software development and exposes the user to the perils of a dynamic software environment. However, the benefits obtained now (and in the future, after the final release) outweigh any temporary glitches.

As of the beginning of November 2002, HEDC contains more than 30 000 data products of over 1500 flares. Having a database to classify all pertinent RHESSI data products instead of a file system-based archive is a big advantage: a single event can warrant so many data products that users get lost trying to sort them again if they rely on a standard file system. HEDC allows for quick, easy searches.

For the astrophysics researcher, the determination of flare position has been found to be most useful. Many small flares not in the catalog are located by users and stored. The demand for this function will greatly increase once the aspect solution is more reliable. Also in high demand are mission-long lightcurves and the visualization of the data in the observing summaries of satellite orbits.

## Acknowledgements

We thank the RHESSI software team for continuous encouragements and support, in particular Brian Dennis, André Csillaghy, Jim McTiernan, Richard Schwartz, and Kim Tolbert for their help, explanations, feedback, and goodwill.

The RHESSI work at ETH Zürich is supported, in part, by the Swiss National Science Foundation (grant No. 20-67995.02) and ETH Zürich (grant TH-W1/99-2).

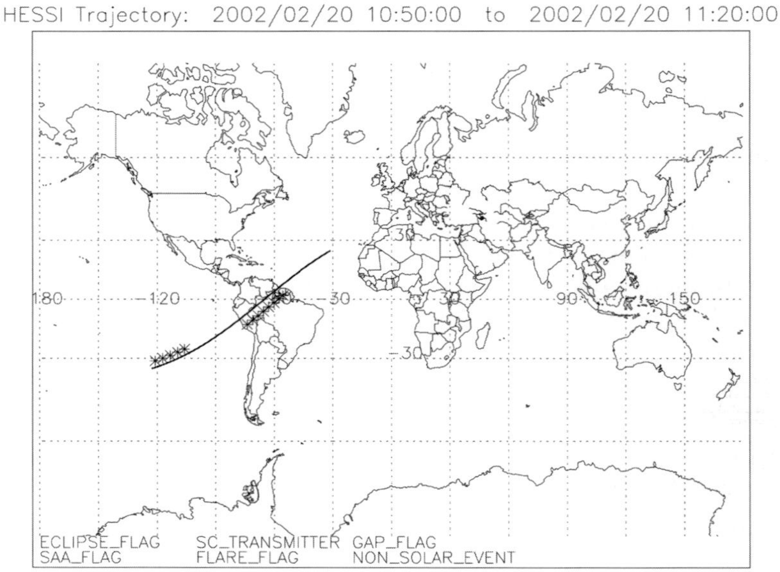


*Figure 5.* For each event, HEDC has a projection of RHESSI's subpoint on a mercatorian view of the Earth. A list of Observing Summary flags is shown at the bottom of the plot. Those are the flags that are being checked for and displayed with stars along the trajectory. The color-coded version is available with the CD-ROM material accompanying this volume.

## Appendix A. HEDC Extended Catalog contents

This appendix describes the current state of what is generated and stored on HEDC. It is liable to change. Consult the on-line documentation for the latest updates. Currently, the generation of the HEDC extended catalog is done about a week after observation by RHESSI. Later reprocessings will occur periodically and incrementally, following improvements or additions to the catalog generator, or major modifications to the raw data or the flare list. The newest, reprocessed versions of HEDC events and their associated data products will replace previous versions. Of course, user-made events and data products will never be reprocessed.

### A.1. Detection of events

Currently, only solar flares and some 'other' flares (i.e., with parameters still undefined) are being looked for and generated. Later, this might be extended to gamma-ray bursts and electron events.

Solar flares are given by the flare list attached to the Level-0 data. Basically an increase in photon count rates in the 12–25 keV energy band is looked for. The signal must also be strongly modulated in RHESSI's two coarsest detectors (num-

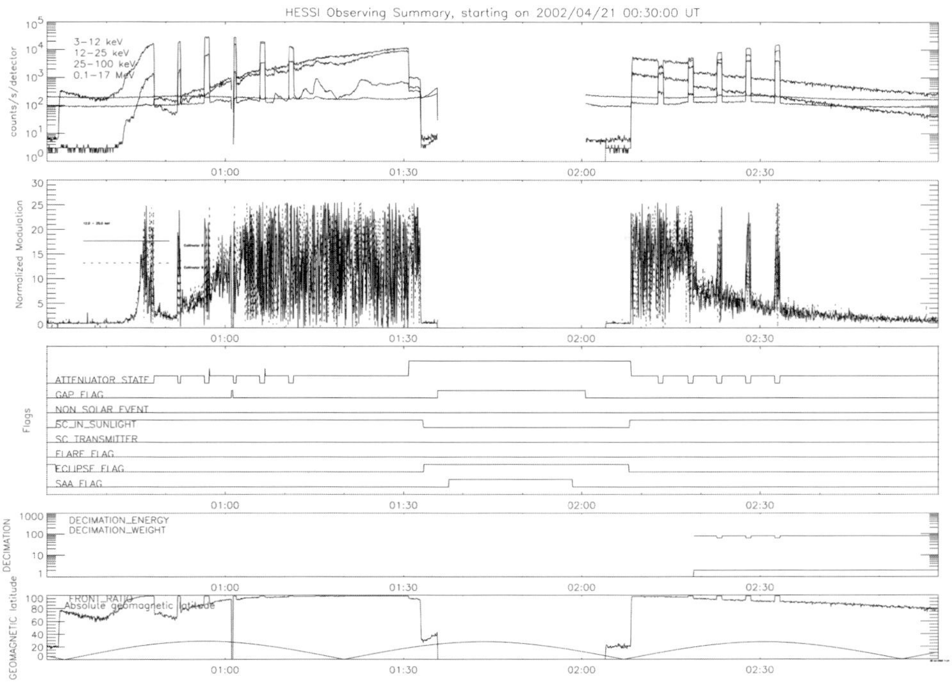


*Figure 6.* For each event, HEDC has an 'Observing Summary page', showing several products available in the Observing Summary, as well as RHESSI's geomagnetic latitude. The color-coded version is available with the CD-ROM material accompanying this volume.

ber 8 and 9). See the RHESSI Data Analysis Software pages[12] for more details on this.

'Other flares' are those enhancements in the count rates as seen by HEDC or the flare list, and for which no other classification was (yet) found.

After an event is detected, and its type determined, a set of attributes is determined which characterize the event for later database queries.

## A.2. Determination of Event Attributes

Attributes for each events (such as start and end times, total counts, peak count rates in certain energy bands, SAA and eclipse flags, etc.) are determined as each event is generated. Those attributes can be used as search fields during database queries.

Appendix B gives a full listing of HEDC event attributes.

[12]http://www.RHESSI.ethz.ch/software/

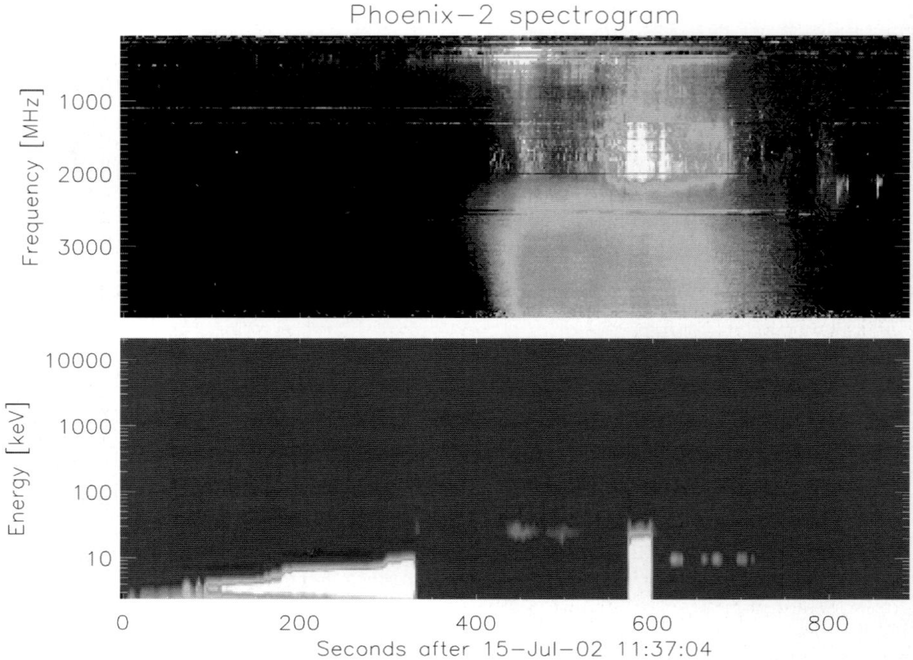


*Figure 7.* For each event, HEDC has a background-subtracted spectrogram from both RHESSI and *Phoenix*-2 radio data. (See the color reproduction on the accompanying CD-ROM.)

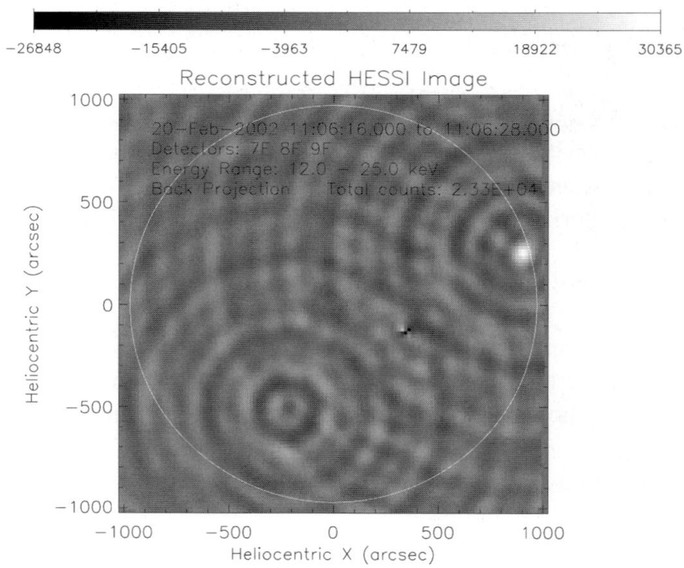


*Figure 8.* For each event, HEDC has a full-Sun back-projected image, at the peak of the 12–25 keV flux. Note flare position (850, 280), spin axis (350, −150) and ghost image (−220, −520).

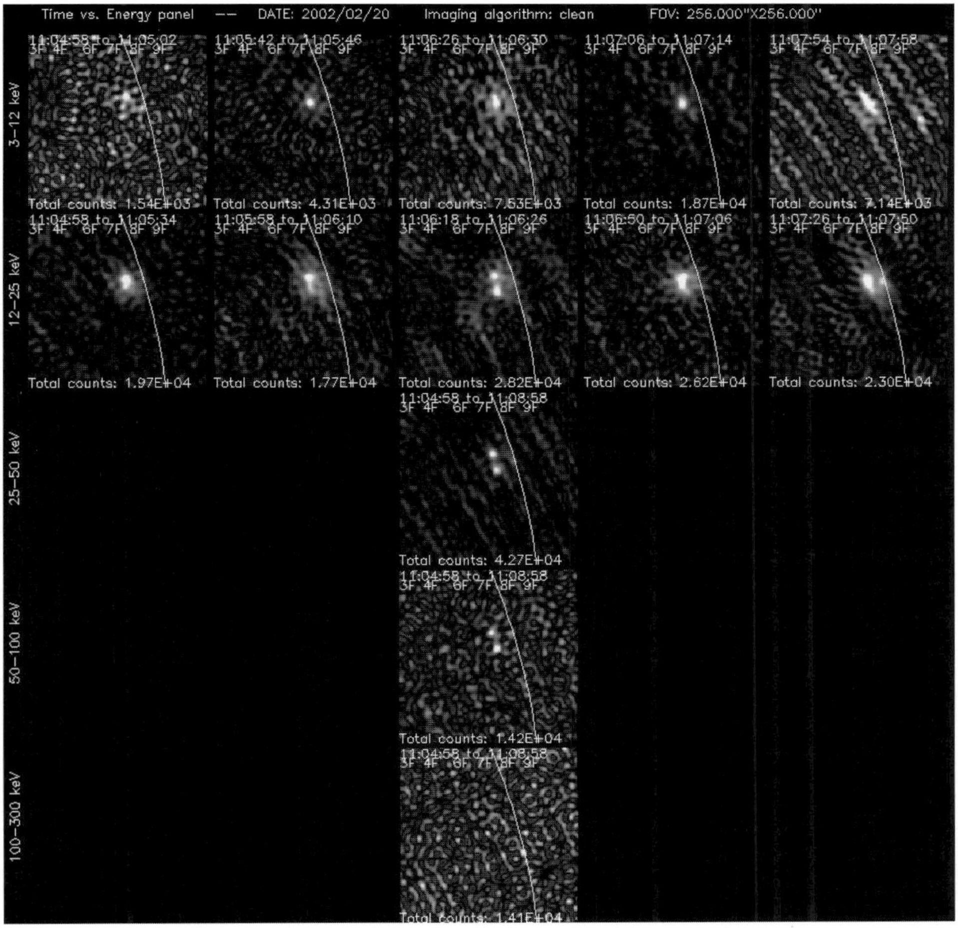


*Figure 9.* For each event, HEDC has a panel of CLEANed images showing the evolution of the flare in time (*horizontal*) and energy (*vertical*). Only images with a minimum number of counts are made. Hence, small flares do not necessarily have five images in every energy band. (See the color reproduction on the accompanying CD-ROM.)

## A.3. Data Products Automatically Generated with Each Event

For all events:

– Lightcurves of the whole event, in different energy bands.

– Three spectra in the 3–2500 keV range, with one minute accumulation time. One done at peak time, one midway between start time and peak time, and one midway between peak and end time.

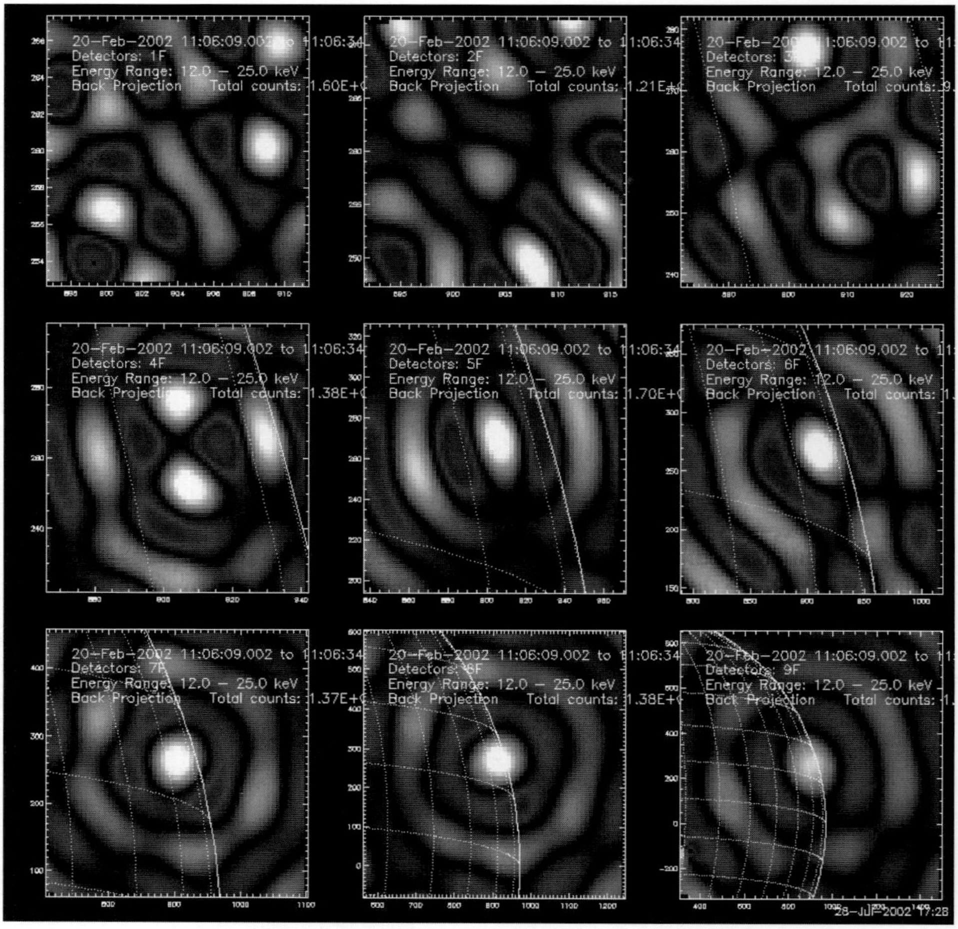

*Figure 10.* For each event, HEDC has a panel of back-projected images at the peak of the 12–25 keV flux, one for each sub-collimator. The field of view (FOV) increases proportionally to collimator resolution. Using the same FOV for all collimators does not allow a proper visual appreciation of each collimator's contribution to the final image. (See the color reproduction on the accompanying CD-ROM.)

– Images made from Observing Summary[13] data: count rates in different energy bands; RHESSI trajectory on a Mercator projection of Earth; modulation variance lightcurves; flags; geomagnetic latitude (Figures 5 and 6).

– RHESSI spectrograms are generated. If possible, they are also superimposed with radio spectrograms from *Phoenix-2* (Figure 7). Both are background-subtracted.

– Background-subtracted time series of the 25–50 keV over 12–25 keV counts ratio for the whole event.

[13]http://www.hessi.ethz.ch/software

TABLE I

List of query fields for HEDC events.

| | |
|---|---|
| Code | The 'name' of an event. Typically 'HXS202261026', where HX is an event made by HEDC, 'S' for solar flare, and '202261026' for the peak time of the February 26th, 2002 10:26 flare). Another possible format is 'hadar', 'hadar001', etc. for an event generated by user 'hadar'. |
| Event ID | An internal, unique ID number for each event. |
| Event type | 'S' for solar flares, 'G' for gamma-ray bursts, 'E' for electron events, 'O' for other flares. |
| Flarelist | Time-concurrent flare list number for a solar flare. |
| Minimum energy | Lower edge of highest energy band where flare counts were seen. |
| Maximum energy | Upper edge of highest energy band where flare counts were seen. |
| Total counts | Total counts of the flare, in the 12–25 keV energy band. |
| Distance to Sun | Solar flare's offset from Sun center, in arc sec. |
| X pos | Solar flare's west-east offset on the Sun, in arc sec. |
| Y pos | Solar flare's north-south offset on the Sun, in arc sec. |
| Creation date | Creation date of the event. |
| Start DATE+TIME | Date and time of the start of the flare, 12–25 keV band. |
| End DATE+TIME | Date and time of the end of the flare, 12–25 keV band. |
| Start time-of-day | Time, in seconds since midnight, of the start of the flare. |
| End time-of-day | Time, in seconds since midnight, of the end of the flare. |
| Duration | Time between flare's start and flare's end, in seconds. |
| Peak D+T (3–12 keV) | Date and time of the peak of the flare, 3–12 keV band. |
| Peak t-o-d (3–12 keV) | Peak time, in seconds since midnight, 3–12 keV band. |
| Total counts (3–12 keV) | Total counts of the flare, in the 3–12 keV band. |
| Peak rate (3–12 keV) | Count rate at peak time, in the 3–12 keV band. |
| Peak D+T (12–25 keV) | Date and time of the peak of the flare, 12–25 keV band. |
| Peak t-o-d (12–25 keV) | Peak time, in seconds since midnight, 12–25 keV band. |
| Total counts (12–25 keV) | Total counts of the flare, 12–25 keV band. |
| Peak rate (12–25 keV) | Count rate at peak time, 12–25 keV band. |
| Peak D+T (25–100 keV) | Date and time of the peak of the flare, 25–100 keV band. |
| Peak t-o-d (25–100 keV) | Peak time, in seconds since midnight, 25–100 keV band. |
| Total counts (25–100 keV) | Total counts of the flare, 25–100 keV band. |
| Peak rate (25–100 keV) | Count rate at peak time, in the 25–100 keV band. |
| Ratio 25–50/12–25 | Ratio of counts in the 25–50 keV and 12–25 bands at peak time. |
| Source multiplicity | Number of sources in a solar flare. Not operational yet. |
| Active region | Where the flare occurred, as given by the flare list. |
| Is simulated data | 0/1 or NO/YES flag. |
| S/C in SAA flag | 0/1 or NO/YES flag. S/C stands for spacecraft (i.e., RHESSI). |
| S/C in night flag | 0/1 or NO/YES flag. S/C stands for spacecraft (i.e., RHESSI). |
| Background rate | Background count rate. Not operational yet. |
| Comments | Made automatically by HEDC (e.g., highest geomagnetic latitude during an event), or by a user for a user-made event. |
| Reserves | unused yet. |

TABLE II

List of query fields for data products.

| | |
|---|---|
| Code | The ‘name’ of a data product. For a data product associated with an HEDC-made event, the data product’s code is usually the same as the HEDC event’s (e.g., HXS202261026). For user-made data products, any combination of 12 characters is possible. |
| Product ID | An internal, unique ID number for each data product stored on HEDC. |
| Product type | ‘IM’ for images, ‘SP’ for spectra, etc. See the online documentation for a complete listing. |
| Imaging algr | ‘BACK’ for back projection, etc. See the online documentation for a complete listing. |
| Movie code | Most of the RHESSI images made on HEDC are meant to be viewed in sequence, i.e., they share energy bands, imaging algorithm, etc., and differ only by their time ranges. All those images have the same movie code. |
| Movie frame | The order in which an image which is part of a movie appears. |
| Creation date | Creation date of the data product. |
| Start DATE+TIME | Date and time of the start of the accumulation time. |
| End DATE+TIME | Date and time of the end of the accumulation time. |
| Start time-of-day | Time, in seconds since midnight, of the start of the accumulation time for the data product. |
| End time-of-day | Time, in seconds since midnight, of the end of the accumulation time for the data product. |
| Duration | Accumulation time for the data product. |
| Min energy | Lower edge of the energy bands used for the data product. |
| Max energy | Upper edge of energy bands used for the data product. |
| Time resolution | Time binning for lightcurves (corresponds to LTC_TIME_RES). |
| Front segments used? | 0/1 or NO/YES flag. |
| Rear segments used? | 0/1 or NO/YES flag. |
| Subcollimator used | example: 101111100. |
| Distance to sun center | Angular offset (in arc sec) of the center of an image with respect to Suncenter (image data products only). |
| Xpos | Angular $x$-offset from suncenter of the center of an image. |
| Ypos | Angular $y$-offset from suncenter of the center of an image. |
| Xdimension | Number of horizontal pixels in an image (images only). |
| Ydimension | number of vertical pixels in an image (images only). |
| Xpixel size | Horizontal size (in arcseconds) of a pixel (image data products only). |
| Ypixel size | Vertical size (in arcseconds) of a pixel (image data products only). |
| Data quality | Unused yet. |
| Is simulated data | 0/1 or NO/YES flag. |
| Is background-subtracted | 0/1 or NO/YES flag. Not used yet. |
| Other alg. params | Information on some other parameters of the data product. |
| Comments | Text added by HEDC or by users, for their own data products. |
| Reserves | Unused yet. |

Additionally, for 'solar flare' events only:

– Full-Sun image (Figure 8), using back-projection.

– Movies, i.e., series of images in the following energy bands: 3–12, 25–50, 50–100, and 100–300 keV.

– 'Quicklook' images and spectra (i.e., those that are included with the raw data) are also extracted and inserted in the database.

– Panel of up to 5 × 5 images (up to 5 different time intervals, in 5 different energy bands) of the region of interest (Figure 9), using CLEAN.

– Panel of 3 × 3 images of the region of interest, one for each RHESSI subcollimator, using the back projection imaging algorithm (Figure 10) and photons in the 12–25 keV energy band.

Appendix B gives a full listing of data product database attributes. The time taken to generate a single 'solar flare' event and its associated data products is less than one hour for the above list of data products. More images per event will certainly be generated later on, increasing the processing time accordingly.

### A.4. Others

RHESSI mission-long daily lightcurves in different energy bands are available through the home page.

## Appendix B. Attributes Used for Browsing Queries

Tables I and II are lists of the attributes that may be used by users to query for data on the HEDC using the Web interface's 'expert' query form. The on-line documentation provides an up-to-date listing, as well as additional details.

## References

Handy B. *et al.*: 1999, *Solar Phys.* **187**, 229.

Hurford, G. *et al.*: 2002, *Solar Phys.*, this volume.

Lin, R. P. *et al.*: 2002, *Solar Phys.*, this volume.

Mendiboure, C.: 1998, *Second Advances in Solar Physics Euroconference. ASP Conf. Series* **155**, 302.

Messmer, P., Benz, A. O., and Monstein, C.: 1999, *Solar Phys.* **187**, 335.

Schwartz, R. A. *et al.*: 2002, *Solar Phys.*, this volume.

Stolte, E. and Alonso, G.: 2002a, Optimizing Scientific Databases for Client-Side Proccessing. *Proceedings of the VIII Conference on Extending Database Technology (EDBT)*, Prague, Czech Republic.

Stolte, E. and Alonso, G.: 2002b, Efficient Exploration of Large Scientific Databases. *Proceedings of the 28th International Conference on Very Large DataBases (VLDB)*, Hong Kong, China.

Szalay, A. S., Gray, J., Thakar, A., Kunszt, P. Z., Malik, T., Raddick, J., Stoughton, C., and van den Berg J.: 2002, The SDSS SkyServer – Public Access to the Sloan Digital Sky Server Data. *ACM International Conference on Management of Data*, SIGMOD.

Szalay, A. S., Kunszt, P. Z., Thakar, A., Gray, J., and Slutz, D. R.: 2000, Designing and Mining Multi-Terabyte Astronomy Archives: The Sloan Digital Sky Survey. *ACM International Conference on Management of Data*, SIGMOD.

# RHESSI DATA ANALYSIS SOFTWARE: RATIONALE AND METHODS

R. A. SCHWARTZ[1], A. CSILLAGHY[2,3,4], A. K. TOLBERT[1], G. J. HURFORD[2], J. McTIERNAN[2] and D. ZARRO[5]

[1]*NASA Goddard Space Flight Center/SSAI, Greenbelt, MD 20771, U.S.A.*
[2]*Space Sciences Laboratory, University of California, Berkeley, CA 94720, U.S.A.*
[3]*Institute of Astronomy, ETH Zurich, CH-8092 Zürich, Switzerland*
[4]*University of Applied Sciences, CH-5210 Windisch, Switzerland*
[5]*NASA Goddard Space Flight Center/L3 Corp., Greenbelt, MD 20771, U.S.A.*

(Received 14 September 2002; accepted 16 September 2002)

**Abstract.** The Reuven Ramaty High-Energy Solar Spectroscopic Imager (RHESSI) performs imaging spectroscopy of the Sun with high spatial and spectral resolution from 3 keV to 17 MeV using indirect Fourier-transform techniques. We review the rationale behind the RHESSI data analysis software, and explain the underlying structure of the software tools. Our goal was to make the large data set available within weeks after the RHESSI launch, and to make it possible for any member of the scientific community to analyze it easily. This paper describes the requirements for the software and explores our decisions to use the SolarSoftWare and Interactive Data Language programming packages, to support both Windows and Unix platforms, and to use object-oriented programming. We also describe how the data are rapidly disseminated and how ancillary data sets are used to enhance the RHESSI science. Finally, we give a schematic overview of some of the data flow through the high-level analysis tools. More information on the data and analysis procedures can be found at the RHESSI Data Center website, *http://hesperia.gsfc.nasa.gov/rhessidatacenter*.

## 1. Introduction

The Reuven Ramaty High-Energy Solar Spectroscopic Imager (RHESSI) is a NASA Small Explorer (SMEX) mission to study the acceleration and transport of high-energy particles in solar flares (Lin *et al.*, 2002). Observationally this is accomplished by high-resolution imaging spectroscopy of X-rays and gamma-rays from 3 keV through 17 MeV. This energy range requires the use of indirect imaging techniques. In this case a set of rotating modulation collimators encodes the imaging information into a time-modulated lightcurve (Hurford *et al.*, 2002). A primary function of the data analysis software is to decode this modulated waveform back to a quantitative estimate of the original image. An additional challenge is the recovery of high-resolution spectra that require background subtraction and the deconvolution of the instrument response to yield the incident spectrum as a function of time with $\sim 1$ keV resolution up to 100 keV (increasing to 5 keV resolution at 17 MeV). Additional data products include the generation of conventional light curves (parameterized by energy), and imaging spectroscopy in the form of feature-based light curves and spectra.

*Solar Physics* **210:** 165–191, 2002.

This type of imaging spectrometer poses several distinct challenges and opportunities in the design of a data analysis system. The first challenge arises from the large volume of information that must be processed to make either the basic images or spectra. The imaging and spectroscopy both start from a Level-0 database of 4-byte 'photon-tagged events' that encode the detector ID, arrival time, and energy for each detector count, as well as some live-time information (Curtis *et al.*, 2002). (We refer to these events as photon events in this paper, although they include particle events as well.) The data set for some flares consists of hundreds of millions of these photon-tagged events, with rates up to a half million per second. Once calibrated using the aspect solution, the modulated count rates contain information that is equivalent to a set of Fourier components of the source distribution. From that stage, the image reconstruction task is comparable to that in radio interferometry. One helpful factor is that (unlike the case of radio interferometry) the RHESSI instrument response does not vary significantly with time. However, imaging spectroscopy, achieved by the interpretation of data cubes resolved into spatial, energy, and temporal coordinates, does place stringent requirements on the photometric accuracy of the resulting images.

A second challenge is posed by supporting the fraction of the RHESSI user community not familiar with either indirect imaging techniques or with many of the instrumental issues associated with X-ray or gamma-ray spectroscopy. This is particularly important since the RHESSI data are made freely available online within a few days of observation and it is imperative that interested users be able to analyze the data without waiting for secondary databases to be generated. At a minimum, they should be able to obtain lightcurves and to reconstruct images and photon spectra with minimum knowledge about the inner workings of the software. Therefore, it is essential that the user interface provide the option of performing the image reconstruction or spectral acquisition in a manner that does not burden such users with the details of those processes.

A third challenge is that the RHESSI mission relies on data obtained in other wavelength regimes by other instruments to provide the context information on which much of the RHESSI science is dependent. Therefore, special emphasis is placed on enabling convenient comparisons with external data sets. This latter task is eased somewhat by the intrinsic absolute accuracy and stability of RHESSI image locations to $\sim 1$ arc sec and times to one millisecond.

A unique opportunity is also afforded by the photon-tagged nature of the RHESSI data set. In most imaging-spectroscopy instruments, difficult tradeoffs must be made during the instrument design phase or during operations to optimize the allocation of finite telemetry resources in order to meet the conflicting requirements of imaging field-of-view and resolution, energy range and resolution, temporal coverage and cadence, etc. Since the RHESSI telemetry includes the arrival time, detector, and energy for each detected photon, all of these spectrometer choices can be made during the data analysis phase. Therefore, such decisions can be made iteratively and on a case-by-case basis in response to the unique characteristics of the solar event under study and in response to the user's particular scientific objectives. This unique capability greatly enhances the scientific return from the observations. *Therefore, a key driver of our data analysis approach is the preservation of this flexibility for use by the data analyst.*

A further opportunity arises because the RHESSI imaging data are self-calibrating to a significant extent. Because of this and the expectation that the calibrations and imaging algorithms will become more refined over time, we anticipate significant improvements to the RHESSI image reconstruction during the course of the mission. The photon-tagged nature of the data set then permits these improvements to be applied retroactively to the entire mission data set.

To maintain the flexibility to choose time, energy, and imaging parameters and to conveniently exploit the anticipated improvements in calibration and algorithms, the generation of extensive secondary databases is minimized. Instead, most scientific analysis begins with the primary database with the most current calibration information. The software applies such calibration data to yield images and spectra with the most appropriate time, spatial, and spectral resolution and spectral range.

Other factors that define the data analysis approach are that the user interface and analysis kernel should support both interactive analysis and multi-event batch processing. It was also deemed a requirement to support both Unix and Windows platforms. Finally, event-driven science, such as flare observations, places an additional premium on the rapid dissemination of both data and software.

## 2. Meeting the Science Requirements

### 2.1. Database Preparation

Before we discuss the top-level and underlying structure of the RHESSI software, we briefly describe the data products that we archive online to facilitate scientific study. These data products provide the entry point to any scientific study, and help guide the more detailed analyses required by most users. Another category of data products is created dynamically according to the needs of the analyst and we will describe them together with the software architecture in the sections that follow.

The first component of the archived data products is a database of observations and calibrations that can be used to meet the scientific objectives. The primary database includes the following principal elements: packed photon-tagged event lists; Solar Aspect System (SAS) and Roll Aspect System (RAS) data; Monitor Rates; Fast Rate Counter output (when appropriate); and housekeeping data. The Monitor Rates consist of one-second readouts of various detector events that give the operators a picture of the detector status and environment. The Fast Rate Counters contain broad energy-band event rates suitable for imaging during the largest solar flares that might paralyze the normal detector electronics. The SAS and RAS packets contain the aspect sensor data that allow us to know the modulation response at any location on the Sun as a function of time (Zehnder *et al.*,

2002). All of these data are included within our Level-0 data files. Except for some header information to make the primary database appear as a sequence of time-ordered, non-duplicated, and quality-flagged data, the primary database has no 'value-added' content over the raw telemetered data. To ensure prompt preparation, the primary database is generated without routine operator intervention in the Mission Operations Center and Science Operations Center (MOC/SOC). The Level-0 packets are organized into FITS files up to $\sim 110$ Mbytes in length covering no more than a single orbit between local midnights. During flares, with their higher data rate, there are often multiple files per orbit. The FITS format, for which there is a large and well-tested library of low- and high-level routines, is used to facilitate the documentation and the addition of useful tables within FITS extensions and completely preserves the telemetry packets. The first extension is a table of the packet headers, containing the packet number, packet type, and time range included. As each FITS file is completed by the MOC/SOC, the file is copied to the archive centers at Goddard Space Flight Center and the Swiss Federal Institute of Technology in Zürich, Switzerland (ETHZ) over the Internet and made available there for anyone to download.

Another archival component is the 'quicklook products' (QLP). These allow a quick survey via low-resolution lightcurves, spectra, and images. These are the Level-1 data products created by the MOC/SOC autonomously using the same software used for higher-level analyses. These are also referred to as catalog or summary data. Summary quantities describing the instrument status and spacecraft position, flare positions, flare meta-data, and data quality are included, too. All QLPs are created both as FITS files (or extensions) and browser-viewable image formats such as GIF, PNG, and text files. The SOC posts the most recent QLPs on the RHESSI web pages as they are created. On the web page, any user can easily access the entire QLP archive for either the graphic format or FITS format files. Additionally, a simple summary event catalog will be available for browsing and for searching via a web page. The browser-viewable products are well suited to public access, since no special analysis skills or RHESSI-specific software are required. We have also developed a software interface to the QLPs for use during analysis sessions. (As of August 2002, the QLPs are included with the Level-0 data, but in the near future, the QLPs will only be written to separate files and not to the Level-0 files. This difference should be transparent to anyone using the summary data on the Web or through the analysis software.)

### 2.2. Software to Meet the Science Requirements

Conceptually the solar X-ray and gamma-ray input to RHESSI is in the form of a data cube dimensioned along four axes: two spatial, one for energy, and the other for time. The action of the instrument transforms this incident data cube of real photons into a time-ordered list of photon-tagged events. In this list, the spatial (more accurately angular) dimensions have been convolved into a temporal modulation, and the true photon energy distribution has been convolved with the detector response. Although some of the spatial structure of the true data cube can be inferred from the temporal modulation, without the use of high-level software it basically appears only as a temporal modulation. In contrast to this, much of the spectral information can readily be seen in the photon-event lists binned in time and energy. This contrast suggests that the requirement for complete flexibility in the energy dimension for building images makes the data management task even more difficult for imaging. Thus, it is the full imaging spectroscopy task that has driven much of the structure of the data analysis software in order to meet our goal of a robust system capable of producing results for the general scientific user.

One of the first decisions the software team made was to build our system around SolarSoftWare (SSW) (Freeland and Handy, 1998) and the Interactive Data Language (IDL). Both SSW and IDL have a long heritage for solar data analysis, both are used extensively throughout *Solar Maximum Mission* (SMM), *Yohkoh*, and *Solar Heliospheric Observatory* (SOHO). SSW is a package of IDL programming utilities and libraries to facilitate the data input/output, manipulations, and comparisons needed for a data set obtained with a solar instrument. Moreover, most of the scientific software team for RHESSI had extensive experience with both the IDL language and the programming environment created using standard SSW setup procedures. SSW includes installation and update procedures to ensure that users will have the correct and current software environment for RHESSI analysis. Also, because of the wide usage of IDL for other missions, we believe that the commercial license required for IDL imposes a mild constraint on the community compared with the effort that would be required by the team and the scientific community to use any alternative. IDL provides tools to build Graphical User Interfaces (GUIs) which we believe are critical to creating a wide community of users. These are simpler to operate and do not have the burden of syntactic structures. From the beginning, we decided to fully support two of the operating systems supported by IDL, Unix/Linux and Windows, for several reasons. There has been and continues to be a large installed base of Unix servers, there is an increasing number of institutions deploying personal computers using Linux, and there is a large and growing base of users analyzing data on personal computers, in particular laptops, using all variants of Windows since Windows 95. At the time we started, there was minimal support for Windows within SSW. The RHESSI team has been a dominant contributor in making the joint utilization of SSW, IDL, and Windows routine seamless since then. We have found that personal computers running either Windows or Linux make the most cost-effective platform for RHESSI analysis. At present, we do not provide explicit support for the Macintosh operating system because it lacks the environment variables relied on by SSW. We expect that RHESSI software will work on the newest Macintosh operating systems (very Unix-like) supported by IDL version 5.6 and beyond.

The approach of the RHESSI data analysis scheme is to create a robust system that allows the routine production of Level-2 photon-calibrated images and spectra by any scientist given the Level-0 FITS files, the QLPs, the calibration database, and the software installation. The RHESSI team members interact with the data set in the same way. Recalling our incident data cube of solar photons, the software's task is to take the analyst's specification of the resolution and range for the component desired, i.e., spectrally resolved image, spectrum, or lightcurve, and complete the processing required for the associated data product. Here we list the specific requirements of our software:

(1) The software must deal with large volumes of data, potentially processing nearly four Gigabytes of photon-tagged events to obtain a spectrogram covering a large flare, and do this with the relatively modest resources found on common notebook computers.

(2) The software must unerringly guide the user through the various processes needed to complete a task based only on the properties of the result such as range and resolution along the data cube axes.

(3) The software must be efficient. Some operations are very costly in time and other resources. The results of costly operations should be saved and reused wherever possible.

(4) The software must allow the recovery of intermediate data products to facilitate testing and debugging. This feature also allows building complex chains of processing from smaller and simpler elements.

(5) The action of the software must be highly configurable by user-selected parameters and algorithms, yet require only a few changes from its default configuration to produce meaningful results.

(6) The software must be callable from a command line interface (CLI) to facilitate scripting of repetitive actions.

(7) There must be a graphical user interface (GUI) to reduce the need for an analyst to learn our programming syntax.

## 3. Object-Orientation in the RHESSI Data Analysis Software

The architecture of the RHESSI data analysis software is based on object-oriented design concepts. In this section we explain what we mean by object-orientation, show how we implement this design technique in the RHESSI software, and show why we believe that the object-oriented approach helps us meet our software requirements.

### 3.1. Object-oriented software

Object-oriented programming describes a standard way of designing software in which the software is divided into independent parts – the objects – that communicate through well-defined interfaces. Object-oriented design can be seen as a realization of the 'divide and conquer' rule that has been driving software development for the past two decades: complex problems are easier to solve by first dividing them into smaller, isolated problems. We refer to Booch (2002) for a discussion of object-orientation, or to Schach (2001) for a software engineering approach to the topic. An online resource on object-orientation is given by Appendix Item 11.

Object-oriented design has been used successfully in numerous programs, from operating systems to text editors. Today, it is the main support for the deployment of Web applications. Surprisingly, object-oriented design has been used little for developing data analysis systems in solar physics, even though this field has traditionally been in the lead for innovative computer technologies. We see three reasons for this: (1) Scientific data analysis systems are complex, and object-oriented code is harder to develop than 'procedural' code. These two characteristics collide. The rewards of object-oriented code come in the long term, but short-term considerations often prevail in scientific software development. (2) The vendor of IDL, the main data analysis programming language in solar physics, only recently introduced the object-oriented software syntax, delaying the build-up of experience in this domain. (3) There is a large amount of legacy, non-object-oriented software, available. There is a (false) tendency to think that object-orientation prevents the reuse of this existing code.

Object-orientation does not replace procedural programming, it extends it. Objects are containers for the data types used in an application and for the operations – also called methods – that apply to them (Gamma *et al.*, 1995). The data within an object can only be accessed through a limited number of access methods, which define the interface for that object. This is the principle of *encapsulation*: all operations on a data type are associated with the object that contains this data type. Furthermore, new objects can extend already existing objects. This is the principle of *inheritance*: an object that extends another object inherits all its definitions, i.e., the extended object can use all data types and methods defined in the basis object.

### 3.2. Object-Oriented Design Applied to RHESSI

The RHESSI software was built to take advantage of the many strengths of object-oriented programming. One drawback is the current unfamiliarity of the syntax to many members of the user community. However, subsequent experience has shown that this is quickly overcome and of course it does not apply at all to those using the RHESSI graphical user interface. Other drawbacks for developers unfamiliar with the programming techniques are discussed in Section 6.

The advantages gained include minimizing the amount of processing required by keeping track of and reusing intermediate products, shielding the user from internal data analysis details (although all intermediate products are available), reusing of code for generic object-handling functions, and supporting both interactive and batch processes easily. The next section describes these advantages more fully. All of these features could be implemented with standard procedure-driven code, but it would be clumsier, require more coding and testing, and probably

TABLE I
A list of the main RHESSI data products.

| Type | Object name | Product |
|---|---|---|
| Main RHESSI data products | hsi_image | Images |
| | hsi_lightcurve | Lightcurves |
| | hsi_spectrum | Spectra |
| Data products associated with a specific image algorithm | hsi_clean | Images processed by the CLEAN algorithm |
| | hsi_forwardfit | Images processed by the Forward Fitting algorithm |
| | hsi_pixon | Images processed by the Pixon algorithm |
| | hsi_mem_sato | Images processed by the Maximum Entropy algorithm (MEM SATO) |
| | hsi_memvis | Images processed by the Maximum Entropy algorithm using visibilities |
| | hsi_bproj | Back-projection images |
| Intermediate data products | hsi_psf | Point spread function |
| | hsi_modul_profile | Modulation profiles |
| | hsi_modul_pattern | Modulation patterns |
| | hsi_calib_eventlist | Calibrated event lists |
| | hsi_binned_eventlist | Binned event lists |
| | hsi_eventlist | Event lists |
| | hsi_packet | Telemetry packets |

burden the analyst with keeping track of parameter changes and how that affects the reuse of intermediate products.

Now we discuss how we apply the concepts of objects to accomplish the goals of the RHESSI software. There are three top-level objects to handle the three primary RHESSI data products – images, spectra, and lightcurves. These three objects employ a set of intermediate data products for each step of the processing cycle: data retrieval, accumulation, time binning, energy binning, aspect calculation, image reconstruction, etc. These data products are listed in Table I. Each data product is associated with an object. Thus, there is a packet object, an event list object, etc. These intermediate objects are created when necessary and stored within the three primary objects. The RHESSI data analysis software can be considered as a chain of objects. To arrive at the final product, a chain of transformation of data products occurs.

Every object in the chain has identical access methods: a *Set* method for setting parameters in the object, a *Get* method for extracting parameters from the object, and a *GetData* method for processing and retrieving the primary data of the object.

A standard interface object called Framework was written to handle these methods so that access to every object (and hence data product) is the same. The framework object is reused (i.e., inherited) by each individual data product object to handle the generic parts of the access methods. The data product-dependent code (the transformation algorithm itself) is embedded into the process method of the specific object.

First, we will outline the steps an analyst would take to create an object and retrieve data from it, and then we will discuss the operation of the objects. We will use image reconstruction as an example (lightcurve and spectrum generation are similar). We proceed as follows:

(1) An instance of an image object is created. This creates a reference (a variable name) to an object and initializes all of the control parameters in the object to default values. At the SSW IDL command prompt, the command used is

```
obj = hsi_image()
```

(2) Parameters that need to be changed from the default are set via the *Set* method. We will usually want to set values for the time interval, energy range, spatial resolution, and position of the field of view on the Sun. For example, to set the position of the field of view, the command used is

```
obj − > Set, xyoffset = [−900., −250.]
```

(3) The image that corresponds to the current setting of the parameters is retrieved using the *GetData* method. If necessary, the image is processed, following the data transformation chain mentioned above.

```
data = obj − > GetData()
```

(4) Parameters that give information about what happened during image generation can be retrieved using the *Get* method:

```
params = obj − > Get()
```

There are two general classes of parameters – control parameters that the user sets to specify the data requested, and information parameters that are set by the object to show what happened during the processing of the data. Any of these parameters can be retrieved using the *Get* method. An important feature of getting and setting parameter values in RHESSI objects is that because objects are chained together with framework, the parameters can be set and retrieved from any object in the chain. For instance, we can set the image dimension (associated with the modulation pattern object) in the same way that we set the time range (associated with the event list object), both with the *Set* method of the image object.

Now we return to the chain of transformations of data products initiated by the *GetData* method. In this chain, telemetry packets are transformed into event lists, which are transformed into binned event lists, which are transformed into

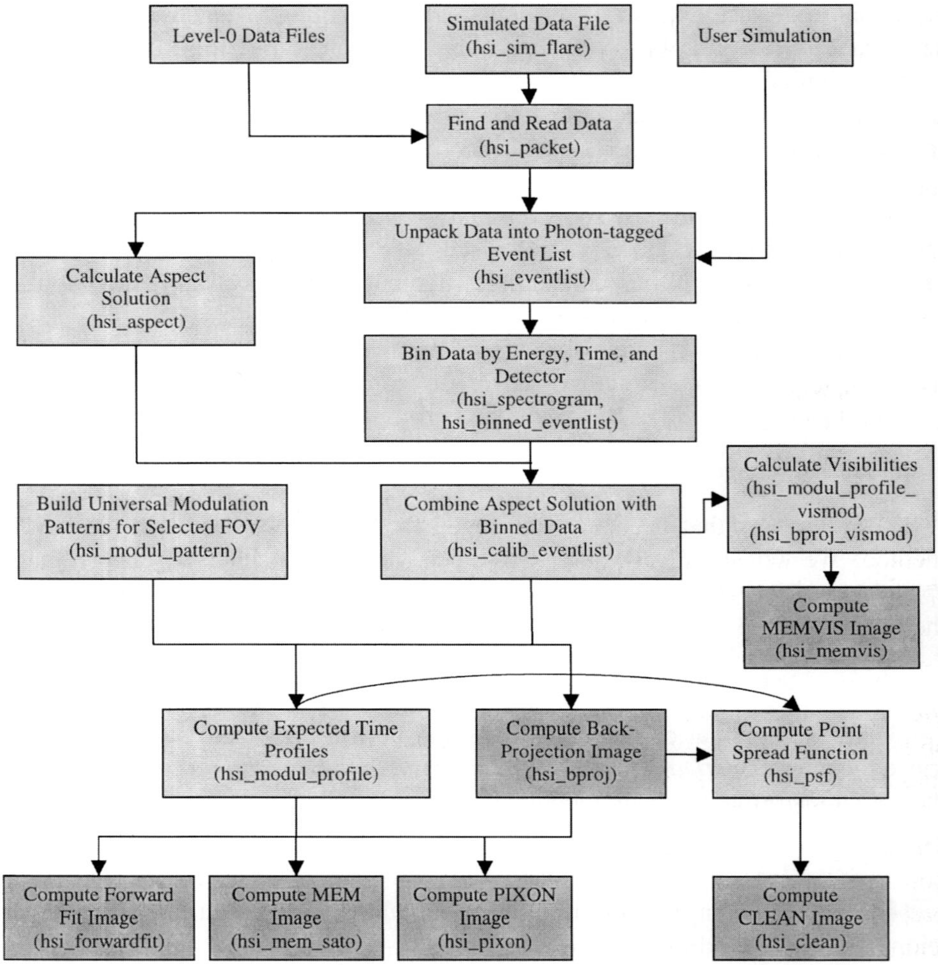


*Figure 1.* Image creation diagram. This figure shows the chain of transformations for an image reconstruction. The name of the object (or in some cases, program) corresponding to each task is shown in parentheses. The final products are in the darker gray boxes.

calibrated event lists, back-projections after including the modulation patterns, and finally images. This chain is shown in Figure 1.

When an image is requested, the image object checks whether it already contains an image corresponding to the current parameter setting. If it does, this image is returned to the user without any further processing. If it does not, the object needs to process a new image. To do this, it relies on several intermediate data objects. One way an image can be processed is by summing up back-projection maps, the basis of the RHESSI image reconstruction method (Hurford *et al.*, 2002). Thus, the image object relies on a back-projection object, which is responsible for building back-projection maps. Other objects of interest for the image object include those implementing the different image algorithms (Forward Fit, MEM, Pixon, Clean, MemVIS), as shown in Figure 1.

The back-projection object now proceeds in a similar fashion. It first checks whether it contains back-projection maps corresponding to the parameter settings. If it does, these data are returned to the image object without further processing. If it does not, the object needs to process new back-projection maps. Back-projection maps are processed by building a set of modulation patterns weighted by the binned photon counts. Thus, the back-projection object relies on a modulation pattern object, which is responsible for processing the modulation patterns specific to a spatial and spectral resolution.

This process continues along the chain of objects, from the image to the Level-0 data files, with each object returning either previously-processed or newly-processed data products as required.

As for the access interface, the framework in which these transformations occur is the same for all steps in the chain. It is 'only' the data product in consideration that changes. Thus, the Framework object implements the generic part of the software (therefore, the name Framework). Extensions of the Framework implement the data product-dependent code.

### 3.3. Benefits of object-oriented design for RHESSI

We believe that the data analysis software is more structured, efficient, maintainable, and extendable by using object-oriented techniques than it would be with traditional procedural software for the following reasons:

*Minimized reprocessing:* The Framework object also implements the reprocessing policy. Whenever a parameter is set, the Framework checks the parameter value passed. According to the reprocessing policy, the Framework decides whether the data already generated in the object are consistent with the parameter settings. If not, the Framework sets a flag to regenerate the data at the time of the next request. This strategy allows not only for updating the data only when it is necessary, but also for waiting until a request comes to generate any new data product.

*Shielded details:* Shielding the user from the burdensome details of data analysis is accomplished by enabling the user to access data in a simple and obvious way. Access to a data product should involve as few commands as possible and the analyst should not have to be concerned about the intermediate data products. Objects provide the necessary tools by allowing the simplest user interface one can imagine. Not only are the *Set, Get* and *GetData* access methods available in all objects, all objects use the same implementation of those methods. This technique allows a straightforward access to virtually any byte in the data, while keeping the interface in its simplest form. And though the data analyst is shielded from many of the specific details, every detail of every process can be examined. All of the intermediate data products can be examined, and due to the interactive nature of IDL any line in any procedure can be found and examined during processing.

*Support of interactive and batch processing:* Using objects also allows us to accomplish our goal to support both interactive and batch processing easily. The access interface described above can clearly be used for interactive processing in IDL's command-line interface (CLI) and in batch processing through the use of scripts. In addition, we have built a graphical user interface as a layer over the CLI. GUI development was significantly simplified because of the object design of the underlying software. With objects, parameter values and data products are accessed through a single object reference. Within the GUI, the object reference is used to pass parameters and data to individual widgets. Thus, widgets specialize in presenting options for the data analysis task, and delegate data and parameter management to the objects. This architecture allows a clear separation between the widget-based event processing and the data processing. This, on the one hand, avoids duplicating parts of the software, and, on the other hand, makes the localization of software bugs easier; hence, it contributes significantly to the robustness of the system. It also contributes to the power of the GUI by giving every widget access to the entire functionality of the data analysis software.

In addition to the support of our primary objectives, objects also provide support for further development and management of the software during the mission. The object-oriented design is an adequate response to issues in reuse, maintenance, extensions, and combination with other software packages.

*Software reuse:* The definition of a single interface for accessing any data product provides a strong potential for reusing code. The management of data products (getting and setting parameters, reuse policy) is written only once, in the Framework object.

*Software maintenance:* The principle of encapsulation allows us to locate bugs efficiently and correct the software with minimal effort. Furthermore, since changes remain localized, bug corrections in one area will not generate an avalanche of other changes in the software. Finally, specific objects can be assigned to specific programmers and the interactions in developing software are considerably easier to manage. This is an important consideration when a software team is spread around the world.

*Software extension:* Here again, encapsulation helps. New functionality for specific data products can be added by adding new methods to the objects. In addition, completely new objects can be added with relatively few changes in other parts of the software.

*Combination with other software packages:* Object-oriented software and procedural software are compatible. The RHESSI Graphical User Interface (GUI) is an example; the GUI is not object-oriented itself, but relies on object-oriented software. Conversely, an object-oriented interface can be built on top of legacy (procedural) software. This allows the definition of higher-level objects that can be designed to combine data from several instruments. Software development using this approach includes the Map Objects (Appendix Item 13) and is being considered for new projects such as the Virtual Solar Observatory (Hill *et al.*, 2002).

## 4. Using the RHESSI Data Analysis Software

Analysts use the RHESSI software either by typing commands through IDL's command-line interface (CLI), or by entering the RHESSI GUI. Using the CLI, analysts create and manipulate the RHESSI objects directly. The CLI offers the full power of the objects, but requires the analyst to be familiar with IDL syntax and RHESSI procedure and parameter names. The GUI shields the analyst from those kinds of details and makes it easy to use most of the functionality available in the objects, but has its own limitations.

### 4.1. Using the RHESSI Objects from the CLI

As mentioned above, there are three main RHESSI objects that provide access to the main RHESSI data products – images, spectra, and lightcurves. They are listed along with their intermediate data products in Table I and are discussed more fully in Sections 5.1 and 5.2.

Another object is provided to access the observing summary and quicklook data. This object and its underlying objects are listed in Table II, and discussed in more detail in Section 5.3.

Because of the structure of the RHESSI objects, every object in either list can be used directly *or* as an invisible part of the chain of objects needed to produce a data product. To use an object directly it can be created interactively by the user as a stand-alone object, or it can be extracted from a higher-level object. The discussion that follows demonstrates the types of operations that can be used on any of these objects.

#### 4.1.1. *Initialization and Methods*

Every object is created with the command obj = hsi_objectname() and has a *Set, Get*, and *GetData* method as discussed earlier. Many objects have additional methods that are specific to the object. For example, most of the top-level objects and some of the underlying objects have a *Plot* method. Appendix Item 2 provides a complete list.

#### 4.1.2. *Object Parameters*

Every object has control and information parameters that are specific to it (Appendix Item 7). As mentioned earlier, the user sets control parameters to specify the data requested; the object sets information parameters to show what happened during the processing of the data. During initialization of the object, all control parameters are set to default values. Users only need to set parameters they want to be different from their default value. When using objects that are part of a chain of objects, users do not need to be concerned about which particular object a parameter applies to. The Framework manages setting the parameter in or getting the parameter from the correct object in the chain. Any number of parameters may be set or retrieved in one command. Some examples of setting and getting parameters follow:

```
obj -> set, obs_time_interval=['29-mar-02 21:20','29-mar-02 21:30'],
     energy_band=[6.,12.]

params = obj -> get(/time_range, /energy_band, /pixel_size)
```

Setting parameters can be combined with initialization of the object, or with *GetData, Plot*, or other method calls. When any method is called, it first calls the *Set* method to set any object parameters that were passed in the call, then retrieves whatever data are necessary to accomplish the function of the call.

#### 4.1.3. *Keywords*

In addition to control parameters, control over the objects is implemented by using keywords in the method calls. Keywords allow us to refine the method calls. It is important to note the distinction between keywords and control parameters, because they are syntactically similar. Keywords apply only to the immediate method call – they do not persist in the object; control parameters are stored in the object and, once set, persist until changed. Some keywords are data product-specific, others are generic. Refer to Appendix Item 2 for a complete description of available keywords. For instance to retrieve all of the information parameters from an object, we use the **INFO_ONLY** keyword:

```
all_info_parameters = obj -> get(/info_only)
```

or to retrieve a specific parameter, we can specify that parameter as a keyword. For example, to retrieve the control parameter specifying the image dimensions we want in our reconstructed image, we type:

```
image_dim = obj -> get(/image_dim)
```

### 4.2. Graphical User Interface

We have constructed a graphical user interface (GUI) as a layer on top of the objects (Appendix Item 8). The GUI provides an easy way to use the RHESSI software without requiring any knowledge of the RHESSI data objects, or the IDL command names and syntax. To start the GUI, we simply type *hessi* at the SSW IDL command prompt. The main GUI window is essentially a control panel – it provides entry points into more specialized widget interfaces to access the various forms of RHESSI data, while managing the storage, display, and user interaction with plots of retrieved data. It also provides access to the Synoptic Data Archive (see Section 5.4), and manages overlaying those data with RHESSI data.

Because the GUI is built around the RHESSI objects, it has access to the entire functionality of the data analysis software. The GUI manages the user input to the objects and the data extraction from the objects, but all of the data processing takes place within the objects. Therefore, there is no duplication of code, and there should be no difference in results obtained through the GUI and command line.

There are some limitations of the GUI: (1) While the GUI can *potentially* access all the functionality of the objects, it is limited in practice to the options programmed into each widget interface. All of the primary options are currently available, and with time, as we learn which additional options analysts use most frequently, we will add new features. (2) There is currently no scripting capability available in the GUI. Analysts must enter repeated sequences by clicking buttons every time.

A powerful feature of the GUI that allows the GUI and command line to share objects mitigates the first limitation. A combined approach of using the GUI for operations most easily done through the GUI and using the command line for purposes that are more esoteric offers endless possibilities. We will address the second limitation in the future.

The GUI employs a Plot Manager object (Appendix Item 9) written for the RHESSI project that has potential for general usage. The Plot Manager object provides most of the generic plot-handling features of the GUI. All of the different data types retrieved in the GUI (spectra, time plots, images, contours, etc.) are registered in the Plot Manager object, which then uses the same code, regardless of the origin of the data, to allow the user to interactively manipulate the plot, print it, save it in a plot file, etc. This code is not RHESSI specific, and could be applied to data from other missions.

## 5. Elements of RHESSI Data Analysis Software

Analysts can access all of the data elements discussed below through the IDL CLI or through the RHESSI GUI. Refer to Appendix Items 2 and 7 for a table of methods that apply to each object and the parameters that apply to each object.

### 5.1. Images and Imaging Spectroscopy

Appendix Items 3 and 5 provide step-by-step guides to RHESSI imaging. There are six image reconstruction algorithms available: Back-Projection, Clean, Forward Fit, Maximum Entropy, Maximum Entropy with Visibilities, and Pixon. Hurford *et al.* (2002) discuss these algorithms in detail. Figure 1 shows the chain of transformations required to produce an image with each algorithm starting with the Level-0 data files or a user simulation.

We recall the example given in Section 3.2 to generate an image. Here in addition we set the time range and use the plot method to display the result:

```
obj = hsi_image()
obj -> set, time_range= ['2002/7/23 00:30:00','2002/7/23 00:30:04'],
        xyoffset= [-900, -250]
image = obj -> getdata()
obj -> plot
```

As shown in this example, setting just two parameters – the time range and the position of the field of view on the Sun – and leaving all other parameters at their default setting will produce a reasonable image. Many other parameters can be set however, such as image dimensions, pixel size, detectors, energy range, and image algorithm.

There is an important feature of the objects that we can take advantage of in imaging. Since the different algorithms are each implemented as their own strategy in the same object, the results from each algorithm are stored simultaneously in the object and can be retrieved easily or processed further. This allows us to compare images using different algorithms with full confidence that all other parameters in the object are identical.

There are several tools for performing feature-based imaging spectroscopy. The GUI has options to specify multiple time and energy bins and produce a 4-dimensional image cube FITS file. Once we have created the image cube file, there are several options for examining it: (1) The GUI has an interface for selecting regions of interest and computing quantities such as the flux, centroid, and peak through the image cube. (2) The GUI has an option to show the image cube as a movie either in the time or energy dimension. (3) There is a standalone program called IMSPEC, which computes spectra through the cube for regions of interest. Imaging spectroscopy capabilities within the RHESSI objects are under development.

### 5.2. Spectra and Lightcurves

Smith *et al.* (2002) discuss RHESSI spectroscopy in detail. Figure 2 shows the chain of transformations required to produce a spectrum or lightcurve starting with the Level-0 data files or a user simulation.

The lightcurve object is based on the same objects and methods as the spectrum object; in fact, the lightcurve object *inherits* the spectrum object. The main difference is that the data returned by the *GetData* method are ordered by time rather than energy.

The RHESSI software offers complete flexibility in selecting energy and time bins for spectra and lightcurves. The bins can be of fixed or arbitrary size.

An important feature of the RHESSI software is that an event list of any size can be processed; the computer should not run out of memory. This means the user can select any duration, even during periods of high photon counts, for accumulating spectra. This is accomplished by reading the data in bunches (a fixed number of data packets), and reusing the memory after processing a bunch. The only limitations are therefore the final size of the spectrum array (which depends on the number of energy and time bins) and the amount of time it takes to read and process the data (which can be considerable).

Normally the spectrum object returns a count rate spectrum. There is also an option to return a 'semi-calibrated' spectrum. A semi-calibrated spectrum does not include a background subtraction and uses only the diagonal elements of the response matrix. It is not a complete solution, but gives a first-order approximation below 100 keV. Another element is needed for complete analysis of the spectra – the spectral response matrix (SRM). There is an SRM object incorporated into the spectrum object that offers varying levels of complexity in computing the SRM, ranging from just the diagonal elements to the full response matrix including all off-diagonal elements.

Using the *FileWrite* method in the spectrum object, we can export the accumulated spectra and the SRM to FITS files. This serves two purposes – it saves results that may have taken a long time to accumulate, and it is the first step in performing complete spectral analysis. Currently we do not fully analyze spectra in the RHESSI object software (and hence the GUI). After exporting the data to FITS files, we proceed with external spectral analysis software to produce more accurate photon spectra and compute best-fit function parameters to the spectral data. Implementing spectral analysis within the object environment (and hence the GUI) is under development.

The external packages we use to analyze spectra are SPEX (see Smith *et al.* (2002) for a description of the process) and XSPEC (Arnaud, 1996). SPEX is a multi-mission SSW/IDL tool written by Schwartz 8 years prior to the RHESSI mission; XSPEC is a FORTRAN-based package written by the Laboratory for High Energy Astrophysics at GSFC. Both packages provide an interface suitable for X-ray spectral analysis of data from a number of astronomical instruments. The RHESSI spectra and SRM are imported into SPEX via FITS files we create from the RHESSI objects. We have not yet built the software to complete the conversion of the spectrum and response-matrix files to XSPEC formats, but this is straightforward using the tools in SSW and could be done on demand. Appendix Item 4 provides a guide to getting started with SPEX.

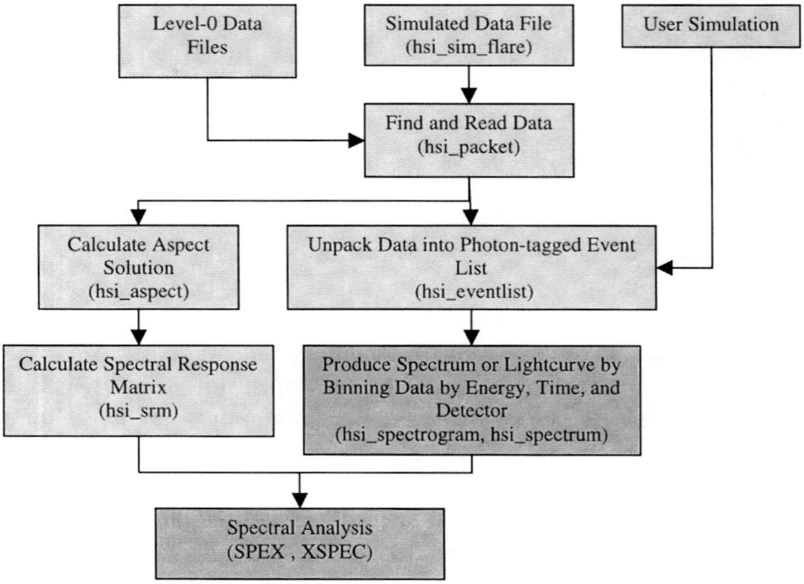


*Figure 2.* Spectrum and lightcurve creation diagram. This figure shows the chain of transformations for calculating a spectrum or lightcurve.

An example of accumulating a spectrum and writing spectrum and SRM FITS files follows:

```
obj = hsi_spectrum()
obj -> set, obs_time_interval='23-jul-2002'+[' 00:10:00', ' 01:20:00']
obj -> set, sp_energy_binning=14, sp_time_interval=10.
obj -> filewrite, /buildsrm
```

### 5.3. Observing Summary Including Quicklook Products

The observing summary consists of the Level-1 data products listed in Table II. Each type of data in the observing summary is stored in a separate binary extension of a FITS file (originally the Level-0 FITS files, but as of September 2002, a separate daily catalog file). Refer to Appendix Item 6 for a complete description of the contents of each observing summary data type, and Appendix Item 10 for documentation on using the hsi_obs_summary object.

### 5.4. RHESSI Synoptic Data Archive

The RHESSI Synoptic Data Archive (hereafter referred to as the Synoptic archive) has its origins in the SOHO Synoptic Archive that was constructed to meet the

TABLE II

A list of the Observing Summary and Quicklook Data Products.

| Type | Object name | Product |
|---|---|---|
| Main object | hsi_obs_summary | All obs. summ. data |
| Individual Data Products | hsi_obs_summ_rate | Observing summary rates |
| | hsi_mod_variance | Modulation variance |
| | hsi_ephemeris | Spacecraft position |
| | hsi_qlook_pointing | Quicklook pointing |
| | hsi_qlook_roll_angle | Roll angle |
| | hsi_qlook_roll_period | Roll period |
| | hsi_obs_summ_flag | Obs. summ. flags |
| | hsi_flare_list | Flare list |
| | hsi_qlook_image | Quicklook images |
| | hsi_qlook_spectra | Quicklook spectra |
| | hsi_qlook_summary_page | Summary page |
| | hsi_qlook_monitor_rate | Quicklook monitor rates |
| | hsi_qlook_packet_rate | Quicklook packet rates |
| | hsi_qlook_soh | Quicklook state of health data |

planning and joint observation requirements of experiments onboard the SOHO mission. The SOHO archive is a central database of selected datasets from cooperating space- and ground-based observatories. The database is updated daily and is accessible via Web and IDL interfaces developed and maintained at the Solar Data Analysis Center (SDAC). The RHESSI Synoptic archive is an extension of the SOHO archive to include concurrent observations of RHESSI-observed solar flares at different wavelengths. (Appendix Item 12.)

#### 5.4.1. *The Synoptic Archive*

The purpose of the Synoptic archive is to complement RHESSI observations by providing additional context information (e.g., magnetic morphology and geometry) and physical parameters (e.g., temperature, density, velocity, etc.) that cannot be obtained directly from RHESSI observations. Synoptic archive datasets are selected according to the following requirements. They must

- overlap spatially and temporally with RHESSI-observed flares;
- include different datatypes (image, spectra, lightcurves, spectroheliograms);
- cover a wide range of wavelengths (optical, radio, EUV, soft X-ray);
- be in a standard data format (e.g., FITS);
- be available online for easy access.

Table III lists datasets and their corresponding sources that have been identified to meet the above requirements.

TABLE III

Synoptic archive datasets that complement RHESSI observations. SXI observations are expected to become available January 2003.

| Datasets | Source |
|---|---|
| Magnetogram (including vector) | Mees, NSO, BBSO, MSFC, SOHO/MDI |
| H-alpha imaging/spectroheliogram/polarization | Mees, Meudon |
| Optical imaging/spectra | Mees, Meudon, NSO, BBSO, Kanzelhöhe, Kiepenheuer |
| Radio imaging/spectra/lightcurves | OVSA, Nobeyama, Phoenix, Nancay |
| Soft/hard X-ray/EUV imaging | Soft X-ray Imager (SXI), SOHO/EIT, TRACE |
| Soft/hard X-ray lightcurves | GOES, HXRS (Czech) |

The task of obtaining and disseminating Synoptic datasets is performed by mirroring data files from remote sites around the world to a central server at the SDAC where they are made available for searching, browsing, and download (via ftp or http) to individual user systems. In order to facilitate the mirroring of datasets, each site is required to store its data in FITS-compliant files using date-based naming conventions. For example, a Big Bear Solar Observatory H-alpha image file with an observation start date of 01:00 UT on 12-December-2002 would have the filename: *bbso_halph_20021202_00100.fts*. To facilitate mirroring, remote files are stored on an anonymous FTP server in directories that are organized by date.

#### 5.4.2. *The Synoptic Archive Software*

The bulk of the original IDL software that supports the Synoptic archive was written several years prior to the launch of RHESSI. The software is fully integrated into SSW and has been tested on current versions of Unix and Windows operating systems using the most recent version of IDL. The software consists of general routines to read FITS files, specific readers for particular instruments (e.g., SOHO/Extreme Ultraviolet Telescope [EIT]), and tools for coaligning and displaying images.

The task of integrating the Synoptic and RHESSI data analysis software systems has been guided by the requirement that the two systems be capable of passing data and relevant parameter information in a seamless fashion. Why is this requirement important? Consider the typical task of overlaying a reconstructed RHESSI image on a ground-based H-alpha image or magnetogram to identify sites of thick-target electron precipitation. This task is accomplished by first creating a RHESSI image using available techniques, finding the relevant H-alpha or magnetic dataset that is nearest in time and spatial location to the RHESSI event, and finally reading and plotting the two images on a common spatial scale.

A typical user cannot be expected to know all the details of module names and procedure calls for different Synoptic datasets in order to perform the above overlay task easily. The solution that we have adopted is to develop object-oriented IDL wrappers for each different dataset stored in the Synoptic archive. For example, the SOHO/EIT data reader is embodied as a read method of an object class named ‘EIT’, while a SOHO/MDI data reader is embodied as a read method of an object class ‘MDI’. The advantage of this formalism is that different synoptic data types can be read into memory using the same *Read* method name. To enhance the utility of synoptic data objects further, we have developed *Set, Get*, and *GetData* interface methods analogous to those of RHESSI objects. With these methods, one can retrieve the underlying image data and associated parameter information such as solar pointing, observation time, etc. In addition to image datasets, we have developed corresponding object-oriented wrappers to lightcurve datasets such as GOES soft X-ray, and *Phoenix*, OVSA, and RSTN radio observations. The lightcurve objects use the same interface method names as RHESSI and, accordingly, can be used to overlay with RHESSI lightcurve observations.

The development of IDL object wrappers for Synoptic archive datasets and software has resulted in several benefits. First, it has minimized the development of new software. Existing procedural code that has already been written and debugged for particular synoptic datasets can be reused as methods of the corresponding object. Second, because they share similar interface method names, synoptic data objects have the same ‘look and feel’ as RHESSI data objects. Consequently, RHESSI analysis software such as the GUI can conveniently operate on different synoptic datasets using the same method calls as RHESSI objects.

## 6. Discussion

In the previous sections, we have described the RHESSI data analysis software in terms of its successes. Real software development projects encounter problems, especially when new software methods are employed. In this section, we discuss some of these problems in an effort to help development teams for other projects plan their strategies, as well as to preview some of the improvements we are bringing to our own software.

### 6.1. Management of the Software Development

To meet the software challenges posed in Sections 1 and 2.2, we employed object-oriented design methods. Nevertheless, this solution had its own costs, most of which were related to IDL’s meager support for objects, and the software team’s lack of expertise with objects.

IDL’s support for objects is mainly syntactic, so many utilities were needed for our chain of RHESSI objects to function optimally. Consequently, the Framework and other management objects were being developed and enhanced throughout the software development period, sometimes hindering progress in other areas.

Initially, only one member of the software team, A. Csillaghy, was versed in object-related methods and philosophy so all of the object-specific development flowed through him. There was clearly a learning curve as additional team members became familiar with some of the principles so they could successfully code and debug modules without his assistance.

An area where the lack of experience hurt was in defining the correct object structure and determining when objects should be used instead of ordinary procedures or functions. As an example of the former, we have found in hindsight that the SRM (spectrum response matrix) object and the SPECTRUM object have the wrong relationship in the processing chain. As written, any change in SRM parameters would cause reprocessing of the SPECTRUM object, when in fact the two objects are mostly independent. We have modified the SPECTRUM object's reprocessing policy by writing special cases to prevent the extra reprocessing; however, if the objects were properly related that would not be necessary.

The example above illustrates the need to build software not from a functional point of view but from a data-modeling point of view. We believe that the best way to design software is by first understanding the data products that are involved at every data processing step. Defining the correct data products makes it possible to infer the software functionality from the data. Nevertheless, many software developers already think in terms of data products, but program in a procedural way. For them, the shift to object-oriented programming is reduced to the adoption of a few syntactic structures that will *help* them in their development, and increase their efficiency. Had we understood this at an earlier stage our software development would have been faster and more robust.

Programmers also need experience to decide when an object class is required to supplant procedural methods. For this we have found that a figure of merit is obtained in the product of the number of times a result is accessed by the time it takes to produce a result. As that product becomes a larger proportion of the overall processing time, an object with clear rules that determine reprocessing should probably be utilized. We have identified several procedures within the SPECTROGRAM classes related to energy and time binning that would perform better using our Framework object.

For future projects requiring complex data analysis methods, we anticipate that object methods will become a necessity rather than a choice. In this context, our experiences should prove educational. Studying several concepts and techniques used in our software might be beneficial: (1) Our basic template for object development (Appendix Item 1) should be examined for its simplicity and reusability. Using that template, a simple processing chain could be built to gain experience. (2) We use several techniques to encode variety and exceptions within the object classes. For example, we use the STRATEGY_HOLDER class to achieve differentiation in our image processing. This technique allows us to process images using different strategies (image algorithms) but with identical input and output structures, and with the maximum reuse of intermediate data products. Another example is the variety achieved in the SPECTROGRAM objects by using a post-processing routine. The SPECTROGRAM objects (SPECTRUM, LIGHTCURVE, and BINNED_EVENTLIST) start with identical data and are processed through the inherited SPECTROGRAM process method, but then each apply their own additional processing routines. These techniques are based on standard software engineering methods, the design patterns, described by Gamma *et al.* (1994). Our innovation is in applying them to scientific data analysis. We believe that design patterns have a significant potential of application in scientific data.

### 6.2. Lessons Learned Post-Launch and Anticipated Changes

How well were our pre-launch expectations for our software and the deployment of our databases met? As expected, the RHESSI hardware behaved somewhat differently in space than on the ground. The main unanticipated problems that required immediate software solutions were data dropouts (Smith *et al.*, 2002) and the need to use the PMTRAS data for roll aspect (Hurford and Curtis, 2002), both of which required non-trivial solutions. In addition, our original plan for organizing the data files was flawed. We originally attached Level-1 Quicklook Products (QLPs) to the Level-0 FITS files. This meant that any change in the QLPs necessitated rewriting the entire Level-0 database, and propagating all of the rewritten files to mirror sites and individual local archives. The files are large and the time to regenerate and propagate them became prohibitive. As of September 2002, we are removing the QLPs from the Level-0 FITS files, and appending them instead to a much smaller daily catalog file.

It was necessary to address the issue of compute-time, particularly since users were starting from a photon-based data set and since many algorithms are intrinsically iterative. (Maximizing re-use of intermediate data products is valuable in this context.) Our initial efforts were directed at speeding the image reconstruction task, where one to two orders of magnitude were saved by algorithm innovations (Hurford *et al.*, 2002). Speed is not a dominant issue with the current package except for large flares where hundreds of millions of photons must be processed. We believe that we can significantly reduce the processing time needed to build the spectrograms by processing the event list once and saving the results in a file. The size of this file will depend on the flare X-ray intensity and the user-selected time range. Further processing in the same time and energy range will then use this file instead of the Level-0 file.

Simulation software for RHESSI played several roles during the development phase. Simulation output could be expressed either as an event list, a simulated Level-0 data file or as a series of timed 'photon' stimuli into the flight hardware. This permitted testing of data handling and image reconstruction software as well as end-to-end testing of hardware/software systems. After launch, simulation software is expected to provide a tool for evaluating the robustness of image reconstruction. Documentation on how to use the simulation tools can be found in Appendix Item 5.

The simulation software was based on many of the same geometry modules used for image reconstruction. While this enabled significant savings in the effort required to develop the simulation capability, it compromised the value of simulations for the discovery of some classes of 'geometrical' bugs since errors in simulation were exactly compensated during analysis.

Another area in which our preparations were incomplete is the tracking of instrument configuration changes. The software team was unaware that complete housekeeping information was not included in the telemetry stream, and consequently made no provision for compiling logs of changes in housekeeping information. We are now compiling these logs, but a better solution would have been to have this information encoded in telemetry.

A series of ongoing hands-on workshops provided a combination of presentations by software developers and hands-on experience for potential users of RHESSI software. This not only represented an additional channel by which to reach potential users, it also provided useful feedback to the software development team. The workshops were an integral part of our documentation strategy and provided the impetus for the creation for many of documents found on the RHESSI datacenter website. We believe these workshops to have been successful and to have enabled many scientists to begin scientific analysis with RHESSI data within a few weeks after the launch.

Finally, we expect to improve the RHESSI analysis software through the life of the mission. Improvements will be made to the structure of our objects, to the speed of our process methods, and to the quality of our scientific algorithms. The focus of our efforts will be, as always, to maximize the scientific return from the RHESSI project.

## Acknowledgements

We are grateful to M. Aschwanden, P. Bilodeau, A. J. Conway, M. Fivian, S. Krucker, T. Metcalf, E. Schmahl, J. Sato, and D. M. Smith for significant contributions to the RHESSI software package. We also thank R. Bentley, B. R. Dennis and C. Johns-Krull for their valuable insights and roles in documenting the software. Without contributions such as these, the package would not exist. We also thank R. P. Lin for his support in allocating sufficient resources and priority to this task. This work is funded by NASA grant NAS5-98033-05/03. A. Csillaghy is partially funded by the Swiss National Science Foundation (Grant No. 2000-061559) and UAS Grant No. FHA-02-07-007.

## Appendix

The main RHESSI Software Web site is located at http://hesperia.gsfc.nasa.gov/rhessidatacenter.

The following is a list of Internet documents referred to in this paper. The URLs given for the references are relatively stable, but in the nature of Web documents, could change. The CD included with this paper contains the top-level page of the current versions of these Web documents.

1. Csillaghy, A.: Object-Oriented Data Analysis Software Concepts
   URL: http://hessi.ssl.berkeley.edu/software/hessi_oo_concept.html
   A description of object-oriented concepts, especially as applied to HESSI Data Analysis.

2. Csillaghy, A.: RHESSI Objects Reference Manual
   URL: http://hessi.ssl.berkeley.edu/software/reference.html
   Tables of the classes, keywords, and methods used in the RHESSI object construct, as well as descriptions of each object class.

3. Dennis, B.R.: RHESSI Imaging – First Steps
   URL: http://hesperia.gsfc.nasa.gov/~dennis/imaging/first_steps.htm
   A beginner's tutorial-style guide to using the GUI to create images. Includes instructions for each step starting with generating a lightcurve through creating a back-projection image, and provides screen snapshots of the widget interfaces and the expected output.

4. Dennis, B.R.: RHESSI Spectroscopy – First Steps
   URL: http://hesperia.gsfc.nasa.gov/~dennis/spectroscopy/first_steps.htm
   A beginner's tutorial-style guide to using the GUI to generate spectra, and using SPEX to analyze spectra. Provides screen snapshots of the widget interfaces and the expected output.

5. Johns-Krull, C.M.: An Overview of the Command Line Interface for RHESSI Data Analysis Software
   URL: http://hessi.ssl.berkeley.edu/~cmj/hessi/doc.html
   A thorough tutorial-style guide to using IDL's command line interface to run the RHESSI software including imaging, spectroscopy, simulations, and access to quicklook data. Numerous examples demonstrate the syntax and commands to use and output to expect.

6. McTiernan, J.: RHESSI Quicklook Data Products
   URL: http://sprg.ssl.berkeley.edu/~jimm/hessi/hsi_obs_summ_soc.html
   A description of the data and info structures returned for each class of observ-

ing summary data.

7. Tolbert, K.: RHESSI Data Object Parameters
   URL: http://hesperia.gsfc.nasa.gov/ssw/hessi/doc/hsi_params_all.htm
   A table listing all control and information parameters for all RHESSI objects. Includes a brief description of each parameter, its default value, and the name of the object that uses it.

8. Tolbert, K.: HESSI GUI Guide
   http://hesperia.gsfc.nasa.gov/ssw/hessi/doc/gui_help.htm
   User's Guide to the HESSI Graphical User Interface.

9. Tolbert, K.: Plot Manager (PLOTMAN)
   URL: http://hesperia.gsfc.nasa.gov/ssw/hessi/doc/plotman_help.htm
   A user's guide to the Plot Manager, a generic plot-handling package.

10. Tolbert, K.: Using the Observing Summary and Quicklook Data
    URL: http://hesperia.gsfc.nasa.gov/ssw/hessi/doc/obs_summ_access.htm
    A guide to using the hsi_obs_summary object to retrieve and plot the observing summary and quicklook data.

11. Usenet: Object FAQ
    URL: http://www.cyberdyne-object-sys.com/oofaq2/
    A complete explanation of the concept of object-orientation, its background, why it should be used, and how to implement it.

12. Zarro, D. M.: The RHESSI Synoptic Data Archive
    URL: http://orpheus.nascom.nasa.gov/~zarro/synop
    A description of the RHESSI Synoptic Data Archive, an example of a database input and search engine, a tutorial on overlaying images from different sources, and a user's guide to reading and plotting GOES data.

13. Zarro, D.M.: IDL Software for Analyzing Solar Images
    URL: http://orpheus.nascom.nasa.gov/~zarro/idl/maps.html
    An explanation of map objects, how to create and manipulate them, and how to apply them to solar images.

## References

Arnaud, K. A.: 1996, in G. Jacoby and J. Barnes (eds.), *Astronomical Data Analysis Software and Systems V, ASP Conf. Series*, Volume 101.

Booch, G.: 2002, *Object-Oriented Analysis and Design with Applications*, 3rd edition, Addison-Wesley, New York.

Curtis, D., Berg, P., Gordon, D., Harvey, P. R., Smith, D. M., and Zehnder, A.: 2002, *Solar Phys.*, this volume.

Gamma, E., Helm, R., Johnson, R., and Vlissides, J.: 1995, *Design Patterns: Elements of Reusable Object-Oriented Sofware*, Addison-Wesley, New York.

Hill, F., Csillaghy, A. *et al.*: 2002, in A. Szalay (ed.), *EGSO in Need for a Global Schema, Proc. SPIE* **4846**, in press.

Hurford, G. J. and Curtis, D. W.: 2002, *Solar Phys.*, this volume.

Hurford, G. J., Schmahl, E. J. *et al.*: 2002, *Solar Phys.*, this volume.

Freeland, S. L. and Handy, B. N.: 1998, *Solar Phys.* **182**, 497.

Lin, R. P. *et al.*: 2002, *Solar Phys.*, this volume.

Schach, S.: 2001, *Object-Oriented and Classical Software Engineering*, 5th edition, McGraw Hill, New York.

Smith, D. M. *et al.*: 2002, *Solar Phys.*, this volume.

Zehnder, A. *et al.*: 2002, in Szalay (ed.), *Proc. SPIE* **4853**, in press.

# RECONSTRUCTION OF RHESSI SOLAR FLARE IMAGES WITH A FORWARD FITTING METHOD

MARKUS J. ASCHWANDEN[1], ED SCHMAHL[2] and the RHESSI TEAM[1,2,3]
[1]*Lockheed Martin ATC, Solar & Astrophysics Laboratory, Dept. L9-41, Bldg.252, 3251 Hanover St., Palo Alto, CA 94304, U.S.A. (e-mail: aschwanden@lmsal.com)*
[2]*NASA Goddard Space Flight Center, Solar & Astrophysics Laboratory, Greenbelt, MD 20771, U.S.A.*
[3]*Space Sciences Laboratory, University of California Berkeley, Berkeley, CA 94720; U.S.A.*

(Received 7 August 2002; accepted 19 August 2002)

**Abstract.** We describe a forward-fitting method that has been developed to reconstruct hard X-ray images of solar flares from the Ramaty High-Energy Solar Spectroscopic Imager (RHESSI), a Fourier imager with rotation-modulated collimators that was launched on 5 February 2002. The forward-fitting method is based on geometric models that represent a spatial map by a superposition of multiple source structures, which are quantified by circular gaussians (4 parameters per source), elliptical gaussians (6 parameters), or curved ellipticals (7 parameters), designed to characterize real solar flare hard X-ray maps with a minimum number of geometric elements. We describe and demonstrate the use of the forward-fitting algorithm. We perform some 500 simulations of rotation-modulated time profiles of the 9 RHESSI detectors, based on single and multiple source structures, and perform their image reconstruction. We quantify the fidelity of the image reconstruction, as function of photon statistics, and the accuracy of retrieved source positions, widths, and fluxes. We outline applications for which the forward-fitting code is most suitable, such as measurements of the energy-dependent altitude of energy loss near the limb, or footpoint separation during flares.

## 1. Introduction

Image reconstruction from RHESSI data can be performed with a number of methods, including *backprojection*, CLEAN, the *maximum entropy method*, *pixon mapping*, *polar mapping*, and *forward-fitting*. In this paper we describe the principles of a *forward-fitting algorithm* that is currently implemented in the software. We simulate RHESSI observations, perform image reconstructions, and compare the test results with the input models, in order to quantify the fidelity of the reconstructed images as a function of the count rate. These tests provide error bars on the accuracy of retrieved source positions, sizes, or relative displacements in energy and time.

This paper includes a brief definition of relevant instrumental parameters (Section 2), the parameterization of images (Section 3), a description of the forward-fitting code, numerical test results of image reconstruction from simulated data (Section 4), and a summary (Section 5). More detailed information to some sec-

*Solar Physics* **210:** 193–211, 2002.

CD ROM

tions is provided on the attached CD-ROM, also available at *http://www.lmsal.com/ ~aschwand/ eprints/2002_fwdfit.*

## 2. Instrument

### 2.1. Instrumental description of RHESSI

Technical descriptions of the RHESSI instrument can be found in Lin *et al.* (1993, 1994, 1998) and Dennis *et al.* (1996). The RHESSI webpages (*http://hessi.ssl.berkeley.edu, http://hesperia.gsfc.nasa.gov/ rhessidatacenter/*), contain the most up-to-date information. Here we summarize the instrumental specifications that are most relevant for image reconstruction. For a detailed description of the imaging capabilities of RHESSI, the grid characteristics, the modulation patterns, and the modulation profiles we refer to Hurford *et al.* (2002).

RHESSI employs the technique of Fourier-transform imaging with rotating modulation collimators, using a set of 9 bi-grid collimators with a grid period (pitch) ranging from 34 $\mu$m to 2.6 mm, separated by a distance of 1550 mm between front grids and rear grids. The grid pitch increases by a factor of $\sqrt{3}$ from one grid to the next coarser one, yielding a FWHM resolution ranging from 2.26$''$ to $2.26'' \times \sqrt{3}^8 = 183''$. This defines the range of spatial scales over which RHESSI is able to map. The full angular resolution of $\approx 2''$ is achieved at the lowest energies, up to $\lesssim 100$ keV. At higher energies, where the thickness of the tungsten grids (Nos. 2–9) or molybdenum (Mb) grid (No. 1) is not sufficient to completely absorb high-energy photons, the modulational efficiency is thus reduced, restricting the spatial resolution to the spacing of the thicker, fully-absorbing grids, which is $\approx 35''$ above 1 MeV. The angular coverage of an image is thus $2''-183''$, while the field of view includes the full Sun, as long as the spacecraft spin axis points within $\approx 12'$ of the Sun center. The temporal resolution for highest quality images is half a spacecraft rotation period (providing maximum uv-coverage in Fourier space), i.e., $\approx 2.0$ s, but 'snapshot images' with poorer uv-coverage can be obtained down to tens of ms. The pointing of the spacecraft is determined by the Solar Aspect System (SAS) which measures the direction to the Sun center to sub-second accuracy, and by two redundant systems that measure the roll angles to $< 3'$ – a CCD system and a PMT-based Roll Angle Aspect System (PMTRAS). Great care has been taken to warrant that the relative twist of the finest grid pair is maintained to within $1'$ by the metering tube structure. Spectroscopy of hard X-rays and gamma rays is achieved with internally segmented germanium detectors (GeDs), placed behind the bi-grid collimators, having diameters of 71 mm and a length of 85 mm. The germanium detectors are cooled down to 75 K, providing $< 1$ keV FWHM energy resolution at 10 keV, increasing to $\approx 5$ keV at 7 MeV (see Smith *et al.*, 2002).

### 2.2. Data structure of observations

RHESSI records every photon above a threshold energy of $\gtrsim 3$ keV as a time-tagged event up to a maximum rate of $\approx$ 50 000 $s^{-1}$, while attenuation and decimation schemes take place at higher rates to avoid pulse pileup effects. Thus, the telemetry data contain photon event information for each of the 9 detectors, tagged with a time accuracy of 1 $\mu$s (called Photon Event List).

For standard image processing, the time-tagged photon events are binned into time intervals that resolve the grid-induced modulation due to the spinning spacecraft. The number of modulations per rotation depends on the angular distance $r$ of the observed source from the spacecraft spin axis, divided by the angular pitch $p_l = 4.52''\sqrt{3}^{(l-1)}$ of the collimator ($l = 1, \ldots, 9$),

$$N_l^{\mathrm{mod}} = \frac{4r}{p_l} = \frac{0.88\, r''}{\sqrt{3}^{(l-1)}}\,, \quad l = 1, \ldots, 9\,. \tag{1}$$

For a maximum field of view of $r \leq 1°$ we record $N_1^{\mathrm{mod}} = 3185$ modulations for the finest grid ($l = 1$) and $N_9^{\mathrm{mod}} = 39$ modulations for the coarsest grid ($l = 9$). The required time binning to resolve at least 2 data points per modulation period is thus

$$\Delta t_l^{\mathrm{mod}} = \frac{T^{\mathrm{rot}}}{2N_l^{\mathrm{mod}}} = 0.6\left(\frac{r''}{3600''}\right)^{-1}\sqrt{3}^{(l-1)}\ \mathrm{ms}\,, \tag{2}$$

ranging from $\Delta t_1^{\mathrm{mod}} \gtrsim 0.6$ ms for the finest grid to $\Delta t_9^{\mathrm{mod}} \gtrsim 50$ ms for the coarsest grid, for a spacecraft spin period of $T^{\mathrm{rot}} = 4.0$ s. A typical binning of the data is done by the nearest numbers of powers of 2, i.e., the default number of binning intervals varies from 128 (Detector 9) to 8192 (Detector 1), and the corresponding time intervals range from 0.5 ms (Detector 1) to 32 ms (Detector 9). The values of the detector pitch angles $p_l$ and the default time binning $\Delta t_i$ are listed in Table I. Thus the typical data structure used for image processing consists of 9 arrays of *modulation profiles* $N_l^{\mathrm{obs}}(t_k)$ for the detectors $l = 1, \ldots, 9$ with variable number of time bins, $t_k =, 1, \ldots, N_l$, ranging from $N_1 = 8192$ to $N_9 = 128$, amounting to a total of $\sum_{l=1}^{9} N_l = 17,792$ data points per spacecraft rotation. This data set of binned photon counts, along with the associated aspect variables, instrumental transmission parameters, amplitudes and phases, livetimes and other quantities required for mapping, is called the Calibrated Event List in the RHESSI jargon.

## 3. Parameterization of Maps

The simulation of maps as well as the forward-fitting method require a parameterization of the geometry of source morphologies. Hard X-ray sources in solar flares are known to appear as (often unresolved) point sources, double sources, multiple sources, or parts of flare loops (Sakao, 1994; see also *Yohkoh/HXT Image*

TABLE I
RHESSI data structure.

| Detector No. $l$ | Angular pitch $p_l$ | FWHM resolution $\alpha_l$ | Modulation per rotation[a] $N_l^{mod}$ | Data time bins[b] $N_l^{bin}$ | Time bin $\Delta t_l$(ms) | Minimum count rate[c] $R_l^{min}$ ($s^{-1}$ SC$^{-1}$) |
|---|---|---|---|---|---|---|
| 1 | 4.52″ | 2.26″ | 3168 | $2^{13} = 8192$ | 0.512 | 900 |
| 2 | 7.84″ | 3.92″ | 1829 | $2^{12} = 4096$ | 1.024 | 450 |
| 3 | 13.57″ | 6.79″ | 1056 | $2^{11} = 2048$ | 2.048 | 220 |
| 4 | 23.5″ | 11.8″ | 609 | $2^{10} = 1024$ | 4.096 | 110 |
| 5 | 40.7″ | 20.4″ | 352 | $2^{10} = 1024$ | 4.096 | 110 |
| 6 | 70.5″ | 35.3″ | 203 | $2^9 = 512$ | 8.192 | 54 |
| 7 | 122.2″ | 61.1″ | 117 | $2^9 = 512$ | 8.192 | 72 |
| 8 | 211.6″ | 105.8″ | 68 | $2^8 = 256$ | 16.384 | 38 |
| 9 | 366.5″ | 183.2″ | 39 | $2^7 = 128$ | 32.768 | 17 |

[a]For a source with a distance of 1° from spin axis.
[b]Default values for time binning.
[c]Minimum count rate to produce detectable modulation, producing r.m.s. fluctuations of the modulation profiles that exceed the Poisson fluctuations.

*Catalogue* by Kosugi *et al.*, 1995, and Sato *et al.*, 1998). We adopt the philosophy that such maps can mathematically be represented as a superposition of a few elementary geometric shapes.

### 3.1. Geometry of Elementary Source Structures

The simplest shape is a point source, placed at location $(x_i, y_i)$, with a flux amplitude $f_i$. Every point source has also a finite width, regardless whether this width represents the physical extent of the source or the apparent size of an unresolved source, as rendered by the instrumental point-spread function. Thus, it is most natural to characterize a point source with a gaussian width $w_i$, so that its 2-dimensional distribution in a map is defined by 4 parameters $(x_i, y_i, f_i, w_i)$,

$$F(x, y) = f_i \ \exp\left[-\frac{(x - x_i)^2 + (y - y_i)^2}{2w_i^2}\right] . \quad (3)$$

In the next step, we generalize a circular gaussian geometry to an elliptical shape, which can be defined by two additional parameters, the eccentricity $e_i$ and the tilt angle $\alpha_i^{tilt}$. We define the departure from circularity by the ratio of the ellipse half-axis $w_i^x$ along the $x$-axis to the half-axis $w_i^y$ along the $y$-axis (Figure 1, top), and subtract the value of one to render the parameter zero for circular symmetry $(w_i^x = w_i^y)$, i.e.,

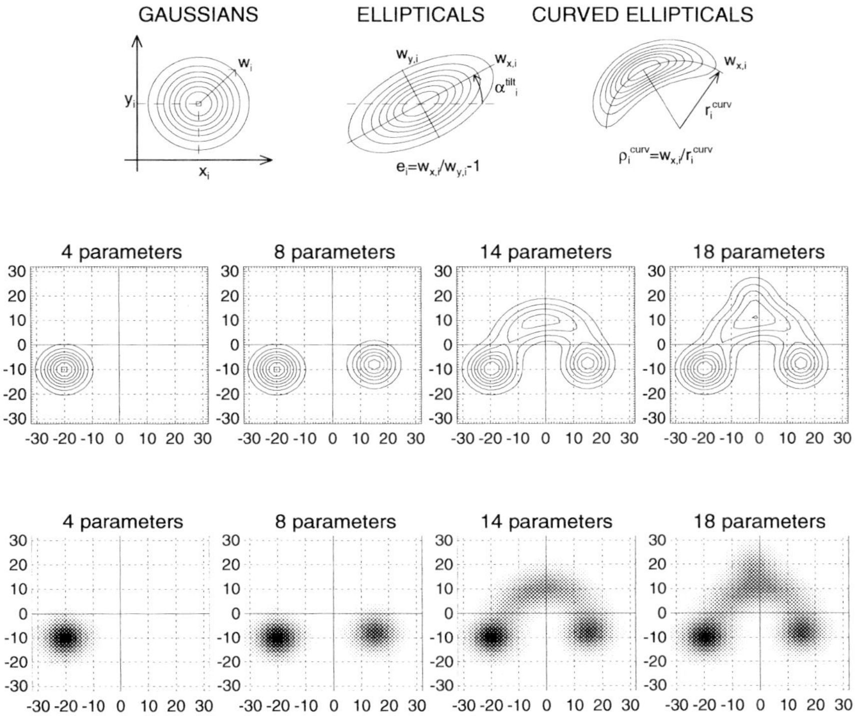


*Figure 1.* The parameterized model maps, using circular gaussians (4 parameters, $x_i$, $y_i$, $w_i$, $f_i$), elliptical gaussians (6 parameters, ... , $e_i$, $\alpha_i^{\mathrm{tilt}}$), and curved ellipticals (7 parameters, ... , $\rho_i^{\mathrm{curv}}$), are defined in the *upper panels*. The *middle* and *lower rows* show 4 examples with superpositions of the basic geometric elements defined above (with 1, 2, 3, 4 elements, requiring 4, 8, 14, 18 free parameters, respectively).

$$e_i = \left( \frac{w_i^x}{w_i^y} - 1 \right) . \tag{4}$$

The 2-dimensional distribution of an elliptical gaussian is then

$$F(x'', y'') = f_i \ \exp\left[ -\frac{(x'' - x_i)^2}{2w_i^2} - \frac{(y'' - y_i)^2}{2w_i^2}(e_i + 1)^2 \right] . \tag{5}$$

To obtain the 2-dimensional distribution of an elliptical gaussian in an arbitrary orientation, the coordinates $(x'', y'')$ aligned with the elliptical major and minor axes have to be rotated by the tilt angle $\alpha_i^{\mathrm{tilt}}$ (defined anti-clockwise with respect to the $x$-axis),

$$x = x'' \ \cos(\alpha_i^{\mathrm{tilt}}) - y'' \ \sin(\alpha_i^{\mathrm{tilt}}), \tag{6}$$

$$y = x'' \ \sin(\alpha_i^{\mathrm{tilt}}) + y'' \ \cos(\alpha_i^{\mathrm{tilt}}). \tag{7}$$

In order to provide geometric elements that can represent loop-like shapes we go a step further and introduce curved gaussians, which require only one additional parameter, the curvature radius $r^{curv}$. It is mathematically more convenient to choose the reciprocal curvature radius, because this way we have no singularity $r^{curv} = \infty$ for uncurved ellipticals or gaussians, and this additional parameter is then zero if no curvature is wanted. We define a dimensionless parameter $\rho_i^{curv}$ by the ratio of the gaussian width $w_i$ along the $x$-axis to the curvature radius $r_i^{curv}$,

$$\rho_i^{curv} = \frac{w_i}{r_i^{curv}} \, . \tag{8}$$

For coordinate transformations from curved ellipse coordinates $(x'', y'')$ into cartesian coordinates $(x, y)$, it is most convenient to use the rotation angle $\alpha^{curv}$ that measures the $x$-position on the curved $x$-axis with respect to the curvature center (Figure 1, top right),

$$\alpha^{curv} = \frac{x''}{r_i^{curv}} = \frac{x''}{w_i} \rho_i^{curv} . \tag{9}$$

The transformation from curved ellipse coordinates $(x'', y'')$ into cartesian coordinates $(x', y')$ is then simply

$$x' = (r_i^{curv} + y'') \sin(\alpha^{curv}), \tag{10}$$

$$y' = (r_i^{curv} + y'') \cos(\alpha^{curv}). \tag{11}$$

With a subsequent rotation by the orientation angle $\alpha_i^{tilt}$, the cartesian ellipse coordinates $(x', y')$ can then be rotated into the Cartesian reference system $(x, y)$ (by substituting $(x', y')$ from Equations (10) and (11) into $(x'', y'')$ of Equations (6) and (7)).

### 3.2. Map Parameterization

We have now defined the general geometry of a single source with 7 free parameters, which has the morphology of a curved gaussian in the most general form, but can be reduced to 6 free parameters (in the form of an elliptical gaussian without curvature), or to 4 parameters (in the form of a circular gaussian). The geometry of these three shapes can be described with the same general 7-parameter formalism $(x_i, y_i, f_i, w_i, e_i, \alpha_i^{tilt}, \rho_i^{curv})$, while setting a subset of the last three parameters to zero will automatically reproduce the simpler shapes.

We define a map simply by a superposition of the 2-dimensional distributions $F_i(x, y; P_i)$ of the elementary shapes, defined by the parameter sets $P_i = [x_i, y_i, f_i, w_i, e_i, \alpha_i^{tilt}, \rho_i^{curv}]$,

$$F(x, y) = \sum_{i=1}^{n^g} F_i(x, y; P_i), \tag{12}$$

with $n^g$ the number of gaussian sources (or elementary shapes). The total number of free parameters of such a parameterized map is

$$n^{\text{free}} = n^g \times n^{\text{par}} - 1, \tag{13}$$

where $n^{\text{par}}$ represents the number of parameters per source (i.e., $n^{\text{par}} = 4$ for circular gaussians, $n^{\text{par}} = 6$ for elliptical gaussians, and $n^{\text{par}} = 7$ for curved gaussians). The total number of free parameters is reduced by 1, because the total flux of the map can be normalized to unity $[I^{\text{norm}}(x, y) = I(x, y)/\int I(x, y)\,\mathrm{d}x\,\mathrm{d}y]$, having the physical unit of 1 photon count per map. The total number of photons can be obtained by summing the time profiles, $N_l^{\text{phot}} = \sum_{k=1}^{N_l} N_l^{\text{obs}}(t_k)$ for each detector $l = 1, \ldots, 9$. A map in physical units of photons per pixel and per detector can then simply be retrieved by scaling the normalized map with the averaged photon number per detector, i.e., $I(x, y) = I^{\text{norm}}(x, y) \times \langle N_l^{\text{phot}} \rangle$. The standard unit used in RHESSI software is also normalized by time, the detector area, and arc sec$^2$ of a pixel, i.e., photons s$^{-1}$ cm$^{-2}$ arc sec$^{-2}$.

Some examples of such parameterized maps are shown in Figure 1, including a single gaussian source (4 parameters), double gaussian sources (8 parameters), a double source with a loop (14 parameters), and a Masuda-type flare (Masuda *et al.*, 1994) with an additional above-the-loop-top source (18 parameters).

## 4. Forward-Fitting Algorithm

### 4.1. Pixelized and Unpixelized Approach

We describe now the mathematical and numerical formalism of the basic *forward-fitting method*, as currently implemented in the RHESSI software (since release on 4 January 2000), a package in the Interactive Data Language (IDL)-based Solar SoftWare (SSW) (Freeland and Handy, 1998). This approach is also called the *pixelized forward-fitting method* because the spatial distribution of the photon flux $F(x, y)$ is modeled as a function of a coordinate grid $(x, y)$ with regular spacing, either in cartesian $(x, y)$ or polar coordinates $(r, \vartheta)$. An *unpixelized forward-fitting method* has been proposed, where an image of the photon flux can be reconstructed with a set $(i = 1, \ldots, n)$ of $n$ point or gaussian sources at positions $(x_i, y_i)$, without using the parameterization of a regular grid $(x, y)$. This second method is theoretically more efficient for image reconstruction, but is currently not implemented in the RHESSI software.

### 4.2. The Basic Steps of the Algorithm

The algorithm is basically defined by the input, which consists of a list of time-tagged photon events from the 9 detectors (packed in telemetry files) and the desired output in the form of a reconstructed map $I(x, y)$. This map should be

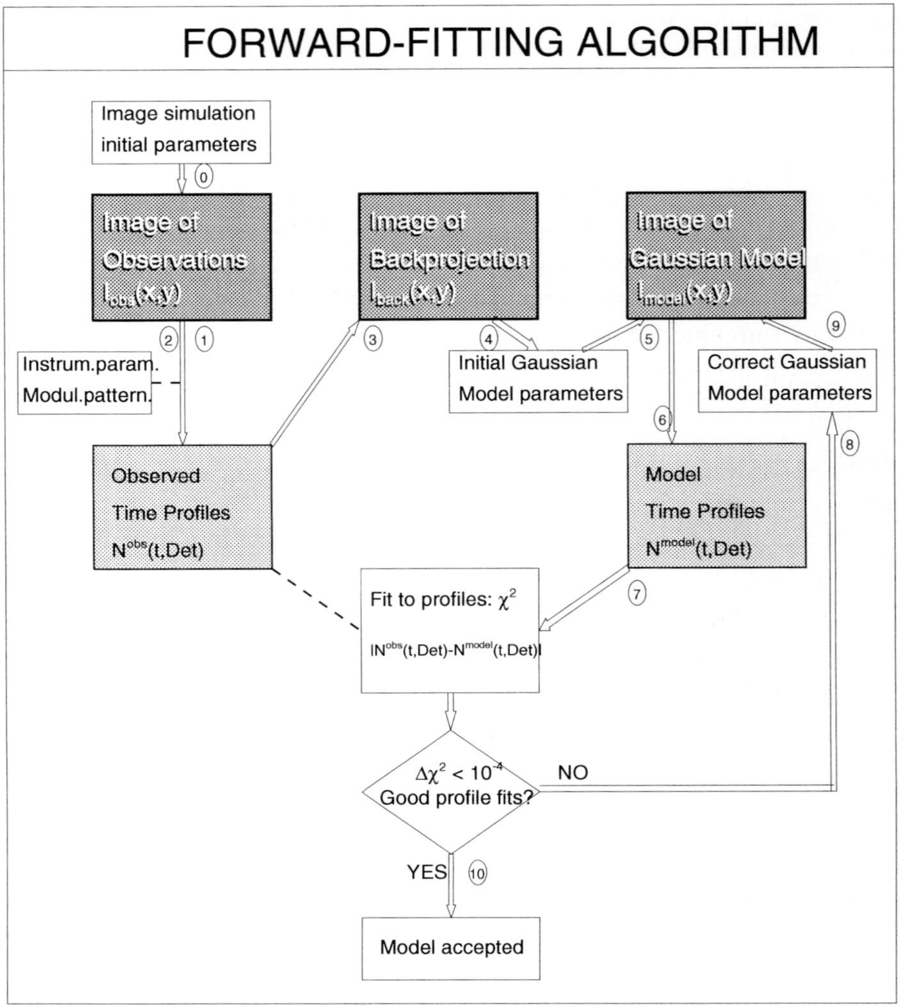


*Figure 2.* Flow chart of the numeric algorithm of forward fitting (see description of 10 basic steps in Section 4.2).

consistent with the observed count-rate time profiles, within the expected Poisson fluctuations. The algorithm consists of 10 basic steps (Figure 2). For the simulations here, an additional step is required beforehand, which includes the simulation of telemetry files from a model map (numbered as 'step 0' in the following scheme).

*Step 0:* selection of a set of model parameters $P_i, i = 0, \ldots, n^{\text{free}}$ that define a model map $I^{\text{model,sim}}(x, y)$. Simulation of a list of photon events that are randomly distributed in time and reproduce the spatial 2D distribution prescribed by the model map. The time-tagged photon events are written out into telemetry files that have the same format as observations telemetered down from the spacecraft.

*Step 1:* time intervals $[t_1, t_2]$, energy intervals $[E_1, E_2]$, image size $[N_x, N_y]$, image offset from Sun center $[X_{\rm offset}, Y_{\rm offset}]$, and pixel resolution $\Delta x$ are selected for desired image reconstruction.

*Step 2:* from the telemetry files, a binned event list is created, and then from the aspect solution, livetime calculations and instrumental calibration tables, a calibrated event list is created, $N^{\rm obs}(t_k, D_l)$ (for time bins $t_k$ and detectors $D_l, l = 1, \ldots, 9$), and the corresponding modulation patterns are calculated, $M_{ijkl}$ (for each pixel $x_i, y_j$).

*Step 3:* a *back-projection* (Hurford *et al.*, 2002) is performed from the calibrated event list to obtain an initial first-guess map $I^{BP}(x, y)$ for the forward-fitting procedure.

*Step 4:* the back-projection map $I^{BP}(x, y)$ is decomposed into circular source components (see Figure 4 and description on CD-ROM) in order to determine a first-guess value of the number ($n_g$) of gaussian components and their spatial positions. These are used to specify an initial set of gaussian parameters $P_i, i = 0, \ldots, n^{\rm free}$ (for a small number of gaussian source components).

*Step 5:* based on the first-guess gaussian parameters $P_i, i = 0, \ldots, n^{\rm free}$, a first model map $I^{\rm model1}(x, y)$ is composed. It provides the input for the first iteration of the forward-fitting procedure. We normalize the model map to 1 photon count per image, i.e., by $I^{\rm model1,norm}(x, y) = I^{\rm model,1}(x, y) / \int I^{\rm model1}(x, y)\, {\rm d}x\, {\rm d}y$.

*Step 6:* the first model map $I^{\rm model1}(x_i, y_j)$ is multiplied with the modulation pattern $M_{ijkl}$ to obtain the modulation time profile $P_{kl}$,

$$P^{\rm model1}(t_k, D_l) = M_{ijkl}\, I^{\rm model1,norm}(x_i, y_j)\ . \tag{14}$$

This is then scaled into the same physical units as the observed time profiles, i.e., counts per time bin,

$$N^{\rm model1}(t_k, D_l) = R^{\rm phot} \Delta t(D_l) q^{\rm trans}(D_l) P^{\rm model,1}(t_k, D_l)\ , \tag{15}$$

where $R^{\rm phot}$ represents the photon counts per s and per subcollimator (s$^{-1}$ SC$^{-1}$), $\Delta t_k(D_l)$ is the time bin used for detector $D_l$, and $q^{\rm trans}(D_l)$ is the bi-grid transmission factor of detector $D_l$.

*Step 7:* the values of the reduced $\chi^2$ (or C-statistic, which is currently implemented in the code, see Appendix A for a description) is calculated between the observed time profiles $N^{\rm obs}(t_k, D_l)$ and the model time profiles $N^{\rm model1}(t_k, D_l)$,

$$\chi^2 = \frac{1}{n_{t'}} \sum_{k'=1}^{n_{t'}} \frac{[N^{\rm obs}(t_{k'}) - N^{\rm model}(t_{k'})]^2}{[\sigma^{\rm model}(t_{k'})]^2}\ , \tag{16}$$

where the time series $t_k$ from each detector $D_l$ are concatenated $[t_{k'} = t_k\ (D_1), t_k\ (D_2), \ldots, t_k\ (D_9)]$, $k' = 1, \ldots, n_{t'}$ to obtain a single $\chi^2$ value for a given image.

The currently implemented method gives each spatial scale the same weight, by averaging the $\chi^2$'s from each detector.

*Step 8:* the parameters $P_i, i = 0, \ldots, n^{\text{free}}$ are varied in all ($n^{\text{free}}$) dimensions and the change of $\chi^2$ (or C-statistic), $\Delta C = C(P_i + \Delta P_i) - C(P_i)$, is calculated. A new parameter set $P_i^{\text{new}} = P_i - \Delta P_i(\partial C/\partial P_i)$ is calculated from the steepest gradient of $\chi^2$ in the $n^{\text{free}}$-dimensional parameter space, so that the new parameter set $P_i^{\text{new}}$ points in the direction where $C$ decreases most rapidly. The search for a local minimum in the direction of the steepest gradient is performed by increasing the steps $\Delta P_i$ by powers of 2 and using a cubic spline fit of $C(P_i^{\text{new}})$. The set of parameters $P_i^{\text{new}}(\min[C])$ associated with the local minimum of $C$ is used as the starting point for the next iteration cycle of the forward-fitting procedure.

*Step 9:* the next model map $I^{\text{model2}}(x_i, y_j)$ is calculated based on the new parameter set $P_i^{\text{new}}, i = 0, \ldots, n^{\text{free}}$ and steps 6–8 are repeated iteratively until the change in $C$ converges, e.g., $\Delta C < 10^{-4}$ per iteration cycle.

*Step 10:* after $n$ iterations, the final map $I^{\text{model},n}(x, y)$ becomes the desired reconstructed map of the sources. In the case of the simulations, the final map $I^{\text{model},n}(x, y)$ can be compared with the initial model $I^{\text{model,sim}}(x, y)$ used for creating the telemetry files (Step 0) to verify the fidelity of the reconstruction. In the case of real observations, the final solution may be compared with maps reconstructed using other methods such as CLEAN or the maximum entropy method (MEM). A self-consistency criterion between the reconstructed map and the observations is the convergence value of $\chi^2$ or $C$-statistic, which should have values near unity, say within $\lesssim 10\%$. If the $\chi^2$ or $C$-values are significantly above unity (provided there is no convergence problem to the global minimum), this would indicate that the model parameterization is not adequate to represent the data, and thus requires a more complex model.

This outline is only a formal description of the basic steps of the forward-fitting code (Figure 2). To understand the most efficient implementation, the accuracy and limitations of the solutions, and the convergence behavior of the algorithm, we provide more detailed information in the following subsections.

### 4.3. Limits Set by Poisson Fluctuations

It is a property of RHESSI image reconstruction algorithms that parameter changes are most significant in the first iteration steps, while later iterations near the convergence limit are less efficient and essentially are dominated by Poisson fluctuations. (For the purposes of this paper, we ignore background, cosmic rays and other types of noise, which in many cases have been found to be smaller than the Poisson fluctuations.) In the case of forward-fitting in particular, it is desirable to suppress those computation-expensive parts of the calculations that bluntly fit the Poisson fluctuations. A first important efficiency consideration is therefore to select only those detectors with time bins large enough to have a reasonable count-to-fluctuation ratio to be sensitive to the fitting algorithm.

The time profiles of the finest collimators have the greatest Poisson fluctuations, because with the standard RHESSI time-binning, the same number of photons (per detector) is spread out over a larger number of time bins (by the requirement of resolving the rotational modulation). Ignoring background, the amplitude of the fluctuations is entirely Poisson, i.e., count/fluctuations $= \sqrt{N^{\mathrm{phot}}}$, and the ratio (for standard binning) is smaller for the finer grids, because it decreases as $C/F \sim \sqrt{\Delta t_i^{\mathrm{bin}}}$, where $\Delta t_i^{\mathrm{bin}}$ is the time bin length for detector $i$. For instance, if the coarsest grid gives a count/fluctuations ratio of 100:1, it will be 12:1 for the finest grid, because the default time bins vary by a factor of $\Delta t_9^{\mathrm{bin}}/\Delta t_1^{\mathrm{bin}} = 64$.

In order to decide which time profiles can be ignored in the forward-fitting procedure without significant loss of information, we define a significance criterion that is expressed by the ratio of its standard deviation $\sigma^{\mathrm{r.m.s.}}$ to the expected Poisson fluctuations $\sigma^{\mathrm{Poisson}}$. The r.m.s. fluctuation of a time profile $N^{\mathrm{obs}}(t_k, D_l)$ is composed of the variability due to the rotational modulation $\sigma^{\mathrm{mod}}$ and the photon count statistic $\sigma^{\mathrm{Poisson}}$,

$$(\sigma^{\mathrm{r.m.s.}})^2 = (\sigma^{\mathrm{mod}})^2 + (\sigma^{\mathrm{Poisson}})^2. \tag{17}$$

For instance, when the variability due to rotational modulation, $\sigma^{\mathrm{mod}}$, falls below 20% of the counting fluctuations, $\sigma^{\mathrm{Poisson}}$, the combined r.m.s. variability, $\sigma^{\mathrm{r.m.s.}} = \sigma^{\mathrm{Poisson}} \times \sqrt{1+0.2^2} = 1.02\,\sigma^{\mathrm{Poisson}}$, exceeds the Poisson fluctuations by 2% only. Any fitting method based on a $\chi^2$-criterion becomes quite insensitive at this level, because $\chi^2 \sim (\sigma^{\mathrm{Poisson}})^{-2}$ (Equation (16)) varies then by 4% only, which has little significance in judging the goodness of a fit. Therefore, we adopt a threshold criterion of

$$\left(\frac{\sigma^{\mathrm{r.m.s.}}}{\sigma^{\mathrm{Poisson}}}\right) \geq 1.02, \tag{18}$$

to select detectors that have time profiles with modulation large enough to be usable by the forward-fitting procedure. This criterion is equivalent to

$$\left(\frac{\sigma^{\mathrm{mod}}}{\sigma^{\mathrm{Poisson}}}\right) \geq \sqrt{(1.02^2-1)} = 0.2\,. \tag{19}$$

The average counts per time bin is $N_l^{\mathrm{obs}} = R^{\mathrm{phot}} \Delta t_l q_l^{\mathrm{trans}}$ (see Equation (15)), while the total number of events in the time interval $(t_2 - t_1)$ is $N_l^{\mathrm{tot}} = R^{\mathrm{phot}}(t_2 - t_1) q_l^{\mathrm{trans}}$. For Poisson statistics, with $n = N_l^{\mathrm{tot}}$ events and a mean of $\langle x \rangle = N_l^{\mathrm{obs}}$ the standard deviation from the mean is $\sigma = \sqrt{\langle x \rangle}$ (Bevington, 1969, p. 33), and thus

$$\sigma_l^{\mathrm{Poisson}} = \sqrt{N_l^{\mathrm{obs}}} = [R^{\mathrm{phot}} q_l^{\mathrm{trans}} \Delta t_l]^{1/2}. \tag{20}$$

In Figure 3 we compare the r.m.s. $\sigma_l^{\mathrm{Poisson}}$ with the fluctuations due to rotational modulation $\sigma_l^{\mathrm{mod}}$, which is

$$\sigma_l^{\mathrm{mod}} \approx 0.5\, \varepsilon_l^{\mathrm{mod}} N_l^{\mathrm{obs}} \tag{21}$$

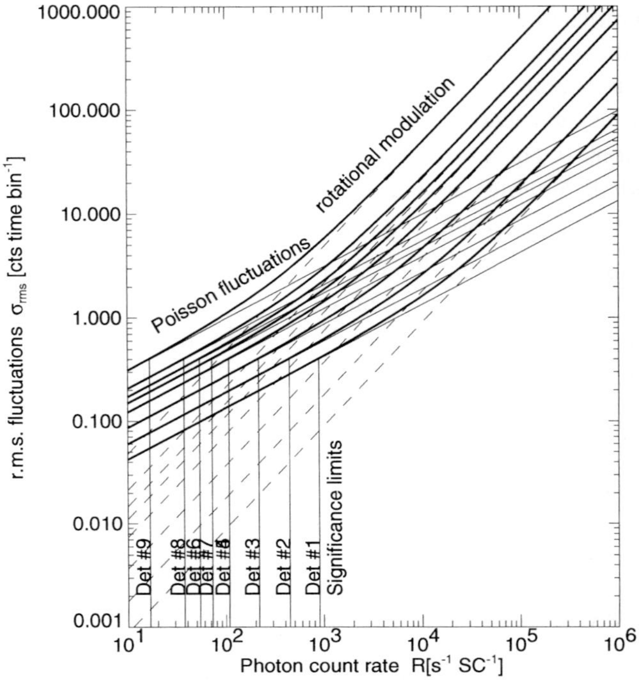


*Figure 3.* The r.m.s. fluctuations of observed time profiles $N^{\mathrm{obs}}(t_k, D_l)$ as a function of the photon count rate $R$ (*thick lines*), modeled in terms of two components: the photon counting r.m.s. fluctuations $\sigma^{\mathrm{Poisson}} \sim \sqrt{R}$ (*thin lines*), and the r.m.s. fluctuations due to the rotational modulation $\sigma^{\mathrm{mod}} \sim R$ (*dashed lines*). Each component is shown for each detector $l$, $D_1, \ldots, D_9$, where the coarsest detector $D_9$ represents the topmost lines. Note that the photon Poisson fluctuations dominate for small photon count rates $R_l$. The limit $R_l^{\mathrm{min}}$ where the rotational modulation exceeds 20% of the Poisson fluctuations is indicated with a *vertical line* for each detector. These values of $R_l^{\mathrm{min}}$, which represent lower limits for the photon rate where rotational modulation is detectable in the time profiles, are also listed in Table I.

for the 9 detectors ($l = 1, \ldots, 9$) and modulation efficiency $\varepsilon_l^{\mathrm{mod}} (\leq 1.0)$. From the requirement that the variability due to rotational modulations exceeds the Poisson fluctuations by 20% (Equation (19)), we can infer a requirement for a minimum photon rate to produce significant modulations in the detectors. Combining Equations (19)–(22) we find

$$N_l^{\mathrm{obs,min}} \geq \left( \frac{0.2}{0.5\varepsilon_l^{\mathrm{mod}}} \right)^2 = \frac{0.16}{(\varepsilon_l^{\mathrm{mod}})^2} , \tag{23}$$

which yields a minimum photon rate $R_l^{\mathrm{min}}$ of

$$R_l^{\mathrm{min}} \geq \frac{N^{\mathrm{obs,min}}}{q_l^{\mathrm{trans}} \Delta t_l} \approx \frac{10^3}{\sqrt{3}^{\,l-1}} . \tag{23}$$

These limits are indicated in Figure 3 and tabulated in Table I, ranging from $R_1^{min} \geq 900$ counts s$^{-1}$ SC$^{-1}$ for the finest detector No. 1 to $R_9^{min} \geq 17$ counts s$^{-1}$ SC$^{-1}$ for the coarsest detector No. 9. These count rate limits are derived for 100% modulation efficiency, but would be higher for source sizes that are comparable or larger than the collimator resolution, where the modulation efficiency drastically drops. They would also be higher at higher energies because the grids become transparent, and at lower energies for the finest grids because of diffraction effects.

The count-rate limits $R_l^{min}$ for different detectors $l$ (indicated in Figure 3 and listed in Table I) save a lot of computational effort. For instance, if a weak flare with a relatively low count rate of $R^{phot} = 100$ counts s$^{-1}$ SC$^{-1}$ is observed, we see immediately (Table I) that this count rate is below the requirement $R_l^{min}$ of significant modulation for detectors Nos. 1–5. We need therefore only to consider detector Nos. 6–9 for image reconstruction. The number of time bins from detectors Nos. 6 –9 is $\sum_{i=6}^{9} N_i^{bin} = 1408$, and thus a factor of $(17\,792/1408) = 12.6$ smaller compared with the number of time bins from all detectors Nos. 1–9, making the computation a factor of $\approx 12$ faster (since the computation of modulation profiles scales linearly with the number of time bins).

### 4.4. Decomposition of Backprojection Maps

For the first guess of source positions the forward-fitting algorithm makes use of a backprojection map (Hurford *et al.*, 2002), by iterative decomposition of the strongest peaks in the backprojection map. The exact peak location $(x_i, y_i)$ is determined with sub-pixel accuracy by parabolic interpolation of the maximum in both $x$- and $y$-direction. An example with a detailed description of the procedure is given in Figure 4 on the CD-ROM.

### 4.5. Parameter Optimization in Forward-Fitting

The forward-fitting method requires the optimization of $n^{free}$ parameters until the modulation time profiles resulting from the model map are consistent with the observed time profiles $N^{obs}(t_k, D_l)$. For simple maps, e.g., the examples shown in Figure 1, this task consists in the optimization of 3–18 free parameters. The minimization of the $\chi^2$ or $C$-statistic of a parameter set $(P_i), i = 1, \ldots, n^{free}$ requires the calculation of a modulation profile $N^{model}(t_k, D_l)$ for each new trial of a model map $I(x, y; P_i)$, in order to calculate the $\chi^2$ or $C$-statistic of a new parameter set $(P_i)$. We developed a variant of the multi-dimensional *downhill simplex method* (Press *et al.*, 1986, p. 289), which is described in some detail in Figure 6 on the CD-ROM.

### 4.6. Numerical Simulations and Tests

In this study we performed tests using some 500 simulations with image reconstructions by the forward-fitting method. To obtain statistics on the accuracy of reconstruction we simulated basic maps that contain one or two circular gaussian sources, which can be quantified by 4 or 8 free parameters. A few simulations with more complex source topologies were also performed as a feasibility test, i.e. with triple, quadruple, elliptical, and curved elliptical sources. The results of the tests are reported in more detail in the accompanying CD-ROM (Figures 5–13). We tested mainly the accuracy $\Delta r_i$ of the reconstructed source positions $(x_i, y_i)$, widths $(w_i)$, and flux amplitudes $(a_i)$ for model maps with 1 or 2 gaussian sources, and obtained uncertainties that scale with the square root of the count rate $R$,

$$\frac{\Delta r_i}{w_i} = \frac{\sqrt{(x_i^{\rm fit} - x_i^{\rm sim})^2 + (y_i^{\rm fit} - y_i^{\rm sim})^2}}{w_i^{\rm sim}} = \sigma_r \left(\frac{R}{10^4}\right)^{-1/2} , \tag{24}$$

$$\frac{\Delta w_i}{w_i} = \frac{(w_i^{\rm fit} - w_i^{\rm sim})}{w_i^{\rm sim}} = \sigma_w \left(\frac{R}{10^4}\right)^{-1/2} . \tag{25}$$

The results can be summarized as follows. For double-source simulations, a forward-fitting solution with two gaussians was found in 90 cases, using a fully automated computation mode without human interaction. For uncontaminated statistics of errors, we excluded cases with insufficient separation and cases with weak secondary sources. For a count rate of $R = 10^4$ counts $s^{-1}$ $SC^{-1}$, two gaussian centroid positions could be reconstructed with automated forward-fitting with a typical positional accuracy of $\sigma_r \approx 10$–$20\%$ of their gaussian width, and a width accuracy of $\sigma_w \approx 5\%$. The total photon flux was typically retrieved with an accuracy of $\approx 1\%$. Overlapping double sources or weak sources with a contrast of less than 1:20 are less reliably reconstructed. The details of the accuracy tests are provided in Figures 8–13 on the accompanying CD-ROM.

### 4.7. Applications

The six different image reconstruction algorithms currently supported by the RHESSI software each have different advantages and disadvantages. Image reconstruction with forward-fitting has the specific property that it places all the flux into the prescribed number of gaussian source components, and thus has a high photometric accuracy because no flux is spread into noisy residuals (ignoring the detector background). Another specific property is that the determined source positions are more stable than in other imaging algorithms. This is because the fitted gaussians are sensitive to the centroid position. For these reasons, forward-fitting is especially well suited to measure small differential changes of source positions in time or energy. A possible application for small differential changes in source positions as a function of time is the separation of flare loop footpoints. This is expected to increase systematically in the Kopp–Pneuman model (Kopp and Pneuman, 1976) due to the rise of the X-type reconnection point. Previous measurements, based on *Yohkoh* HXT images, showed a mixed trend, where only half of the flares showed evidence for the predicted increase in footpoint separation with time (Sakao, 1998). Another possible application for small differential changes in source positions is the energy-dependent altitude of footpoint sources, which can be measured most favorably near or above the limb. Such accurate height measurements can constrain the energy loss of precipitating electrons in the chromosphere and can be used to infer a chromospheric density model (Brown, Aschwanden, and Kontar, 2002; Aschwanden, Brown, and Kontar, 2002).

## 5. Summary and Conclusions

We described a newly-developed forward-fitting method to reconstruct images from the RHESSI telescope for simple source morphologies that can be represented by a superposition of multiple gaussians. We performed some 500 numerical simulations of model source geometries and compared the forward-fitted images with the model images in order to quantify the fidelity of the image reconstruction, the accuracy of the retrieved physical parameters, the maximum spatial resolution, and the convergence behavior. These tests, covered in detail in the accompanying RHESSI CD-ROM, represent a basic reference for all other RHESSI image reconstruction algorithms, providing error bars and quantitative uncertainties that are important for the scientific interpretation of hard X-ray images from RHESSI. The results of the forward-fitting method can be summarized as follows:

(1) Using circular, elliptical, and curved gaussians, most of the so far observed hard X-ray maps of solar flares can be represented by a relatively small number of free parameters, e.g., $n^{\text{free}} = 3$ for a single source, $n^{\text{free}} = 7$ for double sources, $n^{\text{free}} = 14$ for double footpoint flares with loop-top hard X-ray emission, or $n^{\text{free}} = 18$ for Masuda-type flares. The current implementation of the forward-fitting code is designed to reconstruct hard X-ray images for this variety of source morphologies.

(2) The minimum count rate required for a detector to produce modulation large enough for the current forward-fitting algorithm to work is approximately $R_l^{\min} = 1000/\sqrt{3}^{\,(i-1)}$ counts $\text{s}^{-1}$ $\text{SC}^{-1}$ for detector $l = 1, \ldots, 9$, e.g., $R_1^{\min} \approx 1000$ counts $\text{s}^{-1}$ $\text{SC}^{-1}$ for the finest grid (detector No. 1), and $R_9^{\min} \approx 10$ counts $\text{s}^{-1}$ $\text{SC}^{-1}$ for the coarsest grid (detector No. 9). For a given flare event with count rate $R$, detectors that have a count rate below this limit, $R_l^{\min} < R$, should be discarded for image reconstruction by this method. This saves considerable computational effort.

(3) Test runs of forward-fitting with model maps containing a single gaussian source show that the relative accuracy of most gaussian parameters scales with the count/Poisson-fluctuation ratio, i.e., $\sim \sqrt{R}$, with $R$ being the photon count rate. The average accuracy in the reconstruction of single gaussian sources at a count rate of $R = 10^4$ counts $\text{s}^{-1}$ $\text{SC}^{-1}$ is $\approx 5\%$ (of the source width) for the position, and $\approx 3\%$ for the gaussian source width. The accuracy depends primarily on the photon count rate and not at all on the chosen pixel size (as long as the source

is larger than the half pixel size). Subpixel accuracies in source position down to 1% can be achieved for count rates of $R \approx 10^5$ counts $\mathrm{s}^{-1}$ $\mathrm{SC}^{-1}$, slightly above the value at which the RHESSI attenuators automatically switch in. This allows efficient mapping to be performed with relative large pixel sizes without losing accuracy.

(4) Test runs of forward-fitting with model maps containing double gaussian sources demonstrate that the positions and widths can be reconstructed with almost the same accuracy as single source maps, unless there are source confusion problems or strong contrast ($\lesssim$ 1:20 for the weaker source). We find that the forward-fitting code converges to a $\chi^2$ or $C$-statistic value that is equally consistent with the data as the model.

(5) The maximum spatial resolution that can be obtained with the forward-fitting technique for high count rates is $\mathrm{FWHM}^{\mathrm{min}} \approx 0.8''$, which is about a factor of 3 below the nominal resolution of the finest grid, i.e., $\mathrm{FWHM}_1^{\mathrm{res}} = 2.3''$.

(6) We find that the forward-fitting algorithm virtually always converges to the same solution as the model (used for the simulation of the telemetry data). The final $\chi^2$- and $C$-statistic values reached at the convergence limit, which are found in the range of $0.95 \lesssim C \lesssim 1.1$, are constrained by the Poisson fluctuation statistics of a particular event, and not by the convergence behavior of the forward-fitting code.

(7) Forward-fitting is especially well suited to measure small differential changes of source positions in time or energy. Possible applications are energy-dependent altitude measurements of energy loss near the limb, or footpoint separations during flares.

Future work with the forward-fitting method will include (1) the simulation of RHESSI telemetry data with more complex source morphologies (i.e., curved loops, Masuda-type flares), (2) combined spatial-spectral imaging, (3) combined spatial-temporal imaging, (4) fast imaging on time intervals shorter than a half spacecraft spin period, and quantitative comparisons with other RHESSI imaging methods.

## Acknowledgement

Support for this work was provided by the NASA SMEX grant NAS5-98033 through University of California, Berkeley (subcontract SA2241-26308PG).

## Appendix
## $\chi^2$ and *C*-Statistic for Sparse Sampling

An observed time series of photon counts, $f^{\mathrm{obs}}(t)$, which is binned into time intervals of length $\Delta t$, so that the number $n_i = f^{\mathrm{obs}}[t_i < t < t_i + \Delta t]$ represents the photon counts in bin $[t_i, t_i + \Delta t]$, is modeled by an analytical function, $f^{\mathrm{model}}(t)$, which predicts an expectation value $e_i = f^{\text{model}}[t_i < t < t_i + \Delta t]$ of photon counts in time bin $i$. In the limit of large-number counts, $n_i \gg 0$, Poisson statistics can be approximated by a gaussian distribution, which predicts an r.m.s. fluctuation of $\sigma_i = \sqrt{e_i} \approx \sqrt{n_i}$. The goodness of fit of a model can then be evaluated by the well-known $\chi^2$-statistic,

$$\chi^2 = \frac{1}{N} \sum_{i=1}^{N} \frac{(n_i - e_i)^2}{\sigma_i^2} = \frac{1}{N} \sum_{i=1}^{N} \frac{(n_i - e_i)^2}{e_i} , \tag{A1}$$

yielding a normalized value of $\chi^2 \approx 1$ in the case when the model is consistent with the data, within the uncertainties of the expected fluctuations.

For sparse sampling, in particular when the observed time series $n_i$ contains zero's or few counts, say $n_i \lesssim 10$, e.g., as it is the case for the default timebins of the finest RHESSI collimators, where the total photon counts are binned into some $2^{12} = 4096$ time bins (for the finest grids) to resolve the rotational modulation, the gaussian $\chi^2$-statistic is not appropriate anymore. Photon counting statistics for sparse sampling can be derived from the general probability function $P$ as outlined by Cash (1979), which is

$$P = \prod_{i=1}^{N} \frac{e_i^{n_i} e^{-e_i}}{n_i!} \tag{A2}$$

for a particular result $n_i$ given the correct set of $e_i$. The likelihood ratio, called *C-statistic* by Cash (1979), can be expressed in logarithmic form from Equation (A2),

$$C_{\text{Cash}} = -2 \ln P = -2 \sum_{i=1}^{N} (n_i \ \ln e_i - e_i - \ln n_i!) . \tag{A3}$$

Let the theoretical model $e_i(\Theta_1, \ldots, \Theta_p)$ be defined by $p$ free parameters $\Theta_i$, $i = 1, \ldots, p$. The best-fitting model is found by varying all $p$ parameters $\Theta_1, \ldots, \Theta_p$ until the $C$-statistic reaches a minimum, which we denote by $(C_{\text{min}})_p$. Now, during the iterative fitting procedure, only a partial subset of model parameters, say $q$ parameters (with $q < p$), may have already converged to the true solution, $\Theta_1^T, \ldots, \Theta_p^T$, so that $\Theta_1, \ldots, \Theta_q$ are set to $\Theta_1^T, \ldots, \Theta_q^T$, while the remaining $p - q$ parameters, $\Theta_{q+1}, \ldots, \Theta_p$, still need to be varied until a global minimum of $C$ is reached. We denote this partial solution, where $p - q$ parameters have already converged to the true solution $\Theta_i^T$, with the value $(C_{\text{min}})_{p-q}^T$. According to Cash, quoting the theorem of Wilks (1938, 1963), the difference

$$\Delta C = (C_{\text{min}})_{p-q}^T - (C_{\text{min}})_p \tag{A4}$$

will be distributed as $\chi^2$ with $q$ degrees of freedom. Therefore, the quantity $\Delta C$ can be used to establish a confidence criterion of the model $e_i$ to the data $n_i$. Following Cash (1979), the term $\ln(n_i!)$ of Equation (A3) drops out in the difference $\Delta C$, because only the parameters $e_i(\Theta_1, \ldots, \Theta_p)$ are varied during the fitting procedure. Thus, it is more convenient to use the simplified statistic

$$C_{\text{simplified}} = -2 \ln P = -2 \sum_{i=1}^{N} (n_i \ \ln e_i - e_i) \ . \tag{A5}$$

For the evaluation of the difference $\Delta C$ in Equation (A4) we have the partially optimized term $(C_{\text{min}})_{p-q}^{T}$,

$$(C_{\text{min}})_{p-q}^{T} = -2 \sum_{i=1}^{N} (n_i \ \ln (e_i)_{p-q}^{T} - (e_i)_{p-q}^{T}) \ , \tag{A6}$$

and the absolute minimum $(C_{\text{min}})_p$, which represents the asymptotic limit when the data $n_i$ perfectly match the model, i.e., $e_i = n_i$,

$$(C_{\text{min}})_p = -2 \sum_{i=1}^{N} [n_i \ \ln (n_i) - n_i] \ . \tag{A7}$$

The combination of Equation (A4)–(A6) yields then

$$\Delta C = -2 \sum_{i=1}^{N} [n_i \ \ln (e_i) - e_i - n_i \ \ln (n_i) + n_i] \ . \tag{A8}$$

Instead of the reduced $\chi^2$-statistic (A1), we can therefore use the equally simple C-statistic (where we drop the symbol $\Delta$ for brevity),

$$C := \frac{1}{N} \Delta C = \frac{2}{N} \sum_{i=1}^{N} \left[ n_i \ \ln \left( \frac{n_i}{e_i} \right) - (n_i - e_i) \right] \ . \tag{A9}$$

This form has the advantage that, in addition to being asymptotic to $\chi^2$, C vanishes identically when the model fits the data exactly. Numerical care has to be taken for the time bins that contain zeros in the data, $n_i = 0$, in which case the mathematical relation $n_i \ln (n_i) = 0$ has to be used to avoid the singularity $\ln (n_i) \mapsto \infty$. On the other hand, a singularity could arise when the model predicts zero counts, $e_i = 0$, but the observed counts are not zero, $n_i \neq 0$, because the term $-n_i \ \ln (e_i)$ yields then infinity. It seems therefore to be recommendable to restrict the model fit to time intervals with a finite probability for photon counts, i.e., $e_i > 0$.

## References


Aschwanden, M. J., Brown, J. C. and Kontar, E. P.: 2002, *Solar Phys.*, this volume.

Bevington, P. R.: 1969, *Data Reduction and Error Analysis for the Physical Sciences*, McGraw-Hill Book Co., New York.

Brown, J. C., Aschwanden, M. J., and Kontar, E. P.: 2002, *Solar Phys.*, this volume.

Cash, W.: 1979, *Astrophys. J.* **228**, 939.

Dennis, B. R. *et al.*: 1996, *Proc. SPIE* **2804**, 228.

Freeland, S. L. and Handy, B. N.: 1998, *Solar. Phys.* **182**, 497.

Hurford, G. J. *et al.*: 2002, *Solar Phys.*, this volume.
Kopp, R. A. and Pneuman, G. W.: 1976, *Solar Phys.* **50**, 85.
Kosugi, T., Sawa, M., Sakao, T., Masuda, S., Inda-Koide, M., Yaji, K., and Sato, J.: 1995, *The Yohkoh HXT Databook October 1991 – December 1994*, Natl. Astron. Obs. Tokyo, Japan.
Lin, P. R., Dennis, B. R., Emslie, A. G., Ramaty, R., Canfield, R., and Doschek, G.: 1993, *Adv. Space Res.* **13**, 401.
Lin, P. R., Dennis, B. R., Ramaty, R., Emslie, A. G., Canfield, R., and Doschek, G.: 1994, *Geophys. Monogr.* **84**, 283.
Lin, P. R. *et al.*: 1998, *Proc. SPIE* **3442**, 2.
Masuda, S., Kosugi, T., Hara, H., Tsuneta, S., and Ogawara, Y.: 1994, *Nature* **371**, 495.
Press, W. H., Flannery, B. P., Teukolsky, S. A., and Vetterling, W. T:. 1986, *Numerical Recipes, the Art of Scientific Computing*, Cambridge University Press, Cambridge.
Sakao, T.: 1994, Ph.D. Thesis, University of Tokyo.
Sakao, T., Kosugi, T., and Masuda, S.: 1998, in T. Watanabe, T. Kosugi, and A. C. Sterling (eds.), *Observational Plasma Astrophysics: Five Years of Yohkoh and Beyond*, Kluwer Academic Publishers, Dordrecht, Holland; p. 273.
Sato, J., Sawa, M., Masuda, S., Sakao, T., Kosugi, T., and Sekiguchi, H.: 1998, *The Yohkoh HXT Image Catalogue, October 1991 – August 1998*, Nobeyama Radio Observatory/NAO, Japan.
Smith D. M. *et al.*: 2002, *Solar Phys.*, this volume.
Wilks, S. S.: 1938, *Ann. Math. Stat.* **9**, 60.
Wilks, S. S.: 1963, *Mathematical Statistics*, Ch. 13, Princeton University Press, Princeton.

# TIME-DOMAIN DEMODULATION OF RHESSI LIGHT CURVES

KASPAR ARZNER
*Paul Scherrer Institut, CH-5232 Villigen PSI, Switzerland*

(Received 5 August 2002; accepted 11 September 2002)

**Abstract.** This paper presents an algorithm to decompose the modulated RHESSI light curves into periodic functions and a smooth function, representing the true (demodulated) time profile of an impulsive source. The decomposition is achieved by optimizing a trade-off between the Poisson likelihood, a smoothness constraint, and conditions on the average grid transmission and the (modulating or non-modulating) background. The algorithm, which operates on the level of count rates and does not require imaging information, is verified by numerical simulations and applied to some early RHESSI data, where – as a preliminary result – several impulsive features on time scales $< 4$ s may have been identified.

## 1. Introduction

The RHESSI instrument uses the rotational modulation principle in combination with germanium detectors to obtain spatially, spectrally, and temporally resolved images of solar HXR outbursts. The interpretation of the observed counts raises a series of inverse problems, such as the deconvolution of the spectral response (Smith *et al.*, 2002), and the image reconstruction from the modulated count rates (Hurford *et al.*, 2002). In this article, the focus is on the 'reverse' problem, i.e., on the estimation of the true time dependence of the source on time scales below the RHESSI spin period $T_s \sim 4$ s. The motivation for this stems from earlier HXR observations, which revealed impulsive features on sub-second time scales in some 10% of all flares with sufficient count rates to see such variations (Dennis, 1985). The shortest features found so far were called 'HXR spikes'; they have typical durations of order of 400 ms (Kiplinger *et al.* (1984) from HXRB/SMM; Machado *et al.* (1993) from BATSE/CGRO), and rise times in the order of $\tau_s \sim 100$ ms; occasionally, even shorter time scales of a few 10 ms have been observed (see Dennis, 1985). The sub-second evolution of the HXR light curves is believed to be in close relation with the energy release process, and is therefore interesting from an acceleration physics point of view (Benz *et al.*, 1994; Miller *et al.*, 1997), and relevant for the comparison with radio data (e.g., Gary, 2000; Nakajima, 2000).

The modulational contributions can be eliminated by different methods such as non-recursive (time-domain) filters ($O_n = \sum_{i<n} a_i I_i$ with $I$ the input and $O$ the output signals), recursive filters ($O_n = \sum_{i<n} a_i O_i + \sum_{j\leq n} b_j I_j$), autoregressive models (e.g., Rosen and Porat, 1989), or (frequency-domain) Wiener filters (Kailath, 1974; Mallat, 1998). The latter option has been closely examined, and a

demodulation method based on it will be presented in a separate communication. The classical Wiener filter minimizes the deviation of the restored and true signals on the basis of the power spectral densities of the signal and noise.

Singular spectrum analysis (SSA) (Vautard and Ghil, 1989; Varadi *et al.*, 1999, 2000) offers another, more general, approach to the detection of periodic structures in noisy data. SSA disentangles deterministic (periodic or chaotic) and stochastic contributions of an observed time series, and identifies the dimension of an underlying dynamical system. This is achieved by considering the singular values of the covariance of vectors of consecutive (or time-ordered) observations, with the motivation that a single differential equation of order $d$ (involving $d$ discrete values) is equivalent to a $d$-dimensional system of first-order equations. SSA does not require an independent estimate on the period of the deterministic contribution, and could therefore apply to situations where the RHESSI spin period is unknown. In what follows, however, we focus on the (usual) case where $T_S$ is known.

The modulation involves two time scales. The first one is the fine-scale modulation which would result from a pair of well separated grids ('modulation amplitude'), and the second one is the slower modulation due to the internal shadowing of each individual grid ('grid transmission'). Together, these modulational contributions yield a 'chirp' covering the frequency range from $\sim 1$ Hz up to $k^i \rho / T_S$ Hz, with $k^i = 2\pi / p^i$ the wave vector of the $i$th subcollimator $i$ ($p^i$ being its angular pitch), and $\rho$ the distance between the imaging axis and the source(s). As a consequence of their broad-band nature, the modulation frequencies tend to interfere with the signal frequencies, and a proper separation requires the inclusion of phase information into the Wiener filter, or the use of non-uniform representations such as Bessel functions or wavelets (Mallat, 1998). This insight is consolidated if one keeps in mind that the modulation is deterministic rather than stochastic. However, phase information requires a preceding estimate of the source position(s) or an equivalent phase identification from the count rates, and thus a higher level of pre-processing. A similar caveat holds for non-recursive (time domain) filters, which are equivalent to convolutions. In order to effectively suppress spin period artifacts the convolution kernel must decay on a time scale $> T_S$, which yields an undesirably vigorous smoothing of the demodulated signal.

When the RHESSI subcollimators sweep an unknown brightness distribution they create, in general, a light curve of complex shape. However, one statement can be made without any detailed knowledge of the brightness distribution: the observed intensities must be $T_S$-periodic if precession and angular acceleration can be neglected, and if the source distribution did not change with time. An intuitive way to see this projects the transmission probability to the solar disk ('modulation pattern') (Figure 1). In the absence of precession, the rotation axis remains stationary, and the imaging (optical) axis $\mathbf{P}(t)$ (dotted circle) rotates around the rotation axis with rigidly co-rotating wave vector $\mathbf{k}(t)$ of the subcollimator (Figure 1 top). By convention, $\mathbf{P}(t)$ is measured with respect to the Sun center (Fivian *et al.*, 2002). The angle between the imaging axis and the rotation axis is called the

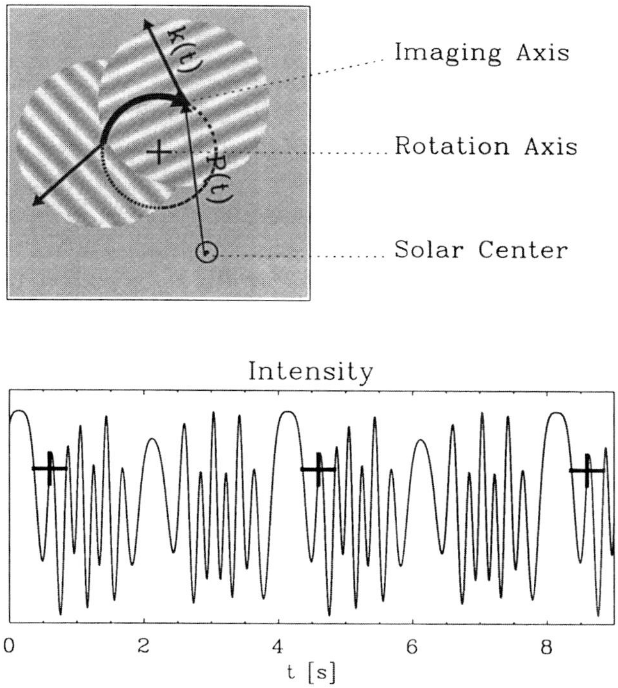


*Figure 1.* Illustration of the rotational modulation in the absence of precession but presence of coning. *Top panel*: in the inertial (solar) frame, the imaging axis $\mathbf{P}(t)$ rotates clockwise around the rotation axis, and carries rigidly along the subcollimator's wave vector $\mathbf{k}(t)$. Two subsequent snapshots are shown. If the solar brightness distribution is time-independent, then the observed light curve is periodic with the RHESSI spin period $T_S \sim 4$ s (*bottom*: crosses denote sample intensities for better clarity.)

coning angle, and RHESSI is 'coning' if this angle is different from zero. Clearly, the modulation pattern reproduces itself after $T_S$ (and possibly also after $nT_S/2$ for favourable $\mathbf{P}(nT_S/2)$), implying equal Poisson intensities for a stationary (or $T_S$-periodic) brightness distribution. This fact is exploited by the demodulation method proposed in the present paper. The method essentially detects deviations from periodicity, and assigns them to the intrinsic time dependence of the source distribution. It is thus limited to time intervals with a (relatively) stable spin axis.

The organization is as follows: Section 2 outlines the modulation principle, Section 3 defines the demodulation (inverse) method, Section 4 describes numerical issues and tests, Section 5 shows some examples of early RHESSI data, and Section 6 contains a brief discussion and preliminary conclusions.

## 2. Forward Model

The method invokes the following assumptions on the ordering of relevant time scales. First, the time $\tau_b$ over which the solar (gradual flare) and non-solar (cosmic, magnetospheric) backgrounds can be considered as constant satisfies $\tau_b \gg T_S$, where $\tau_b$ is of the order of a minute. This rules out, for instance, times just before South Atlantic Anomaly crossings. Secondly, the RHESSI spin period can be considered as constant during $\tau_b$. The source variability itself, as outlined above, is expected to involve time scales down to $\tau_s \sim 100$ ms. Throughout this paper, the analysis is restricted to time intervals of duration $\leq \tau_b$, and to a fixed energy band which is the same for all subcollimators.

The photons arriving at detector $i$ are supposed to be Poisson distributed with time-dependent intensities $\lambda^i(t) = \int M^i(\mathbf{x}, t)\, B(\mathbf{x}, t)\, \mathrm{d}\mathbf{x}$, where $\mathbf{x}$ are coordinates in the (solar) image plane, $M^i(\mathbf{x}, t)$ is the $i$th modulation pattern, and $B(\mathbf{x}, t)$ is the brightness distribution, which, during $\tau_b$, is modeled as $B(\mathbf{x}, t) = B_0(\mathbf{x}) + B_1(\mathbf{x}, t)$. Here, $B_1(\mathbf{x}, t)$ represents an 'impulsive' source and $B_0(\mathbf{x})$ is a stationary background. The short rise time of HRX spikes indicates a – presumably unique – compact source ($c\tau_s \lesssim 1'$ by causality; $v_A \tau_s < 1''$ by the Alfvén travel time), so that one may set $B_1(\mathbf{x}, t) = s(t) B_1(\mathbf{x})$ with some localized function $B_1(\mathbf{x})$ and $s(t)$ the true time profile. The temporal modulation due to a point source located at $\mathbf{x}_s$, $M^i(t) = \int M^i(\mathbf{x}, t)\delta(\mathbf{x} - \mathbf{x}_s)\, \mathrm{d}\mathbf{x}$, is called the 'modulation function'. Its average over time and over the RHESSI field of view is denoted by $a_0^i$ ($\sim$ average grid transmission), and the fundamental Fourier coefficient of $M^i(t)$ is denoted by $a_1^i$ ($\sim$ modulation amplitude). These quantities are taken from the grid transmission software (Hurford *et al.*, 2002b)

Under the above assumptions, the Poisson intensities have the form $\lambda^i(t) = b^i(t) + m^i(t)s(t)$, with an arbitrary positive function $s(t)$, and where the functions $b^i(t)$ and $m^i(t)$ are positive and $T_S$-periodic (including the constant case). The functions $s(t)$, $m^i(t)$ and $b^i(t)$ represent the true time profile, the modulation, and the background, respectively. More precisely, the functions $b^i(t)$ and $m^i(t)$ return to comparable values at times $\{t\}$ with comparable aspect data, in the sense that

$$\left(\frac{\pi a_1^i}{p^i}\right)^2 \left\langle r_\odot^2 (\Phi(t) - \langle\Phi\rangle)^2 + |\mathbf{P}(t) - \langle\mathbf{P}\rangle|^2 \right\rangle < \epsilon^2 , \tag{1}$$

where $\Phi(t)$ is the roll angle, $\mathbf{P}(t)$ is the imaging axis in heliocentric coordinates (Figure 1), $r_\odot$ is the solar radius, and averages are over the time set $\{t\}$. Equation (1) is a sufficient (but not necessary) condition for the rms deviation of the modulation functions being less than $\epsilon$, under the 'worst-case' assumption that the distance between the imaging axis and the source is one solar radius. In practice, $\epsilon < 0.1\, a_0^i$ is found to suppress artifacts due to violation of the periodicity assumption. On times scales $\lesssim 10\, T_S$ and for the coarser grids, Equation (1) is dominated by $\Phi(t)$, whereas on longer terms, the non-recurrent motion of $\mathbf{P}(t)$ cuts off the comparable times. This long-term cut off varies widely from orbit to orbit, and may completely inhibit periodicity for the finest subcollimators. It is assumed that for times satisfying Equation (1), and within the accuracy required by Equation (1), $\Phi(t)$ can be replaced by a linear function of time. This reduces the aspect recurrence to time periodicity, and admits a simple numerical implementation. The roll angle is taken from the PMTRAS instrument (Hurford and Curtis, 2002), with a relative (period-to-period) angular accuracy of $\sim 1'$. The imaging axis is provided by the SAS instrument (Fivian *et al.*, 2002), the accuracy of which is $\Delta\mathbf{P} < 1''$. The term $b^i(t)$ also accounts for a time-independent non-solar background, which – as long as it passes the RHESSI detectors – is $T_S$-periodic.

Soon after the RHESSI launch it became evident that the observed count rates suffer from so-called dropouts where the count rate is zero (Smith *et al.*, 2002; Schwartz *et al.*, 2002). The dropouts are presumably due to cosmic rays hitting the detectors. Empirically, they occur at random with frequency $\sim 1$ Hz and exponentially distributed durations of decay time $\sim 0.25$ s, as if they were triggered and terminated by subsequent points of a Poisson process of intensity $\sim$ 4 Hz. The dropouts, together with the detector dead time, are unified in a lifetime flag $L_k^i$ (Schwartz *et al.*, 2002), which measures the fraction of operational time in each time bin (i.e., $L_k^i = 0.8$ indicates that detector $i$ was operational during 80 percent of the time interval $t_k \ldots t_{k+1}$). For the present purpose, count rates with $L_k^i < 0.5$ are ignored, and corrected for lifetime if $0.5 < L_k^i < 1$.

## 3. Demodulation

Consider a time interval $\tau_b$ with the count rates binned at an integer factor of the spin period, or at a sufficient number of spacecraft binary $\mu$s bins, such that round off does not affect Equation (1). Since $\tau_b > T_S$, this requires somewhat finer time bins ($\Delta t$) than the SSW default. Assume that the corresponding discrete times satisfy Equation (1). The forward model for the count rates is then

$$\lambda_t^i = b^i_{(t \bmod N_S)} + r_t\, m^i_{(t \bmod N_S)}\,, \tag{2}$$

where $N_S = T_S/\Delta t$, $r_t$ is the retrieved 'impulsive' time profile (the estimate for $s(t)$), and we have adopted the convention that superscripts label subcollimators, and subscripts label time. Clearly, the decomposition with Equation (2) is ill-posed or ill-conditioned. A unique solution may, however, be found by assigning as much time variation to modulation as possible. To this end, the goal function

$$F = \underbrace{\sum_{i,j,k}^{*} \left\{ -(b^i_j + m^i_j r_{j+N_S k}) + c^i_{j+N_S k} \left( 1 + \ln \frac{b^i_j + m^i_j r_{j+N_S k}}{c^i_{j+N_S k}} \right) \right\}}_{\ln L/L_0} - \underbrace{-\frac{\alpha}{2} \sum_t (r_{t+1} - r_t)^2}_{A} - \underbrace{\frac{\beta}{2} \sum_i \left( a^i_0 - \frac{1}{P} \sum_j m^i_j \right)^2}_{B} + \underbrace{\gamma \sum_{i,j} \ln b^i_j}_{C} \tag{3}$$

is maximized with respect to $r$, $m$ and $b$ (omitted indices indicate the full index set). Above, $c$ stands for the observed count rates, and $\alpha$, $\beta$, $\gamma$ are positive numbers defined below. The different terms in Equation (3) represent different design goals of the demodulation algorithm. The first term is the Poisson maximum log likelihood ratio, which reduces to minus one half times the usual chi square expression in the limit of large count rates (Eadie *et al.*, 1971; Cash, 1979). The asterisk indicates that the sum is restricted to valid times and subcollimators (no dropouts) and that the lifetime correction is applied if necessary. The second term ($A$) represents a smoothness constraint for minimum derivatives; if data are lacking (dropouts), this term attempts to linearly interpolate $r$ across the data gap. The third term ($B$) ties down the mean values $\langle m^i \rangle$, which are relatively insensitive to the exact source shape and -position (except for spin axis artifacts in the absence of coning), and can be predicted from the grid geometry alone. The fourth term ($C$), finally, resolves the overall ambiguity between $b$ and $mr$ in favour of $b$, such as that periodic contributions from a constant offset of $s(t)$ are attributed to the background, and $m$ (and hence $r$) is not contaminated by $b$. In a Bayesian (Eadie *et al.*, 1971) interpretation, $(C - A - B)$ is proportional to the logarithm of the a priori probability of $(r, m, b)$.

Numerical experience shows, and analytical arguments (Appendix) support, that Equation (3) has a unique solution (global maximum) for sufficiently large $\alpha$, $\beta$, $\gamma$. The current choice is as follows. First, $\alpha = N N_t^{-1} (\tau^2/\Delta t)^2 \bar{r}^{-1}$ with $N$ the number of valid count rates, $N_t$ the number of time bins, $\bar{r} = \sum^*_{it} c^i_t / \sum^*_{it} a^i_0$, and $0.1 \text{ s} < \tau < 1$ s. This allows $r$ to vary by $\sim \sqrt{\bar{r}}$ over the *a priori correlation time* $\tau$. The second regularization parameter is $\beta = N N_c^{-1} (0.2\, a_0)^{-2}$ with $N_c$ the number of subcollimators, which admits a 20% deviation of the retrieved versus theoretical average grid transmissions. The third parameter is $\gamma \lesssim \frac{1}{2} N / (P N_c \ln \bar{c})$ with the average observed (valid) count rate. The upper limit for $\gamma$ is needed in the presence of a dominant background; otherwise, $\gamma$ may be smaller without losing its regularizing function.

At low count rates ($< 1000$ counts s$^{-1}$ subcollimator$^{-1}$), there may not be enough information available to reliably determine the full model $b + mr$. In this situation, the background degrees of freedom are ‘frozen in’ ($b^i_t = 0$). This applies especially to energies above 20 keV, whereas the full model becomes important at energies $\sim 10$ keV with significant background contribution. Note that a frozen background still allows for a *constant* offset, $\lambda^i_t = b^i_0 + m^i_t s_t$, which is included in the periodic functions $m_t^i$ and the constant part of $s_t$. The retrieved and true time profiles are then connected by $r_t = s_t + \sum_i b_0^i / \sum_i a_0^i$.

## 4. Numerical Tests

Equation (3) is solved in an iterative way. The state vector $(r, m, b)$ is updated by a simplified Newton/Marquardt method (Press *et al.*, 1998) with the Hessian approximated by its diagonal. The diffusion equation resulting from the '$A$' term is treated by an implicit (Crank–Nicholson) scheme. The range of $m$, which is given by the slit-to-slot ratio of the grids, is enforced rigorously in the numerical code, and $r$ and $b$ are enforced to be positive. The initial guess for $m_j^i$ is $a_0^i$. If $b$ is frozen to zero then the initial guess for $r$ is $r_t = \bar{r}$; otherwise, the initial guesses for $r$ and $b$ are $r_t = \bar{r}/2$ and $b_j^i = \bar{c}/2$. If $b$ is frozen to zero then convergence is reached after some 30 iterations; if $b$ is varied as well, then convergence is somewhat slower.

The algorithm has been tested by numerical simulations with known modulation functions, 'impulsive' time profiles, and backgrounds. A general example is shown in Figure 2. The true brightness distribution is of the form $B(\mathbf{x}, t) = B_0(\mathbf{x}) + s(t)B_1(\mathbf{x})$, where $B_1(\mathbf{x})$ is a point source 0.75 $r_\odot$ away from the imaging axis, and $B_0(\mathbf{x})$ is an extended (200″) stationary source at distance 0.3 $r_\odot$. The imaging axis is assumed to be constant (no coning), and the sources are not aligned with the imaging axis. The modulation functions involve the first harmonics only. The time profile $s(t)$ of the 'impulsive' source (Figure 2 top middle) consists of a linear rise with a superimposed 'spike' at $t = 30$ s. Both the impulsive and stationary sources contribute about the same number of photons, so that the total average count rate is 1600 counts/s/subcollimator, simulated dropouts being included. The simulation involves the subcollimators 7–9. The spin period is 4 s, the time bin size is 8 ms, the total time interval is 1 min, and an error of 0.5% has been added to the spin period in order to account for uncertainties and binning effects. The a priori correlation time of the retrieved signal is $\tau = 0.3$ s. The input to the demodulation algorithm is the Poisson distributed count rates $c$ (Figure 2, left column). Its output is the decomposition into the 'impulsive' time profile $r$ (top left), the functions $m$ (middle column, black: retrieved, gray: true), and the background $b$ (right column, black: retrieved, gray: true). Since the 'impulsive' source is pointlike, $m$ must coincide with the true modulation functions. As can be noticed, the 'spike' is correctly recovered, although its presence might not be obvious from the count rates by eye. Owing to the term $(C)$ in Equation (3), the constant offset of $s(t)$ is (partially) shifted to the background. The numerical convergence is demonstrated in the top right panel, showing the evolution of the figures of merit $\ln L$, $A$, $B$, and $C$ as a function of the iteration step. All figures of merit are normalized to the number of valid count rates. In the example of Figure 2, the final reduced log likelihood indicates slight over-resolution, which is tolerated for illustration purposes.

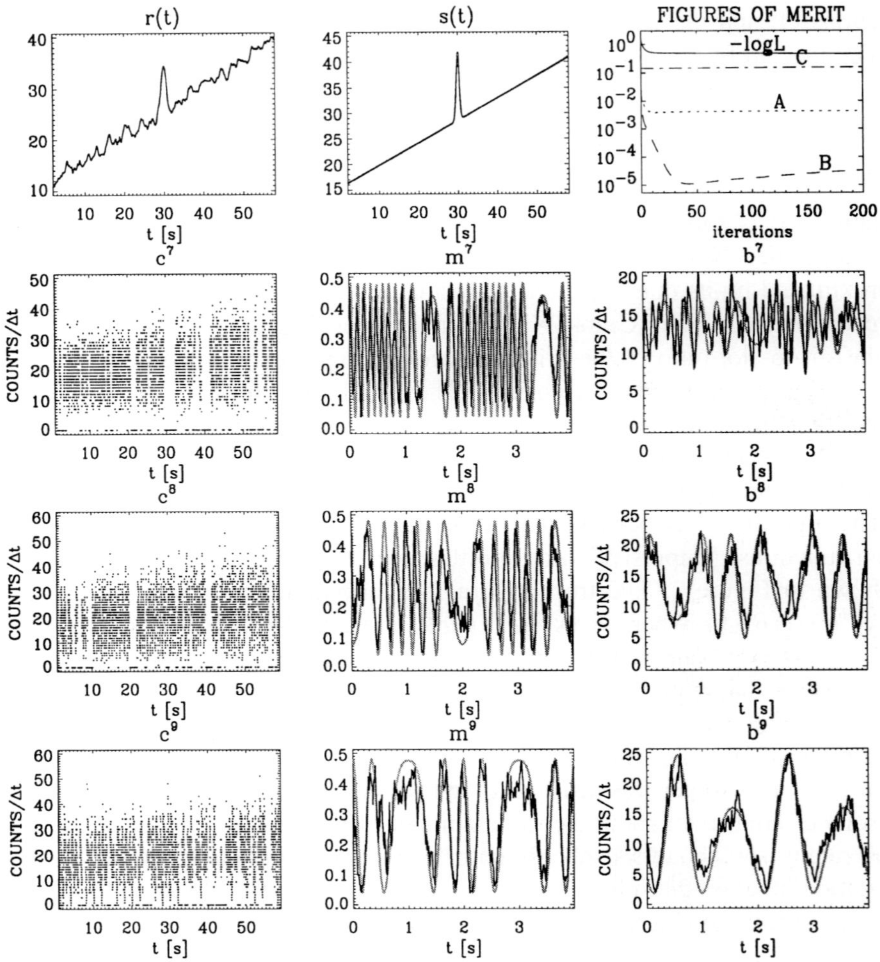

*Figure 2.* Computer simulation of the demodulation problem. *Top line*: retrieved $(r(t))$ and true $(s(t))$ time profiles, and the figures of merit of Equation (3). *Lines 2 to 4*: simulated counts (*left column*); retrieved/true modulation functions (*middle column, black/gray*); retrieved/true backgrounds (*right column, black/gray*). The time bin size is $\Delta t = 8$ ms, and the *a priori* correlation time is $\tau = 0.3$ s.

## 5. First Results

After testing with the simulated time series, the demodulation algorithm was applied to several flares observed by RHESSI. A prominent one is the flare of 20 February 2002, 09:58 UT (Figure 3), which is distinguished by both a high peak count rate ($\sim 6000$ counts $\mathrm{s}^{-1}$ subcollimator$^{-1}$ in the energy band 12–25 keV) and by a high stability of the spin axis. The motion of the imaging axis $\mathbf{P}(t)$ (Figure 3 middle top) reveals coning but only marginal precession; it is recurrent to better than 10 arc sec over one minute around peak time ($\sim$ 09:58:00 UT). The stability of

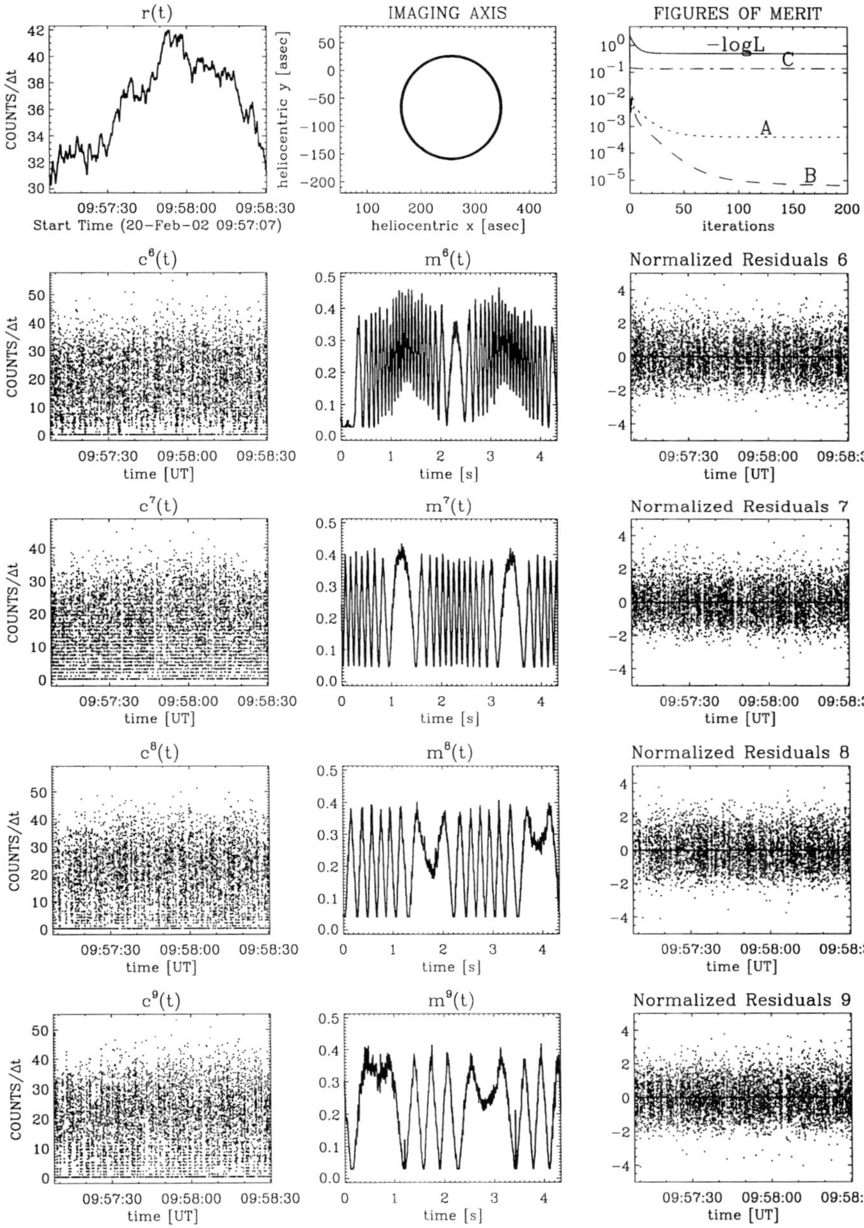


*Figure 3.* Demodulation of the flare of 20 February 2002, 11:06 UT in the energy band 12–50 keV. The time bin size is $\Delta t = 1.9$ ms and the *a priori* autocorrelation time is $\tau = 0.3$ s. Both the retrieved 'impulsive' time profile (*top left*) and an estimate for the functions $m$ (*middle column*) are obtained from the observed counts in detectors 6–9 (*left column*). The 'normalized residuals' are defined by $(c - \lambda)/\sqrt{\lambda}$, with $\lambda$ the predicted count rates (Equation (2)). Not all counts are shown.

the spin axis predicts periodicity for subcollimator 9 during almost the whole flare, and for subcollimators 6–9 during more than one minute at peak time. In fact, with the choice $\Delta t = 0.19$ ms, Equation (1) becomes $(0.005, 0.003, 0.002, 0.001) < 0.03\, a_0^i$ for subcollimators $i = (6,7,8,9)$ in the time interval 09:57:08–09:58:30 UT. The RHESSI spin period, obtained from a linear fit to the aspect solution $\Phi(t)$, is 4.334 s. The observed counts, corrected for lifetime, are shown in the left column, with the demodulated 'impulsive' contribution at the top left panel. The retrieved functions $m^i$, which are a proxy for the modulation functions, are presented in the middle column. The retrieved background (not shown) is approximately proportional to $m$, in agreement with the fact that CLEAN imaging (Högbom, 1974; Lannes, Anterrieu, and Maréchal, 1997) indicates the presence of a single source (same energy band; several 8 s-subintervals). The background contributes about 12 counts$^{-1}$ $\Delta t^{-1}$ subcollimator$^{-1}$; it is constrained by $\gamma = 0.2N/(N_S N_c \ln \bar{c})$. Diminishment of $\gamma$ has a negligible effect on the solution $(r, m, b)$. The final reduced log likelihood is $-0.505$, corresponding to reduced chi squares of (1.1, 0.96, 0.98,0.95) in subcollimators (6,7,8,9). The iterative convergence of the different terms of Equation (3) is shown in the top right panel, where all quantities are again normalized by the number of valid count rates. It can be seen that the trade-off is dominated by the likelihood, in the sense that the absolute change of $\ln L$ is larger than the absolute changes of $A$, $B$ and $C$. (The change of $\ln L$ is more indicative than $\ln L$ itself, because $\ln L$ contains summands which are independent of $(r, m, b)$.) The dominance of $\ln L$ suggests that the qualitative features of the solution $(r, m, b)$ are determined by the agreement between prediction and observation rather than by the regularization.

Another example flare event was observed at 11:06 UT of the same date as the previous example (Figure 4). The energy band is 12–50 keV. Here, the stability of the spin axis predicts periodicity only for the coarsest subcollimators. The time bin size is 8 ms, and the final reduced log likelihood is $-0.507$. The retrieved time profile (Figure 4, top) exhibits several features on time scale $<4$ s; among which the peak at 11:06:44 UT is believed to be real, since it clearly correlates with a type III burst observed by the *Phoenix-2* radiometer (Messmer, Benz, and Monstein 1999) at $\sim 600$ MHz. The three 'peaks' at 11:06:14–24 UT might correspond to similar type III events, but are probably also contaminated by spurious spin periodicity which is not completely removed by the demodulation procedure. Contrary to the 09:58 UT flare event, the retrieved background was found to be nonproportional to the retrieved $m$, which is a hint to the presence of multiple sources. More precisely, the data are not compatible with the assumption that the X-ray brightness distribution is of the form $B(\mathbf{x}, t) = s(t) B_1(\mathbf{x})$. This rules out, for instance, a single time-dependent footpoint of a magnetic loop, or a pair of simultaneously flashing footpoints. Instead, there must be a further (temporally) independent component, which is likely to be driven by different physical causes or configurations.

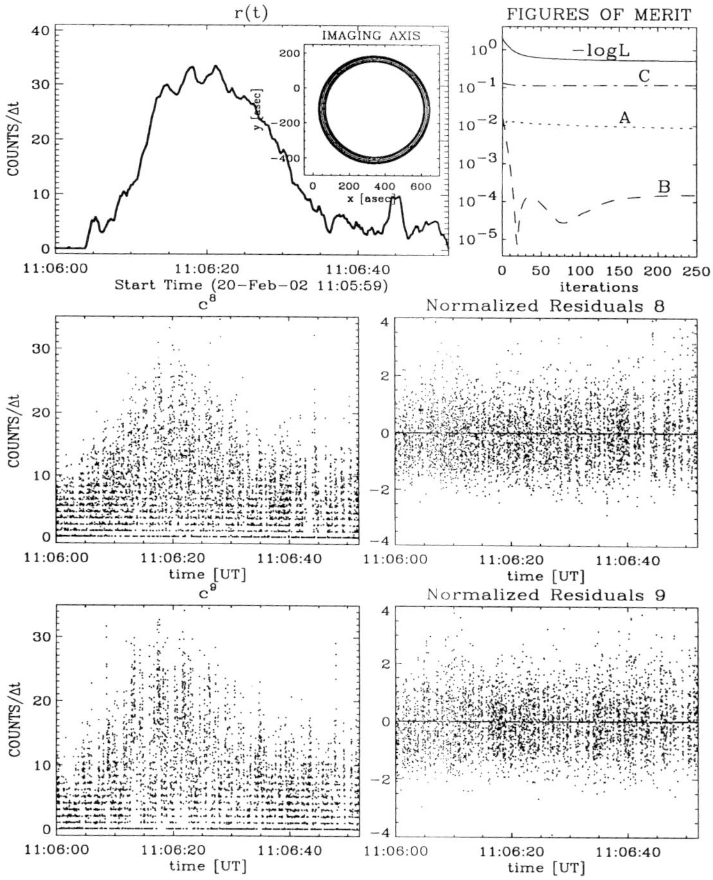


*Figure 4.* Demodulation of the flare of 20 February 2002, 11:06 UT in the energy band 12–50 keV, with $\Delta t = 8$ ms and $\tau = 0.25$ s. All observed counts in subcollimators 8 and 9 (*left column*) are shown.

A third example is provided by the flare of 17 March, 19:28 UT (Figure 5). Here, the energy band is 25–100 keV, and the background is neglected (frozen in). The comparably slow modulation indicates that the source(s) are close to the imaging axis. This, together with a good spin stability, admits the use of subcollimators, 4–9 over the time interval 19:27:25–19:29:10 UT, with $\Delta t = 0.19$ ms and $\tau = 0.25$ s. Subcollimator 5 is excluded, since it gives rise to spurious spin periodicity and an unsatisfactory likelihood ratio. This might be connected to known problems of this subcollimator, which could enter the demodulation through a wrong theoretical mean grid transmission. The retrieved 'impulsive' time profile shows several interesting short-scale features. The retrieved functions $m$ bear close resemblance with the expected modulation functions, suggesting that the source(s) have not been resolved, except for possibly subcollimator 4. In fact, imaging reveals the presence

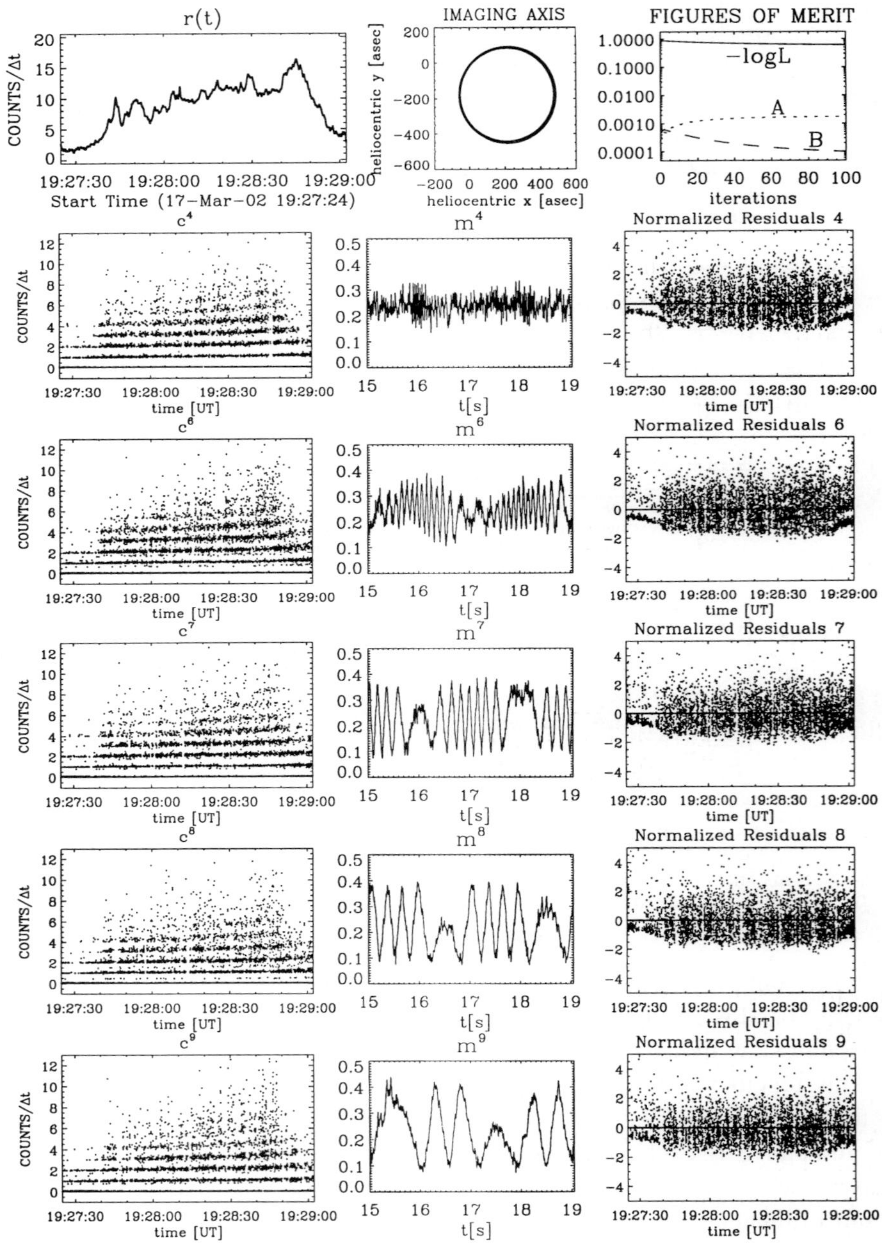


*Figure 5.* The flare of 17 March. 19:28 UT in the energy band 25–100 keV, with $\Delta t = 1.9$ ms and $\tau = 0.25$ s. Here, the background degrees of freedom are frozen in, thus assuming only contributions form the 'impulsive' source. Not all counts are shown.

of two or three sources within 20″, which makes the neglect of the background ($B(\mathbf{x}, t) = s(t)B_1(\mathbf{x})$) questionable for subcollimator 4 ($p^4 = 23''$). Note also the skewness of the residual ($(c - mr)/\sqrt{mr}$) distribution (right column), indicating that a normal approximation would not apply to the low count rates under consideration.

## 6. Summary and Discussion

A time-domain demodulation algorithm is developed and applied to RHESSI light curves, based on the simple observation that the count rates with comparable aspect data should coincide if the source distribution did not change with time. This allows to decompose the count rates into periodic and non-periodic contributions, where the periodic parts are due to modulation and background, and the non-periodic part represents the true time profile of an impulsive source. (Remember that 'periodic' includes the constant case.)

A thorough error analysis of the demodulated light curve $r(t)$ is not elementary and out of the scope of the present article. As an order-of-magnitude estimate, one may however argue as follows. By construction, $r(t)$ has correlation time $\tau$, so that $\{r_t\}$, $t_0 \leq t \leq t_0 + \tau/\Delta t$, represents one degree of freedom. By coherently perturbing $r_t$ in intervals of duration $\tau/\Delta t$ one can find that perturbation $\Delta r$ which causes the (non-reduced) log likelihood, in average, to change by unity. This corresponds to a change of about 2 in the (non-reduced) chi square ($\sim 95\%$ confidence). For the flare of 20 February, 2002, 09:58 UT (Figure 3) one obtains $\Delta r_t \simeq 1.25$; for the flare of 20 February 2002, 11:06 UT (Figure 4) one finds $\Delta r \simeq 1.5$, and for the flare of 17 March 2002, 19:28 UT (Figure 5) $\Delta r \simeq 0.2$. The above estimates neglect other than Poisson errors, as well as errors due to $m$ and $b$. The latter are not independent of $r$, but better known than $r$ if several spin periods available. In practice, the reliability of $r$ should always be verified by comparison with the observed counts, and by exploring different time intervals and sets of subcollimators.

Besides an intrinsic physical interest in short-time structures, the demodulation may also have implications for imaging, since it allows to verify (or disprove) the stationarity assumption underlying the current imaging algorithms. As a side product, it provides an independent estimate on the modulation functions, whenever the source is well-resolved ($m_t^i \simeq M^i(t)$), which is the case if the $m_t^i$ covers the full theoretical range of $M^i(t)$.

The basic advantage of the presented demodulation method is its weak assumptions, which are likely to fail only if several 'impulsive' sources are present at the same time. Its principal limitation is the required aspect recurrence, which rules out the finest subcollimators. The recurrence condition (1) may be relaxed if the grid periodicity is taken into account, so that comparable times $\{t\}$ may occur at integer multiples of $T_S/2$ if $(\mathbf{k} \cdot \Delta\mathbf{P})/2\pi \simeq 0, \pm 1, \pm 2, \ldots$. This relaxation will enlarge the set of flares which are amenable to the demodulation method. The finest subcollimators may be included in another way, e.g., by Wiener filtering. A consistent incorporation into the present framework is one of the major future projects, together with imposing a (computationally more expensive) non-periodicity constraint on $r$, instead of merely asking for smoothness. This is hoped to cure the spurious spin periodicity which is sometimes present in the demodulation. A – more theoretical – caveat is the use of Poisson statistics which is not properly defined for the non-stationary case. The Poisson assumption in Equation (3) is with the understanding that is represents a convenient and sensible approximation to the underlying binomial statistics.

**Acknowledgements**

The author thanks A. Benz, G. Hurford, R. Schwartz, A. Zehnder, and M. Fivian for helpful discussions and A. Benz for providing synoptic radio data.

## Appendix
## Existence of a Global Maximum of *F*

Here are some analytical arguments on the extremal problem $F = \max!$. A sufficient condition for a unique solution is the (strict) convexity of $F$. Let $x \doteq (r, m, b)$ and $F(x)$ be defined as in Equation (3). Remember that all elements of $x$ are positive, and that $0 < m < 1$. Applying the triangle inequality to the sums over spin periods and subcollimators it can be shown that the Hessian $H_{ij} = \partial^2 F/\partial x_i \partial x_j$ satisfies $H_{ii} < 0$ and $H_{ii}H_{jj} > |H_{ij}|^2$, which is a necessary condition for negative definiticity. In order to characterise a sufficient condition, consider the function $\Delta F \doteq F(\frac{1}{2}(x_1 + x_2)) - \frac{1}{2}(F(x_1) + F(x_2))$. The discussion is restricted to a single subcollimator, since the inclusion of several subcollimators is not expected to degrade the conditioning. Convexity requires $\Delta F > 0$. Itemized by $(r, m, b)$, one has that

$$\Delta F = \tfrac{1}{4}\sum(m_1 - m_2)(r_1 - r_2) + \tfrac{\alpha}{8}\sum(r_1 - r_2)\mathcal{D}(r_1 - r_2) + \\ + \tfrac{\beta}{8}\left(\sum(m_1 - m_2)\right)^2 + \sum c \ln \tfrac{(b_1+b_2)/2+(m_1+m_2)(r_1+r_2)/4}{(b_1+m_1r_1)^{1/2}(b_2+m_2r_2)^{1/2}} + \gamma \sum \ln \tfrac{(b_1+b_2)/2}{(b_1b_2)^{1/2}},$$

with $\mathcal{D}_{ij} = -\delta_{i-1,j} + 2\delta_{i,j} + \delta_{i+1,j}$ the 2nd order difference operator, and where the sums run over the full index sets. The 2nd, 3rd and 5th term of $\Delta F$ are non-negative definite (by positivity of $\mathcal{D}$ and $\beta$, and the arithmetic-geometric inequality), while the 1st and 4th term may be negative. Assume now that $b$ is frozen to zero ($b_1 = b_2 \to 0$). The 5th term of $\Delta F$ then vanishes, and each summand of the 4th term possesses the (weak) lower bound

$$c \ln \tfrac{(m_1+m_2)(r_1+r_2)/4}{(m_1r_1)^{1/2}(m_2r_2)^{1/2}} = -\tfrac{1}{2}c \ln[1 - (\tfrac{m_1-m_2}{m_1+m_2})^2] - \tfrac{1}{2}c \ln[1 - (\tfrac{r_1-r_2}{r_1+r_2})^2] \geq \\ \geq \tfrac{1}{2}c(\tfrac{m_1-m_2}{m_1+m_2})^2 + \tfrac{1}{2}c(\tfrac{r_1-r_2}{r_1+r_2})^2 \geq \tfrac{1}{8}c(m_1 - m_2)^2 + \tfrac{1}{8\max(r)^2}c(r_1 - r_2)^2,$$

where it was used that $-\ln(1-\xi) \geq \xi$ for $0 \leq \xi \leq 1$, and that $\max(m) < 1$. Therefore, $\Delta F \geq \frac{1}{4}(\Delta r, \Delta m)^T M(\Delta r, \Delta m)$, where the symmetric matrix $M = \begin{pmatrix} M^{rr} & M^{rm} \\ M^{mr} & M^{mm} \end{pmatrix}$ has the blocks $M^{rr}_{ij} = \frac{c_i}{\max(r)^2}\delta_{ij} + \alpha \mathcal{D}_{ij}$, $M^{rm}_{ij} = \delta_{i \bmod P, j}$, and $M^{mm}_{ij} = \delta_{ij} \sum_k c_{j+Pk} + \beta$. It may (numerically) be shown that $M$ is positive definite for suitable $\alpha, \beta > 0$; increasing count rates favour the dominance of the diagonal blocks over the off-diagonal blocks, and allow decreasing regularization. The case $b_1 \neq b_2 > 0$ can be treated by similar lower bounds on the 4th and 5th terms of $\Delta F$. Analytical regularity bounds on $(\alpha, \beta, \gamma)$, which are both rigorous and tight, were not obtained so far.

## References


Benz, A. O. *et al.*: 1994, *Solar Phys.* **153**, 33.

Cash, W.: 1979, *Astrophys. J.* **228**, 939.

Dennis, B.: 1985, *Solar Phys.* **100**, 465.

Eadie, W. T., Drijard, D., James, F. E., Roos, M., Sadoulet, B.: 1971, *Statistical Methods in Experimental Physics*, Elsevier Science Publ., Amsterdam.

Fivian M., Hemmeck, R., Mchedlishvili, A., and Zehnder, A.: 2002, *Solar Phys.*, this volume.

Gary, D.: 2000, in R. Ramaty and N. Mandzhavidze (eds.), *High Energy Solar Physics – Anticipating HESSI, ASP Conference Series* **206**, 297.

Högbom, J. A.: 1974; *Astron. Astrophys. Suppl. Series* **15**, 417.

Hurford, G. J. and Curtis, D.: 2002, *Solar Phys.*, this volume.

Hurford G. J. *et al.*: 2002, *Solar Phys.*, this volume.

Kailath, T.: 1974, A view of three decades of linear filtering theory, *IEEE Trans. Info. Theory*, **IT-20**, 146.

Kane, S. R., Chupp, E. L., Forrest, D. E., Share, G., and Rieger, E.: 1986, *Astrophys. J.* **300**, L95.

Kiplinger, A. L., Dennis, B. R., Frost, K. J., and Orwig, L. E.: 1984, *Astrophys. J.* **28**, L105.

Lannes, A., Anterrieu, E., and Maréchal, P.: 1997, *Astron. Astrophys. Suppl. Ser.* **123**, 183.

Machado, M. E., Ong, K., Emslie, A. G., Fishman, G. J., Meegan, C., Wilson, R., and Paciesas, W. S.: 1993, *Adv. Space Res.* **13**, 175.

Mallat, S., 1998, *A wavelet tour of signal processing*, Academic Press, London.

Messmer, P., Benz, A. O., and Monstein, C.: 1999, *Solar Phys.* **187**, 335.

Miller, J. A., Ramaty, R., and Murphy, R. J.: 1987, *Proc. 20th Cosmic Rays Conf.* **3**, 33.

Miller, J. A., Cargill, P. J., Emslie, A. G., Holman, G. D., Dennis, B. R., LaRosa, T. N., Winglee, R. M., Benka, S. G., and Tsuneta, S.: 1997, *J. Geophys. Res.* **102**, 14631.

Nakajima, H.: 2000, in R. Ramaty and N. Mandzhavidze (eds.), *ASP Conference Series* **206**, 313.

Press, W., Teukolsky, S., Vetterling, W., and Flannery, B.: 1998, *Numerical Recipes in C*, 2nd ed., Cambridge University Press, Cambridge.

Rosen, Y. and Porat, B.: 1989, *IEEE Trans. Info. Theory* **IT-35**, 342.

Schwartz, R. *et al.*: 2002, *Solar Phys.*, this volume.

Smith, D.: 2002, *Solar Phys.*, this volume.

Vautard, R. and Ghil, M.: 1989, *Physica* **D35**, 395.

Varadi F., Pap, J. M., Ulrich, R. K., Bertello, L., and Henney, C. J.: 1999, *Astrophys. J.* **526**, 1052.

Varadi F., Ulrich, R. K., Bertello, L., and Henney, C. J.: 2000, *Astrophys. J.* **526**, L53.

# RELATIVE TIMING AND SPECTRA OF SOLAR FLARE HARD X-RAY SOURCES

SÄM KRUCKER [1] and R. P. LIN[1,2]
[1] *Space Sciences Lab, University of California, Berkeley, CA 94720-7450, U.S.A.*
[2] *Physics Department, University of California, Berkeley, CA 94720-7450, U.S.A.*

(Received 7 August 2002; accepted 2 September 2002)

**Abstract.** Hard X-ray lightcurves, spectrograms, images, and spectra of three medium-sized flares observed by the Reuven Ramaty High-Energy Solar Spectroscopic Imager (RHESSI) are presented. Imaging spectroscopy of the 20 February 2002, 11:06 UT flare at $10''$ spatial resolution, comparable to the best previous hard X-ray imaging from *Yohkoh*, shows two footpoints with an $\sim 8$ s delay of peak emission between footpoints. Subsequent imaging at $\leq 4''$ shows three sources consistent with two separate loops and simultaneous brightening in connected footpoints. Imaging for the simple two footpoint flare of 2 June 2002 also shows simultaneous footpoint brightening. The more complex 17 March 2002 flare shows at least four different sources during the main peak of the event, and it is difficult to clearly demonstrate simultaneous brightening of connected footpoints. Non-thermal power laws are observed down to $\sim 12-13$ keV without flattening in all these events, indicating the energy content in energetic electrons may be significantly greater than previously estimated from assumed 25 keV low energy cutoff. Simultaneously brightening footpoints show similar spectra, at least in the three flares investigated. Double-power-law spectra with a relatively sharp break are often observed.

## 1. Introduction

During the flare impulsive phase, a double footpoint structure is often observed with the two footpoints brightening simultaneously to within a fraction of a second (Sakao *et al.*, 1994, 1996). These coincide, spatially and temporally, with H$\alpha$ and white-light brightenings. For some flares occurring near the solar limb, the *Yohkoh* Hard X-ray Telescope (HXT) has detected a co-temporal, weaker, hard X-ray source in the corona (Masuda *et al.*, 1994; Alexander and Metcalf, 1997) above the soft X-ray loop linking the hard X-ray footpoints. This source has been interpreted as evidence for energy release by magnetic reconnection in a region above the soft X-ray loop.

Very rapid transport of energy from the release site down to the footpoint interaction regions is required. This can only be achieved by fast electrons streaming down the loop and depositing their energy in lower coronal and chromospheric footpoints, implying that nonthermal, tens-of-keV electrons are interacting with a cool ($T < 10^5$ K) environment. The resulting heating of the ambient gas leads to evaporation of chromospheric material and flare radiation in the visible, UV/EUV and soft X-rays. Hard X-ray studies of electron time-of-flight (Aschwanden *et al.*,

1996) further support this nonthermal thick-target model. RHESSI's hard X-ray imaging, with much wider energy range (3 keV to 17 MeV), higher dynamic range, and finer spatial ( 2.3 arc sec) and spectral resolution (1 keV FWHM) than *Yohkoh* HXT, is designed to study these sources and electron streaming in detail. Here we present the first results for three flares, concentrating on the timing, imaging, and spectral signatures in hard X-rays to provide information on the acceleration and propagation of the energetic electrons in the flares.

## 2. Data Analysis

Overview plots (Figures 1, 6, 8) are used to select flare events for study. To produce these plots, the RHESSI counts are binned in energy in two ways: (1) The photons are binned into 4 energy bands to make time profiles at 1 second resolution (second panel from top); or (2) into 32 logarithmically spaced energy channels to make count spectrogram plots (third panel from top). The decrease in counts below $\sim 10$ keV in these plots is due primarily to absorption by the thin shutter (cf., Lin *et al.*, 2002) that is in front of the detectors during these observations. The logarithmic binning produces a clearer representation by flattening out the otherwise steep spectrum. Overview plots of this kind are available online at *http://sprg.ssl.berkeley.edu/~krucker/hessi_plots*.

### 2.1. Imaging

For all events, cleaned images (Hurford *et al.*, 2002) averaged over one rotation period ($\sim 4$ s) are constructed every second around the time of the HXR peak. Hence, there is time overlap in these time series and short lived sources could be smeared out. The standard RHESSI software corrects these images for efficiency using the diagonal elements of the spectral response matrix only (cf., Hurford *et al.*, 2002). Generally, grid 2 and higher are used giving a resolution down to 4 arc sec. Including grid 1 (2.3 arc sec) did not enhance the image quality for these events at the present state of the calibration (Figure 2, bottom). These time series of images allowed the event evolution to be studied in detail, and the time profiles of individual sources to be derived. In particular, it allows the determination of when sources appear and disappear, and the selection of the time interval to integrate over to get the best possible image of a particular source.

### 2.2. Imaging spectroscopy

To determine the spectrum of an individual source, images at 20 logarithmically spaced energy bands are made. Individual sources are then selected and the flux within each source is summed separately at each energy. An error estimate is approximated from the residual map provided by the CLEAN algorithm. The maximum in the residual map outside the source is taken as a very rough approximation of a $3\sigma$ error. The same error is used to determine the uncertainties of the time profiles mentioned above. In the graphical representation, the derived errors are divided by 3 to represent a $1\sigma$ error bar.

As a first attempt, a very simple fitting of the derived flux spectra is used that will be improved in future work. A power law or broken power law is first fitted to the spectra above $\sim 15$ keV. The cutoff energy of the power law is chosen normally between 10 to 15 keV; at lower energies the power law contribution is neglected. This power law fit is then subtracted from the observed spectrum before fitting a thermal spectrum to the values below $\sim 10$ keV. The thin shutter (cf., Lin *et al.*, 2002) was in front of the detectors for all the flares presented here. This shutter absorbed most of the photons below $\sim 10$ keV. Therefore, for the events presented here, fitting below 10 keV depends critically on the calibration.

Fitting of the thermal component in the HXR footpoint spectra has an additional, more severe problem. It is very difficult to separate the thermal emission of the footpoint from the much brighter thermal emission of the loop often seen between the footpoints. Contamination from the thermal loop source (due to sidelobes) might contribute significantly to the flux in the selected source region of the HXR footpoints. Therefore the fitted thermal emission of footpoints presented in this work are preliminary and need more investigation. The derived values appear reasonable, between 10–20 MK; however, because of the above mentioned difficulties, the derived footpoint temperature and emission measure are not discussed in detail.

## 3. Observations

Three events are presented here: 20 February 2002, 11:05 UT; 2 June 2002, 11:44 UT; and 17 March 2002, 19:28 UT. All these events were selected because they show prominent hard X-ray peaks above 30 keV. The 20 February flare is presented in detail, the 2 June flare is selected as an example of a simple two footpoint source, and the 17 March flare is an example of a very complex flare.

### 3.1. 20 February flare

The 20 February, 11:05 UT flare presented here is a smaller flare (GOES class C7.7) occurring on a day when 5 GOES M-class flares were observed. Nevertheless, this event was selected since only a single, simple HXR peak is seen. This event is also discussed by the images, by Aschwanden *et al.* (2002) studying the height at which the HXR emission originates, and by Sui *et al.* (2002) modeling the HXR emission. Additionally, this flare produced intense radio bursts that are compared with the HXR emission by Vilmer, Krucker, and Lin (2002).

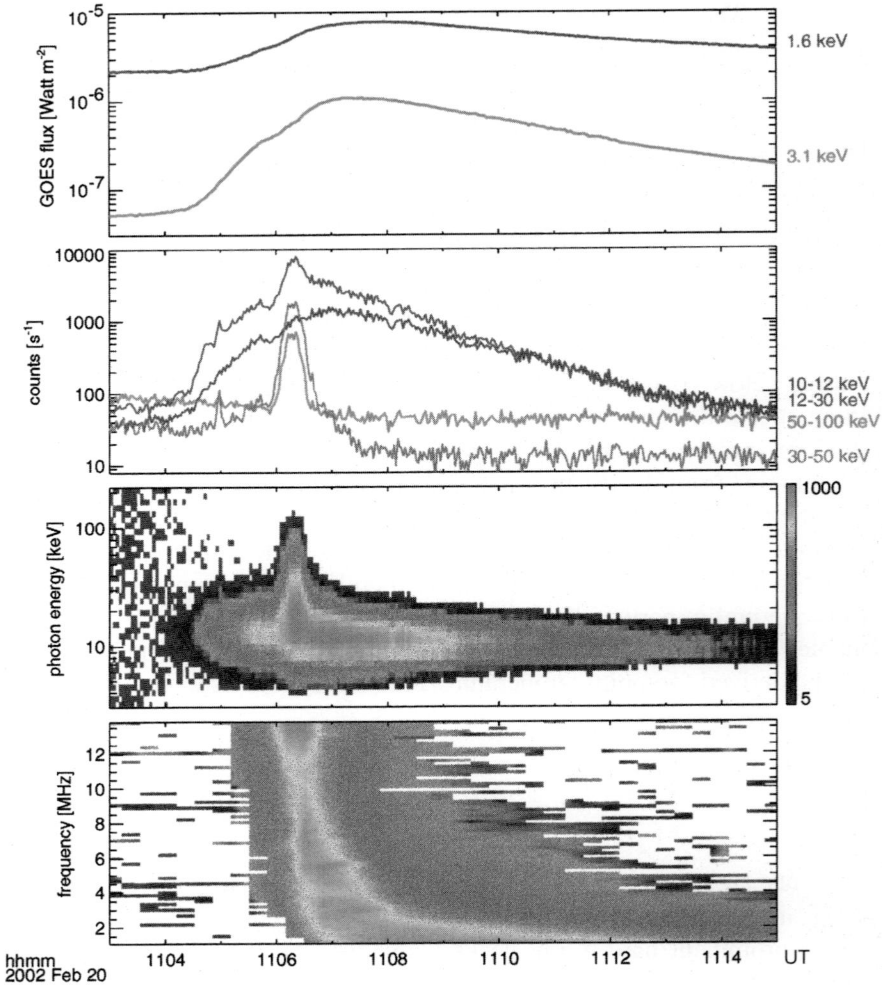


*Figure 1.* X-rays and radio waves overview plot of the 20 February 2002 flare. *Top*: GOES 1–8 Å (1.6 keV) and 0.5–4 Å (3.1 keV) lightcurves. *Second from top*: RHESSI lightcurves at four energies. *Third from top*: RHESSI spectrogram plot, the background has been subtracted. *Bottom*: WIND/WAVES RAD2 radio spectrogram. See the accompanying CD-ROM for color versions of this and other figures.

### 3.1.1. *Imaging*

Imaging around the hard X-ray peak reveals a flare location close to the west limb (N14 W79). At a spatial resolution of 10″, comparable to that of the *Yohkoh* HXT, two main HXR sources could be identified (Figure 2, top). Surprisingly, the maxima of the two sources did not occur simultaneously (Figure 2, bottom) contrary to the general behavior seen by Sakao *et al.* (1996). The southern source peaked about ∼ 8 seconds later than the northern source. However, imaging at resolution

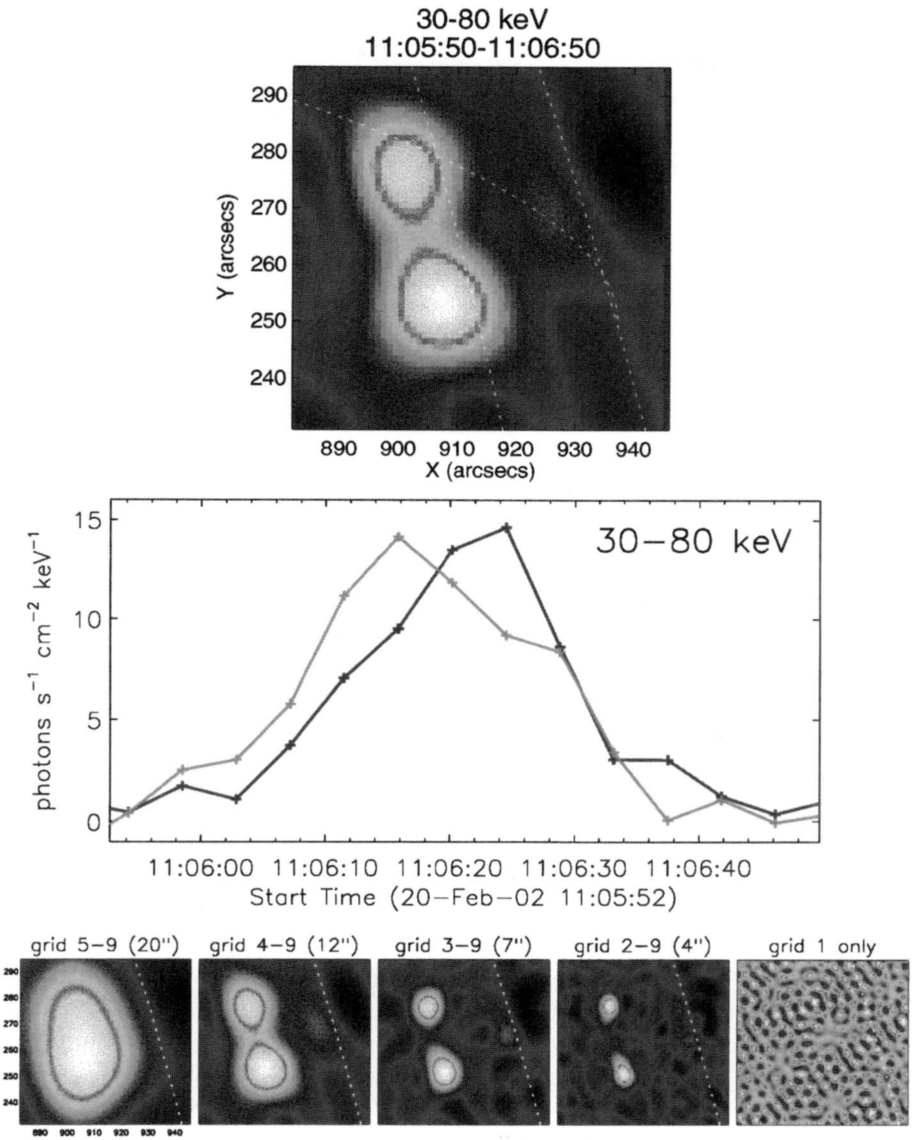


*Figure 2.* Non-thermal X-ray image (30–80 keV) of the 20 February flare with a resolution of 12″ (*top*). At this resolution, only two sources are seen that show different time profiles (in gray for the northern source) suggesting non-simultaneous brightening of footpoints (*middle*). Cleaned images with succesively better resolution are shown below. The last image is computed using the finest grid only (grid 1, 2.3″) and does not show significant emission. All times in UT.

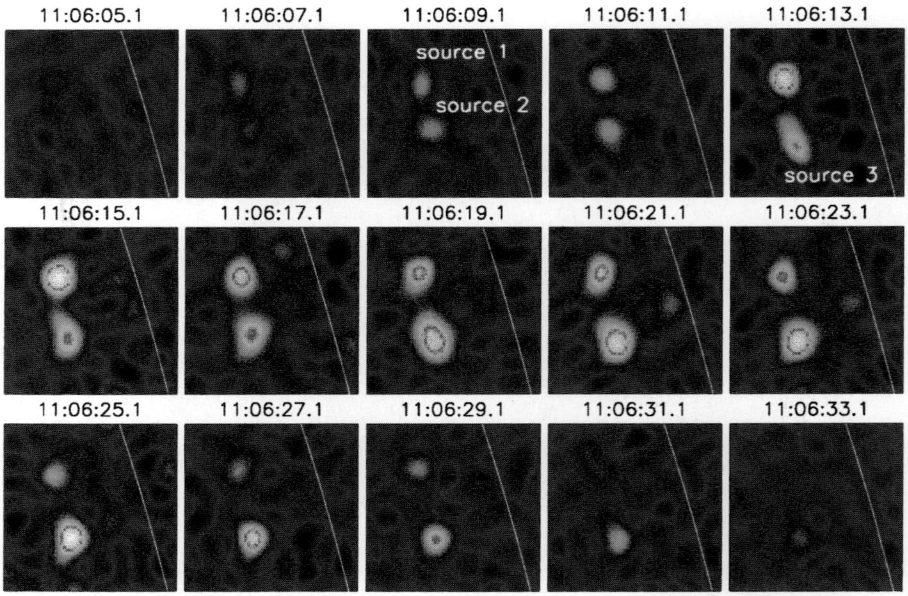


*Figure 3.* HXR images (30–80 keV, 7″ resolution) taken every 2 s with ~4 s integration time of the 20 February flare are shown. The field of view is 64″ by 64″, the *white lines* give the location of the solar limb. All times in UT.

below 4″ and with a cadence of 1 second (with an integration time over ~4 s, one spin period) showed three sources: Source 1 to the north, source 3 to the south, and a weaker, earlier peaking source 2 in-between but much closer to source 3 (Figure 3). With this third source, the observed temporal evolution could be made consistent with simultaneous brightening of connected footpoints in two loops; one loop connecting source 1 and 2, the second connecting source 1 and 3.

In the beginning of the HXR peak (around 11:06:08 UT), source 1 and 2 are seen brightening simultaneously, reaching only moderate fluxes. Then, source 3 appears when source 1 peaks, while source 2 is slowly disappearing. In a third and last phase, source 3 peaks with a weaker simultaneous brightening of source 1.

Figure 4 shows X-ray images of the three different phases. The thermal emission seen in the 6–9 keV images suggest a hot loop connecting source 1 and 2, but there seems to be no hot loop connection to source 3. However, loop connection to source 3 might be much fainter and therefore be lost in the residuals of the cleaned images. Difference images produced from EIT observations (Delaboudinière *et al.*, 1995) showing possibly the loop connecting source 1 and 2 some minutes later, but the uncertainty of the roll angle of EIT images, of the order of 10″ (private communication by Gallagher), makes it difficult to relate the HXR sources and the post flare loops seen in EIT.

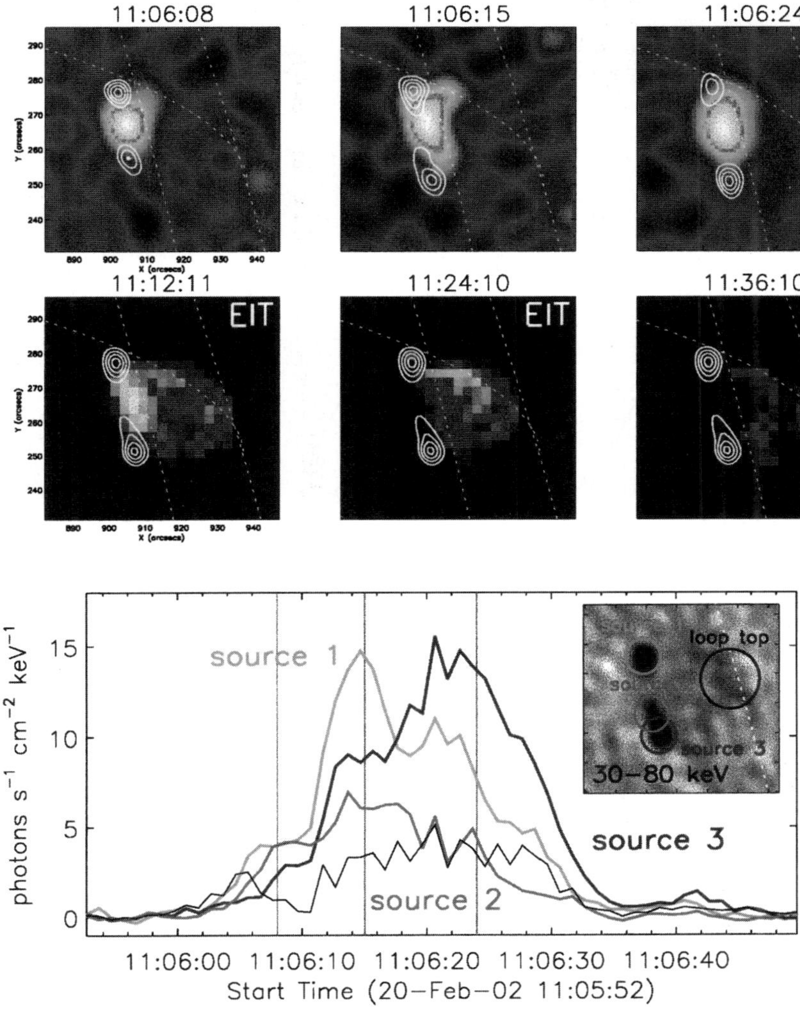


*Figure 4.* The 3 different HXR sources appearing in the 20 February flare are shown in the *first row* of figures. The images show the thermal emission (6–9 keV) with the 30–80 keV HXR contours overlayed (levels are 30, 50, 70, 90%). Later in the event EIT 195 Å difference images show the post flare loops (*second row* of figures). To outline the position of the earlier occurring HXR sources, contours of the time averaged 30–80 keV emission are overplotted. The *bottom panel* shows the temporal evolution of the different sources, including a possible source above the EIT loop top (*black*). The circles in the insert mark the source locations, but do not represent the source sizes. The three *gray vertical lines* give the center time of the interval of the above shown images; time intervals are $\sim 8$ s, $\sim 8$ s, and $\sim 16$ s, respectively. All times in UT.

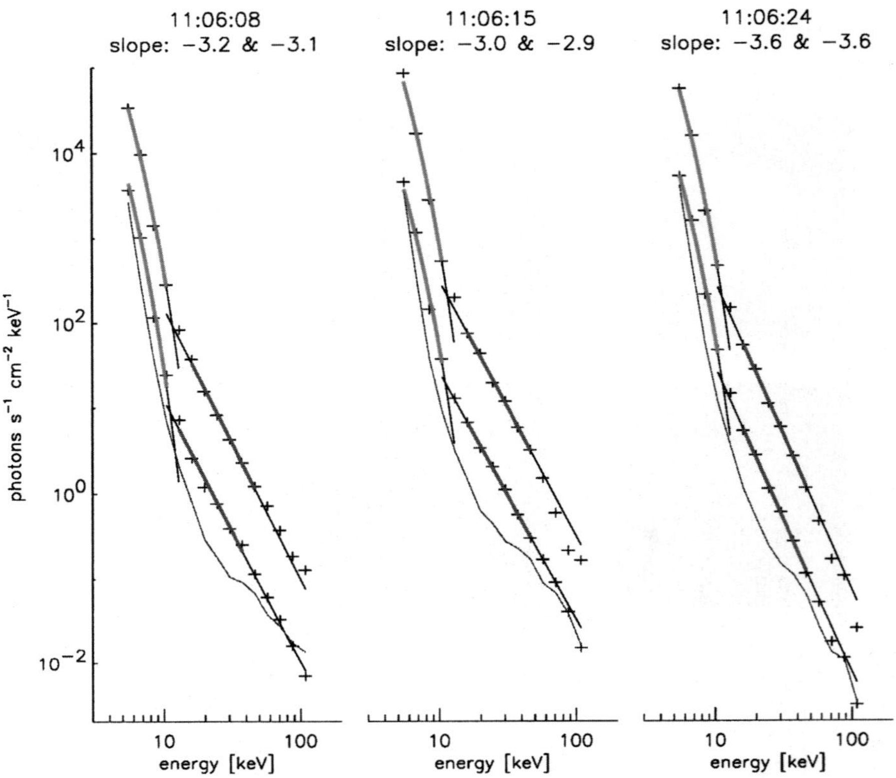


*Figure 5.* Spectra of the different sources outlined in Figure 4. Each plot contains two spectra, source 1 is always plotted on top and shifted by a decade, source 2 is only shown at the bottom of the first plot, source 3 in the remaining two spectra. Thermal fits are shown in *thick, medium gray* and power-law fits in *thick, dark gray*. The *gray lines* give rough estimates of the uncertainties. The *thick lines* give the range in energy used to make the fit, values above 50 keV are questionable.

Next to the three main sources, there may be a fourth source that might just be slightly above the EIT loop; this is discussed in Sui *et al.* (2002). The time variation of this possible loop top source is shown in Figure 4 (bottom) in black.

### 3.1.2. *Imaging Spectroscopy*

For the three different phases of the HXR peak, images reconstructed in 16 narrow energy ranges were used to provide energy spectra for the three sources seen (Figure 5). All spectra show power laws at high energies, extending down to $\sim 12$–$13$ keV. Since the thermal emission is of similar magnitude around 10 keV, only the energy range above $\sim 15$ keV is used for the power law fit. At energies above $\sim 50$ keV, the uncertainties are large and these points are not used in the fit. No break in spectra is seen in this energy range: This is not incompatible with spatially integrated spectra that show a break at higher energies (between 45 to 60 keV).

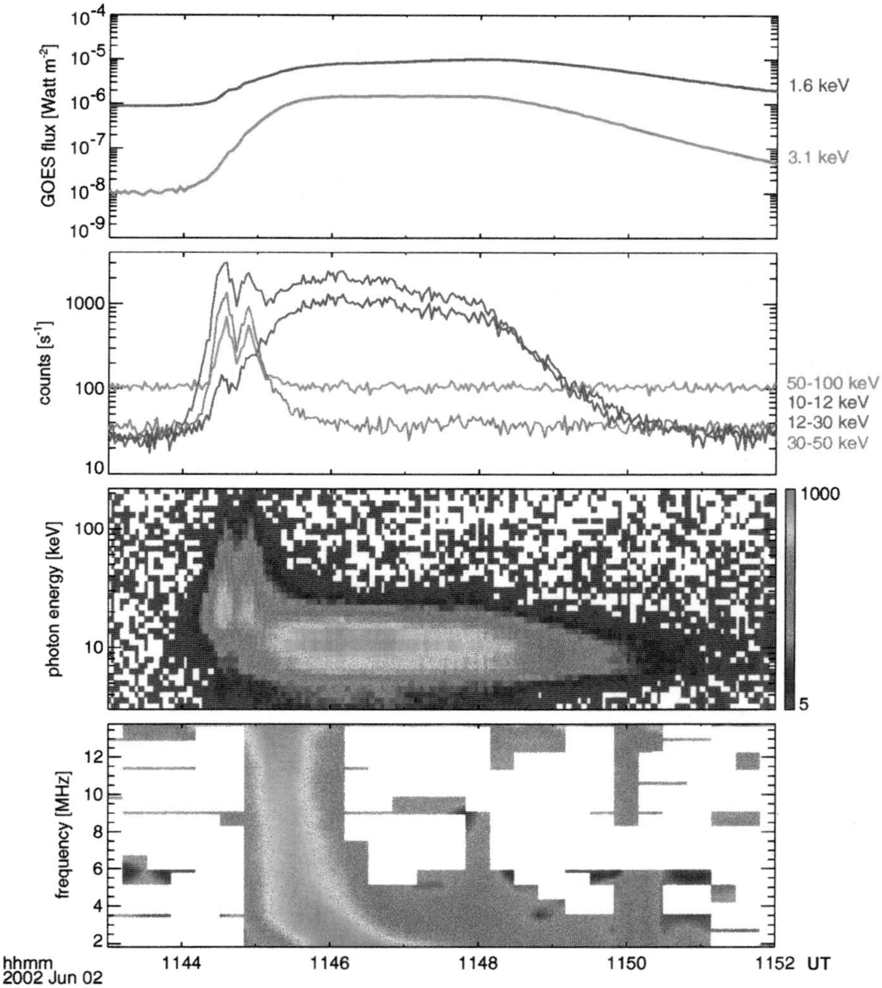


*Figure 6.* X-ray and radio wave overview plot of the event on 2 June 2002.

Spectra of loop-connected footpoints show very similar spectral slopes for all the three intervals. Power-law exponents between −2.9 to −3.2 are calculated for the first two phases, while the last peak gives a softer spectral exponent around −3.6. Spatially integrated spectra show power-law exponents between −3.1 to −3.4 for the first two phases, and around −3.6 for the last peak. Considering the early stage of data analysis, the agreement is good.

### 3.2. 2 June flare

The 2 June, 11:44 UT flare (GOES class M1.0) occurred on a day with several low-M-class flares (Figure 6). It was selected for its impulsive onset with a rise

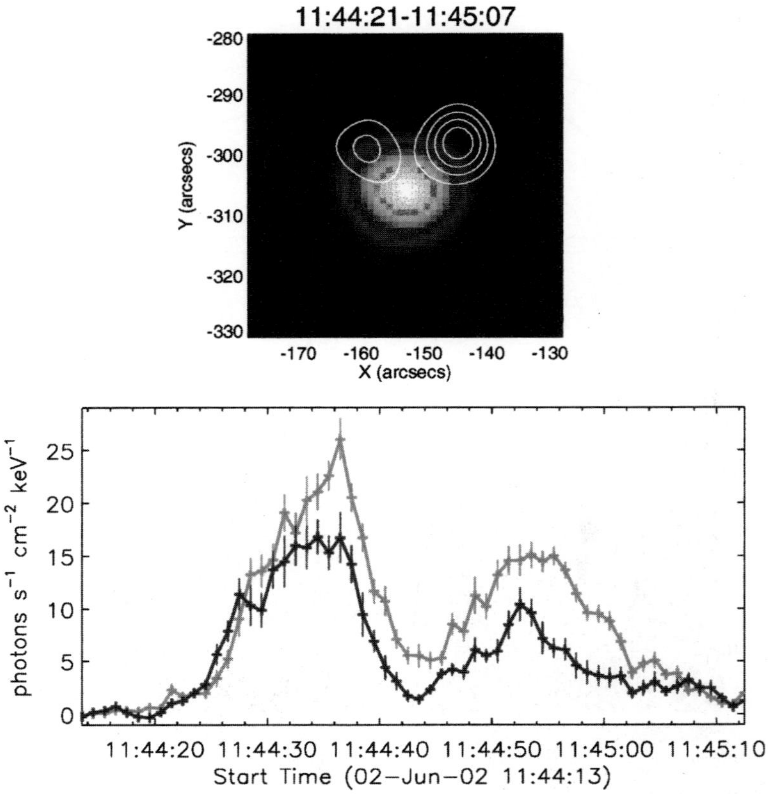


*Figure 7. Top*: thermal X-ray image (6–9 keV) and non-thermal X-ray contours (30–80 keV, levels are 30, 50, 70, 90%) averaged of the HXR peak of the 2 June flare are shown. *Bottom*: the time profile of the two sources are given. Times in UT.

phase of 3 min and its simple double-peaked HXR time profile. The images around the peak times show only two footpoints occurring near S19 E10 (Figure 7), and even after careful investigation, no other sources could be found. This simple flare geometry is a good test for simultaneity: first results (Figure 7, bottom) show that the two footpoints brighten simultaneously within the uncertainties.

### 3.3. 17 March flare

The GOES M4.4 flare occurring at S21 E17 on 17 March 2002 is an example of a more complex flare. The RHESSI data show three distinct groups of HXR peaks above 30 keV: rise (19:26:30–19:27:30 UT), main peak (19:27:30–19:29:00 UT), and decay (19:29:00–19:33:00 UT), which continues after the GOES flux is already decaying (Figure 8). The spatial distribution is also very complex with at least 4 different sources appearing during the main peak alone (Figure 9). With this complexity of the source structure, it is difficult for the source temporal evolution

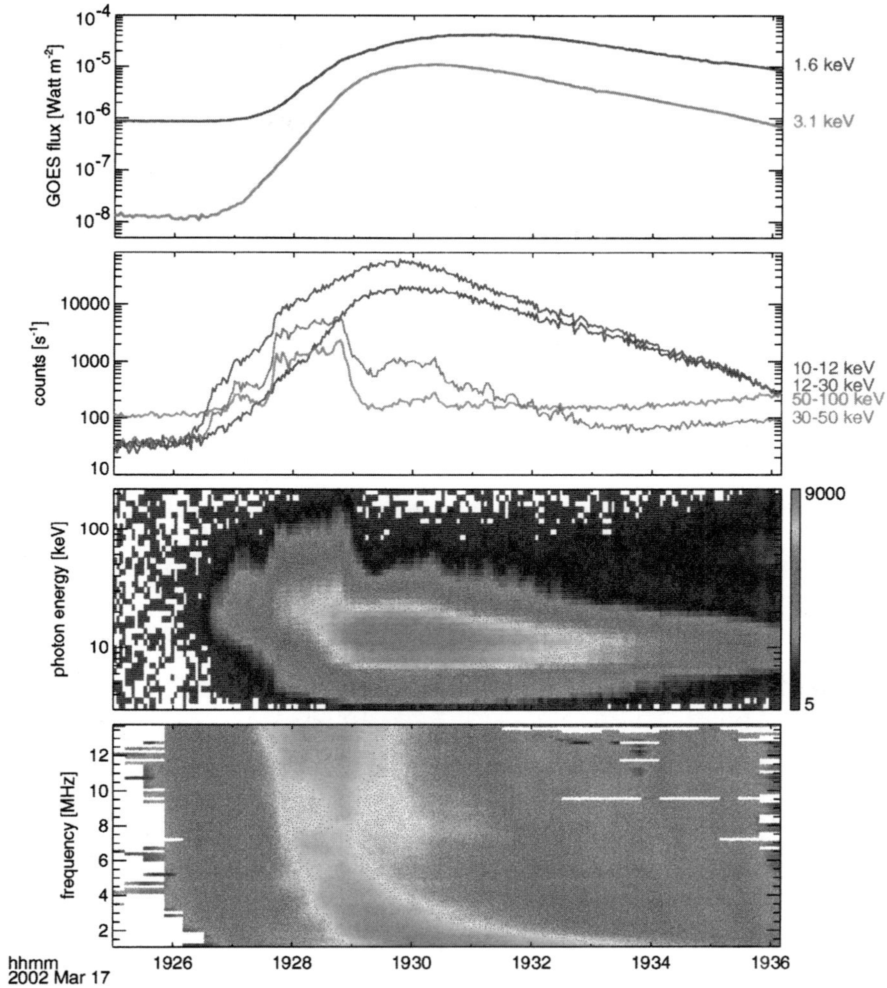


*Figure 8.* X-ray and radio wave overview plot of the event on 17 March 2002. Note that the enhanced emission starting late in the event around ~ 19:33 UT seen mostly above 30 keV is not of solar origin, but produced locally around the spacecraft by magnetospheric particles.

to be clearly interpreted as simultaneous brightening of loop-connected footpoints. Alexander and Metcalf (2002) analyze this event using the Pixon algorithm.

The main HXR peak (19:27:30–19:29:00 UT) was divided into two time intervals that were studied separately. The most prominent sources are two footpoints connected by a thermal loop seen between 19:28:38 and 19:28:50 UT (Figure 10, right). Earlier (19:27:36–19:28:36 UT), the source structure is more complex, showing an elongated source in the east (Figure 10, left) that most likely is composed of 3 sources (see time series of images shown in Figure 9). The source seen to the west seems to be the same source that later shows up during the main peak.

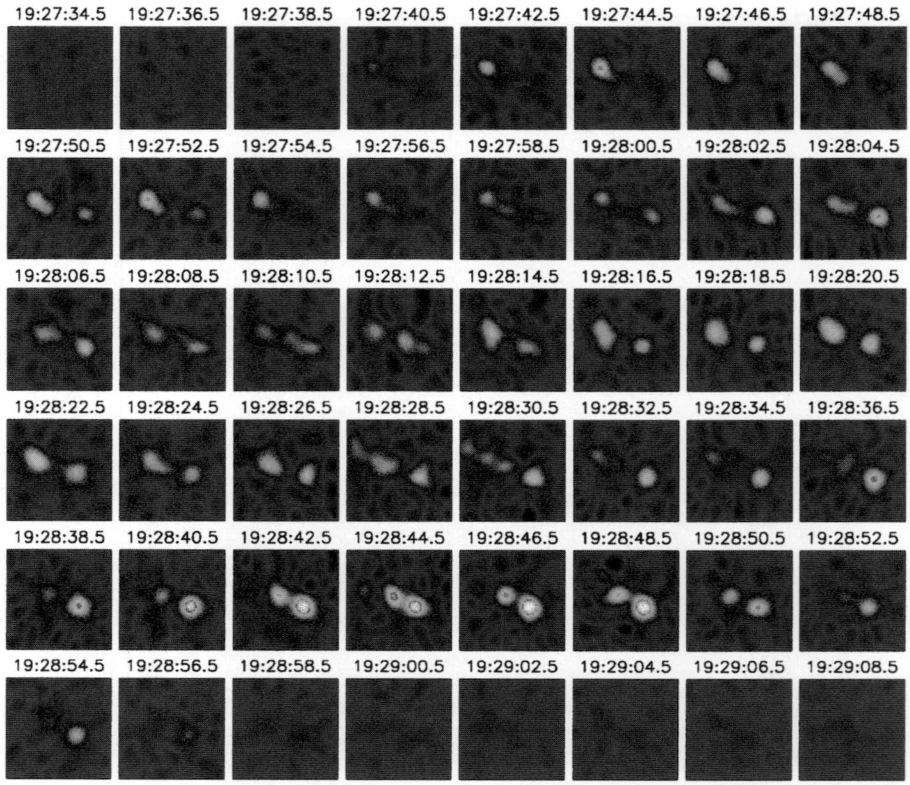


*Figure 9.* HXR images (30–80 keV, 7″ resolution) taken every 2 s with $\sim$4 s integration time of the 17 March event. Time in UT.

However, the second footpoint seen during the main peak is missing (or much fainter) during this earlier phase.

Interestingly, the fits to the spectrum of the sources seen in the earlier phase and the main peak give different results: the spectra in the earlier phase can only be fitted with a broken power law with a steeper spectra above $\sim$30 keV, whereas the main peak can be fitted with a single power law up to 60 keV. The similarity of the spectra for the two main sources in each time interval suggests a possible spatial (loop) connection.

## 4. Discussion

First results of HXR images and imaging spectroscopy obtained by the RHESSI mission reveal the great potential of this mission. Three medium-sized flares with rather short and impulsive onsets in X-rays are studied revealing the complexity of X-ray sources seen during a flare. Even if the HXR time profile shows only one,

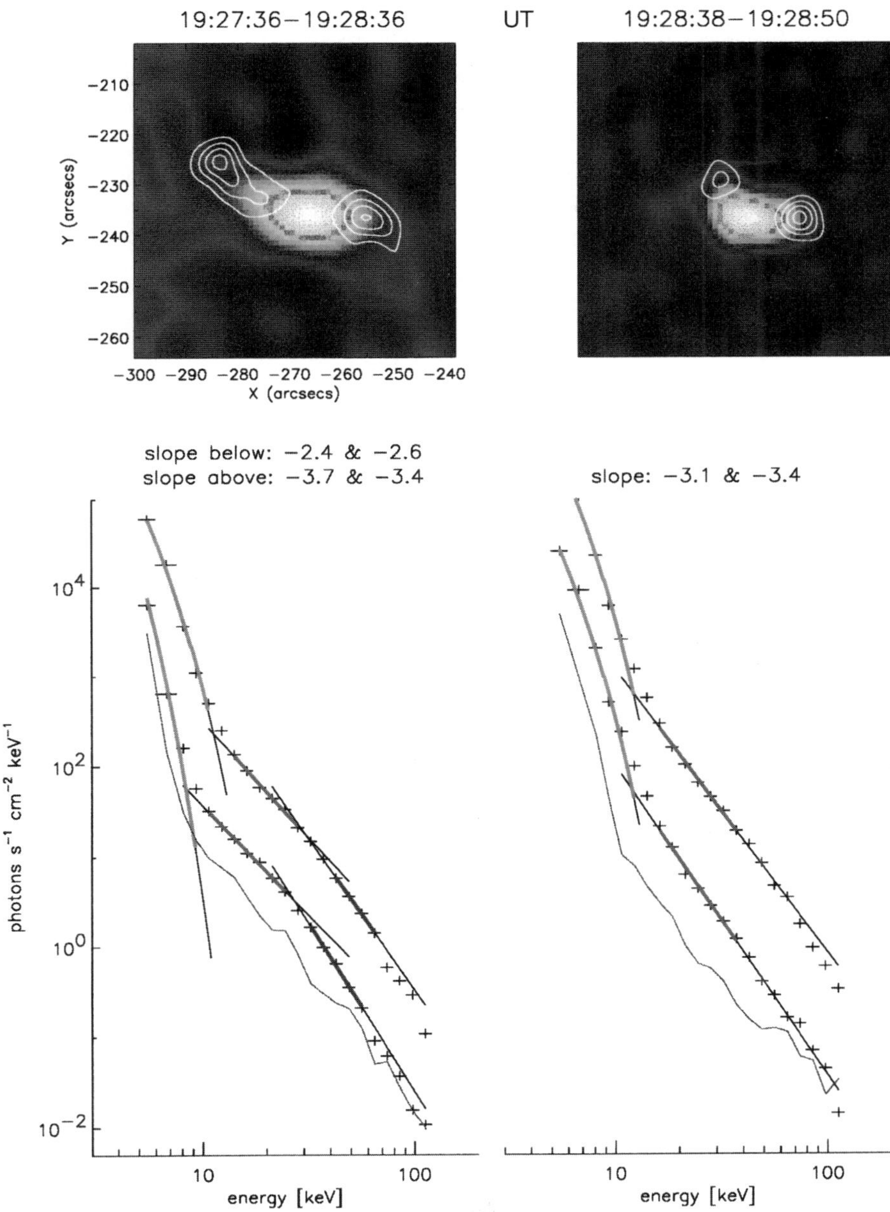


*Figure 10.* Images and spectra of the inital HXR phase and the main HXR peak of the 17 March flare. The images show again the 6–9 keV thermal emission and the contours are the non-thermal 30–80 keV hard X-rays (levels are 30, 50, 70, 90%). *Below*, the spectra taken during these two time periods are shown. The spectrum of the western source is shown on top shifted up by a decade from the eastern source. The *shades* represent the same as in Figure 5 except that additionally a broken power law is fitted in *thick dark gray* to the earlier phase of the flare. Values above 65 keV are questionable.

simply shaped peak, several sources can be involved. Without adequate spatial resolution, these observations can be mistakenly interpreted as simple two-footpoint flares. By wrongly combining two or more sources into one source, the time history then shows non-simultaneous brightening of footpoints. Imaging with resolution down to 4″, however, reveals the complexity of source structure. The time evolution and the spectra of these sources are then found consistent with simultaneous brightening of connected footpoints. More detailed studies are need to give quantitative values and error estimates of the simultaneity. Thermal emission is observed to be coming from compact regions often connecting HXR footpoints.

Imaging spectroscopy with RHESSI gives for the first time the possibility to distinguish footpoints and other features in the images and to compute their spectra independently down to 3 keV with a spectral resolution of 1 keV. Generally, non-thermal power-law spectra are observed down to at least $\sim 12$ keV. This indicates that the lower energy cutoffs used in previous years of 20–25 keV (e.g., Crosby *et al.*, 1994) for the calculation of the total energy in energetic electrons are likely underestimating the energy content in non-thermal electrons by a factor of at least $> 4$ (assuming a power law index of 3). For microflares where the thermal component is weaker and at lower energies, making it easier to detect non-thermal emission below 10 keV, power laws extending even down to 7 keV are reported (Krucker *et al.*, 2002).

These first imaging spectroscopy results also show that simultaneously brightening footpoints have similar non-thermal power law indices, even if the absolute values of the spectra can be different. This similarity of the spectra provides independent support for the loop-connection of the two footpoints.

In the event of 17 March where statistics are most favorable, a break above 35 keV is observed in the spectra, with a steeper slope at higher energies. Later in the event during the main peak of the flare, the spectra, however, are best represented with a single power law fit. In events where the statistics are good, double-power-laws with a relatively sharp break appear to be common. Such spectra may have significant implications for the acceleration and/or energy loss processes.

## Acknowledgements

This research is supported by NASA contract NAS 5-98033. We acknowledge the RHESSI team's outstanding effort in making this mission a success under adverse conditions.

## References

Alexander, D. and Metcalf, T. R.: 1997, *Astrophys. J.* **489**, 442.
Alexander, D. and Metcalf, T. R.: 2002, *Solar Phys.*, this volume.

Aschwanden, M. J., Wills, M. J., Hudson, H. S., Kosugi, T., and Schwartz, R. A.: 1996, *Astrophys. J.* **468**, 398.
Aschwanden, M. J. *et al.*: 2002, *Solar Phys.*, this volume.
Crosby, N. B., Aschwanden, M. J., and Dennis, B. R.: 1993, *Solar Phys.* **143**, 275.
Delaboudinière J.-P. *et al.*: 1995, *Solar Phys.* **162**, 291.
Sui, L., Holman, G. D., Dennis, B. R., Krucker, S., Schwartz, R. A., and Tolbert, K.: 2002, *Solar Phys.*, this volume.
Hurford, G. J. *et al.*: 2002, *Solar Phys.*, this volume.
Lin, R. P. *et al.*: 2002, *Solar Phys.*, this volume.
Krucker, S., Christe, S., Lin, R. P., Hurford, G. J., and Schwartz, R. A.: 2002, *Solar Phys.*, this volume.
Masuda, S., Kosugi, T., Hara, H., Tsuneta, S., and Ogawara, Y.: 1994, *Nature* **371**, 495.
Sakao, T., Kosugi, T., Masuda, S., Yaji, K., Inda-Koide, M., and Makishima, K.: 1994, *Proceedings of Kofu Symposium*, Kofu, Japan, p. 169.
Sakao, T., Kosugi, T., Masuda, S., Yaji, K., Inda-Koide, M., and Makishima, K.: 1996, *Adv. Space Res.* **17**, 60.
Vilmer, N. *et al.*: 2002, *Solar Phys.*, this volume.

# MODELING IMAGES AND SPECTRA OF A SOLAR FLARE OBSERVED BY RHESSI ON 20 FEBRUARY 2002

LINHUI SUI[1], GORDON D. HOLMAN[2], BRIAN R. DENNIS[2], SÄM KRUCKER[3], RICHARD A. SCHWARTZ[4] and KIM TOLBERT[4]

[1]*CUA & Laboratory for Astronomy and Solar Physics, NASA Goddard Space Flight Center, Greenbelt, MD 20771, U.S.A. (e-mail: lhsui@stars.gsfc.nasa.gov)*

[2]*Laboratory for Astronomy and Solar Physics, NASA Goddard Space Flight Center, Greenbelt, MD 20771, U.S.A. (e-mail: holman@stars.gsfc.nasa.gov; Brian.R.Dennis.1@gsfc.nasa.gov)*

[3]*Space Science Laboratory, University of California at Berkeley, Berkeley, CA 94720, U.S.A. (e-mail: krucker@ssl.berkeley.edu)*

[4]*SSAI & Laboratory for Astronomy and Solar Physics, NASA Goddard Space Flight Center, Greenbelt, MD 20771, U.S.A. (e-mail: richard.schwartz@gsfc.nasa.gov; kim.tolbert@gsfc.nasa.gov)*

(Received 6 August 2002; accepted 30 August 2002)

**Abstract.** We have analyzed a C7.5 limb flare observed by RHESSI on 20 February 2002. The RHESSI images appear to show two footpoints and a loop-top source. Our goal was to determine if the data are consistent with a simple steady-state model in which high-energy electrons are continuously injected at the top of a semicircular flare loop. A comparison of the RHESSI images with simulated images from the model has made it possible for us to identify spurious sources and fluxes in the RHESSI images. We find that the RHESSI results are in many aspects consistent with the model if a thermal source is included between the loop footpoints, but there is a problem with the spectral index of the loop-top source. The thermal source between the footpoints is likely to be a low-lying loop interacting with the northern footpoint of a higher loop containing the loop-top source.

## 1. Introduction

The study of hard X-ray emission is important for understanding electron acceleration and transportation in flares. Most observations of solar hard X-rays have been obtained with poor spectral and spatial resolution. Years of *Yohkoh* HXT observations indicate that many types of hard X-ray sources exist, even in a single flare (Masuda, 2002). For this reason, imaging spectroscopy is very important for investigating individual sources in flares, demanding X-ray data with high spatial, spectral and temporal resolution. The Ramaty High-Energy Solar Spectroscopic Imager (RHESSI), launched on 5 February 2002, is now obtaining high spatial ($\sim 2$ arc sec), spectral ($\sim 1$ keV) and temporal (tens of milliseconds) resolution data in 3 keV–17 MeV X-rays and $\gamma$-rays. All these new features offer a dramatic improvement in studying the hard X-ray emission from flares.

In order to interpret RHESSI images and spectra, we established flare models (Holman *et al.*, 2001, 2002) based on *Yohkoh* HXT observations of loop-top

*Solar Physics* **210:** 245–259, 2002.

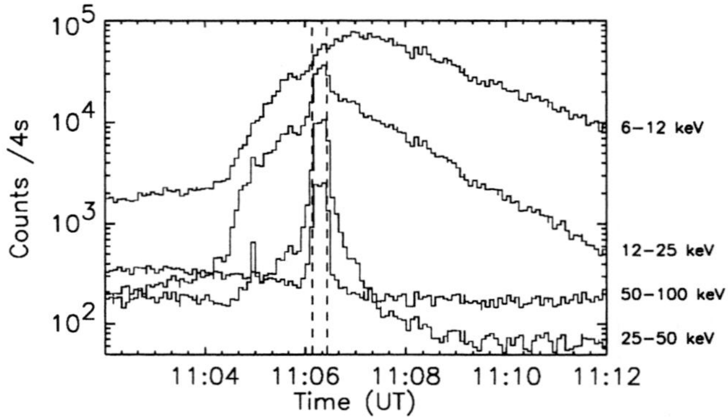


*Figure 1.* RHESSI light curves in four energy bands. The count rate (4-s time integration) is summed over detectors 1 through 9, excluding detectors 2 and 7. For clarity, we have scaled the count rates by 5.0 (6–12 keV), 1.5 (12–25 keV), 1.0 (25–50 keV), and 1.0 (50–100 keV). The two *vertical lines* show the integration time interval for the RHESSI images and spectra in this paper.

sources (Masuda, 1994). We apply one of these models to a flare observed by RHESSI on 20 February 2002. We first compute model images with flare parameters derived from the observational data. We then obtain simulated RHESSI images and spectra using the computed model images as input. Finally, we compare the simulated images and spectra with RHESSI results. This method not only tests our model, but enables us to test the RHESSI imaging software as well.

## 2. Observational Results and Data Analysis

On 20 February 2002, RHESSI observed a C7.5 flare in NOAA active region 9825, located near the northwest limb of the Sun at N16 W80. The GOES-8 soft X-ray flare started at 11:02 UT and ended at 11:12 UT. The RHESSI hard X-ray peak time was at 11:06:20 UT, as shown in Figure 1.

The spatially integrated and background-subtracted photon spectrum at the time of the hard X-ray peak, 11:06:10–11:06:24 UT (see Figure 1), is shown in Figure 2. This spectrum was obtained using the RHESSI spectral executive (SPEX) software (see Smith *et al.*, 2002). We have integrated a program into SPEX that computes the thick-target bremsstrahlung flux spectrum from a single or a double power-law distribution of electrons. The photon spectrum in Figure 2 was obtained by fitting the RHESSI count data to the bremsstrahlung spectrum from an isothermal plasma plus a double power-law electron distribution.

The background varied with time during this flare. It was subtracted from the flare data by obtaining a linear fit to measurements of the background flux before and after the flare. Below 100 keV uncertainties in the fluxes resulting from the background subtraction are less than 20%.

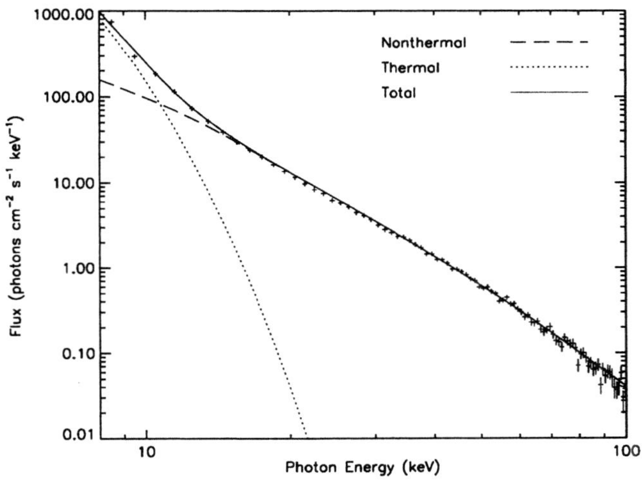


*Figure 2.* RHESSI spatially integrated, background-subtracted photon spectrum for the time interval 11:06:10–11:06:24 UT, the same as that for all RHESSI images shown later. The bremsstrahlung spectrum from an isothermal plasma plus a double power-law electron distribution (both shown separately on the plot, in addition to the total spectrum) were fit to the RHESSI count data to obtain this spectrum.

The RHESSI thin shutters were in the field of view of the detectors for this event. When the thin shutters are in, the effective area of the detectors drops rapidly as the photon energy falls below 10 keV. The current uncertainty in the correction factor is large at these energies, so fluxes below 10 keV were not included for spectral fitting. The X-ray background flux becomes significant at photon energies above 100 keV, so fluxes at energies above 100 keV were also not included.

The emission measure and temperature determined from our fit to the spatially integrated RHESSI spectrum in Figure 2 are $2 \times 10^{48}$ cm$^{-3}$ and 15 MK, respectively. We obtained an emission measure and temperature from the GOES-8 soft X-rays data at 11:06:20 UT of $3 \times 10^{48}$ cm$^{-3}$ and 14 MK. These were determined using the program GOES_TEM.PRO in the Solar Software Tree, which was developed from the concepts of Garcia (1994) and Thomas, Crannell, and Starr (1985). The GOES-8 results are consistent with those from our spectral fit to the RHESSI data.

We have found several functions that provide a good fit to the non-thermal part of the spatially integrated spectrum. The double power-law fit to the photon spectrum included with SPEX gives a spectral index of $-3.3$ below 56 keV and $-4.3$ above 56 keV. The normalization of the double power-law fit is 0.6 photons s$^{-1}$ cm$^{-2}$ keV$^{-1}$ at 50 keV. We have verified the existence of this break in the photon spectrum by analyzing the data from each individual detector.

The fit shown in Figure 2 is the bremsstrahlung spectrum from a double power-law electron density distribution. This gives a power-law index for the electron distribution of $-4.4$ below 100 keV and $-5.5$ at higher energies. With a low-energy

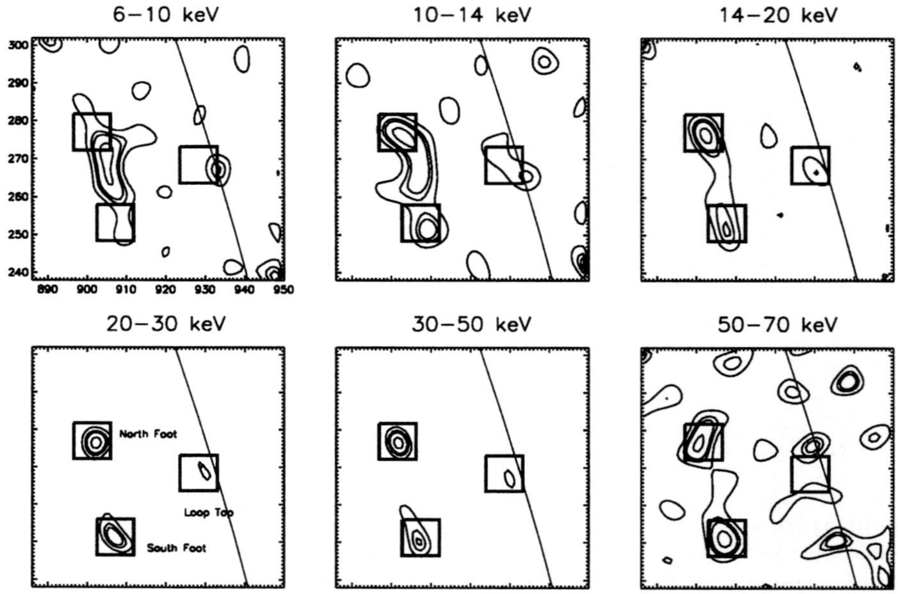


*Figure 3.* Images obtained with the MEM-Sato algorithm. The time interval is 11:06:10–11:06:24 UT. The contour levels are 0.08, 0.2, 0.3, 0.7. The contour levels are normalized to the peak flux of each image. The *smooth line* shows the location of the solar limb. The three *boxes* indicate the pixels that were summed to obtain the footpoint and loop-top spectra in Figure 5.

cutoff in the distribution at 15 keV, the normalization gives the total supra-thermal electron density times source area to be $nA = 4.6 \times 10^{25}$ electrons cm$^{-1}$. The low-energy cutoff was chosen to be 15 keV because this minimizes the total number of supra-thermal electrons while still providing a good fit to the photon spectrum. The bremsstrahlung from a single power-law density distribution with a high-energy cutoff also provides a good fit to the data. This gives a power-law index of −4.5 and a high-energy cutoff of 224 keV. The value of $nA$ is the same as before.

A single power-law electron distribution with a low-energy cutoff at 47 keV also provides an acceptable fit to the spectrum. The power-law index is −5.3 and $nA = 1.24 \times 10^{24}$ electrons cm$^{-1}$. But this fit requires a much higher temperature for the thermal plasma, 40 MK, with an emission measure of $4.6 \times 10^{46}$ cm$^{-3}$. Since this temperature and emission measure do not correspond to the results from GOES, this case will not be pursued further here, but will be explored in future work.

RHESSI images (see Hurford *et al.*, 2002) in six energy bands at the time of the hard X-ray spike are plotted in Figures 3 and 4. Each image is $64 \times 64$ arc sec in size. At the time of this analysis, the relative alignment was not known for collimators 1 and 2, so the spatial resolution of the images is limited to 7 arc sec. Collimator 9 was also not included, since all source structure was well below 180 arc sec in extent.

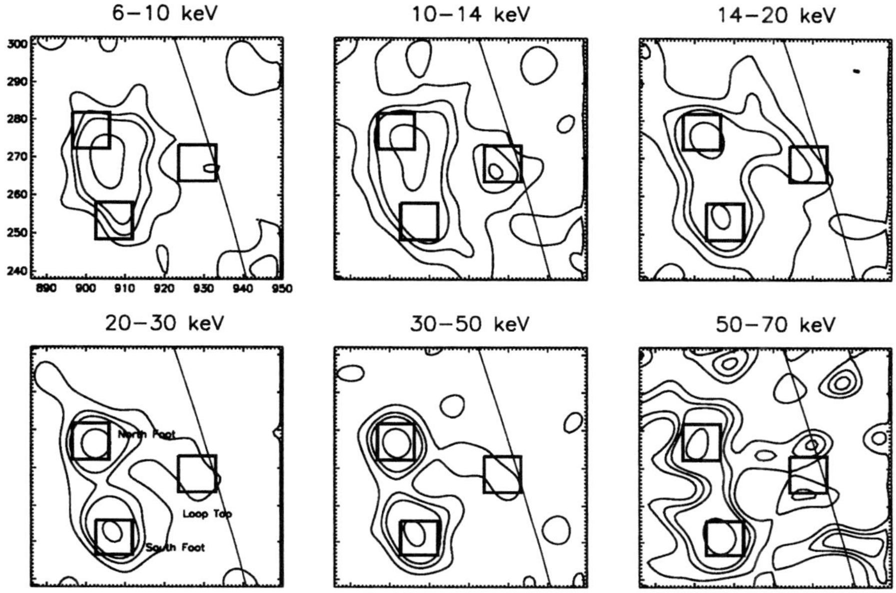


*Figure 4.* Images obtained with the CLEAN algorithm. All other aspects are the same as in Figure 3.

In order to check the reliability of different image reconstruction algorithms, images using both the Maximum Entropy (MEM-Sato – Figure 3) and the CLEAN (Figure 4) reconstruction techniques were obtained. MEM-Sato is an image reconstruction algorithm utilizing the maximum entropy method (MEM) described by Sato, Kosugi, and Makishima (1999), modified for RHESSI. The clean algorithm is an iterative procedure that takes the dirty map, made using back projection (Hurford *et al.*, 2002), finds the brightest pixel, and uses some fraction of its intensity and point spread function to subtract side lobes from the dirty map. This process is then repeated until the brightest pixel is negative, or a maximum number of iterations has been reached. The CLEAN and MEM-Sato images agree with each other in coarse structure. In the low-energy band (6–10 keV), the images show a single source which lies between the two footpoints observed at higher energies. This is identified to be a thermal source with a temperature of 15 MK. In the 10–14 keV band, the images show the thermal source and a possible loop-top source. In the energy bands between 14 keV and 50 keV, the images show two separate footpoints and the loop-top source. In the 50–70 keV band the loop-top source is not present in the images. The quality of the images in this energy band is marginal, however, because of the low count rate.

In order to obtain the spectral characteristics of the footpoints and loop top, we integrated the flux within an 8 arc sec $\times$ 8 arc sec box around each of the two footpoints and the loop top (see Figures 3 and 4). This box size was chosen to encompass the total emission from each feature.

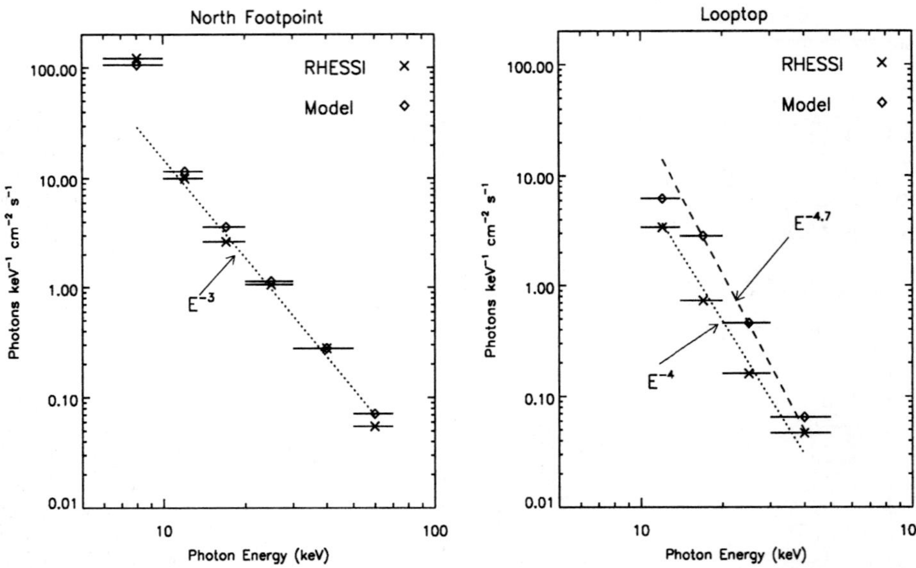


*Figure 5.* Spectra for the north footpoint (*left panel*) and loop top (*right panel*). The horizontal bars marked with an 'x' are from the CLEAN images shown in Figure 4. The bars marked with a diamond are from the model images reconstructed with CLEAN (Figure 9). The time interval is the same as that for Figures 3 and 4.

We found the results from MEM-Sato and CLEAN to be similar. Spectra of the footpoints (left panel) and the loop top (right panel) obtained from the CLEAN maps are shown in Figure 5. Since the two footpoints have similar spectra, we just plot the north footpoint and the loop-top source. The results from the data are shown as horizontal bars marked with an 'x'. The width of the bar indicates the width of the energy band. The bars marked with diamonds are from the flare model (see Section 4 below).

The immediately apparent result from these spectra is that the loop-top spectrum is softer than the footpoint spectra. A rough power-law fit to these spectra gives a spectral index of $-3$ for the footpoints and $-4$ for the loop top. This difference of one in spectral index is consistent with a recent statistical analysis of limb flares observed by *Yohkoh* (Petrosian, Donaghy, and McTiernan, 2002). At 6–10 keV, the flux from the north footpoint is higher than the extrapolated power law. This is because the thermal source between the footpoints is closer to the north footpoint, and the thermal source contributes to the flux from the north footpoint in this energy band. Since the loop-top source does not appear in the images at 50–70 keV, the spectral fitting does not include this energy band.

## 3. Flare Model

In the model, we assume that electrons with a power-law energy distribution and an isotropic pitch-angle distribution are injected at the top of a single semicircular flare loop. The steady-state electron spatial and spectral distributions within the loop were obtained with a Fokker–Planck code (McTiernan and Petrosian, 1990). The code includes Coulomb scattering and energy losses and magnetic mirroring. We computed the hard X-ray emission from the coronal loop and footpoints using thin-target and thick-target bremsstrahlung radiation codes. Hot plasma is assumed to uniformly fill the flare loop. Based on the flare images shown in Figures 3 and 4, we added a thermal source between the two footpoints.

The radius of the semicircular loop in the model is 17 arc sec, and the diameter of the magnetic tube is 4 arc sec. The magnetic field is assumed to be uniform (constant) along the length of the flare loop so that magnetic mirroring of the energetic electrons does not occur. The plane of the loop must be rotated 53° from the plane of the sky to match the flare geometry in the RHESSI images.

The density distribution of the non-thermal electrons injected at the loop top is determined by the fit to the spatially integrated photon spectrum: the non-thermal electron power-law spectral index is $-4.4$ between 15 keV and 100 keV and $-5.5$ above 100 keV, and the total electron density is $3.5 \times 10^8$ cm$^{-3}$. The plasma density and temperature within the loop are $3.5 \times 10^{10}$ cm$^{-3}$ and 10 MK, respectively. The loop would be too bright, i.e., would be seen with RHESSI, if its temperature were much greater than 10 MK at this density.

The thermal source between the footpoints is taken to be 20 arc sec long, 4 arc sec wide, and 4 arc sec deep. The thermal source is shifted 5 arc sec toward the north footpoint from the center of the two footpoints. This configuration was chosen to provide images that agree reasonably well with the RHESSI images. The plasma density and temperature of this thermal source were chosen to agree with the results of the spectral fit: $1.4 \times 10^{11}$ cm$^{-3}$ and 15 MK.

Images at 6, 15, and 50 keV from the model are shown in Figure 6. These images are shown with one arc sec spatial resolution. At 6 keV, the thermal source between the two footpoints is very strong compared with the footpoints and the loop-top source. The flare loop is nearly invisible at this energy because emission from the loop is weaker than that from the central thermal source. At 15 keV, the thermal source between the footpoints disappears. The loop top and footpoint sources are visible. Only the two footpoint sources are visible at 50 keV.

Loop top and footpoint spectra from the model are plotted in Figure 7. The footpoint fluxes are obtained by summing over all pixels that show emission from one footpoint. The loop-top fluxes are obtained by summing over 32 pixels at the top of the model loop. The thermal emission at low energies (dotted curve) is from the model loop, not the thermal source between the footpoints. The spectral index of the footpoints is $-3.1$ between 15 keV and 56 keV, and $-4.4$ above 56 keV. The spectral index of the loop top is $-5.0$ between 15 keV and 56 keV, and $-5.7$

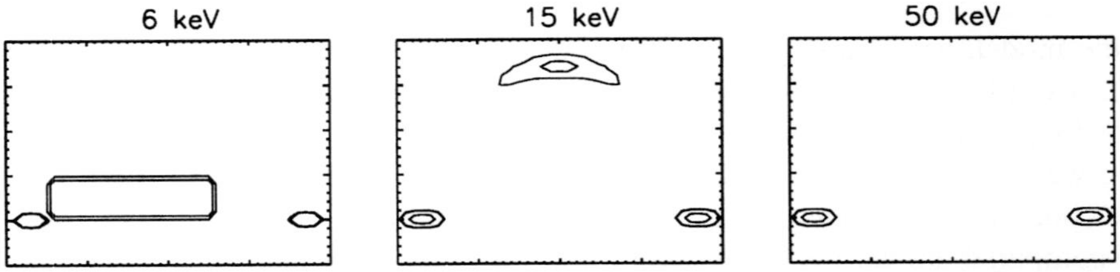


*Figure 6.* Model flare loop images at 6 keV, 15 keV, and 50 keV. Contour levels are 0.05, 0.07, and 0.5 times the peak flux in each image. The spatial resolution is 1 arc sec. These images are not rotated and scaled to the RHESSI images.

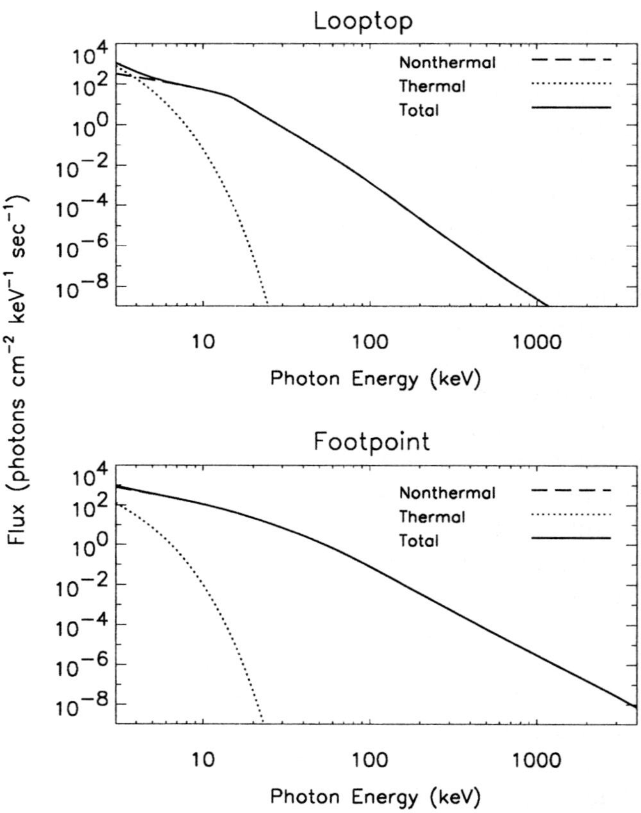


*Figure 7.* Spectra from the model loop images. The *top panel* shows the spectrum from the loop top and the *bottom panel* shows the spectrum from a footpoint. The *dashed curve* is the total non-thermal bremsstrahlung from a footpoint or the loop top, the *dotted curve* is the thermal bremsstrahlung from the same region, and the *solid curve* is the total spectrum.

above 56 keV. Comparing the model spectra with the data, the spectral index of the footpoints from the model below 56 keV agrees with our estimated index, −3, from the RHESSI imaging spectra. The spectral index of the loop top is higher than the value of −4 estimated from the data, however.

## 4. Simulation Results and Analysis

The RHESSI imaging software allows us to input our flare model to obtain simulated images. This allows us to compare simulated images with RHESSI images to test our model. It also allows us to check the imaging software itself to identify possible artifacts from the image-reconstruction process. We input each model image as a $64 \times 64$ array into the imaging software. All imaging parameters are the same as those used for the RHESSI images. As shown in Figures 8 and 9, the simulated MEM-Sato and CLEAN images roughly agree with the images of Figures 3 and 4. At 6–10 keV, only the thermal source between the two footpoints is visible. At 10 –14 keV, the thermal source still dominates. Note that in the MEM-Sato images, the uniform thermal source appears to be a double source. There is a weak loop-top source visible in both the MEM-Sato and CLEAN images. At 14–20, 20–30, and 30–50 keV, two clear-cut footpoints and a loop-top source are visible. This agrees with the RHESSI images. At 50–70 keV, the loop-top source vanishes and only the two footpoints are still visible.

With this simple simulation process, we have found that both the MEM-Sato and the CLEAN imaging algorithms can alter the relative brightness of the flare components. An example of this is shown in Figure 10. An unprocessed image from the model is shown in the top left panel. This 14–20 keV model image was not used in simulating the RHESSI data because the loop top is too bright. This same image processed with MEM-Sato and with CLEAN are shown in the bottom left and bottom right panels. It was processed in the same way as the other images, including only grids 3 through 8. The upper right panel shows the model image smoothed with a conical response function with a full width at half maximum of 7 arc sec. This simulates the lower resolution of the processed images. In the unprocessed image, the peak flux from each footpoint is about 2 times higher than that from the loop top. In the other three images, however, the peak flux of the loop top is about 2 times higher than that of the footpoints. This is because the loop-top source is more extended than the footpoints, and the spatial averaging at the lower spatial resolution picks up flux from many pixels in the loop top while the footpoint emission originates from only a few pixels.

As is the case for the unprocessed model image, the north and the south footpoints have about the same peak flux in the smoothed and the CLEAN images. In the MEM-Sato image, however, the peak flux of the north footpoint is 30% higher than that of the south footpoint. For all the images, however, the integrated flux from the north footpoint is about the same as the integrated flux from the south foot-

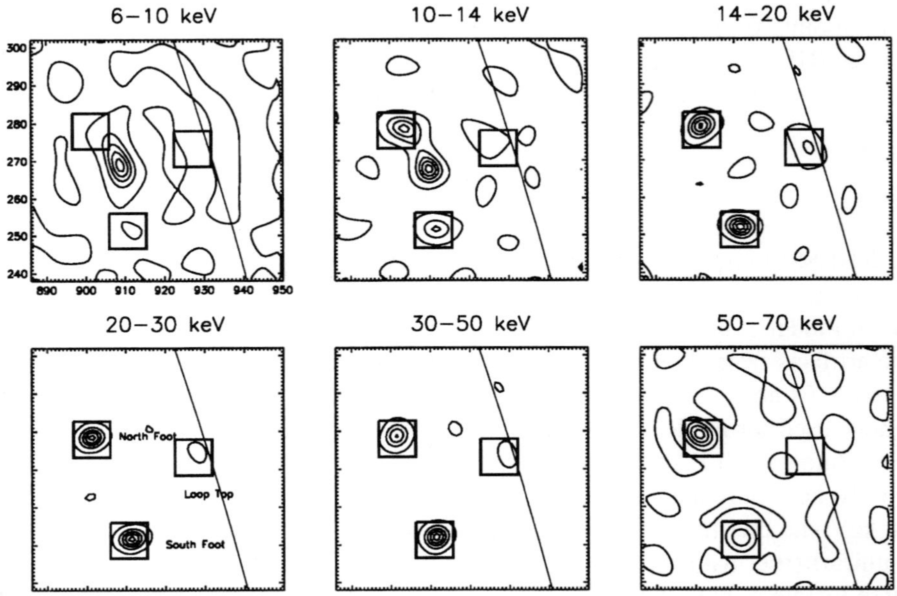


*Figure 8.* MEM-Sato images of the model flare loop. Contour levels are 0.1, 0.3, 0.5, 0.7, and 0.9. The contours are normalized to the peak flux of each image. As in Figures 3 and 4, the *smooth line* is the solar limb.

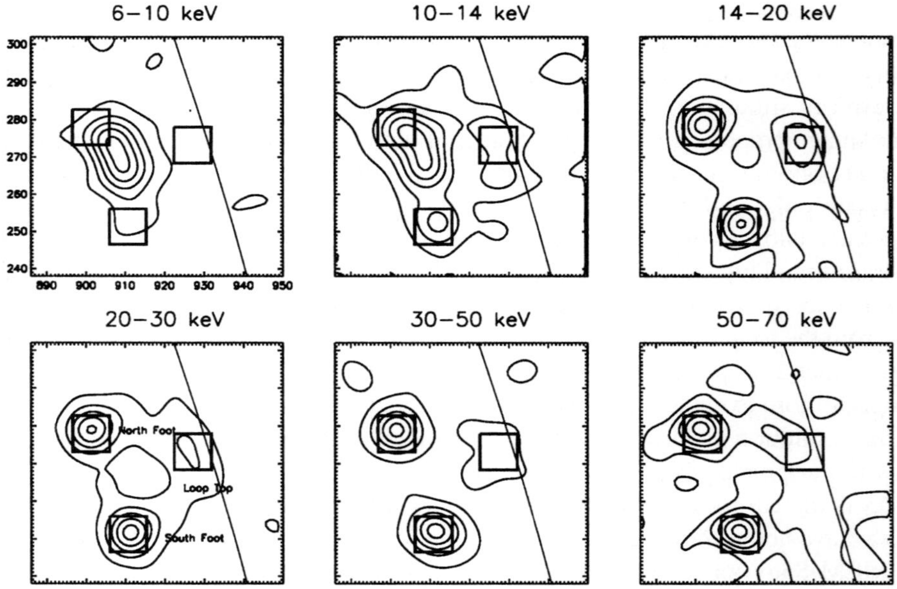


*Figure 9.* CLEAN images of the model flare loop. Contour levels are the same as Figure 8.

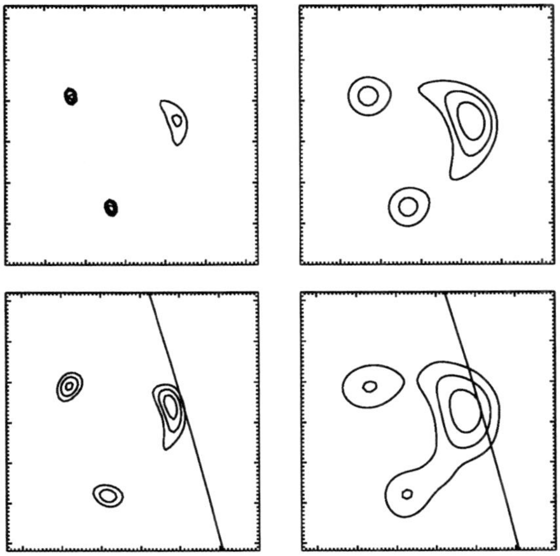

*Figure 10.* An unprocessed image from the flare model (*top left*), the same image convolved with a 7 arc sec FWHM conical response function (*top right*), the same image processed with MEM-Sato (*bottom left*), and the image processed with CLEAN (*bottom right*) are shown. Contour levels are 0.2, 0.4, and 0.7 times the peak flux in each image.

point. MEM-Sato has super-resolved the X-ray sources – they are more compact than the spatial resolution of the instrument. On the other hand, compared to the smoothed image, CLEAN has enhanced the emission from the southern leg of the loop relative to the peak flux from the footpoints by about 130%. Fortunately most of these discrepancies are relatively small, but they could be difficult to recognize in RHESSI images without comparing the images with model images.

Figure 11 shows a simulated image from the model processed with MEM-Sato on the left (Figure 8, 14–20 keV) and the corresponding MEM-Sato flare image on the right (Figure 3, 14–20 keV). For clarity, an additional low-flux contour (0.05) has been added to the RHESSI map from Figure 3. Some artificial sources appear in the simulated image. Comparing the simulated MEM-Sato image with the RHESSI MEM-Sato image, we can see similar patterns in both images. Consequently, we deduce that these 'sources' between the footpoints and the coronal source in the RHESSI flare image are also not real.

Spectra obtained from these simulated images from the model are shown in Figure 5 along with the spectra obtained from the RHESSI flare images. The fluxes from the model are marked with diamonds. Lacking error bars for the fluxes deduced from the flare images, we do not yet have a meaningful, quantitative test of goodness of fit. Nevertheless, the footpoint spectrum can be seen to be closely reproduced by the model. The loop-top spectrum, on the other hand, is poorly reproduced. The model spectrum is too steep relative to the flare data. The loop-

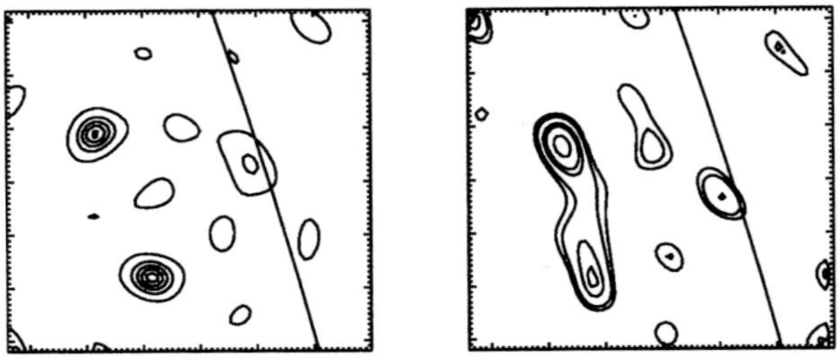

*Figure 11.* MEM-Sato 14–20 keV images of the model (*left*) and of the observed flare (*right*). We can see similar patterns in them. The sources between the loop top and footpoints in the left image are not real. Consequently, we deduce that sources between the footpoints and the loop top in the right image are also not real.

top spectrum for the flare can be fitted with a power-law spectral index of about −4, while the spectral index of the model (15 keV to 50 keV) is about −4.7. We have not found a way to rectify this problem without significantly changing the physical model.

Figure 5 shows that the loop-top flux from the model is too high at low energies. The flux at, say, 14–20 keV could be brought into agreement with the flux from the flare by deceasing the plasma density in the loop. At higher energies the loop-top source would be too weak, however, and would not be visible in the simulated images.

## 5. Discussion and Conclusions

Our results indicate that the simple flare model applied here does not adequately describe the X-ray emission from the 20 February 2002 flare. In this model the loop top and footpoint X-ray sources arise from the injection of supra-thermal electrons at the top of a magnetic loop. The lower energy electrons in the injected particle distribution interact with the loop plasma to produce the loop-top source, while the footpoint sources are produced when the particles reach the high-density footpoints of the loop (Holman, 1996). The problem with the model is that the loop-top flux decreases more rapidly with increasing photon energy than indicated by the spectrum obtained from the RHESSI images.

Another issue that at first appears to be inconsistent with the model is the relative timing of the brightening of the north and south footpoints. Figure 12 shows the time history of the emission from each footpoint in the 20–25 keV energy range. The time of peak emission for the north footpoint is about 8 s earlier than that of the south footpoint. Since the time required for a 25 keV electron with a pitch angle of 45° to travel from the top of the loop to a footpoint is only 0.4 s, our simple model does not provide an explanation for this substantial time delay. Higher spatial resolution maps indicate, however, that the southern footpoint is actually two distinct

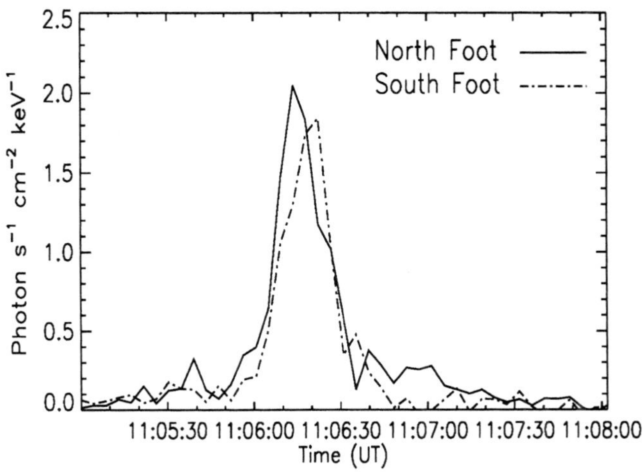


*Figure 12.* Emission history of the two footpoints at 20–25 keV. The start time is 11:05:00 UT. The peak of the north footpoint emission is earlier than the peak of the south footpoint emission by about 8 seconds.

footpoints that reach peak brightness at different times (Krucker and Lin, 2002). The peak emission from the northern footpoint appears to be co-temporal with the peak of the emission from the northern component of the southern footpoint. The peak in the emission from the southern component of the southern footpoint is co-temporal with a secondary peak in the emission from the northern footpoint. Therefore, the timing does not appear to be inconsistent with the electron propagation time. The relative timing and brightness of the footpoint pairs still require an explanation, however.

We have found that the comparison of RHESSI images with images generated from a model provides a valuable check on the image reconstruction process. We found that this comparison allowed us to identify spurious sources and enhanced source fluxes in the reconstructed maps. We also found that the MEM-Sato algorithm misrepresented the uniform thermal source included in the model, displaying it as a double source rather than an elongated single source.

Given the low flux of the loop-top source relative to the footpoints, its reality must be questioned. Our use of two different image reconstruction techniques and comparison with the simulated maps from the model indicate that it is a real source, however. Also, post-flare loops observed with the SOHO EIT indicate the presence of a loop extending from the hard X-ray footpoints to the loop-top source (see Figure 4 of Krucker and Lin, 2002).

The thermal source between the footpoints was an unexpected feature of the flare emission. This source dominated the thermal radiation from the flare. This would not have been unexpected if we had been looking down on a flare loop on the disk of the Sun. It is an unexpected feature, however, for a loop at the limb of the Sun with a loop-top source displaced toward the limb. We found the thermal source to be located closer to the northern hard X-ray footpoint than the southern footpoint. It is interesting that the northern footpoint was also the first to reach peak brightness. It is likely that this thermal source was a low-lying loop interacting with the northern footpoint of the higher loop containing the loop-top source. The northern component of the southern footpoint is likely to have been the southern footpoint of this low-lying loop.

A significant handicap we faced for this analysis was the lack of knowledge of the uncertainties in the fluxes obtained from the RHESSI images (Figure 5). Without a knowledge of these uncertainties, we could not properly evaluate the ability of the model to fit the imaged spectra. Our future work will include a careful evaluation of the imaging and the simulation processes to establish a realistic estimate of these uncertainties. This will include obtaining images using the PIXON method (Metcalf *et al.*, 1996). The PIXON method requires more time to produce an image, but an estimate of the uncertainties is provided by the algorithm and the photometry is expected to be better.

We will also seek out refinements to our model and to other models that may provide a more acceptable description of the observational results from this and other flares observed by RHESSI. Other particle injection and propagation models that include a density enhancement (Wheatland and Melrose, 1995) or magnetic trapping (Fletcher and Martens, 1998) at the top of the loop are likely to suffer the same problem with the loop-top spectrum as the model applied here. If this problem is not resolved, the most likely conclusion is that supra-thermal electrons are accelerated in the loop-top source and/or elsewhere in the observed flare loop.

The temperature of 10 MK or less deduced for the loop containing the loop-top source is rather low for a flare loop. Heating of the loop plasma by the supra-thermal electrons alone, especially in the region of the loop-top source, can be quite high. We will explore models that include a hotter plasma, perhaps confined to the loop-top region, to determine if they can be consistent with the RHESSI results.

We found several functions that provided a good fit to the spatially integrated flare spectrum. For weak to moderate intensity flares such as the flare analyzed here, the flux typically becomes undetectable above the background at energies below 200 keV. These flares are amenable to a wider variety of spectral fits than more intense flares for which the spectrum can be established up to energies of 500 keV or above. Our future studies will include the analysis of these more intense flares that have now been observed by RHESSI.

## Acknowledgements

This work was supported in part by the NASA Sun–Earth Connection Program and the RHESSI Project. This work would not have been possible without the dedication, hard work, and support of the entire RHESSI team. We especially thank

Paul Bilodeau for his help in integrating the thick-target bremsstrahlung code with the SPEX software, and Ed Schmahl for several helpful discussions. We thank the referee, James McTiernan, and the 'unofficial RHESSI referee', Hugh Hudson, for their helpful comments on the paper.

## References

Fletcher, L. and Martens, P. C. H.: 1998, *Astrophys. J.* **505**, 418.
Garcia, H.: 1994, *Solar Phys.* **154**, 275.
Holman, G. D.: 1996, *Bull. Am. Astron. Soc.* **28**, 939; also see http://hesperia.gsfc.nasa.gov/sftheory/loop.htm.
Holman, G. D., Mariska, J. T., McTiernan, J. M., Ofman, L., Petrosian, V., and Ramaty, R.: 2001, *Bull. Am. Astron. Soc.* **33**, 1444; also see http://hesperia.gsfc.nasa.gov/hessi/modelware.htm.
Holman, G. D., Sui, L., McTiernan, J. M., and Petrosian, V.: 2002, Theoretical Model Images and Spectra for Comparison with HESSI and Microwave Observations of Solar Flares, *Yohkoh 10th Anniversary Meeting Proceedings*, in press.
Hurford, G. *et al.*: 2002, *Solar Phys.*, this volume.
Krucker, S. and Lin, R. P.: 2002, *Solar Phys.*, this volume.
Masuda, S.: 1994, *Nature* **371**, 495.
Masuda, S.: 2002, Hard X-ray Solar Flares Revealed with Yohkoh HXT – A Review, *Yohkoh 10th Anniversary Meeting Proceedings*, in press.
McTiernan, J. M. and Petrosian, V.: 1990, *Astrophys. J.* **359**, 524.
Metcalf, T. R., Hudson, H. S., Kosugi, T., Puetter, R. C., and Pina, R. K.: 1996, *Astrophys. J.* **466**, 585.
Petrosian, V., Donaghy, T. Q., and McTiernan, J. M.: 2002, *Astrophys. J.* **569**, 459.
Sato, J., Kosugi, T., and Makishima, K.: 1999, *Publ. Astron. Soc. Japan* **51**, 127.
Smith, D. M. *et al.*: 2002, *Solar Phys.*, this volume.
Thomas, R. J., Crannell, C. J., and Starr, R.: 1985, *Solar Phys.* **95**, 323.
Wheatland, M. S. and Melrose, D. B.: 1995, *Solar Phys.* **158**, 283.

# HARD X-RAY AND METRIC/DECIMETRIC RADIO OBSERVATIONS OF THE 20 FEBRUARY 2002 SOLAR FLARE

N. VILMER[1], S. KRUCKER[2], R. P. LIN[2] and the RHESSI TEAM
[1]*LESIA, Observatoire de Paris, 92195- Meudon-Cedex, France*
[2]*Space Sciences Laboratory, University of California, Berkeley, CA-94720, U.S.A.*

(Received 8 August 2002; accepted 6 September 2002)

**Abstract.** The GOES C7.5 flare on 20 February 2002 at 11:07 UT is one of the first solar flares observed by RHESSI at X-ray wavelengths. It was simultaneously observed at metric/decimetric wavelengths by the Nançay radioheliograph (NRH) which provided images of the flare between 450 and 150 MHz. We present a first comparison of the hard X-ray images observed with RHESSI and of the radio emission sites observed by the NRH. This first analysis shows that: (1) there is a close occurrence between the production of the HXR-radiating most energetic electrons and the injection of radio-emitting non-thermal electrons at all heights in the corona, (2) modifications with time in the pattern of the HXR sources above 25 keV and of the decimetric radio sources at 410 MHz are observed occurring on similar time periods, (3) in the late phase of the most energetic HXR peak, a weak radio source is observed at high frequencies, overlying the EUV magnetic loops seen in the vicinity of the X-ray flaring sites above 12 keV. These preliminary results illustrate the potential of combining RHESSI and NRH images for the study of electron acceleration and transport in flares.

## 1. Introduction

Complementary observations of the hard X-ray (HXR) emitting electrons detected with RHESSI are provided by the radio emission produced in the whole frequency range from a few tens of GHz to a few MHz (see, e.g., Bastian, Benz, and Gary, 1998; White, 1999; Vilmer and Trottet, 1997; Vilmer and MacKinnon, 2002, for reviews of previous multiwavelengths observations of flares). These combined measurements allow to analyse the signature of non-thermal electrons in a whole range of coronal heights. While HXR and centimeter/millimeter wave emissions are the most direct diagnostic of the energetic electrons (see, e.g., Bastian, 1999; Nishio *et al.*, 1999; Lee, Gary, and Shibasaki, 2000; Lee and Gary, 2000), the coherent plasma radiations such as the ones emitted in the metric/decimetric domain at greater coronal heights ($10^4-10^5$ km) represent one of the most sensitive diagnostics of energetic electrons (a few tens of keV) injected in the corona from the flare site. In this paper, we shall focus on the radio observations in this frequency range.

It has been shown in many studies (see, e.g., Trottet, 1994; Vilmer and Trottet, 1997, for reviews) that multiwavelength studies provide a comprehensive approach to understanding particle acceleration and transport in solar flares. Several studies

*Solar Physics* **210:** 261–272, 2002.

have been previously performed combining hard X-ray/$\gamma$-ray spectral measurements with metric/decimetric radio imaging observations (e.g., Raoult *et al.*, 1985; Chupp *et al.*, 1993; Trottet *et al.*, 1994, 1998). They indicate that HXR and radio emissions arise from electrons produced in a common acceleration site and injected respectively in low-height magnetic structures where they produce X-rays and in larger scale, higher magnetic features where they emit decimetric and metric radiations. Many of these observations have also shown that in the course of the flare various large scale coronal structures are successively and/or simultaneously the sites of non-thermal electron radiation. The spectral characteristics of the flare energetic electrons as deduced from HXR observations evolve with time in close temporal association of the 'activation' of these large-scale magnetic structures in which radio emission is produced (Raoult *et al.*, 1985; Chupp *et al.*, 1993; Trottet *et al.*, 1994, 1998; Raulin *et al.*, 2000). It was thus suggested that the electron spectral variability may be linked to the magnetic configuration in which particles are produced. Such variations of HXR emitting electron spectra are also found to correspond to stepwise changes of metric/decimetric spectra (e.g., Trottet *et al.*, 1998). All these previous studies had been performed for flares for which there were either no HXR images or no 2D radio images in the decimetric/metric domains. Such a combination of HXR and radio 2D images was used so far only for the analysis of narrowband decimetric spikes (Benz, Saint-Hilaire, and Vilmer, 2002). The evolution with time of the associated HXR and radio emitting sites was however not investigated. Further probing the link between spectral and spatial variability at all scales is thus a major topic for HXR and possibly $\gamma$-ray imaging observations from RHESSI in combination with spatially and spectrally resolved observations in the whole radio domain.

In this paper, we present a first comparison of spatially resolved HXR and decimetric observations obtained for one of the first flares observed with RHESSI and we investigate the evolution in time and space of the pattern of small and large scale magnetic structures traced by the energetic electrons.

## 2. Instrumentation

The HXR and decimetric observations were obtained respectively by the RHESSI instrument (Lin *et al.*, 2002) and by the Nançay radioheliograph (Kerdraon and Delouis, 1996). At the time of the observations, the RHESSI instrument was operated with the thin shutters in front of the detectors thus cutting out the lowest energy photons (from $\sim 3$ to $\sim 6$ keV). The 2D imaging radioheliograph in Nançay provided observations at 432 MHz, 410 MHz, 327 MHz, 236 MHz, and 164 MHz. In the present study, only the highest frequency images are shown.

The radio spectra in the 150–550 MHz frequency range and in the 1–14 MHz range were obtained respectively by the *Phoenix-2* spectrometer operated by the ETH Zürich (Messmer, Benz, and Monstein, 1999) and by the WAVES experiment aboard the WIND spacecraft (Bougeret *et al.*, 1995). EUV images provided at 195 Å by the Extreme Ultraviolet Imaging Telescope (EIT) aboard SOHO (Delaboudinière *et al.*, 1995) are also used to provide complementary information on the context in which the flare occurred.

## 3. Observations

### 3.1. Temporal Evolution of the Flare at X-ray and Metric/Decimetric Wavelengths

The 20 February 2002 GOES C7.5 flare at 11:07 UT occurred in NOAA region 9825 (N16 W80). It is one of the first solar flares observed by RHESSI at X-ray wavelengths. The flare was seen from the HXR domain (up to around 100 keV) down to the hectometric radio domain. It was indeed simultaneously observed at metric/decimetric wavelengths by the Nançay radioheliograph (NRH) which provided images at five frequencies between 450 and 150 MHz and by the Phoenix-2 spectrometer which provided spectra in the same frequency domain. The WAVES experiment aboard the WIND spacecraft provided observations of the radio emission in the 1–14 MHz bands from electron beams injected in the high corona in association with the flare.

Figure 1 shows the time evolution of the counts (not background subtracted) in 4 energy bands (6–12, 12–24, 24–50, 50–100 keV) and of the X-ray spectrograms (background subtracted) from $\sim 3$ keV to $\sim$ 200 keV measured by RHESSI, together with the radio spectrum observed by *Phoenix-2* in the 150–550 MHz frequency range, the time evolution of the radio flux density at 164 MHz, 236 MHz, 327 MHZ, 408 MHZ , 432 MHz measured with the NRH and the radio spectrum from WIND/WAVES in the 1–14 MHz band.

For these observations thin shutters were placed in front of the RHESSI detectors leading to a decrease in the efficiency for measuring the lowest energy photons. Above 6 keV, the flare started at $\sim$ 11:04 UT and lasted until after 11:15 UT (Figure 1). The X-ray emission observed with RHESSI extends to energies of $\sim$ 100 keV during a shorter time interval during the flare ($\sim$ 40 s around 11:06:20 UT). Although weak radio decimetric/metric emission is observed in the early phase of the flare (i.e., from the start of the X-ray emission in the 10 keV range), there is shortly after 11:06 UT a simultaneous sudden and strong increase of the HXR emission observed up to 100 keV and of the metric/decimetric radio flux. At this time, the radio emission is clearly seen in the metric/decimetric domain starting with groups of fast frequency drifting bursts. Although intense meter wavelength emission is observed, the emission at decimeter wavelengths (i.e., around 300 to 400 MHz) consists of fainter bursts of short bandwidth (see Figure 1). The simultaneous start of intense HXR and metric radio emissions is consistent with the results of previous studies (e.g., Benz *et al.*, 1983; Raoult *et al.*,

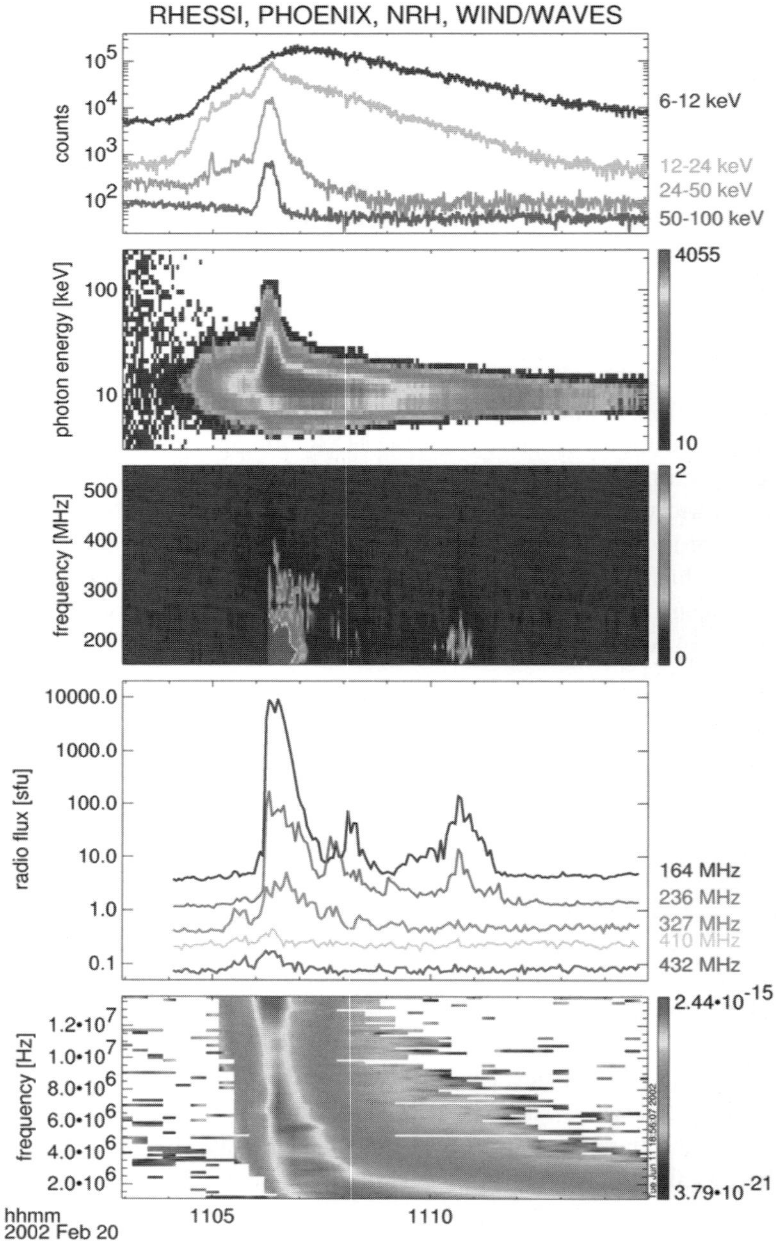


*Figure 1. Top to bottom*: time evolution of the X-ray RHESSI counts accumulated over 1 s in 4 energy channels between 6 keV and 100 keV (the background is not subtracted); RHESSI X-ray spectrogram showing the evolution with time and photon energy of the X-ray emission; decimetric/metric radio spectrum observed between 550 MHz and 150 MHz by *Phoenix-2* (ETH Zürich); decimetric/metric radio flux observed at 5 frequencies (integration time of 4 s) by the Nançay radioheliograph (NRH). The observed fluxes are respectively divided by: 1 (164 MHz), 5 (236 MHZ), 15 (327 MHz), 75 (410 MHz) and 150 (432 MHz); hectometric radio emission observed by the WIND/WAVES experiment. For color versions of this and other figures, see the CD-ROM.

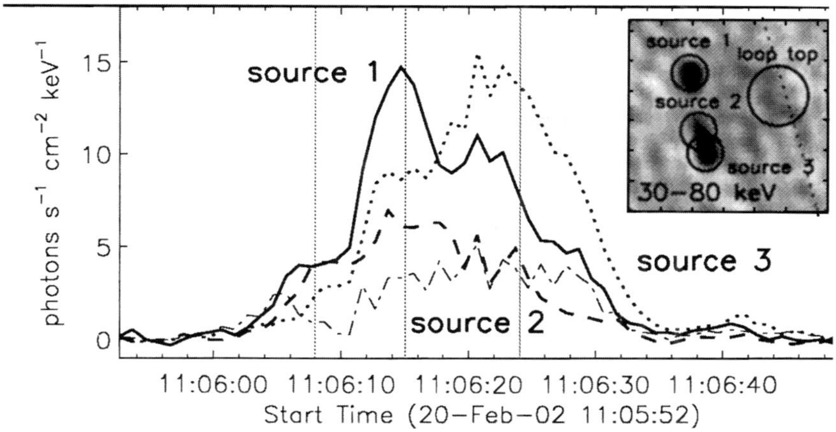


*Figure 2.* Temporal evolution of the different HXR sources observed at 30–80 keV at a resolution below 4″ (adapted from Krucker and Lin, 2002); source 1 (*solid line*), source 2 (*dashed line*), source 3 (*dotted line*), loop top (*dashed-dotted line*).

1985; Trottet *et al.*, 1998; Raulin *et al.*; 2000). The duration of the strongest part of the metric/decimetric radio emission is also found to be similar to the one of the HXR peak above 50 keV. As seen from Figure 1, the radio emission below 12 MHz (i.e., arising from electron beams injected in the high corona) also starts together with the sudden increase of the HXR flux at energies above 50 keV. This shows the link between the injection of electron beams towards the high corona and towards the HXR emitting sites.

## 3.2. HXR AND DECIMETRIC IMAGING OBSERVATIONS

### 3.2.1. *In the Main HXR Burst*

The imaging spectroscopic capabilities of RHESSI are demonstrated for this flare in Krucker and Lin (2002). At energies above 30 keV and a spatial resolution below 4″, the emission comes primarily from 3 compact sources (Figure 2): source 1 to the north, source 3 to the south and a weaker source 2 in between but much closer to source 3. Figure 2 also shows the loop top source discussed in more detail in Sui *et al.* (2002). The time evolution of the photon flux in the different sources is displayed in Figure 2. While a first HXR peak at ~11:06:15 UT is observed coming primarily from the northern component (source 1), a second peak of similar amplitude is observed peaking shortly after 11:06:20 UT originating from the southernmost component (source 3).

The goal of this paper is not to describe in detail the evolution of the X-ray images in different energy bands but to present a first comparison of the time evolution of the radio and of the HXR emitting sites from non-thermal electrons. We shall furthermore focus on the images provided in the highest frequency band by the NRH, since the emissions around 400–300 MHz are more likely to be produced close to the energy release and acceleration sites of X-ray producing electrons (e.g., Benz and Aschwanden, 1991). Indeed, studies based on frequency drifts of decimetric type III bursts or on starting frequencies of pairs of opposite drifting bursts (commonly observed between 300 MHz and 1 GHz) indicate that electron beam acceleration responsible for both downward moving HXR producing electrons and upward radio emitting electrons arise from a medium with densities ranging from $10^9$ to $10^{10}$ $cm^{-3}$ (Benz and Aschwanden, 1991; Aschwanden *et al.*, 1995).

Figures 3 and 4 show combined images of the X-ray emission in the 25–40 keV range and of the decimetric radio emission at 410 MHz superposed on an EIT difference image measured at 195 Å shortly after the HXR flare. While Figure 3 corresponds to a time where X-ray sources 1 and 3 are of equal importance (see Figure 2), Figure 4 corresponds to a time where the southernmost source 3 is predominant. The radio emission sites in both Figures 3 and 4 present a double component. As seen from these figures, the modification of the HXR morphology between the two images correspond to a noticeable change of the relative brightness of the two components of the radio sources at 410 MHz. This shows a causal relationship between the radio and HXR emitting sites which can be attributed to variations in the energy release and electron injection sites. Such variations of the radio emission sources in the development of X-ray flares has been described already in different papers (see, e.g., Chupp *et al.*, 1993; Trottet *et al.*, 1994, 1998). Although it had been suggested that such modifications in the radio emitting sites could be related to variations on the same time period in the morphology of the X-ray emitting sources, a direct comparison such as the one presented here had not been achieved so far.

#### 3.2.2. *In the Late Phase of the HXR Peak*

In the previous section, a comparison of the X-ray and high frequency radio emitting sources was done for two specific time intervals. Figure 5 shows a time series of radio images at 410 and 327 MHz each integrated over 4 s. The field of view is the same for both frequencies and the brightness levels in each image are normalized to the maximum of the image. Figure 5 shows that during most of the HXR main peak (i.e., from 11:06:05 to 11:06:29 UT), the radio emission arises at the two frequencies from two close components with variable relative intensities. At 410 MHz the outermost component of the radio sources is predominant compared to the innermost one (as in Figure 4) from 11:06:21 to 11:06:29 UT (see Figure 5 top), i.e., at the time when the southernmost HXR source (source 3) is the strongest component of the HXR emission.

Figure 5 shows that a radio source is seen (both at 410 MHz and 327 MHz) for a few images around 11:06:33 UT (i.e., in the late phase of the main HXR burst), closest to the limb and to the X-ray flare site. Figure 6 shows combined images at 11:06:33 UT of the X-ray emission in the 12–18 keV range and of the decimetric radio emission observed at 410 MHz superposed on the EIT im-

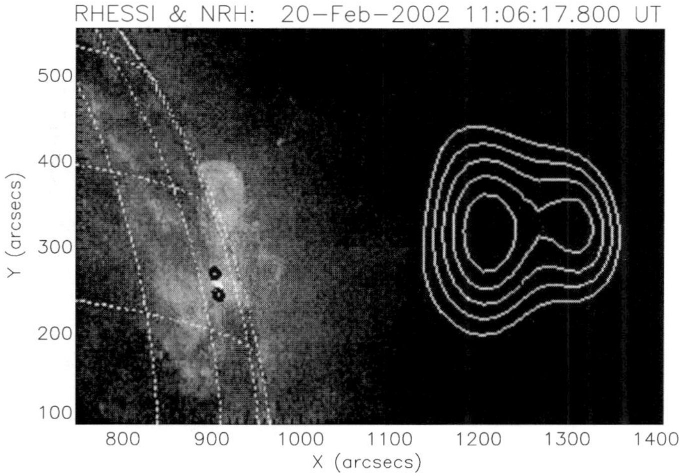


*Figure 3.* RHESSI iso-contours (*black*) (40, 60, 80% of the maximum) at 25–40 keV and NRH contours at 410 MHz (*white*) (50, 60, 70, 80, 90%) observed at 11:06:17.800 at the time when similar flux is radiated from sources 1 and 3 (see text). The RHESSI and NRH contours are superposed on the closest EIT image obtained at 11:12 UT. The RHESSI images are obtained using grids 3 to 9 giving a resolution of 7″.

age obtained shortly after the flare. The X-ray source at 12–18 keV presents a double component, similar to the one observed at higher energies earlier in the flare. At 11:06:33 UT, the northern component is predominant. At 410 MHz, the emission comes predominantly from a new emitting site very close to the limb. The innermost radio component observed previously at the time of the main HXR burst (Figure 4) is still observed, even if its flux has largely decreased. The new radio source is observed in the high frequency part of the NRH observing domain (i.e., 432 MHz, 410 MHz, and 327 MHz) and corresponds to a weak flux of less than 10 s.f.u. It overlays magnetic loops seen with EIT in the close vicinity of the flaring active region. Around 11:06:33 UT, there is still a significant production of non-thermal X-ray emission between 10 and 30 keV (although with a steeper spectrum than during the main phase) (see Figure 1). A detailed discussion of the origin of the radio source overlying the EIT loops is beyond the scope of this paper which presents early science results from RHESSI. A preliminary analysis of the RHESSI spectra shows however that non-thermal electrons between 10 and 30 keV are detected during that period.

## 4. Summary of the Observations and Discussion

This paper presents the results of the preliminary analysis of the 20 February 2002 solar flare at 11:06 UT based on the first comparison of RHESSI and NRH images. The results are summarized below:

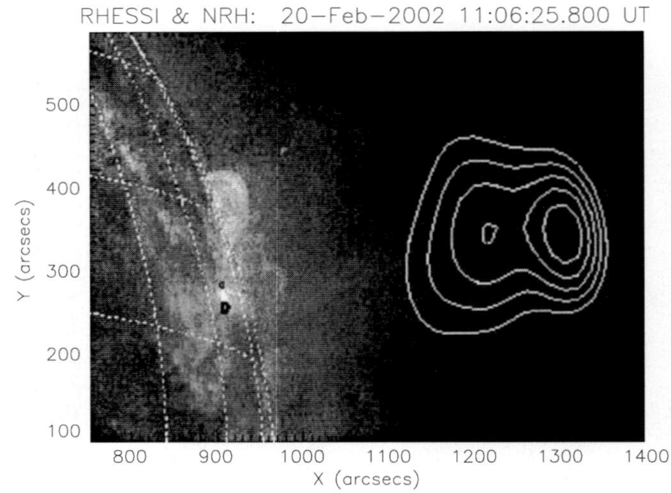


*Figure 4.* Same as Figure 3 at 11:06:25.800 at a time where the southernmost source 3 is predominant (see text). The change of the relative brightness of the two radio components has to be noted.

– Although weak radio decimetric/metric emission is observed in the early phase of the flare in the 10 keV range, intense metric emission starts together with the most energetic X-ray peak observed up to almost 100 keV. This is in good agreement with previous combined observations of spatially unresolved HXR observations and of metric/decimetric observations (e.g., Benz *et al.*, 1983, Raoult *et al.*, 1985; Trottet *et al.*, 1998). The duration of the strongest part of the radio emission in the whole decimetric/metric range is similar to the one of the HXR peak above 50 keV, suggesting as in the previously quoted works a common acceleration site with simultaneous injection of electrons both in low-lying magnetic features where they produce X-rays and in larger scale and higher magnetic structures where radio emission is produced.

– The main results lie in the first comparison of the variations on a few seconds time scale of HXR images above 25 keV and of decimetric radio images. Modifications of the X-ray sources above 25 keV (Figures 3 and 4) are associated with coincident changes in the morphology of the decimetric radio sources. Such changes of the HXR and radio emitting sites occurring on a similar time period had been suggested previously (e.g., Trottet, 1986, 1994; Vilmer and Trottet, 1997) based on several multiwavelength studies combining HXR spectral measurements with metric/decimetric radio imaging observations. Direct comparisons of HXR and radio images obtained on a few seconds time resolution could not be, however, performed. The present observations thus support the previous suggestion of common acceleration/injection sites for HXR and decimetric/metric emitting electrons. Indeed, variations in the HXR and decimetric emitting sites observed over the same time period show that both emissions are closely related to the energy release sites and to their variations during the flare. The link between the X-ray and radio

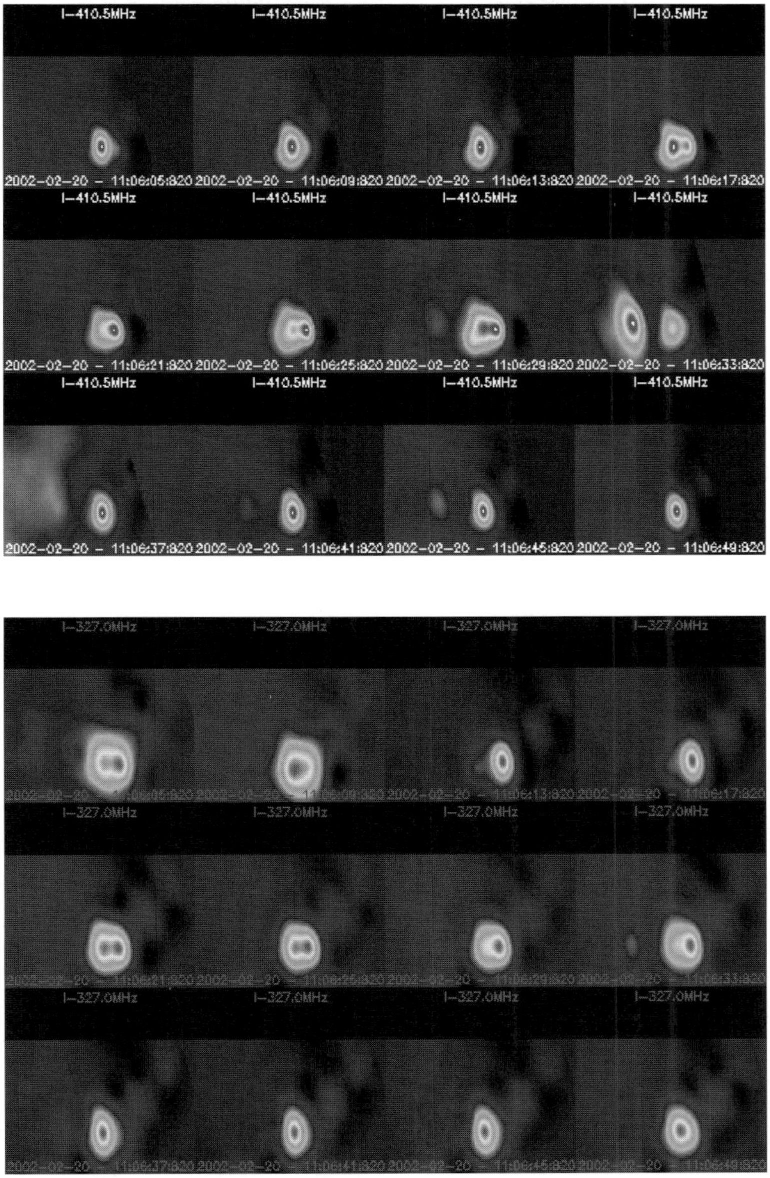


*Figure 5.* Time series of radio images integrated over 4 seconds at 410 (*top*) and 327 MHz (*bottom*) and observed by the NRH. During the time of the HXR peak above 50 keV, the radio sources consist of two close components with variable relative intensities (see also Figures 3 and 4) The appearance of a weak transient radio source closest to the limb and to the X-ray source can also be noted (see Figure 6). (NB: each image is normalized to its maximum.)

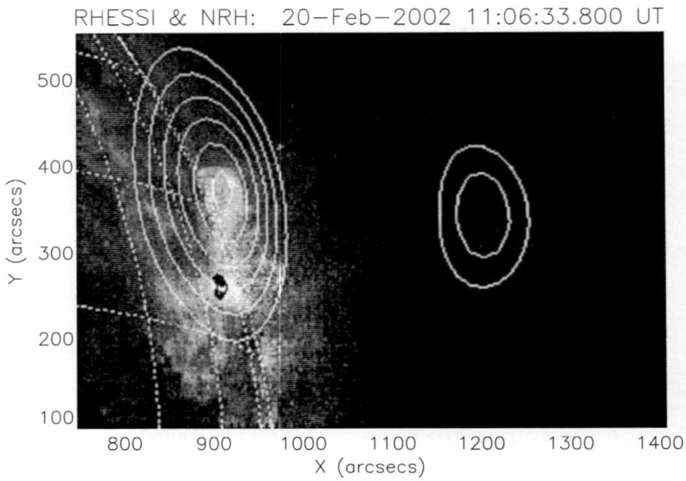


*Figure 6.* RHESSI iso-contours (*black*) (40, 60, 80%) at 12–18 keV and NRH contours (*white*) at 410 MHz (50, 60, 70, 80, 90%) at 11:06:33.800 UT superposed on the same EIT image at 11:12:10:489 UT.

sources and their evolution is, however, not easy to understand in detail without the knowledge of the magnetic connections between the emitting sites. Noticeable changes in the accelerated electron spectra deduced from HXR spectral analysis have been reported in close connection with the changes of the spatial distribution of the electron emitting sites at radio wavelengths (e.g., Chupp *et al.*, 1993; Trottet *et al.*, 1994, 1998; Raulin *et al.*, 2000) similar to the ones shown here in Figures 3 and 4. Spectral variations of the different X-ray sources in this event are presented elsewhere (Krucker and Lin, 2002). A detailed temporal comparison of the spectral evolution of the energetic electrons and of the variations of the pattern of the radio emitting sites is beyond the scope of this preliminary analysis and will be analysed later.

A weak radio source is observed at high frequencies in the late phase of the main HXR burst, overlying EIT magnetic loops and in close proximity of the X-ray flaring sites above 12 keV. The emission comes predominantly from the northern component of the X-ray source. The origin of the weak radio source has not been investigated in detail in this paper and additional work is needed in the future. The observations however suggest that this emission could be linked to the production of non-thermal electrons above 12 keV predominantly observed in the northern part of the flaring active region and subsequently injected towards magnetic field lines overlying the magnetic loops seen at the limb by EIT.

As a conclusion, the present results based on a preliminary comparison between HXR and decimetric/metric images demonstrate the potential of combined RHESSI and NRH observations to understand the properties of electron acceleration/injection and radiation in magnetic structures of all scales. Such comparisons will be performed on many other and more intense flares than this moderate one giving additional possibilities to achieve imaging spectroscopy at higher energies and detailed comparisons between high energy HXR sources and radio emitting sites. The possibility of imaging radio emitting electrons low in the corona (i.e., in the high frequency range of the NRH band) is quite promising in terms of comparison with X-ray emitting sites, since these emissions lie relatively close to acceleration sites. This also demonstrates the potential interest of imaging radio emissions from non-thermal electrons at even higher frequencies (as planned, e.g., by the FASR project) to better image the acceleration sites of the electrons which are responsible for HXR emission and which are now observed with unprecedented imaging-spectroscopy capabilities with RHESSI.

## Acknowledgements

N. Vilmer acknowledges support from the Centre National d'Etudes Spatiales on the RHESSI project. The Nançay Radio Observatory is funded by the French Ministry of Education, the CNRS and the Région Centre. The authors would like to thank Dr T. Bastian for valuable comments on the first version of the manuscript. They are also grateful to Dr H. Hudson for his useful comments.

## References

Aschwanden, M. J., Benz, A. O., Dennis, B. R., and Schwartz, R. A.: 1995, *Astrophys. J.* **455**, 347.
Bastian, T.: 1999, in T. S. Bastian, N. Gopalswamy and K. Shibasaki (eds.), *Proceedings of the Nobeyama Symposium, NRO Report* **479**, 211.
Bastian, T., Benz, A. O., and Gary, D. E.: 1998, *Ann. Rev. Astron. Astrophys.* **36**, 131.
Benz, A. O. and Aschwanden, M.: 1991, in Z. Švestka, B. V. Jackson, and M. E. Machado (eds.), *Eruptive Solar Flares, Lecture Notes in Physics* **399**, 106.
Benz, A. O., Saint-Hilaire, P., and Vilmer, N.: 2002, *Astron. Astrophys.* **388**, 363.
Benz, A. O., Barrow, C. H., Dennis, B. R., Pick, M., Raoult, A., and Simnett, G.: 1983, *Solar Phys.* **83**, 267.
Bougeret, J. L. *et al.*: 1995, *Space Sci. Rev.* **71**, 231.
Chupp, E. L., Trottet, G., Marschhäuser H. *et al.*: 1993, *Astron. Astrophys.* **275**, 602.
Delaboudinière, J. P., Artzner, G. E., Brunaud, J. *et al.* : 1995, *Solar Phys.* **162**, 291.
Kerdraon, A. and Delouis, J. M.: 1997, in G. Trottet (ed.), *Coronal Physics from Radio and Space Observations, Lecture Notes in Physics* **483**, 192.
Krucker, S. and Lin, R. P.: 2002, *Solar Phys.*, this volume.
Lee, J. and Gary, D. E.: 2000, *Astrophys. J.* **543**, 457.
Lee, J., Gary, K., and Shibasaki, K.: 2000, *Astrophys. J.* **531**, 1109.
Lin, R. P. *et al.*: 2002, *Solar Phys.*, this volume.
Messmer, P., Benz, A. O., and Monstein, C.: 1999, *Solar Phys.* **187**, 335.
Nishio, M., Kosugi, K., Yaji, K. *et al.*, 1999, in T. S. Bastian, N. Gopalswamy and K. Shibasaki (eds.), *Proceedings of the Nobeyama Symposium, NRO Report* **479**, 235.
Raoult, A., Pick, M., Dennis, B. R., *et al.*: 1985, *Astrophys. J.* **299**, 1027.
Raulin, J. P., Vilmer, N., Trottet, G. *et al.*: 2000, *Astron. Astrophys.* **355**, 355.

Sui, L., Holman, G. D., Dennis, B. R., Krucker, S., Schwartz, R. A., and Tolbert, K.: 2002, *Solar Phys.*, this volume.

Trottet, G.: 1986, *Solar Phys.* **104**, 145.

Trottet, G: 1994, *Space Sci. Rev.* **68**, 149.

Trottet, G., Chupp, E. L., Marschhaüser, H. *et al.* : 1994, *Astron. Astrophys.* **288**, 647.

Trottet, G., Vilmer, N., Barat, C. *et al.*: 1998, *Astron. Astrophys.* **334**, 1099.

Vilmer, N. and Trottet, G.: 1997, in G. Trottet (ed.), *Coronal Physics from Radio and Space Observations, Lecture Notes in Physics* **483**, 28.

Vilmer, N. and MacKinnon, A. L.: 2002, in K. L. Klein (ed.), *Lecture Notes in Physics*, in press.

White, S. M.: 1999, in T. S. Bastian, N. Gopalswamy, and K. Shibasaki (eds.), *Proceedings of the Nobeyama Symposium, NRO Report* **479**, 223.

# RHESSI OBSERVATIONS OF THE SIZE SCALES OF SOLAR HARD X-RAY SOURCES

E. J. SCHMAHL[1] and G. J. HURFORD[2]
[1] *Astronomy Department, University of Maryland, College Park, 20742, U.S.A.;*
*e-mail: edward.schmahl@gsfc.nasa.gov,*
[2] *Space Sciences Laboratory, University of California, Berkeley, CA 94720-7450, U.S.A.*

(Received 7 August 2002; accepted 3 September 2002)

**Abstract.** The Reuven Ramaty High-Energy Solar Spectroscopic Imager RHESSI telescope produces hard X-ray images by Fourier imaging techniques that are capable of determining the sizes and shapes of sources with spatial scales in the range $\sim 2'' - 180''$. Applying the method of Unpixelized Forward Fitting to RHESSI modulation profiles from simple flares, we have identified the presence of 'halo' sources whose size scale ($\sim 40''$) greatly exceeds the 'core' sizes ($\leq 6'' - 14''$). Although such 'core-halo' structures have been observed at radio wavelengths using a similar technique, the radio and hard X-ray phenomena may be different. These observations raise questions about the nature of these 'halos'. Among the possibilities are that they are albedo sources, thin-target loops, or unidentified diffuse structures.

## 1. Introduction

The primary scientific objective of the Reuven Ramaty High-Energy Solar Spectroscopic Imager (RHESSI) is the study of energy release and particle acceleration in solar flares. This is achieved by imaging-spectroscopy of solar hard X-rays and gamma-rays over a 3-keV to 17-MeV energy range using a set of nine Rotating Modulation Collimators (RMCs) with high-purity germanium detectors. The scientific objectives, instrument and associated imaging techniques are described in detail by Lin *et al.* (2002), Smith *et al.* (2002), and Hurford *et al.* (2002) earlier in this volume.

As described by Hurford *et al.* (2002), the rotation of the spacecraft modulates the detected count rates from the incident hard X-rays, so that the amplitudes and phasing of the observed sinusoid-like time profiles (modulation profiles) provide information on the positions, sizes and shapes of hard X-ray sources. RHESSI's imaging method is precisely analogous to that of earth rotation synthesis in radio interferometry, with collimation replacing interference and spacecraft rotation replacing Earth rotation.

Clearly, the measurement of source size scales and morphology is relevant to determining the properties of the electron acceleration region and for constraining models of the acceleration process. With this in mind, we have selected several RHESSI flares with simple shapes and time profiles for which the size scales can

*Solar Physics* **210:** 273–286, 2002.

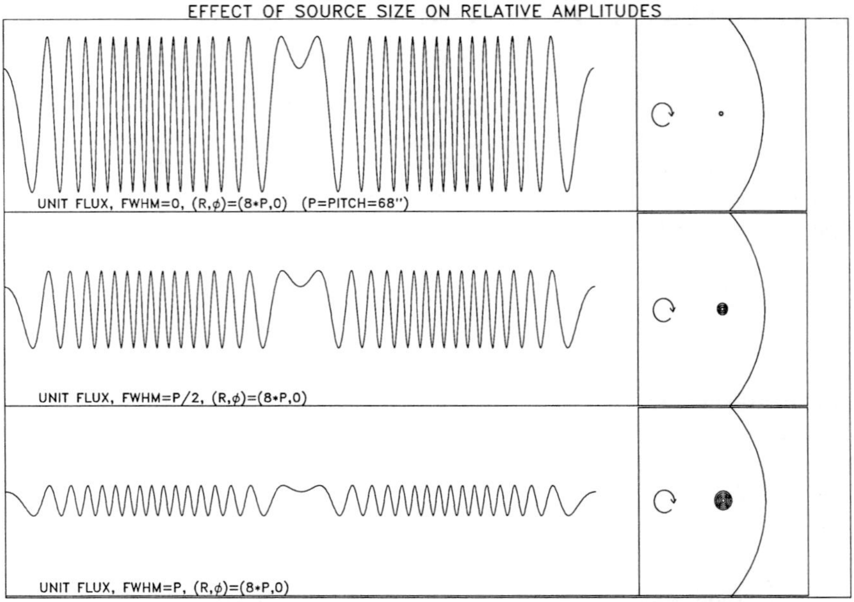


*Figure 1.* Ideal modulation profiles as a function of source diameter.

be most accurately determined, and have applied forward-fitting methods to the light curves to obtain sizes and spatial profiles of the hard X-ray sources.

## 2. Sizes and Relative Amplitudes

The amplitude of the modulation profile depends on the flux and the extent of the source, as well as on instrumental parameters such as the grid transmission and the detector sensitivity. The use of the relative amplitude – defined as the ratio of the modulation amplitude to the mean flux – removes the dependence on flux and detector response and so can directly measure the source extent in the direction perpendicular to the slits. (This assumes that background is negligible.) Wider sources produce smaller relative amplitudes. This is illustrated in Figure 1. For a Gaussian source with FWHM $= \Delta\theta$ the relative amplitude decreases with decreasing resolution $\Delta\theta_{\mathrm{res}}$ as $\exp(-0.89(\Delta\theta/\Delta\theta_{\mathrm{res}})^2)$. In general, for any assumed profile (e.g., Gaussian, pillbox or capped power law), the relative amplitude can be computed by taking the Fourier transform of the profile. Conversely, if the relative amplitude is known at a sufficient number of wavenumbers ($2\pi$/angular pitch), the profile can be found by computing the inverse transform of the relative amplitude.

For an RMC like RHESSI, the fundamental component of the response to a point source is a sinusoidal function of source position, with suitable modifications for the slowly varying grid transmission and effective slit/pitch ratios. The instantaneous response to a point source may be represented by the following equation:

$$C_{pt} = F_0\, \mathcal{G}(\phi)\, [1 + \mathcal{M}(\phi)\, \cos(2\pi(ux + vy) - \Phi)].. \tag{1}$$

In this expression, $F_0$ is the flux of the point source at the time where the slit angle (also the wave number angle) = $\phi$, $\mathcal{G}(\phi)$ is the grid transmission $\times$ livetime at angle $\phi$, $\mathcal{M}(\phi)$ is the maximum modulation amplitude, and $\Phi$ is the instantaneous phase at map center determined by the aspect system. If the instantaneous responses (1) are added together in $(x, y)$ space with appropriate weighting over one rotation, one gets a Bessel function profile ($J_0(kr)$) representing the RHESSI point response as a function of radius, $r$.

By the principle of superposition, an extended source with a brightness distribution $I(x, y)$ can be represented as a sum of point sources around the centroid $(x_0, y_0)$, so that the count modulation profile ($C_{\text{ext}}$) may be represented as a sum of expressions like Equation (2). Therefore the extended instantaneous response becomes:

$$C_{\text{ext}} = F_0 \int\int I(x, y)\, \{\, \mathcal{G}(\phi)\, [1 \;+\; \mathcal{M}(\phi)\, \cos(2\pi(ux + vy) - \Phi)]\}\, \mathrm{d}x\, \mathrm{d}y. \tag{2}$$

Note that any such $(x, y)$ sum over sinusoids, $\cos(ux + vy - \Phi)$, is another sinusoid with different amplitude, and possibly different phase:

$$C_{\text{ext}} = F_0\{\, \mathcal{G}(\phi)\, [1 \;+\; \rho(\phi)\, \mathcal{M}(\phi)\, \cos(2\pi(ux_0 + vy_0) - \Phi')]\}, \tag{3}$$

where

$$\rho(\phi) = \frac{\int\int I(x, y)\, \cos(2\pi(ux + vy) - \Phi)\, \mathrm{d}x\, \mathrm{d}y}{\int\int I(x, y)\, \mathrm{d}x\, \mathrm{d}y}. \tag{4}$$

In this expression, $\rho(\phi)$ is the relative amplitude, which is always $\leq 1$, and $\Phi'$ is a phase-shifted version of $\Phi$. For symmetric distributions $I(x, y)$ whose centroid is at the map center, $\Phi' = \Phi$ (no phase shift).

Figure 2 shows relative amplitudes produced by simple symmetric functions. We have selected three profiles illustrative of a range of behavior: very compact (pillbox), moderately compact (Gaussian), and extended (capped powerlaw):

(1) Pillbox: $I(r) = 1$ for $r \leq a$ and 0 otherwise;

(2) Gaussian: $I(r) = \exp(-\frac{1}{2}(r/a)^2)$;

(3) Capped power law: $I(r) = [(r/a)^2 + 1]^{-3/2}$, $\rightarrow r^{-3}$ for $r \gg a$ and $\rightarrow 1$ for $r \ll a$.

Figure 2 shows that the more compact the source, the narrower the profile of relative amplitudes, and *vice versa*. Note that even though the capped power-law (medium) and Gaussian (light) intensity profiles are similar out to their FWHM,

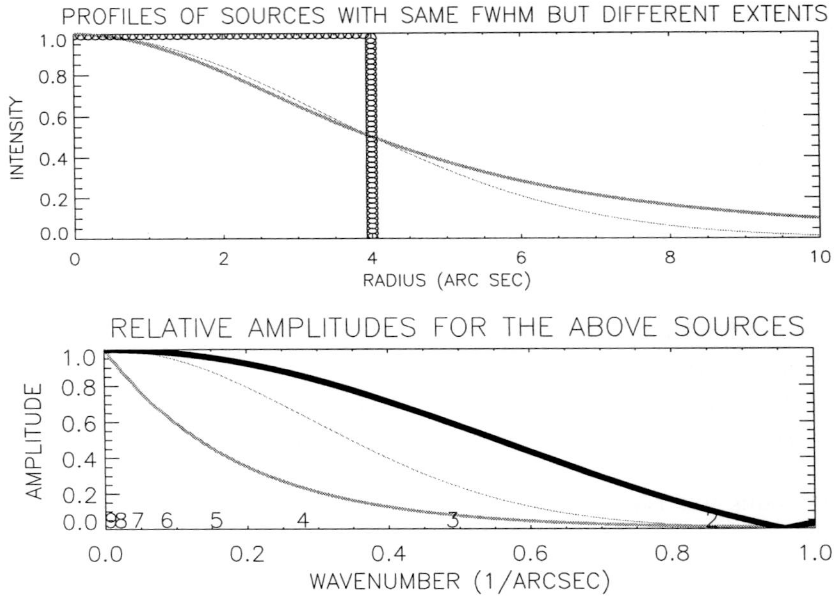


*Figure 2.* Relative amplitudes for representative simple, symmetric functions: a round pillbox (*thick*), a Gaussian (*light*) and a 'capped power-law' intensity profile (*medium width*). The latter corresponds to an exponential relative amplitude profile, ($\exp(-ka)$). The numbers above the abscissa identify the RHESSI subcollimators which measure the corresponding wavenumbers.

the extended tail (representing a 'halo') of the capped power-law intensity profile causes a big difference in the relative amplitude at low wave numbers. This demonstrates the ability of Fourier methods to identify and characterize extended sources.

## 3. Observed Relative Amplitudes

### 3.1. Candidate Flares – Simple, Steady, and Strong

Since our objective is to determine sizes and spatial profiles of HXR flares, we have focused on simple, steady and strong events.

Using the RHESSI CLEAN algorithm (see Hurford *et al.*, 2002, for details), we mapped $\sim 40$ flares, and selected those that satisfy the following criteria:

(1) The source consists of a single component.

(2) The flux is steady to $\lesssim 10$ percent over a rotation ($\sim 4$s).

(3) The source is relatively 'round' (axial ratio $\lesssim 1.25$).

(4) The count rate in the 12–25 keV band is $> 500$ counts $\mathrm{s}^{-1}\mathrm{SC}^{-1}$.

After this filtering of the database, we were left with seven events, whose CLEAN maps are shown in Figure 3. We include only subcollimators 3–9 in our study We excluded 1 and 2, since the threshold of detector 2 is set to $\sim 20$ keV (compared to

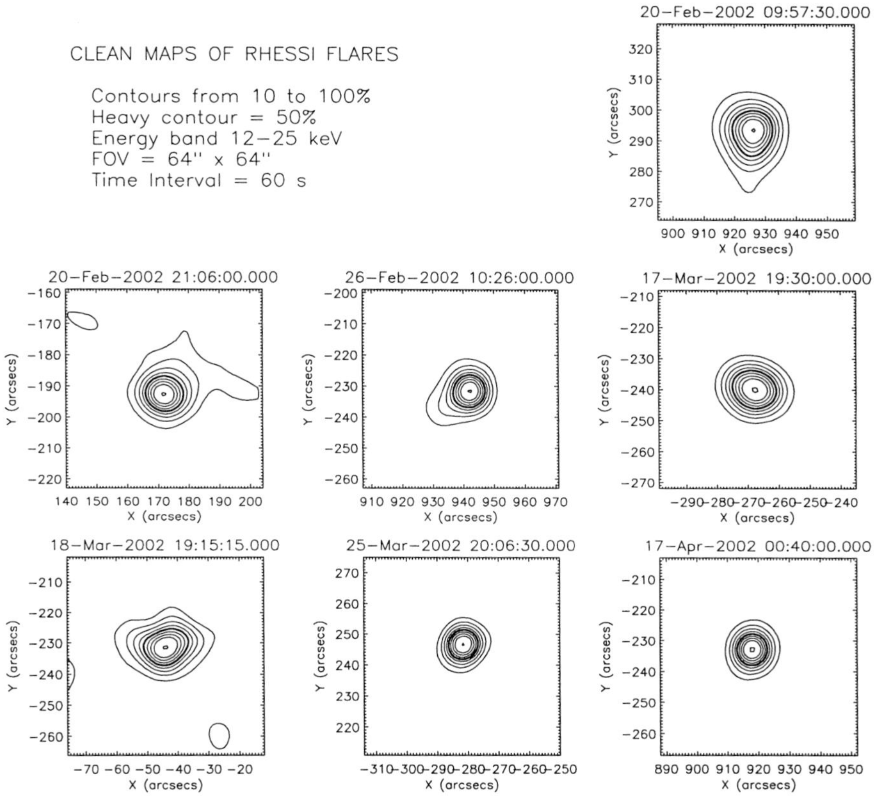


*Figure 3.* CLEAN maps of flares used in this study (UT times). The field of view in each map is $64'' \times 64''$, and the maps were made using subcollimators 3 ($7''$ resolution) through 8 ($105''$ resolution).

our 12–25 keV analysis band) and since subcollimators 1 and 2 have only recently been self-calibrated in phase (June 2002). Subsequent studies will include 1 and 2, since their higher spatial resolutions can provide quite important information.

### 3.2. MODULATION PROFILES

Since the modulation frequency increases systematically with higher spatial frequency of the grids, grouping the observed counts into overly coarse time bins could preferentially reduce the modulation amplitude at higher spatial frequencies. For this reason, we used relatively short time bins of 1 ms so as to reduce the worst-case loss of modulation amplitude to less than 0.5% for grids 5 through 9 and less than 3% for grids 3 and 4. Average reductions would be less than these limits.

With such small time bins, most of the flares in our sample produce mean count rates of only a few counts per time bin, and Poisson fluctuations are quite significant. This is not a problem for our analysis, which is not as sensitive to Poisson fluctuations as other algorithms (e.g., Pixelized Forward Fitting) may be.

It is important to note that the use of the C-statistic is essential to obtaining a proper fit, since there tend to be fewer than 10 counts per time bin (Cash, 1979; and Aschwanden *et al.* this volume.) The $\chi^2$ statistic leads to erroneous results with small time bins, and so it was not used. Valid results were also obtained by minimizing the r.m.s. deviation of the model from the data.

Figure 4 shows four of the modulation profiles during a 4 s interval for one of the events compared to a fit (Equation (3)) for parameters derived from forward-fitting. The method for fitting the model is quite straightforward since the flares are fairly round and constant so that the only time dependence is due to modulation. To start off, guesses are input for $(x, y)$ on a 2-D grid with increments of $\frac{1}{8}$ the current angular pitch. For each $(x, y)$, flux values $(F_0)$ and relative amplitudes $(\rho)$ are found by minimizing the least squares fit of the function (3). After the minimum C-statistic is found, the values for $x$, $y$, $F_0$, and $\rho$ are used for a second-order search using the amoeba method.

Background is not important in these events, as confirmed by inspecting the flux profile. However, there are random data gaps caused by cosmic rays within these profiles, and measures have been taken to exclude the gaps from the analysis. More than $\sim$99% of the gaps have a characteristic signature (Smith *et al.*, 2002) so that once identified, they can be flagged and excluded from subsequent analysis.

## 4. Sizes, Profiles and ‘Core-Halos’

### 4.1. Failure of Single Gaussians

Figure 5 shows the relative amplitude spectra $\rho(k)$ determined for the seven flares. The curves fitted to the data points are functions of the form $\log(\rho) = ak^b$. When $b = 2$, the source is a Gaussian, a traditional shape used in physics and astronomy, and in particular, solar physics, to characterize partially resolved sources (e.g., *Yohkoh*/HXT, microwave, mm-wave observations and the Forward-Fitting method used by the RHESSI software – Aschwanden *et al.*, 2002.) In our case, the fits indicate non-Gaussian shapes, and $b \approx 1$, implying a $e^{-kr}$ dependency. Gaussians that fit the first and last points are shown by the dotted curves, which in most cases are a poor fit to the data.

### 4.2. Profile by Inversion

As mentioned earlier, relative amplitude spectra with sufficient wavenumber coverage may be Fourier back-transformed to provide intensity profiles of the source without making *a priori* assumptions about their functional form. Using interpolation between the data points, we have inverted the spectra to obtain profiles. Since the relative amplitude spectra were obtained under the assumption of azimuthal

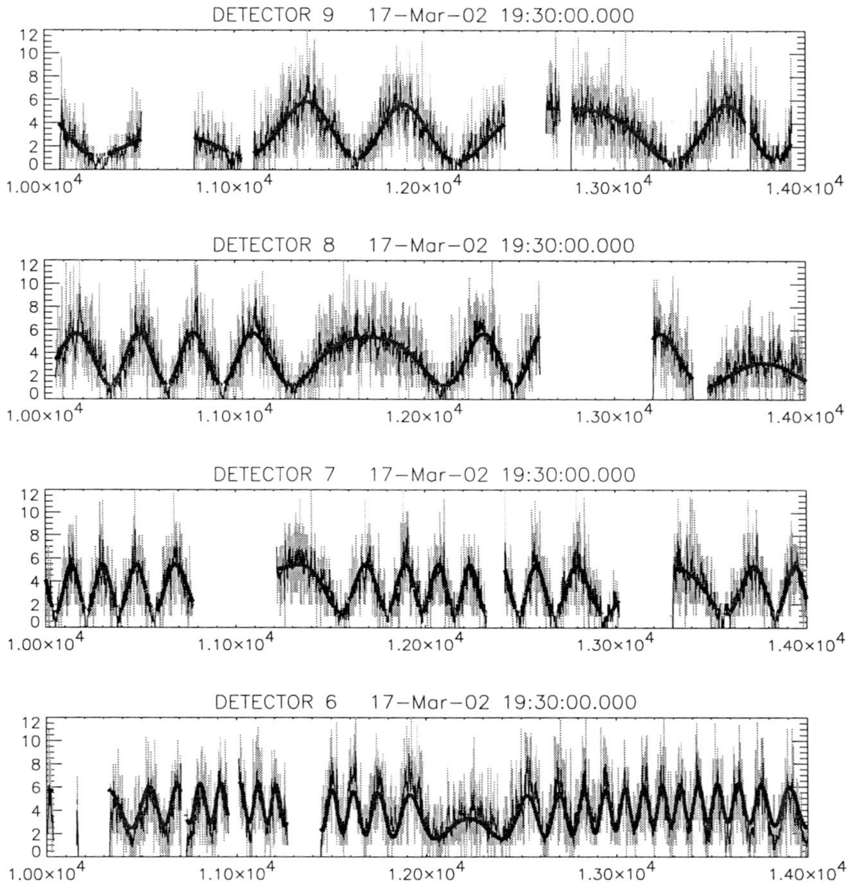


*Figure 4.* Modulation profiles for one rotation of the 17 March 2002 event for the 4 coarsest subcollimators, 6, 7, 8, and 9 whose cycles are sufficiently long that modulation is visible to the eye in the printed format available (UT times). The *thin, gray, jagged curves* are the count profiles (counts per time bin) and the *smooth, dark curves* are the 4-parameter fits using Equation (3). Inspection of the fitted curve for subcollimator 9 shows that the curves drop relatively close to the time axis, which indicates that the relative amplitude is nearly 1. For subcollimators 8, 7, and 6, the ratio of the amplitudes of the modulation to their means decreases as the angular pitch decreases, indicating that the relative amplitude decreases with increasing wave number. (Accurate determination of the relative amplitude requires corrections for the known effects of internal shadowing in the grids, and cannot be done by eye from this figure.) Similar effects may be seen in the plots of all the data and the fitted curves in the CD-ROM provided with this issue.

symmetry, the two-dimensional Fourier transforms reduce to pairs of Bessel transforms:

$$\rho(k) = \int I(r)\ J_0(kr) r\ \mathrm{d}r, \tag{5}$$

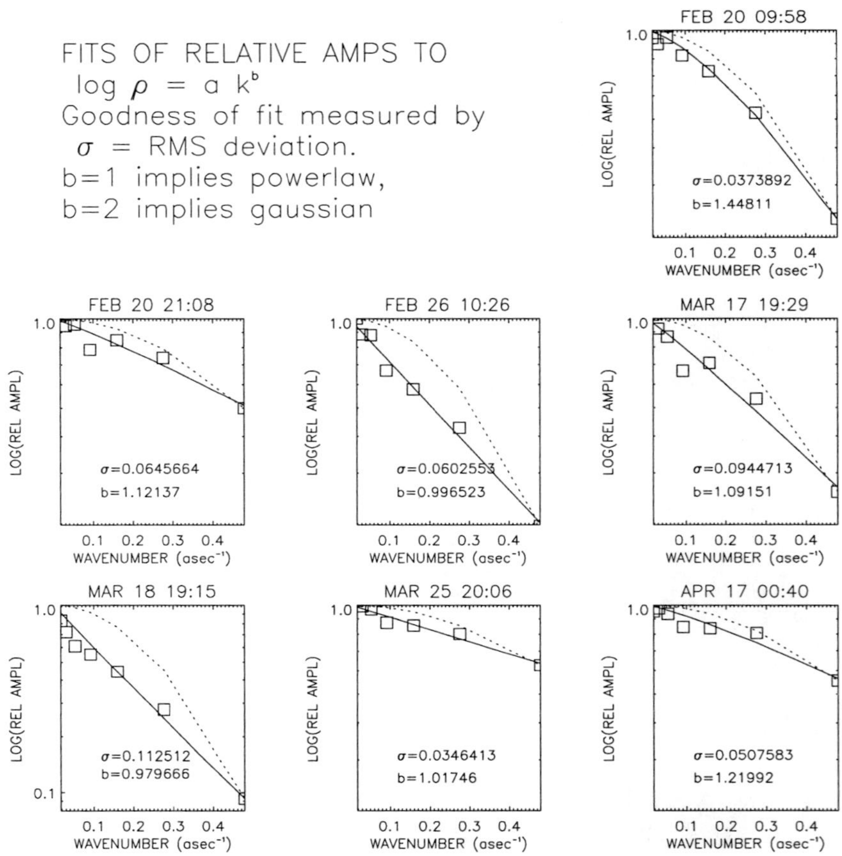


*Figure 5.* Subcollimator-to-subcollimator relative amplitude spectra (UT times). The *squares* are the measured values of $\rho(k)$ and the *solid curves* show fits with a power function $ak^b$. The index $b$, and a measure of goodness of fit, $\sigma$ = r.m.s. deviation, are shown in each plot. The *dotted curves* are for Gaussians, which are poor fits to the data.

$$I(r) = \int \rho(k)\ J_0(kr)k\ \mathrm{d}k. \tag{6}$$

The results of applying Equation (6) to the interpolated $\rho(k)$ spectra are shown in Figure 6. The spectra have been lightly tapered by a hyperbolic tangent function because most of the spectra do not fall off to small values at the largest $k$ ($\approx 0.5$) for subcollimator 3. This smooths out the 'ringing' effect caused by abruptly truncating the spectra. Data for subcollimators 1 and 2 would extend the domain to higher $k$ values, and presumably lower $\rho$ values, and negate the need for tapering, In any event, tapering has no effect on the finding of 'halos'.

Although the profiles shown in Figure 6 appear remarkably like Gaussians, this is simply the result of the use of a linear display. Any broadly extended source several times the diameter of the core sources would have a low surface brightness

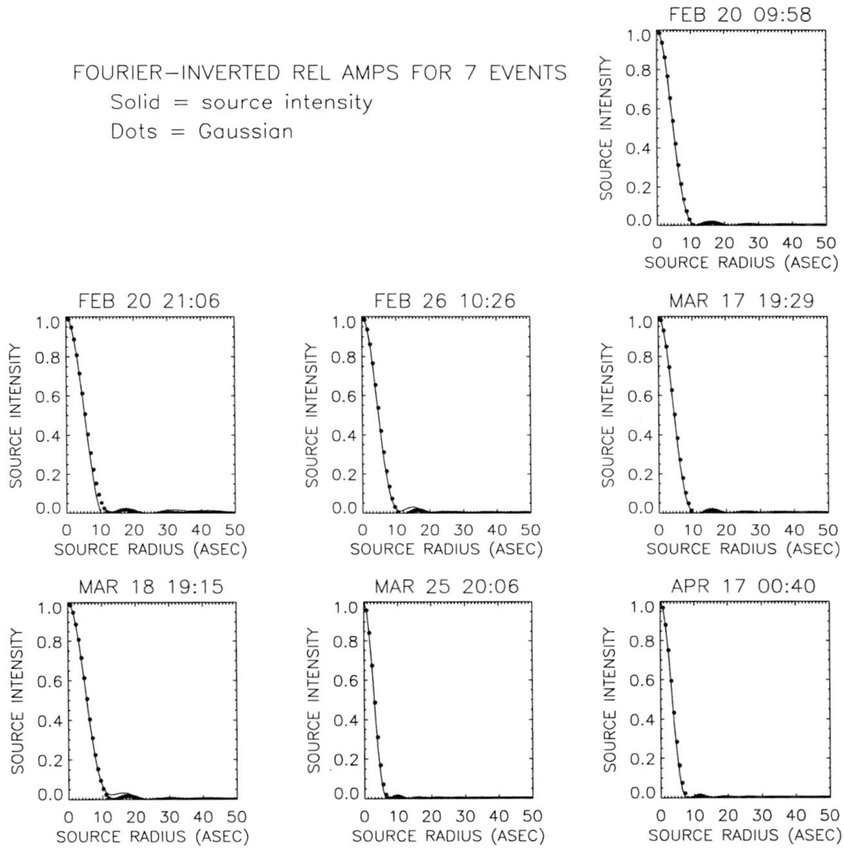


*Figure 6.* Core intensity profiles from inverted relative amplitude spectrum (UT times). The plots show the intensity profiles obtained by Fourier inversion using Equation (6) for the 7 flares. These are well fit by Gaussians (*dotted curves*). But, in common with other pixelized methods, they do not show all the flux. (Compare with the flux curves in Figure 7.)

that cannot be seen either in Figure 6 or in the maps in Figure 3. To get around this difficulty, we plot the cumulative flux from radius, $r$, to $\infty$ as a function of $r$. For a pure Gaussian, this is again a Gaussian. But Figure 7 shows that for most of the 7 events there is substantial flux (10–30%) in an extended region beyond 5–10 arc sec from the center. This appears as a broad 'tail' in the flux profile.

One of the flux profiles, for the 20 February 21:06 event, shows a positive slope near 10–20″, which cannot be real. Inspection of the relative amplitude profile for this flare (Figure 5) shows that it is far from smooth, and the value for subcollimator 6 (at wave number = 0.09) appears to be an 'outlier' from the trend of the data. The low value of relative amplitude in this particular instance causes negative values to appear around 10–20″ in the transform of the profiles, and this leads to a region of positive slope in the cumulative flux curve for this event (Figure 7). The other flares

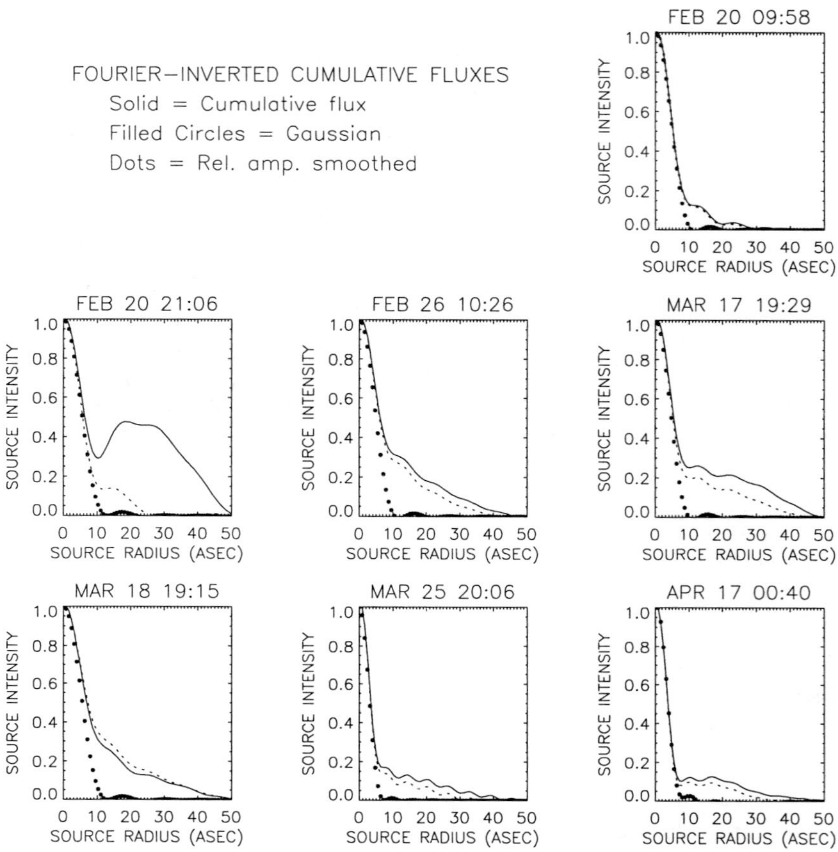


*Figure 7.* Cumulative fluxes from inverted relative amplitude spectra (UT times). The plots show the cumulative flux profiles $\int_r^\infty r\,I(r)\,\mathrm{d}r$ for the same inverted data shown in Figure 6. In all cases the profiles are Gaussian-like in the core, very much like in Figure 6, but extended beyond the FWHM.

do not exhibit this behavior, and it may be significant that the CLEAN map of this particular flare differs from the others in its shape at low levels. The CLEAN map (Figure 3) shows an extended long feature, which either contributes to this anomaly in relative amplitudes at the spatial scale of subcollimator 6 ($\sim 36''$) or illustrates some difficulties in CLEANing at the 10–20% level.

To see how 'noise' (or outliers) in the trends of relative amplitudes can affect the cumulative flux curves, we have smoothed the relative amplitude profiles before doing the transforms. The dotted curves in Figure 7 show the effect of smoothing. In the case of the 20 February 21:06 UT flare, smoothing eliminates the anomalous positive slope and reduces the 'halo' flux by about a factor of 2. For the other cumulative flux curves (none of which have significant positive slopes), the effects of smoothing are minimal, a few percent, either positive or negative.

## 5. Tests and Cross Checks

### 5.1. Analytic Tests

To check that the algorithm used here to find the relative amplitudes works correctly, we have tested it with a simulation in which the RHESSI count rate profiles are replaced by functions of the form

$$C_{\mathrm{sim}} = F_0\, \mathcal{G}(\phi)\, [1 + \mathcal{M}(\phi)\, \cos(2\pi(ux + vy) - \Phi)], \tag{7}$$

where $\Phi$ is the observed array of phases at map center, u and v are the arrays of points (on circles) given by the roll angles and angular pitches for the selected event; $\mathcal{G}$ and $\mathcal{M}$ are found from the standard RHESSI mapping software for the time and location of the flare. Then $F_0$, $\rho$, $x$, and $y$ were found. The resulting fits were exact to machine accuracy, as expected since there was no noise or Poisson fluctuations.

### 5.2. Poisson Simulations

In the next test, a Poisson-distributed function replaced the count rates, where the IDL Poisson-distribution generator created a sample using the mean value of the model in each time bin. The resulting fits were excellent, indicating that Poisson fluctuations do not adversely affect the fits, even when the count rates are as low as $\sim$ 1 count/time bin.

To check that the fits using capped power laws or double superposed Gaussians were better than single Gaussians, we compared C-statistics. Fits with single Gaussian models always have a larger C-statistic.

### 5.3. Possible Instrumental Effects

The presence of the 'halo' components can be directly attributed to the relatively rapid falloff in relative amplitude at the low wavenumbers measured by the coarse subcollimators. Therefore it is appropriate to ask if there are any plausible instrumental effects that could cause such an effect.

#### 5.3.1. *Aspect System Errors*

Imaging with RHESSI is directly dependent on knowledge of the pitch, yaw and roll aspect. Relative aspect errors during 21-s integrations would result in an image which represented a convolution of the source with the characteristics of these errors. To produce a spurious 'halo', such aspect errors would have to have a magnitude of 20 arc sec or larger. This is incompatible with continuous observations of the compact core and internal aspect redundancies that indicate an aspect solution that is self-consistent at the subarcsecond level (Fivian *et al.*, 2002). We conclude that aspect system performance is not an issue for this study.

5.3.2. *Modulation Amplitude Function*
Another possibility is that systematic errors in the modulation amplitude function $\mathcal{M}(\phi)$ (the multiplier of $\rho(\phi)$ in Equation (3)) could cause relative amplitudes to appear low if the values of $\mathcal{M}(\phi)$ were somehow too large. The same method is used to calculate the modulation functions for all grids, based on extensive preflight X-ray and optical calibrations of grid thickness, slat width, etc. There are three reasons for having confidence in the resulting values of $\mathcal{M}(\phi)$. First, the pre-flight calibrations were internally consistent at the $\sim$ 2% level, giving us considerable confidence in the knowledge of the grid parameters. Second, calculations based on these parameters correctly account for the rotational variation of $\mathcal{M}(\phi)$ which is readily seen in the data. Third, for grids with slits and slats that are approximately equal, $\mathcal{M}(\phi)$ is relatively insensitive to the slit/slat ratio, which is itself accurately known.

5.3.3. *Grid Transmission*
Errors in the grid transmission function $\mathcal{G}(\phi)$ could lead to erroneous estimates of the relative amplitudes. The arguments in the foregoing paragraph also apply to this function. We conclude that it is highly unlikely that errors in either the grid transmission or modulation amplitude could account for the falloff in relative amplitude among the coarse grids.

5.3.4. *Background*
One other effect that can produce low relative amplitudes is background. In our flares, the pre-flare and post-flare fluxes in the flare time profiles were used to estimate background, which was at least two orders of magnitude below the peak flux in this energy band. Furthermore, there is no reason to expect the background errors to affect the detectors systematically as a function of the spatial frequency of their grids. Background can therefore be ruled out as a contributor to any errors in relative amplitude.

## 6. Discussion and Conclusions

The main result of this study is that Fourier techniques applied to seven RHESSI flares reveal extended features of low intensity. The surface brightness of these 'halos' is typically only $\sim$ 2% or less than that of the core, so they are difficult to image with other methods. Nevertheless these 'halos' contain a significant fraction (10–30%) of the total flux. Table I shows the flare details, including the core size and the fractional flux of the extended features, as found from the point at which the tails of the profiles in Figure 7 depart from the Gaussian core.

We now consider possible explanations for the 'halo' features, and some ways by which their nature can be explored.

TABLE I
Properties of 'core-halo' events.

| Date 2002 | Time (UT) | X-ray class | Helio. lat./long. | NOAA region | Core FWHM | Halo flux |
|---|---|---|---|---|---|---|
| 20 Feb. | 09:58 | M4.3 | N18 W83 | 9825 | 10″ | 13% |
| 20 Feb. | 21:08 | M2.4 | S18 W11 | 9830 | 12″ | 12% |
| 26 Feb. | 10:26 | C9.6 | S13 W89 | 9830 | 12″ | 30% |
| 17 Mar. | 19:29 | M4.0 | S22 E16 | 9871 | 10″ | 25% |
| 18 Mar. | 19:15 | C8.9 | S21 E03 | 9871 | 14″ | 30% |
| 25 Mar. | 20:06 | C9.8 | N08 E17 | 9878 | 6″ | 16% |
| 17 Apr. | 00:40 | C9.9 | S13 W83 | 9905 | 7″ | 10% |

The flux of Compton solar back-scattered X-rays, called albedo X-rays, can be as large as 25–50% of the primary X-rays in the 12–25 keV band (Alexander and Brown, 2002; Bai and Ramaty, 1978). The spatial distribution of albedo X-rays depends on the height and extent of the source X-rays, so examination of the albedo sources can, in principle, provide information about the height of the source. The center-to-limb dependence of the albedo is predicted by models and there should be discernible differences in the shape of the 'halo' at the limb as compared with near disk center. From the curves of Alexander and Brown, the relative flux of the albedo should be an increasing function of energy from 6 to $\sim$ 40 keV, so the dependence of the relative amplitudes on energy should show this increase if the 'halos' are indeed albedo sources.

In the case of microwave flares containing 'core-halo' features, (e.g., Bastian, Benz, and Gary, 1998), the 'halo' is optically thin and polarized, while the core is optically thick and unpolarized. For gyrosynchrotron emission, this suggests that the core has a strong magnetic field and the 'halo' has a weaker one, compatible with the 'halo's' extended nature. However, for the hard X-ray and radio 'halos' to be the same, the hard X-ray 'halo' would have to be due to thin-target emission from extended loops. In that case, one would expect the hard X-ray 'halo' to be stronger at lower energies. The reverse would be true if the 'halo' were an albedo source.

As RHESSI flares accumulate and the calibration and analysis software improve, this study will be extended to a wider range of energies and source morphologies. It should be possible to use such factors as energy- and limb-dependence to help identify the origin of the 'halo' sources and if they turn out to be albedo sources, this can be exploited to determine the heights of the primary, compact hard X-ray sources. Using the method we have outlined here, it may also be possible to provide good estimates for the the fractional contribution of the albedo component in spectra on a flare-to-flare basis, thereby improving the interpretation of flare source spectra.

## Acknowledgements

The authors thank Alicia Aarnio (NASA Goddard summer intern from Smith College) for mapping and CLEANing numerous candidate RHESSI flares for this study. One of the authors (EJS) acknowledges support by grant NAG-5-10180 from NASA Goddard Space Flight Center to the University of Maryland. The research is supported by NASA grant NAS5-98033-05/03.

## References

Alexander, R. C. and Brown, J. C.: 2002, *Solar Phys.*, this volume.
Aschwanden, M. J., Schmahl, E., and the RHESSI Team: 2002, *Solar Phys.*, this volume.
Bai, T. and Ramaty, R.: 1978, *Astrophys. J.* **219**, 275.
Bastian, T., Benz, A., and Gary, D.: 1998, *Ann. Rev. Astron. Astrophys.* **36**, 131.
Cash, W.: 1979, *Astrophys. J.* **228**, 939.
Fivian, M., Hemmeck, R., Mchedlishvili, A., and Zehnder A.: 2002, *Solar Phys.*, this volume.
Hurford, G. J. *et al.*: 2002, *Solar Phys.*, this volume.
Lin, R. P. *et al.*: 2002, *Solar Phys.*, this volume.
Smith, D. M. *et al.*: 2002, *Solar Phys.*, this volume.

# ENERGY BUDGET AND IMAGING SPECTROSCOPY OF A COMPACT FLARE

PASCAL SAINT-HILAIRE[1,2] and ARNOLD O. BENZ[1]
[1] *Institute of Astronomy, ETH, CH-8092 Zürich, Switzerland (e-mail: shilaire@astro.phys.ethz.ch)*
[2] *Paul Scherrer Institute, CH-5232 Villigen, Switzerland*

(Received 9 August 2002; accepted 26 September 2002)

**Abstract.** We present the analysis of a compact flare that occurred on 26 February 2002 at 10:26 UT, seen by both RHESSI and TRACE. The size of the nearly circular hard X-ray source is determined to be 5.6 ($\pm$0.8)$''$, using different methods. The power-law distribution of non-thermal photons is observed to extend down to 10 keV without flattening, and to soften with increasing distance from the flare kernel. The former indicates that the energy of the precipitating flare electron population is larger than previously estimated: it amounts to 2.6 ($\pm$0.8) $\times$ $10^{30}$ erg above 10 keV, assuming thick-target emission. The thermal energy content of the soft X-ray source (isothermal temperature of 20.8 ($\pm$0.9) MK) and its radiated power were derived from the thermal emission at low energies. TRACE has observed a low-temperature ejection in the form of a constricted bubble, which is interpreted as a reconnection jet. Its initial energy of motion is estimated. Using data from both satellites, an energy budget for this flare is derived. The kinetic energy of the jet bulk motion and the thermal and radiated energies of the flare kernel were more than an order of magnitude smaller than the derived electron beam energy. A movie is available on the CD-ROM accompanying this volume.

## 1. Introduction

The energy of solar flares is commonly assumed to be magnetic in origin, but its release is still unclear. MHD theory of reconnection predicts equal shares of energy for local heating by electric resistivity and the motion of the plasma ejected from the reconnection region (e.g., Priest and Forbes, 2002). However, early observations of centimeter radio and hard X-rays (HXR) from flares made clear that a considerable fraction of the flare energy is initially transferred into energetic electrons (Neupert, 1968; Brown, 1971; Lin and Hudson, 1976) and possibly ions (Ramaty *et al.*, 1995). These 'non-thermal' particles carry a large fraction of the energy away from the flare region and deposit it partially in the chromosphere, where plasma is heated to tens of million degrees, rises into the corona and emits soft X-ray emission (Neupert, 1968). The knowledge of the energy content in the various forms of primary and secondary energy is essential in formulating flare scenarios and modeling flares.

The partitioning of the energy is not well known, as the exact evaluation of the various amounts of energy is hampered by observational limits. Early estimates by Strong *et al.* (1984) of a simple flare using several instruments on SMM find a ratio of 1.3 : 1.7 : 1 for the distribution between electron beam, mass motion,



CD ROM

and thermal energies, where the mass motions were measured in a Ca XIX line and may include also evaporative motions. It may be partially a secondary form of energy, derived from the kinetic energy of energetic particles. De Jager *et al.* (1989) have compared beam and thermal energies of 19 flares with similar results. Observations indicative of reconnection jets have been reported in the literature (e.g., Shibata *et al.*, 1994; Pohjolainen *et al.*, 2001; Zhang, Wang, and Liu, 2000). The identification as a reconnection jet in the corona was often based on the motion of dense material. As the process of reconnection is not predicted to substantially increase the density, these observations suggest that reconnection took place in a high-density region. This may not necessarily be the case in all flares. The heating of reconnection jets is not understood. Although the plasma heated by resistivity is ejected, the jets also contain plasma at preflare temperature in the MHD scenario. Innes *et al.* (1997) have reported reconnection jets in the quiet Sun having a temperature of a few $10^5$ K. On the other hand, stochastic electron acceleration by transit-time-damping of low-frequency waves is currently the most preferred mechanism (Miller *et al.*, 1997). It is expected to take place preferentially in the turbulent plasma of reconnection jets and may also heat them.

With the new generation of solar instruments in EUV lines, soft and hard X-rays, a more accurate determination of energies becomes feasible. Of particular relevance is the Reuven Ramaty High-Energy Solar Spectroscopic Imager (RHESSI) launched on 5 February 2002 (Lin *et al.*, 2002). RHESSI's germanium detectors, flown in space for the first time, register photons in the energy range from 3 keV to 17 MeV with 1 keV resolution at low energies (Smith, 2002). For the first time it is possible to explore the low-energy limit of non-thermal electrons, where most of the beam's energy resides. Nine absorbing grids modulating by satellite rotation provide the basis for imaging. The new method allows reconstructing the image of a flare anywhere on the visible disc of the Sun with a resolution of about 2″ at low energies (Hurford *et al.*, 2002). Thus, RHESSI's spatial resolution also allows determining the size of the high-temperature thermal flare plasma, necessary to estimate its energy content. Furthermore, RHESSI can model the thermal plasma by fitting the low-energy spectrum and determine its temperature and emission measure. The spatial resolution of an instrument like TRACE (Handy *et al.*, 1999) allows measuring the motion of coronal plasma at relatively low temperature predicted by MHD models for parts of the reconnection jets.

In this paper, we use the new capabilities for the first time to estimate the various flare energies in a well-observed, simple flare. The observations and some relevant analyses are presented in Sections 2–6, and in Section 7 the energies are evaluated and compared. A movie is available on the CD-ROM accompanying this volume.

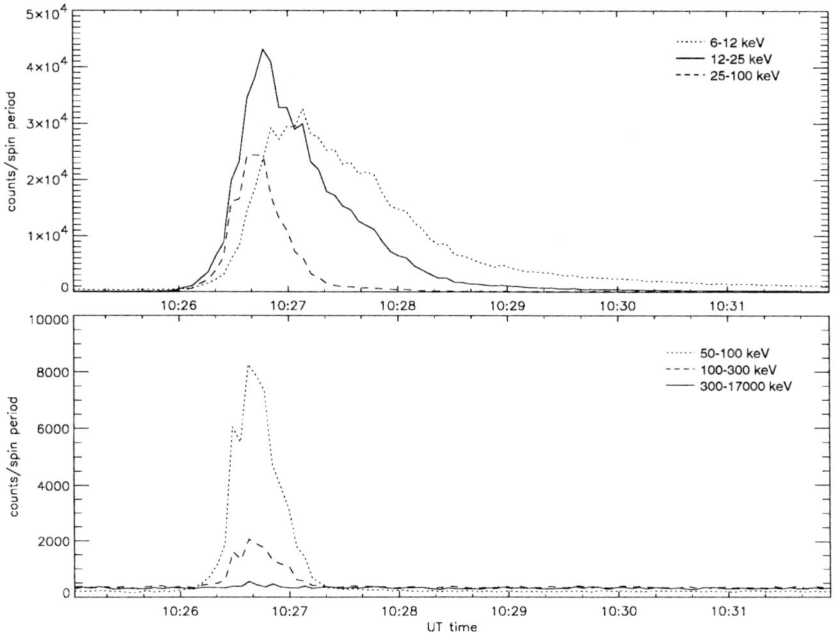


*Figure 1.* Lightcurves of RHESSI observations in uncalibrated counts per 4.35 s rotation period. All RHESSI front detector segments were used.

## 2. Lightcurves and Other Generalities

A flare occurred near the western limb of the Sun on 26 February 2002, and was observed by both RHESSI and TRACE. NOAA/GOES reported it as a C9.7 class flare.

Figure 1 shows that this event, while not the most powerful of flares, did have a non-negligible flux of high-energy photons (i.e., higher than 100 keV): enough to allow images to be reconstructed (Figure 3). The *Phoenix-2* radio spectrometer (Messmer, Benz, and Monstein, 1999) saw gyrosynchroton radiation during the high-energy part of the HXR emission (Figure 2), but very little decimetric emission. It consisted of three reverse drifting type III bursts between 1.2–2.0 GHz and a regular type III burst (620–800 MHz) with possible narrowband spikes at 850 MHz. All these faint coherent emissions (not visible in the overview of Figure 2) occurred after the HXR peak.

RHESSI's attenuator state was '1' (thin in, thick out) at all times. No decimation occurred, RHESSI was 'in the clear' throughout the event (no SAA or eclipse). The flare took place when RHESSI was at high geomagnetic latitude (43 deg).

As the average count rates per detector (total for all energy bands) was less than 6000 counts $s^{-1}$, no pile-up in the detector is expected (Smith *et al.*, 2002). No data gaps were recorded, aside from the fact that detector 2 was turned off at the

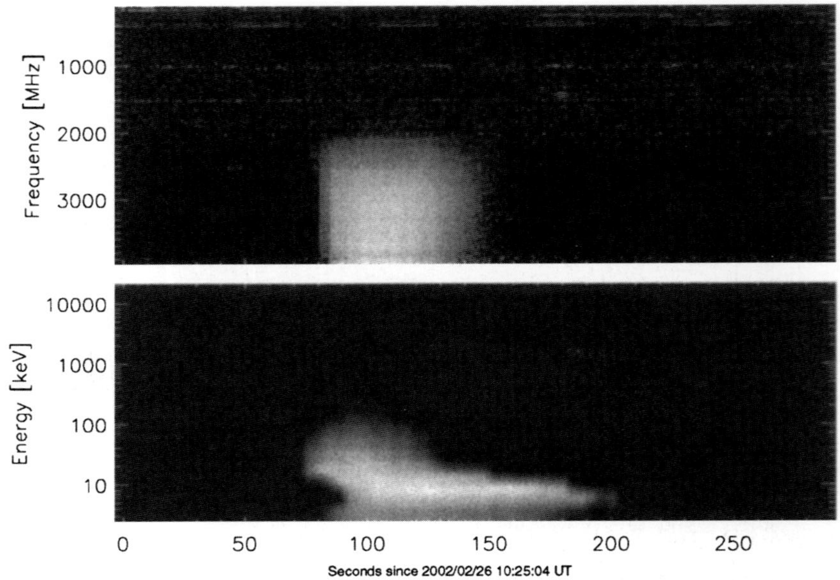


*Figure 2. Phoenix-2* radio (*top*) and RHESSI (*bottom*) spectrograms of the event. Both are background-subtracted.

time of the flare, and the presence of the usual dropouts. Dropouts are short data gaps ($\leq 1$ s) that occur randomly in every RHESSI detector, and are most likely the result of cosmic ray hits that momentarily saturate the detector electronics (Smith, 2002) – the imaging reconstruction software deals with those by ignoring them, i.e., during those times, no weighted modulation pattern contribution is added to the back-projected map.

## 3. Source Size

Figure 3 shows the flare at different times and energies. As the flux diminishes with higher energy, the brightness of the images has been adjusted. It can be seen that its spatial shape is nearly circular and remains practically constant during the HXR emission (special care has been taken to ensure that the *same* aspect solution was used for all the images). The deviation from circular at 100–300 keV is not to be considered statistically significant.

Figure 4 is a close up of the flare, as seen by RHESSI. It is the result of careful elimination of unbeneficial collimators and detectors. The first panel of Figure 5 shows that subcollimator 1 would not have contributed in a useful manner to the overall image: a pattern of minima and maxima of size and spacing comparable to that subcollimators's FWHM. Taking a longer accumulation time, i.e., the whole time interval when HXR counts were above background and the source spatially

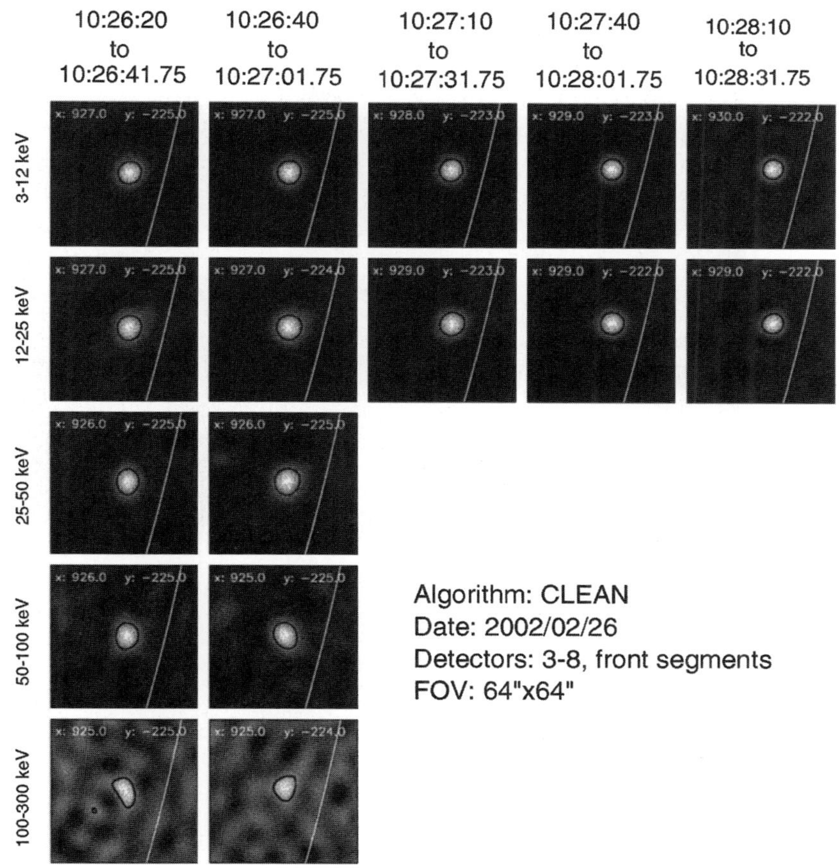


*Figure 3.* Panel of images: time (horizontal) vs. different energy bands (vertical). The *black contour* show the 50% level of the image. The *white line* is the photospheric limb of the Sun. The numbers on each panel refer to the position, in arc sec from Sun center, of the brightest pixel in the image.

stable (10:26:20 to 10:27:10 UT), does not yield a better result, even though the estimated total number of counts above background ($\sim$ 25 000) should have been sufficient. We conclude that subcollimator 1 has over-resolved the source. Detector 2 was unfortunately turned off at that time. Detector 9 (FWHM of 186″) was also removed, as its low resolution does not add any new features to the image. Hence, our imaging capabilities are limited by subcollimator 3, which has a FWHM point spread function of $\sim 6.9''$.

Table I lists the results of several methods used to determine the source size. The 'modulation amplitude' method will be described in this section, whereas the 'imaging spectroscopy' method will be described in Section 4. A source imaged with RHESSI appears larger than its true size $\sigma_s$, the standard deviation of an equivalent gaussian. The observed source size $\sigma_{\rm obs}$ results from convolution with

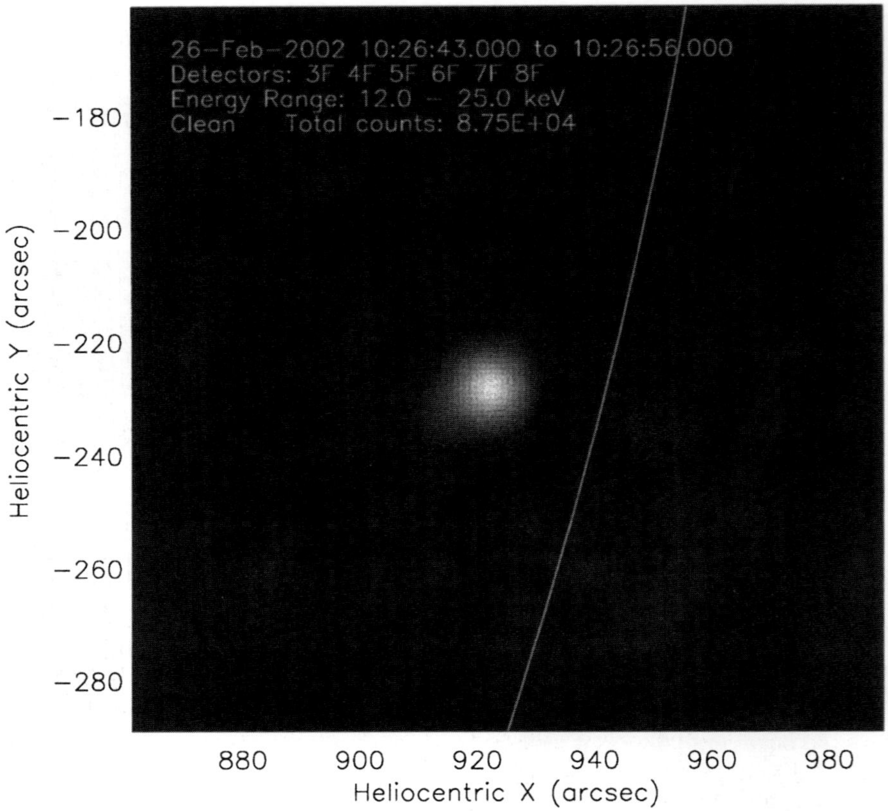


*Figure 4.* RHESSI CLEANed image made at peak 12–25 keV flux time (10:26:43 to 10:26:56 UT).

the point spread function $\sigma_{\rm psf}$ of the instrument, where $\sigma_s = \sqrt{\sigma_{\rm obs}^2 - \sigma_{\rm psf}^2}$, assuming that both source and point spread function have gaussian shapes.

The FWHM of the source was determined from the 50% contour of its image reconstructed by back projection, CLEAN and Maximum Entropy Method (standard RHESSI software)[1]. The FWHM of the point spread function of RHESSI collimators 3–8 was measured the same way. Back projection and CLEAN yield results that agree within error bars. The larger result obtained with forward fit might be due to the presence of an important halo for this flare (Schmahl and Hurford, 2002). The very small FWHM obtained by MEM-Sato may be caused by what has been dubbed 'super-resolution', and is not to be trusted.

Image reconstructions (with both back projection and CLEAN) using the RHESSI simulation tools has been done for different gaussian source sizes. For a true source size of $\sim 6''$, the apparent source size resulting from the simulated map agrees well with that of the observed map, and is close to the value found

[1] http://hesperia.gsfc.nasa.gov/rhessidatacenter/

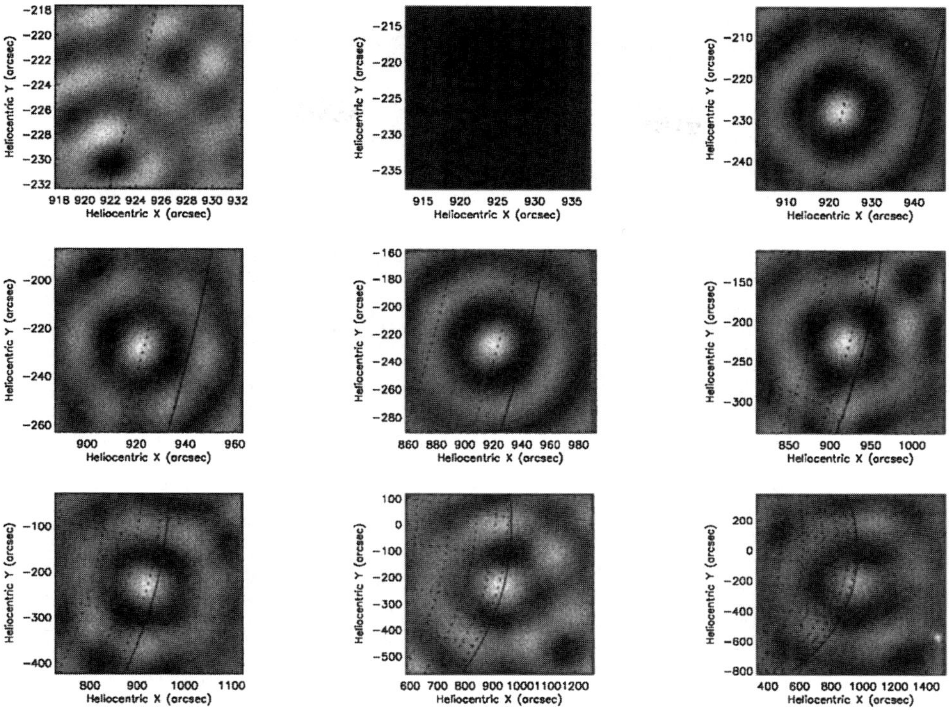


*Figure 5.* Back-projected images for each subcollimator (1 to 9), other parameters are the same as for Figure 4. The field of view and the pixel size are adjusted to the resolution of each subcollimator, such that the spatial sampling is twice the resolution.

TABLE I

Size of non-thermal HXR source (12–25 keV), derived by different methods. The error of the convolved size from RHESSI maps indicate the total range.

| Method | Convolved FWHM | Deconvolved or true FWHM |
|---|---|---|
| FWHM of maps: | | |
| 'Back proj.' using SC 3–8 | $10.5 \pm 0.7''$ | $5.4 \pm 1.4''$ |
| CLEAN using SC 3–8 | $10.6 \pm 0.4''$ | $5.6 \pm 0.8''$ |
| MEM Sato, using SC 3–8 | $2.9 \pm 0.2''$ | – |
| Forward Fit, using SC 3–8 | $17.0''$ | – |
| Pixon, using SC 3–8 | $5.9 \pm 0.7''$ | – |
| Imaging spectroscopy | $11 \pm 0.5''$ | $6.3 \pm 0.9''$ |
| Modulation amplitude | | $4.7 \pm 1.5''$ |

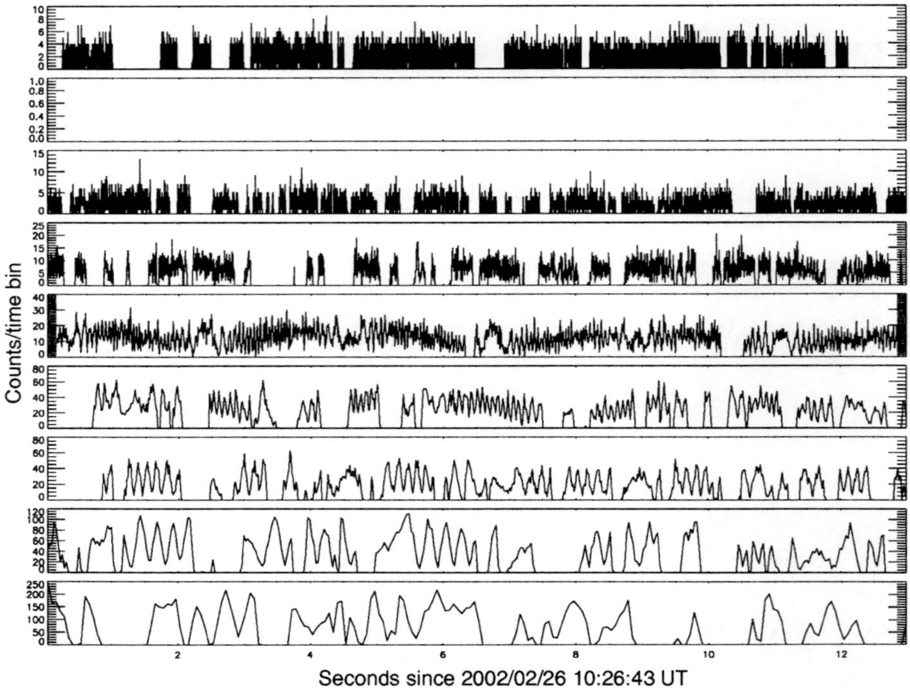


*Figure 6.* Sample of modulation profiles for each subcollimator (1–9: *from top to bottom*) for the time interval used for the image in Figure 4.

in Table I. Another method of determining source size was also attempted using *modulation profiles*.

Figure 6 displays the modulation profiles that were used to produce Figure 4. From those, the *relative modulation amplitudes* (Schmahl and Hurford, 2002) were determined. The relative modulation amplitude $A$ is defined as follows:

$$A = \frac{1}{M}\frac{C_{\text{max}} - C_{\text{min}}}{C_{\text{max}} + C_{\text{min}}} = \frac{1}{M}\frac{C_{\text{max}} - \langle C \rangle}{\langle C \rangle}; \tag{1}$$

$C_{\text{max}}$ and $C_{\text{min}}$ are the maximum and minimum counts per time bin in a modulation cycle. $M$ is the *maximum modulation amplitude*. It is determined by the angle of incidence on the grid as well as the effective slit/pitch ratio. The software calculates it for each time bin. Schmahl and Hurford (2002) give an analytical formula for $A$ in the case of gaussian sources:

$$A = e^{-0.89\left(\frac{\Delta\theta_s}{\Delta\theta_{\text{coll}}}\right)^2}, \tag{2}$$

where $\Delta\theta_s$ is the source's FWHM, and $\Delta\theta_{\text{coll}}$ is the collimator's FWHM. Thus, the modulation disappears gradually as the source dimension exceeds the collimator resolution.

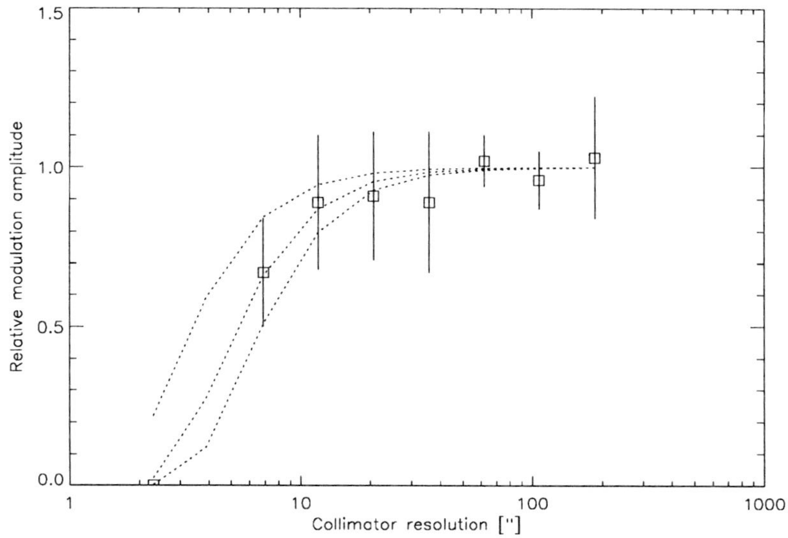


*Figure 7.* Observed relative modulation amplitude vs. subcollimator resolution, from the modulation profiles of Figure 6. The *dotted curves* were computed using Equation (2). The *upper one* for a source size of 3″ FWHM, the *middle one* for a size of 4.7″ FWHM, and the *lower one* for a size of 6″ FWHM.

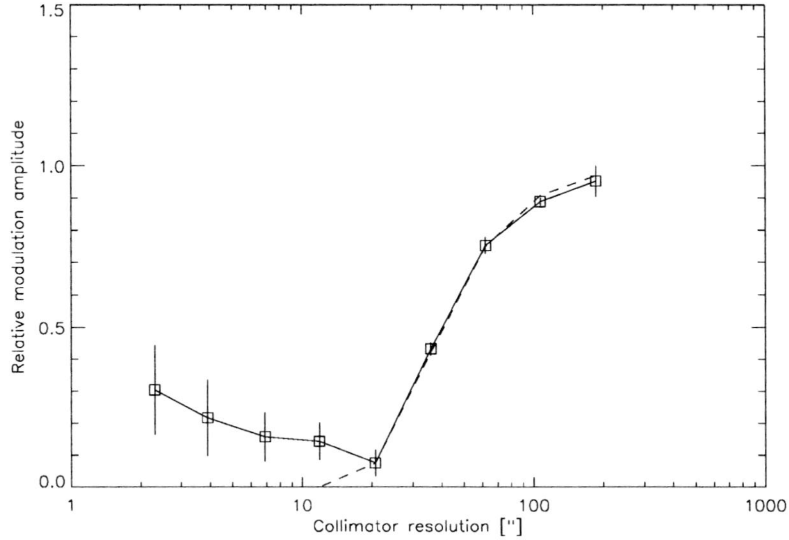


*Figure 8.* Relative modulation amplitude vs. collimator resolution (FWHM), for a simulated 2-D circular gaussian source of size $\sigma = 15''$ ($= 35.3''$ FWHM). The *solid line* links the data points, whereas the *dashed line* was computed using Equation (2).

$C_{max}$ and $C_{min}$ were determined manually and averaged over several modulation cycles (from 3 for subcollimator 9 to 60 for subcollimator 3, also yielding a standard deviation) in the modulation profiles (Figure 6). Thus, the contaminating effects of dropouts have been removed. The results are displayed in Figure 7. The relative modulation amplitude was set to zero for the first (finest) subcollimator, where photon fluctuation was clearly dominant. The relative modulation amplitude of subcollimator 2 was ignored, as it is not available. Few modulation cycles without dropouts were available (Figure 6) for subcollimator 9, resulting in a large standard deviation. Otherwise, the error bars increase with decreasing subcollimator coarseness. This is because the size of the time bins that were used also decreased with decreasing subcollimator coarseness, thus increasing the effects of photon counting noise. Comparing the data points with the theoretical curves, a source size of 4.7 $(\pm 1.5)''$ is assumed, and will be used in the numerical computations of Section 7.

The method was tested on gaussian sources of different sizes using the RHESSI simulation software tools. The match is almost perfect for regions with low photon fluctuations. Figure 8 is one such plot, made with $5 \times 10^5$ photons $s^{-1}$ detector$^{-1}$, for a source size of $35.3''$ FWHM, and in the same 12–25 keV energy band as used previously. The existence of non-zero relative modulation amplitudes at low collimator resolution is due to photon counting noise, and the manual technique for finding the peaks, which does not make any use of the phase (as a forward fitting method would).

The 4.7 $(\pm 1.5)''$ source size derived from the modulation amplitude method is consistent with previous results. As CLEAN yielded the deconvolved result with the smallest error bar, its value of 5.6 $(\pm\,0.8)''$ will be the source size considered for the remainder of this paper.

## 4. RHESSI Imaging Spectroscopy

Imaging spectroscopy is limited by photon-counting noise. Hence, we will simply concentrate on doing imaging spectroscopy during the peak HXR flux of the flare. As the images obtained are not fully calibrated through the spectral response matrix (Schwartz *et al.*, 2002), only energies above 10 keV and below 100 keV were considered.

The flare was imaged (using CLEAN) between 10:26:43 and 10:26:56 UT, from 10 to 100 keV (using 1-keV energy bins from 10 to 20 keV, then 5-keV bands), and using detectors 3, 4, 5, 6, and 8. The image was over-sampled by taking $1''$ pixels. The 26 images thus obtained are not shown here. A 'crosshair' of pixels (10 vertically, 10 horizontally), centered on the flare, was considered. Figures 9 and 10 show results obtained by averaging pixel fluxes at equal distance from the flare's center.

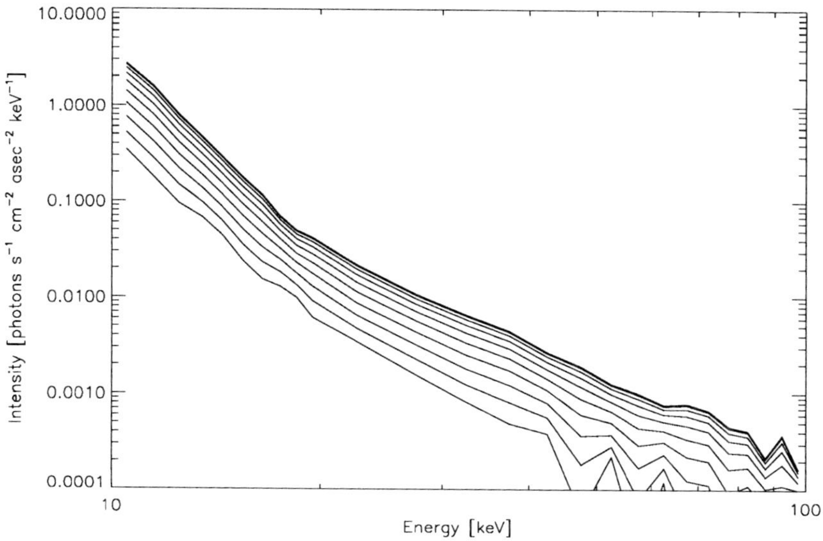


*Figure 9.* Imaging spectroscopy I: spectra at different distances from flare center position at the time of peak HXR flux. Spectra *from top to bottom* were taken at increasing distances, 1″ increments, starting at 0″ for the topmost spectrum. At high energies and large distances, photon fluctuation effects become important.

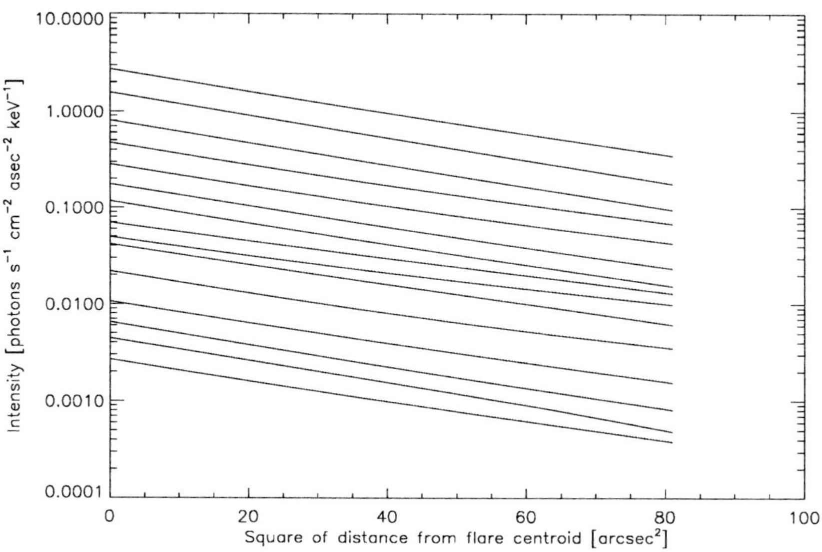


*Figure 10.* Imaging spectroscopy II: flux vs. distance from flare center position, for different energy bands. The flux in different energy bands (from 10 to 45 keV) are ordered with increasing energy, *from top to bottom*. From 10 to 20 keV, 1-keV bands were used. From 20 to 45 keV, 5-keV bands were used.

Figure 9 shows that the portion of the spectrum above the thermal bremsstrahlung component is close to a power law, with photon spectral indices varying from $3.4 \pm 0.2$ (flare center, associated with the brightest pixel in the map) to $3.9 \pm 0.2$ (at $10''$ from flare center), with the fits made in the 20–40 keV energy band.

The hardening of the spectral index towards the center of the flare kernel may be due to a harder electron distribution (more fast electrons) near the center of the flare.

This observation does not support proposed models such as a superposition of thermal distributions (Brown, 1974) mimicking a power law of energy spectral index, e.g., $\gamma = 1/2 + 3/\eta$ from a sum of thermally emitting spherical co-centric shells with temperatures $T(r) = T_0(r_0/r)^\eta$.

The flare had a gaussian shape at all energies (Figure 10). The slope ($= -1/2\sigma^2$) was about $-0.023 \pm 0.002$, implying an apparent source sigma of $4.7(\pm 0.2)''$, or FWHM of $11(\pm 0.5)''$. Note that the apparent source size is again a convolution of the true source size with the point spread function of the imaging instrument.

## 5. Spectral Features

Figure 11 shows a spatially integrated spectrum accumulated during the peak of the HXR flux. As RHESSI's spectral response below 10 keV is not yet completely known (particularly when a shutter is in, as is the case here), spectral fitting (using the full spectral response matrix) has been done on energies above 10 keV. The data points were fitted using the SPEX[2] software, with a thermal free-free bremsstrahlung and double broken power-law model.

The thermal bremsstrahlung component of the fit model yields a temperature of $T = 19.7 \pm 1.0$ MK, and an emission measure of $EM \approx 0.2 \times 10^{49}$ cm$^{-3}$ (accurate to within a factor 2), assuming an isothermal source. The temperature derived from GOES-8 3-second data peaks at 10:26:51 UT at 16.7 MK, with an emission measure of $0.6 \times 10^{49}$ cm$^{-3}$.

The power-law component had a photon flux at 50 keV of $1.5 \pm 0.2$ photons s$^{-1}$ cm$^{-2}$ keV$^{-1}$, and a spectral index of $3.0 \pm 0.1$. This value is not significantly different than the ones derived in the previous section, where a different time interval was used, and where only energies in the 20–40 keV range were considered. A break in the power law is located at $54 \pm 3$ keV. After this break, the photon spectral index is $3.5 \pm 0.1$. Breaks are commonly observed (see, e.g., Lin and Schwartz, 1987) and do not significantly influence the energy budget (Section 7).

An emission volume V can be inferred from the size of the source observed with imaging: $V = (4\pi/3)R^3$, where $R$ is the size of the source. With $R = 2.8\ (\pm 0.4)''$, one finds an emitting volume of $V = 3.5 \times 10^{25}$ cm$^{-3}$, with $(2.2–5.2) \times 10^{25}$ cm$^{-3}$ the range of possible values.

The thermal energy and density can then be calculated, using

[2]http://hesperia.gsfc.nasa.gov/rhessidatacenter/spectroscopy.html

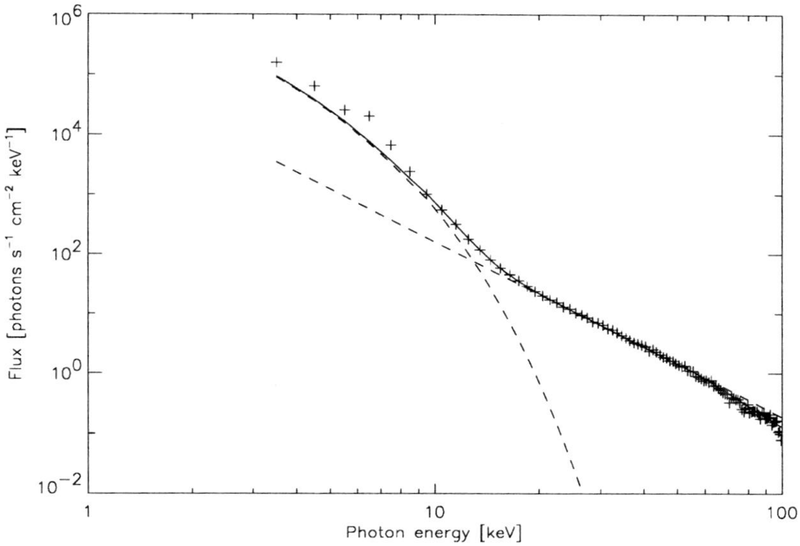


*Figure 11.* Spectrum accumulated from 10:26:05 to 10:27:23 UT, during enhanced HXR > 25 keV flux. Only the front segment of detector 4 was used. The '+' symbols are the data points. A fit with a thermal bremsstrahlung and a power law is also drawn (*solid line*).

$$E_{th} = 3k_B T \sqrt{EMVq}, \tag{3}$$

$$n_e = \sqrt{\frac{EM}{V} q}, \tag{4}$$

where $EM$, the emission measure, and $T$, the temperature, are the ones derived above. The filling factor is represented by a fraction $q$. An inhomogeneous medium possesses less thermal energy than a homogeneous one, for the same temperature and emission measure. $q = 1$ is assumed throughout this paper. Equation (4) yields a density of $n_e = 2.6 \times 10^{11}$ cm$^{-3}$ (range: (2.1–3.3) $\times$ $10^{11}$ cm$^{-3}$), comparable to what is derived from TRACE in Section 7.

For determination of the thermal energy in the flare (Section 7), temperature and emission measure determined at the peak of the soft X-ray (SXR, < 12 keV) flux are needed. Spectral fitting done around 10:27:10 UT, accumulated over three RHESSI rotations of 4.35s, yields $T = 20.8 \pm 0.9$ MK and $EM = 2.9$ $(\pm 0.5) \times 10^{48}$ cm$^{-3}$.

## 6. TRACE Images with RHESSI Overlays

Figure 12 shows TRACE images with RHESSI overlays at different energies. The TRACE images had exposure durations of 20–30 s, except for the fourth one (EUV peak), which had an exposure duration of 8 s. All TRACE images have been translated 10″ northerly, to align with RHESSI. The EUV band pass is dominated by a spectral line of Fe XII (195 Å) having a maximum emissivity at 1.4 MK. At high temperatures (15–20 MK), an Fe XXI line (192 Å) appears and may add significant flux (the filter's response around 15–20 MK is still two orders of magnitude less than at 1.4 MK).

The RHESSI images' accumulation times (as labelled on top of each image) correspond loosely to the time difference between TRACE images (in integer multiples of the spin period of 4.35 s), and were all made using the CLEAN reconstruction algorithm, and subcollimators 3 to 8. Again, care has been taken to use the same aspect solution for all RHESSI images.

The TRACE observations clearly show an ejection occurring with the flare, starting in the second image of Figure 12. Later it develops into the shape of a bubble, which gets constricted at the bottom (best visible in Figure 12, fourth image). It does not rise beyond the TRACE field of view, but becomes turbulent (see movie available on the CD-ROM). The formation, development and constriction of the bubble is suggestive of a reconnection jet scenario speeding from the apparent X-point located at (930,−220) arc seconds from Sun center in Figure 12 (fourth image). The proposed scenario is depicted in Figure 13.

RHESSI overlays show that the HXR (12–25 and 25–50 keV) are emitted in the flare kernel (marked by B in Figure 13). At the peak flux of HXR $> 25$ keV, the peak positions of the sources $> 25$ keV are shifted to smaller radial distances from Sun center ($\sim 1$–$2''$, best seen with Figure 3), consistent with the interpretation that, after being accelerated by a reconnection event, the mildly relativistic electrons precipitate in the lower corona by emitting thick-target HXR radiation. The electrons with the higher energies will lose their energy (mostly via Coulomb collisions) only in the deeper, denser chromosphere. The hot plasma being heated by the precipitation of electrons ($< 25$ keV overlays or images in Figures 12 and 3) was apparently in the same volume. After 10:27:20 UT, when the HXR emission above 25 keV ceases, a slow outward (NW direction) movement of the 3–12 and 12–25 keV (thermal) sources is witnessed, at a speed of about 30 km s$^{-1}$.

## 7. Energy Budget

Table II summarizes the different energies found. In this section, their derivation from the data is described.

*Kinetic energy of precipitating electrons* $E_{\mathrm{kin,beam}}$: assuming that the power-law component in our spectra is related to the HXRs emitted by precipitating electrons in the lower corona (thick target), an electron distribution can be determined from the relations in Brown (1971) and Hudson (1978):

$$\frac{\partial^2 N_e}{\partial E \partial t} = 3.28 \times 10^{33}\, b(\gamma)\, \frac{F_{50}}{50^{-\gamma}} E^{-\gamma-1} \quad [\mathrm{e}^{-1}\ \mathrm{s}^{-1}\ \mathrm{keV}^{-1}], \tag{5}$$

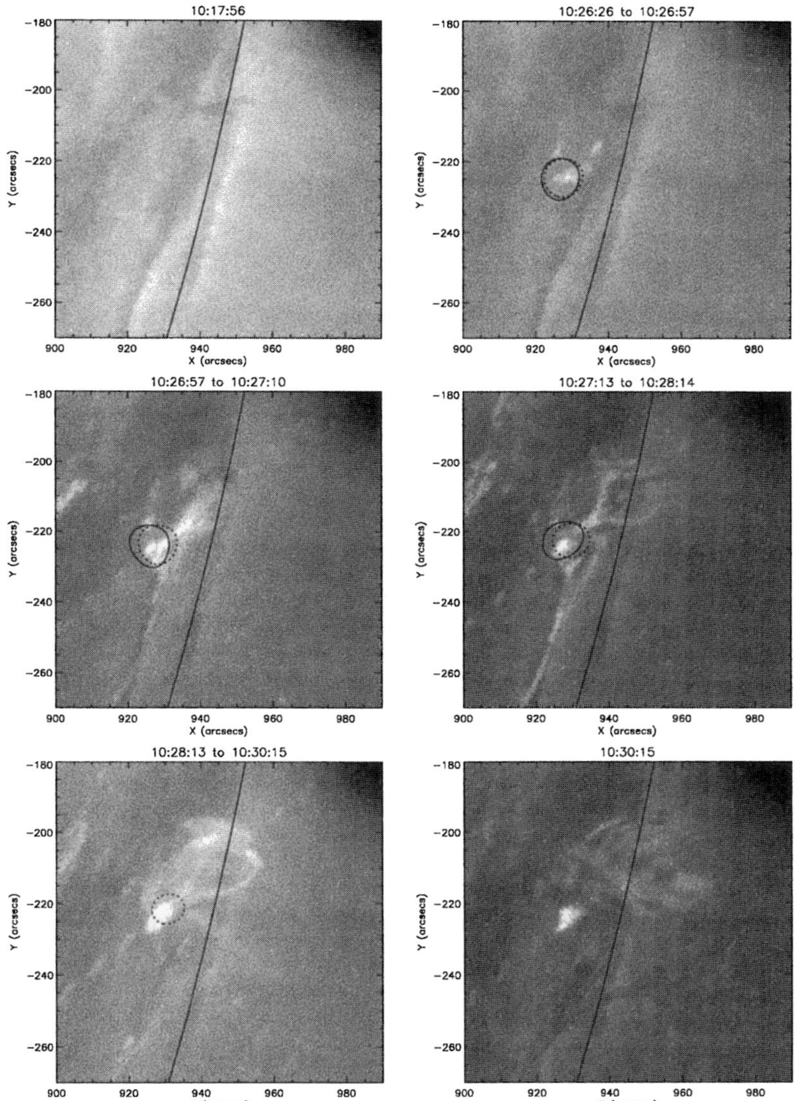


*Figure 12.* TRACE images at 195 Å with RHESSI overlays of different energy bands. The contours correspond to the 50% level. The *dotted black contour* corresponds to the 12–25 keV band and the *full black contour* to the 25–50 keV band. As shown in Figure 3, the 3–12 and 12–25 keV images differ by less than 1″ throughout the flare, as do the 25–50, 50–100, and 100–300 keV images. The first image shows the region of interest before the flare. The second one was taken during the rise of the HXRs. The third one is between HXR and SXR peaks, the fourth one between SXR (3–12 keV) and EUV peak (*flash phase*). The last two were taken during the decay phase.

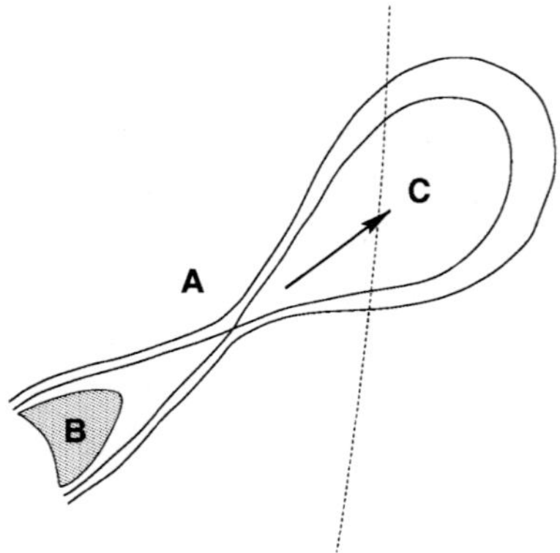


*Figure 13.* Schematic drawing of field lines and interpretation of TRACE image: (A) X-point (reconnection site), (B) flare kernel, (C) upward reconnection jet.

where $\gamma$ is the photon spectral index, $F_{50}$ is the photon flux (in photons $s^{-1}$ $cm^{-2}$ $keV^{-1}$) at 50 keV, $E$ is the electron kinetic energy in keV, and $b(\gamma)$ is equal to 7.05 for $\gamma = 3$.

At low electron energies, the distribution must become flatter, reducing the photon distribution below the cut-off energy. Power-law distributions with low energy cut-offs have been fitted to the spectrum (Figure 11). A cut-off above 10 keV decreases the match with observations, below 10 keV it has little influence on the fit of the photon spectrum. In the absence of effects that enhance the low-energy photon flux (such as non-uniform target ionization, Kontar, Brown, and McArthur, 2002), we conclude that the low-energy cut-off for the photon power-law spectrum is $\leq 10$ keV. In the following, an electron power-law cut-off energy of $10 \pm 1$ keV is assumed.

Determining the number of electrons $\geq 10$ keV that precipitated, as well as their total kinetic energy, is simply done by integrating over time and energy (from $10 \pm 1$ keV to $\infty$). The total kinetic energy in all precipitating electrons is $2.6 \times 10^{30}$ erg. At peak time, an average of $1.4 \times 10^{36}$ electrons $s^{-1}$ precipitated, for a total of about $1.1 \times 10^{38}$ electrons.

*Kinetic energy of ejected plasma (bulk motion)* $E_{\rm mot}$: assuming isothermal plasma, each of TRACE's CCD pixel possesses an accumulated flux

$$F = f_{195}(T) EM, \tag{6}$$

where $F$ is the flux ('data number', in TRACE parlance) in the pixel (CCD dark currents have been subtracted), $EM$ the emission measure observed in that pixel, and $f_{195}(T)$ a known function of the temperature $T$ (the filter response function) (Handy *et al.*, 1999). As TRACE was observing with only one filter band (195 Å, with one aluminium filter in the FOV), temperature and emission measure cannot be determined without additional information or assumptions. $f_{195}(T)$ peaks at $T = 1.4$ MK. Given a certain flux in a pixel, this temperature yields a lower limit of the emission measure (and hence, the density). RHESSI spectral fitting (Section 5) yields an independent measurement, a temperature of 20.8 MK for the flare kernel. Assuming that the ejecta temperature is in the range 1.4–20.8 MK, lower and upper limits can be found for the emission measure, and in particular, the density.

To determine the total thermal energy and the energy of bulk motion of the ejecta, we need to know the number of particles in the ejecta. This was done for the third TRACE image in Figure 12 in the following manner. The shape of the ejecta seen here can be approximated by a truncated cone. The volume of material in a pixel is $V$ = (pixel area) $\times$ $l$, where $l$ is the smallest dimension of the ejecta feature being examined. Combined with Equation (4), and knowing that pixel's emission measure (cf., Equation (6)), a density can be calculated. The average densities $n_{e,\text{ejecta}}$ derived in this manner were $5 \times 10^9$ cm$^{-3}$ for $T = 1.4$ MK and $10^{11}$ cm$^{-3}$ for $T = \;\sim 10$ MK (the temperature where the filter response was lowest). These densities can be compared with those indicated by the observed decimetric type III bursts in the range $1.2 \times 10^9 < n_e < 1.3 \times 10^{10}$ cm$^{-3}$ (assuming emission at the second harmonic).

The volume of all the ejected material was estimated to be $V_{\text{ejecta}} = (1.1 \pm 0.2) \times 10^{27}$ cm$^3$. Hence, $(n_{e,\text{ejecta}} V_{\text{ejecta}})$ yields $4.5 \times 10^{36}$ to $1.3 \times 10^{38}$ electrons. In the flaring kernel, TRACE finds a density of $n_{e,\text{kernel}} \approx 3.5\ (\pm 0.1) \times 10^{11}$ cm$^{-3}$, for temperatures in the range 15–20 MK.

As a self-consistency check, the emission measure of the brightest EUV region outside the flare kernel (the X-point in Figure 13, fourth image in Figure 12) was determined (assuming an upper limit temperature of 20.8 MK). The result of $1.8 \times 10^{47}$ cm$^{-3}$ is below the $EM_{\text{kernel}} = 2.2 \times 10^{48}$ cm$^{-3}$ determined from the RHESSI spectrum for the flare kernel. As RHESSI's dynamic range is currently about 10, this means that this region would have been indeed invisible to imaging, even if it were as hot as the flare kernel.

The bulk motion of the ejecta during the impulsive phase of the flare was determined by TRACE difference images (Figure 12, two and three), and was found to be $v = 290 \pm 70$ km s$^{-1}$. It is used to compute the energy of motion of the ejecta, $E_{\text{mot}} \approx 0.5 m_p (n_{e,\text{ejecta}} V_{\text{ejecta}}) v^2$, where $m_p$ is the proton mass. This yields a result between $2.0 \times 10^{27}$ erg and $1.3 \times 10^{29}$ erg.

*Thermal energy of flaring kernel* $E_{th,\text{kernel}}$: using the RHESSI-derived values from Section 5 and Equation (3), a value of $8.7 \times 10^{28}$ erg is derived. The possible range of values is (6.0–12) $\times\ 10^{28}$ erg.

*Thermal energy of ejected plasma* $E_{\text{th,ejecta}}$: we again use $3 k_B T n_e V$, where $T$ is assumed to be between 1.4 and 20.8 MK, and $n_e$ and $V$ now relate to the ejecta. This yields: $E_{\text{th,ejecta}} = 2.6 \times 10^{27}$ to $1.1 \times 10^{30}$ erg.

*Radiated energy from the flaring kernel* $E_{\text{rad,kernel}}$: assuming a temperature of 19.7 MK, using $EM_{\text{kernel}} = 2.2 \times 10^{48}$ cm$^{-3}$, and integrated between 10:26:05 to 10:27:23 UT, an amount of $3.5(\pm 1.7) \times 10^{27}$ erg has been radiated as the plasma cooled down.

TABLE II

Energy budget of the flare on 20 February 2002 during the impulsive phase (from about 10:26:10 to 10:27:10 UT).

| Type | Best estimate | Range |
|---|---|---|
| $E_{\mathrm{kin,beam}}$ | $2.6 \times 10^{30}$ erg | $(1.8$ to $3.4) \times 10^{30}$ erg |
| $E_{th,\mathrm{kernel}}$ | $8.7 \times 10^{28}$ erg | $(6.0$ to $12) \times 10^{28}$ erg |
| $E_{th,\mathrm{ejecta}}$ | $\sim 10^{30}$ erg | $2.6 \times 10^{26}$ to $1.1 \times 10^{30}$ erg |
| $E_{\mathrm{mot,ejecta}}$ | $\sim 10^{29}$ erg | $2 \times 10^{27}$ to $1.3 \times 10^{29}$ erg |
| $E_{\mathrm{rad,kernel}}$ | $3.5 \times 10^{27}$ erg | $\pm 1.7 \times 10^{27}$ erg |
| $E_{\mathrm{HXR}}$ | $3.2 \times 10^{23}$ erg | $\pm 1.1 \times 10^{23}$ erg |

*Total radiated HXR from precipitating electrons* $E_{\mathrm{HXR}}$: integrating the power-law in Section 5 between 10:26:05 to 10:27:23 UT yields a total of $3.2\ (\pm 1.1) \times 10^{23}$ erg.

## 8. Conclusions

The thermal energy content of the hot flare plasma (flare kernel) is considerably less than the energy in the non-thermal electron beam (Table II). This is consistent with the standard flare scenario where the energy is first released into non-thermal particles and then converted into thermal energy. As some of the target may not be heated to high enough temperatures to radiate X-rays, the energy input by the particle beam can exceed the output visible in soft X-rays. The ratio of beam energy over thermal energy in the kernel is $\sim 30$, much larger than 1, and consistent with results from de Jager *et al.* (1989), taking into account the fact that they only considered electrons above 25 keV. A major uncertainty of the energy budget is the source volume, from which several parameters are derived, such as density, mass and energy. The low-energy cut-off of the non-thermal electron spectrum is the major inaccuracy of the beam energy.

The size of the source was stable in energy and time. The flare kernel contained initially both the thermal and non-thermal electrons. Later, after the HXR emission above 25 keV ended, the thermal source drifted slowly ($\sim 30$ km s$^{-1}$) outwards.

If interpreted by reconnection at point A (Figure 13), the conclusion is that the geometry of energy release and partition was asymmetrical. In the downward jet, not observed by TRACE, the energy was largely transferred to accelerate electrons. The proposed scenario also suggests that the accelerated electrons mostly moved downward from the reconnection site or were accelerated only in the downward reconnection jet. If the observed ejecta is interpreted as the other reconnection jet, this upward jet involved less energy, which showed up mostly as heat. However, the energy estimate of the latter is less accurate.

RHESSI imaging with improved dynamic range may be able to search for the thin-target emission of energetic electrons in the ejecta. Nevertheless, the absence of appreciable decimetric radio emission corroborates the conclusion that acceleration took place mostly below point A and in the downward direction.

The now available high-resolution RHESSI and TRACE observations allow a more quantitative investigation of flare energies. The study of a compact flare yields a detailed scenario (that may not apply to all flares). Based on the above interpretations, we conclude that energy partition is not symmetric about the X-point of reconnection. Most of the initial energy first appears as energetic electrons in the lower, stationary part, and less than half is manifest in thermal energy and even less in bulk motion of the upper part.

More flares need to be analyzed to study the influence of the magnetic field geometry and density on energy partition.

## Acknowledgements

We thank the RHESSI software team for continuous encouragements and support, in particular Gordon Hurford, André Csillaghy, Jim McTiernan, and Richard Schwartz for related help or explanations. Special thanks to Brian Dennis and Ed Schmahl for comments on this paper. The RHESSI work at ETH Zürich is supported by the Swiss National Science Foundation (grant no. 20-67995.02) and ETH (grant TH-W1/99-2).

## References

Brown, J. C.: 1971, *Solar Phys.* **18**, 489.

Brown, J. C.: 1974, in G. Newkirk, Jr. (ed.), *Coronal Disturbances*, *IAU Symp.* **57**, 395.

de Jager, C., Bruner, M. E., Crannell, C. J., Dennis, B. R., Lemen, J. R. and Martin, S. F.: 1989, in M. R. Kundu, B. Woodgate, and E. Schmahl (eds.), *Energetic Phenomena on the Sun*, Kluwer Academic Publishers, Dordrecht, Holland, p. 396.

Handy, B. *et al.*: 1999, *Solar Phys.* **187**, 229.

Hudson, H. S. *et al.*: 1978, *Solar Phys.* **60**, 137.

Hurford, G. J. *et al.*: 2002, *Solar Phys.*, this volume.

Innes, D. E., Inhester, B., Axford, W. I., and Wilhelm, K.: 1997, *Nature* **386**, 811.

Kontar, E. J., Brown, J. C., and McArthur, G. K.: 2002, *Solar Phys.*, this volume.

Lin, R. P. and Schwartz, R. A.: 1987, *Astrophys. J.* **312**, 462.

Lin, R. P. *et al.*: 2002, *Solar Phys.*, this volume.

Messmer, P., Benz, A. O., and Monstein, C.: 1999, *Solar Phys.* **187**, 335.

Miller, J. A. *et al.*: 1997, *J. Geophys. Res.* **102**, 14631.

Neupert, W. M.: 1968, *Astrophys. J.* **153**, L59.

Pohjolainen, S. *et al.*: 2001, *Astrophys. J.* **556**, 421.

Priest, E. and Forbes, T.: 2000, *Magnetic Reconnection: MHD Theory and Applications*, Cambridge University Press, New York.

Ramaty, R., Mandzhavidze, N., Kozlovsky, B., and Murphy, R. J.: 1995, *Astrophys. J.* **455**, L193.

Schmahl, E. J. and Hurford, G. J.: 2002, *Solar Phys.*, this volume.

Schwartz, R. A. *et al.*: 2002, *Solar Phys.*, this volume.

Shibata, K. *et al.*: 1994, *Astrophys. J.* **431**, L51.

Smith, D. M. *et al.*: 2002, *Solar Phys.*, this volume.

Strong, K. T., Benz, A. O., Dennis, B. R., Leibacher, J. W., Mewe, R., Poland, A. I., Schrijver, J., Simnett, G., Smith, J. B., and Sylwester, J.: 1984, *Solar Phys.* **91**, 325.

Zhang, J., Wang, J., and Liu, Y.: 2000, *Astron. Astrophys.* **361**, 759.

# SPECTRAL AND SPATIAL VARIATIONS OF FLARE HARD X-RAY FOOTPOINTS

L. FLETCHER[1] and H. S. HUDSON[2]

[1] *Department of Physics and Astronomy, University of Glasgow, Glasgow G12 8QQ, U.K.*
*(e-mail: lyndsay@astro.gla.ac.uk)*

[2] *Space Sciences Laboratory, University of California, Berkeley CA 94720-7450, U.S.A.*
*(e-mail: hhudson@ssl.berkeley.edu)*

(Received 6 August 2002; accepted 31 August 2002)

**Abstract.** In a sample of strong RHESSI M-class flares we have made a study of the relationship between the 'hardness' of the HXR spectrum and the intensity in the 30–50 keV energy range. In all events we find clear evidence for a 'soft–hard–soft' pattern of correlation between hardness and flux, on time scales as short as 10 s. We investigate whether or not this pattern is intrinsic to the acceleration mechanism. The RHESSI images in this energy range are dominated by footpoint brightenings, and we have searched for a correlation between footpoint separation velocity and spectral hardness, to be compared qualitatively with theoretical flare models. We find quite systematic footpoint motions, and also note that episodes in which footpoint separation varies rapidly often correspond with episodes of significant change in the flare spectral index, though not as the simplest flare models would predict. We report also on one of our events, on 14 March 2002, which exhibits highly sheared HXR footpoint ribbons extending over a scale of 100 arc sec. For this flare we find a correlation between footpoint motion and hard X-ray flux.

## 1. Introduction

The hard X-ray (HXR) emission of a solar flare results mainly from the bremsstrahlung of mildly relativistic electrons accelerated in its impulsive phase. This represents the interval of most intense energy release in a flare, often associated with the time of rapid acceleration of an associated coronal mass ejection and launching of a coronal blast wave. In general, accelerated particles carry a large fraction of the magnetic energy released during the impulsive phase, which itself may dominate the total energy of the flare. However, the mechanism and site of impulsive-phase electron acceleration remain unknown. Because the flare energy resides in the coronal magnetic field prior to its release, the site of particle acceleration is presumably also to be found there. However, the bulk of the hard X-radiation (which marks the location of the collisional energy losses of the energetic electrons) normally occurs in the chromospheric footpoints of the coronal loops. These footpoint regions map magnetically to the coronal regions in which the stored magnetic energy becomes converted into kinetic energy of the accelerated particles.

*Solar Physics* **210:** 307–321, 2002.

CD ROM

With the launching of the Ramaty High-Energy Solar Spectroscopic Imager mission (RHESSI) we have new capability for a systematic exploration, both spectroscopic and imaging, of the flare hard X-ray emissions (Lin *et al.*, 2002). Past observations, almost entirely made with scintillation-counter spectrometers, have typically had energy resolution no better than a few keV in the 10–100 keV energy range; RHESSI uses Ge detectors with resolution on the order of 1 keV and an energy range extending down to a few keV (Lin *et al.*, 2002). The RHESSI imaging throughout this spectral range also provides better sensitivity, dynamic range, and angular resolution than that of the *Yohkoh* Hard X-ray Telescope (HXT; see Kosugi *et al.*, 1991) which has provided the bulk of our current knowledge base in this area.

In this paper we make a first exploration, using RHESSI's new capabilities, of spatial and spectral correlations in the impulsive-phase hard X-ray emission. We choose a sample of strong GOES events and examine the variation of the spectral hardness with (a) HXR intensity and (b) rate of HXR footpoint separation. Past observations have established a 'soft–hard–soft' spectral behaviour throughout flare bursts (Parks and Winckler, 1969; see also Benz, 1977, and the example in Figure 6 of Dennis, 1985). As the burst intensity increases, the photon spectrum hardens, and then softens as the intensity decreases again. This suggests a direct correlation of the flux of accelerated electrons and the 'hardness' of their spectral distribution. These parameters may also be correlated with the rate of separation of flare HXR footpoints, as explained below.

In the reconnection interpretation, the power available for the flare depends on the rate of coronal reconnection. Any reconnection geometry involves the formation of at least one current sheet located at a magnetic separator. Magnetic flux from one magnetic domain reconnects into another domain through this current sheet. The potential drop, $V$, along a reconnecting current sheet is related to the rate at which magnetic flux reconnects through that sheet:

$$V = \frac{\partial A}{\partial t} = \int E_c \, \mathrm{d}l = \frac{\partial}{\partial t}\left(\int B_\perp \, \mathrm{d}a\right)$$

(e.g., Forbes and Lin, 2000; Qiu *et al.*, 2002). Here $E_c$ is the reconnecting electric field in the corona, $l$ is distance measured along the current sheet, $A$ is the magnetic flux, $B_\perp$ is the normal component of the magnetic field and d$a$ is the element of area 'swept out' by the locus of chromospheric points mapping to the reconnecting coronal field. In the simplest symmetric and translationally invariant flare arcade, in which $B_\perp$ remains constant during the flare, the reconnection rate, and the reconnection electric field strength are reflected in the speed $\dot{x}$ at which the locus of footpoints moves. By writing $\mathrm{d}a = L\,\mathrm{d}x$ where $L$ is the current-sheet length, it is found that the electric field strength $E_\circ$ along the X-line above the arcade is given by (e.g., Priest and Forbes, 2002)

$$E_0 = B_\perp(x)\,\dot{x},$$

where $x$ denotes the location of the footpoint.

The events which we examine are clearly not two-dimensional, however the product $B_{\perp}(x) \cdot \dot{x}$ should generally give a measure of the coronal reconnection rate, and hence the power available to accelerate particles. In the simplest interpretation therefore we expect that the peak fluxes, and the highest energy photons, will occur during times of maximum footpoint separation speeds. Clearly there are many model dependences. The flux depends also on the number of particles available to be accelerated which need not depend on the acceleration rate. The geometry of the current sheet, or current sheet turbulence, or the generation of streaming instabilities, could add extra (possibly stochastic) electric field components which act collectively on the distribution of particles to be accelerated.

Recent work from several groups aims at relating the properties of the accelerated particles with footpoint motion or flare ribbon development. Sakao, Kosugi, and Masuda (1998) using *Yohkoh* HXT present the footpoint separation in a single event, finding an overall increase in separation over 30 s of the flare impulsive phase, with superposed small increases and decreases. This event showed no clear time correlation between the burst intensity and the separation rate. However, over a sample of 14 flares it was found that those exhibiting positive separation speeds tended to show upward-breaking HXR spectra, rather than the downward breaks of those events exhibiting zero or negative separation speeds. Qiu *et al.* (2002) look in detail at the motion of the brightest pixels of H$\alpha$ flare footpoints (marking the sites of chromospheric excitation by electron beams) in a flare observed by Big Bear Solar Observatory and *Yohkoh* HXT. Two almost stationary kernels and two H$\alpha$ flare 'ribbons' were present in the event, however these were found to remain more or less fixed in position. Systematic kernel motions during the peak of the flare were *along* the flare ribbons. This is inconsistent with the 'spreading ribbons' pattern expected from an evolving 2-dimensional flare arcade. The coronal electric field calculated from these motions appeared to correlate with HXR flux at the time of the major HXR peak, but the correlation was poor at other times in the flare. This, the authors suggested, may mean (a) that the reconnection geometry is generally not describable as a quasi-2-dimensional structure, (b) that the early phase of the flare does not involve systematic motion, or (c) that particle acceleration is only directly determined by the coronal electric field during a short part of the flare.

Measurements of the reconnecting flux, or electric field during large two-ribbon events have been attempted by Poletto and Kopp (1986), Fletcher and Hudson (2001), Saba, Gaeng, and Tarbell (2001) and Qiu *et al.* (2002). Field strengths range from a few to the order of 100 V cm$^{-1}$, with these being generally highest at the impulsive peak.

One further relevant piece of work is the study of HXR ribbons using *Yohkoh* HXT during the peak of the 14 July 2000 flare (Masuda, Kosugi, and Hudson, 2001). In this rare observation with HXT, a number of discrete HXR sources were observed to brighten at locations along the flare ribbons. By assuming that at any instant in time the brightest kernels were linked in that they mapped the ends of

TABLE I

Event list (times in UT).

| Date | Start | Peak | End | Class | Region |
|---|---|---|---|---|---|
| 14 March 2002 | 01:38 | 01:50 | 02:02 | M5.7 | 9866 |
| 17 March 2002 | 19:24 | 19:31 | 19:34 | M4.0 | 9871 |
| 10 April 2002 | 12:23 | 12:31 | 12:40 | M8.2 | 9893 |
| 17 July 2002 | 06:58 | 07:13 | 07:19 | M8.5 | 0030 |

just-reconnected field, the authors suggested that HXR kernel motions were consistent with the reconnection of progressively less and less sheared field, and that the HXR spectrum resulting from the earlier reconnection (more sheared) is harder than that occurring later on.

In Section 2 we will describe selection of data used in our study. Section 3 presents the time variation of hard X-ray spectra throughout the events studied, and Section 4 the imaging of footpoints in the flare impulsive phase. We focus on more detail on one of our events in Section 5 and discussion and conclusions are presented in Sections 6 and 7.

## 2. Data Selection

We selected a number of RHESSI flares, initially on the basis of their GOES classification. We wanted to make images at a time resolution on the order of the spin period of the spacecraft ($\sim$4 s), in the 30–50 keV spectral range (which should be unambiguously non-thermal emission). Typically, a good back-projected image will require $\geq$2000 counts totaled over all grids being used, corresponding to on the order of 500 counts s$^{-1}$. Flares of GOES class M4 and above tend to have a count rate in the 30–50 keV range sufficient to make 20+ images over the duration of the impulsive phase, from which we can hope to see some systematic motion. Further, we required that the reconstructions must have two footpoints visible throughout the peak – this is not always the case. Sometimes one footpoint disappears into the level of the noise. Or if events are too near the limb, projection effects can make it difficult or impossible to separate footpoints. A further selection based on preliminary reconstructions, and flare positions, results in a short list of events studied, given in Table I.

## 3. Hard X-Ray Spectra

At the time of writing the RHESSI data have only preliminary calibrations, even though the energy resolution is essentially at the design level in eight of the nine Ge detectors, and all are aligned precisely in gain. We have therefore restricted our analysis of the spectra to broad energy bands and to relative variations within each flare, omitting detector G2 (Lin *et al.*, 2002). We do not attempt full imaging spectroscopy, but instead work with these broad-band counting rates independently for imaging and spectroscopy. We accumulate counts in two-second bins, which approximately cancels out any modulations resulting from the spacecraft spin, which is held to the range 14–15 rpm. Residual modulations result in a roughly random fluctuation excess over the photon-counting noise statistics.

Previous observations with Ge resolution have shown that the hard X-ray spectrum does not follow an exact power-law distribution, but may show a rather sharp spectral break between two power laws (Lin and Schwartz, 1987). Spectra obtained at lower energy resolution (Elcan, 1978) suggest that this property may be fairly general, and that the break energy commonly occurs at about 50 keV. In addition to this spectral feature, we know that the soft X-ray spectrum closely matches that expected from a Maxwellian particle distribution, i.e., an exponential bremsstrahlung continuum in the hard X-ray range. The effective temperature of this distribution increases with the photon energy observed, suggesting a multi-thermal plasma (Lin and Schwartz, 1987). The detailed properties of the hard X-ray spectra remain to be delineated by RHESSI data, but the existing data often fit an essentially six-parameter spectral distribution in the 10–100 keV range: temperature and emission measure for a hot or 'super-hot' thermal plasma, with temperature typically not exceeding about 40 MK (3.4 keV); plus the parameters of the broken power law at higher energies.

The first question regards the reality of the soft–hard–soft behavior as seen with RHESSI's greater spectral purity. We define three broad energy bands, with edges at 20, 30, 50, and 70 keV, and plot time profiles of flux (30–50 keV) and the two hardness ratios in Figure 1. Broad energy bands are necessary for good signal-to-noise ratio, and we use the two bands to determine if the same pattern persists both above and below the typical break energy of the double power law. For all four events studied, the soft–hard–soft pattern, and the correlation between flux and spectral hardness is clearly visible in both ratios over bursts down to time scales of $\sim 10$ s.

With broad-band hard X-ray observations there is the clear possibility that the soft–hard–soft pattern might be confused in any two-band comparison, if either band contained appreciable thermal counts. Although the thermal component is expected to be negligible above 20 keV, we test for its presence using flux-flux correlation plots. At low energies (10–20 keV) a flux-flux correlation plot tends to show loops resulting from the slower growth of thermal emission (Figure 2). At higher energies, such as the 30–50 keV band used in Figure 1, the correlation

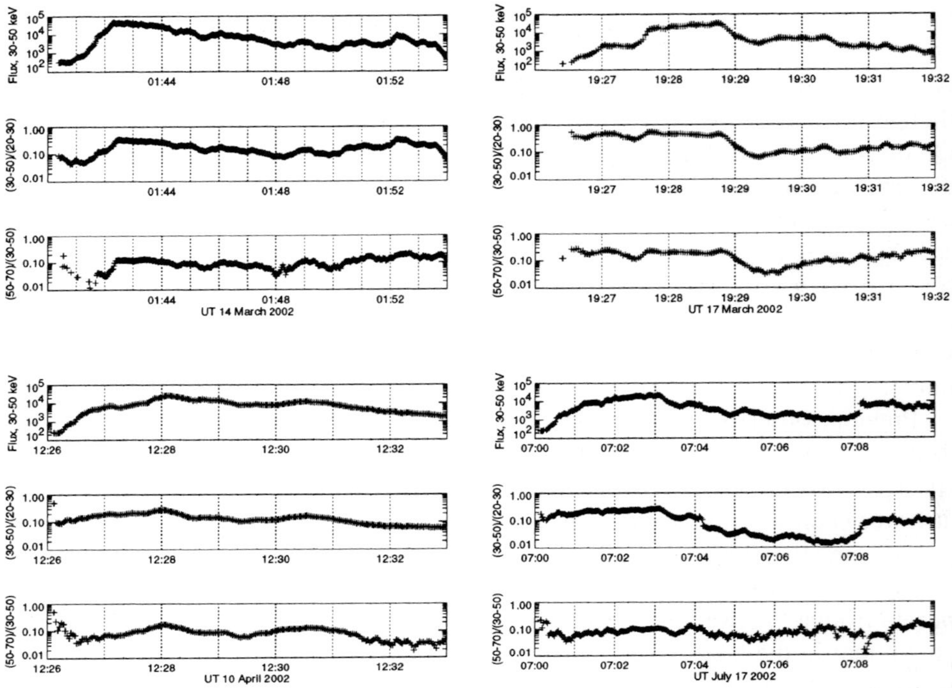


*Figure 1.* Broad-band light curves at 30–50 keV (*upper plots*) and two representative spectral ratios for the four events listed in Table I. The ratios are for adjacent broad bands with edges at 20, 30, 50, and 70 keV. Note the systematic tendency for these hardness ratios to correlate directly with the flux. *Dotted lines* show minutes.

tends to be direct. Because this also rules out appreciable energy-dependent trapping (e.g., Aschwanden *et al.*, 1999), we confirm that the soft–hard–soft pattern originates in the acceleration mechanism itself.

A further test (not shown) of the contribution of thermal sources in our broad RHESSI bands is made by taking advantage of the high spectral resolution of the RHESSI detectors, and correlating 1-keV energy bands against the broad bands defined for Figures 1 and 2. The correlations show that the gradual component is restricted mainly to the energy range below 20 keV, and that each of the correlation coefficients peaks close to the center of the relevant energy band. Thus we are confident that 'contamination' across the edges of the chosen bands is minimal.

## 4. Imaging of Impulsive Phase Footpoints

In each of the events, we have divided up the flare impulsive phase into a number of time bins, which are integer multiples of half of the spacecraft spin period. The criterion for selecting the bin size was simply to get sufficient counts in the bins at the beginning and end of the event; in practice this corresponds to a S/N

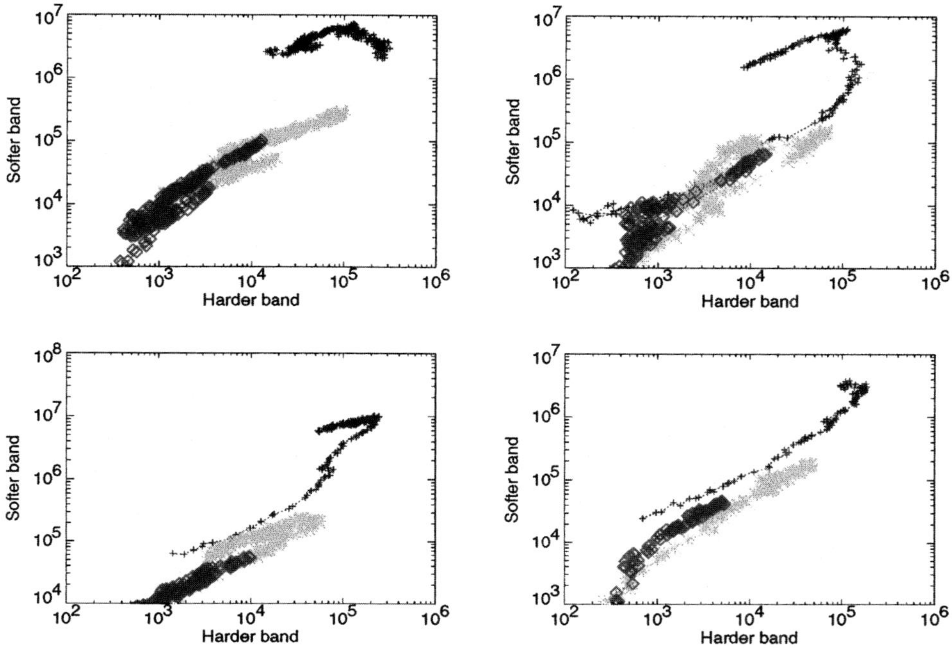


*Figure 2.* Pairwise correlation plots of broad-band flux ratios for the four flares studied, with the harder and softer bands set with boundaries at 10, 20, 30, 50, and 70 keV. The 'hysteresis' loops at the lowest energies (*pluses*) show the presence of an uncorrelated slowly-varying component, identified with the hot flare plasma or the 'super-hot' component reported by Lin and Schwartz (1987). At the highest energies (*diamonds*) the correlation is almost the same for burst rise and fall (see CD-ROM enclosure).

ratio between source (following CLEAN processing, see below) and background of S/N $\geq 3$. This means that in general there are unequal numbers of counts in each bin, and a variable S/N ratio on the reconstructed images. However, we felt that it was important that all components on the $(u, v)$-plane be equally sampled. We have used all grids down to and including grid 3, which has a FWHM of 6.8 arc sec. This provides adequate resolution for centroiding, and is reasonably quick to run for a large number of images. Following back-projection using the standard RHESSI software, the images were further processed using the CLEAN algorithm (e.g., Christiansen and Högbom, 1969), also implemented as part of the RHESSI software. The CLEAN technique assumes that the source distribution can be represented as a number of point sources, and can therefore be poor at imaging large diffuse sources. However, we have good reason to expect the 30–50 keV impulsive-phase chromospheric emission to be well-approximated by a set of point sources.

The footpoints were identified by summing the CLEANed images. Frequently, more than two footpoints are visible; in such cases it is extremely useful to also have TRACE imaging data or MDI magnetograms available to identify all footpoints occurring on the same side of the magnetic neutral line, and hence identify the most likely conjugate pairs. A selection box is drawn around all footpoints on one side of the magnetic neutral line, and the (intensity-weighted) centroid of this box located within each time frame. A continuous motion of a single bright source of emission would thus show up as a continuous line of centroid positions, whereas if sources were brightening and dimming in succession but not moving, a cluster of centroids would be found at each location. These are two extremes of what is generally a rather complex pattern of footpoint brightenings; in reality each flare requires rather detailed study to interpret what is going on. However, in this first paper we are searching only for gross patterns.

Figure 3 shows maps of footpoint centroid location for the flares listed in Table I. These generally reveal a fairly systematic variation in footpoint position as a function of time, often *along* ribbon-like patterns, though the scatter can be considerable. A measure of the rate at which the footpoints separate is given by plotting the distance between conjugate pairs of centroids as a function of time. Figure 4 (lower panels) shows the variation of footpoint separation with time, and the 30–50 keV flux over the same interval.

There are variations in the footpoint separations well above the level of the fluctuations on the centroided positions. Table II shows average separation speeds during the intervals of most rapid change. The fluxes and ratios at the times of these sudden motions are mostly increasing (i.e., the spectrum is hardening) regardless of whether there is a positive or negative footpoint velocity. In the cases of 14 March 2002 and 17 March 2002 a sudden motion matches well with the peak flux (maximum hardness). However, at this time resolution, in general there is no clear correlation between times of peak flux and high separation rates.

We have also looked at times of minimal change in footpoint separation in the four flares, as summarized in Table III. There is a tendency for a dip or decrease in the flux ratios during these times, however, the fluxes themselves may be high (spectra hard) or low (spectra soft).

## 5. The 14th March 2002 Event

In this section we focus on the event of 14 March 2002. If we plot, instead of the centroids of the CLEAN boxes, the locations of the CLEAN sources themselves (i.e., ignoring the contribution of the residuals) we find that the HXR kernels in the early part of the flare are spread out, suggesting systematic motions along a ribbon (Figure 5) that might have been underestimated in Figure 4. Centroiding on the data *including* the residuals, as we do in Section 4, tends to weight the source position towards the center of the centroiding box, so that variations (both systematic, and noise) in source position are generally reduced. The residuals contain information about the side-lobes which should not be discarded, which is why we have chosen in general to centroid on the CLEAN maps in Section 4. However we find, for

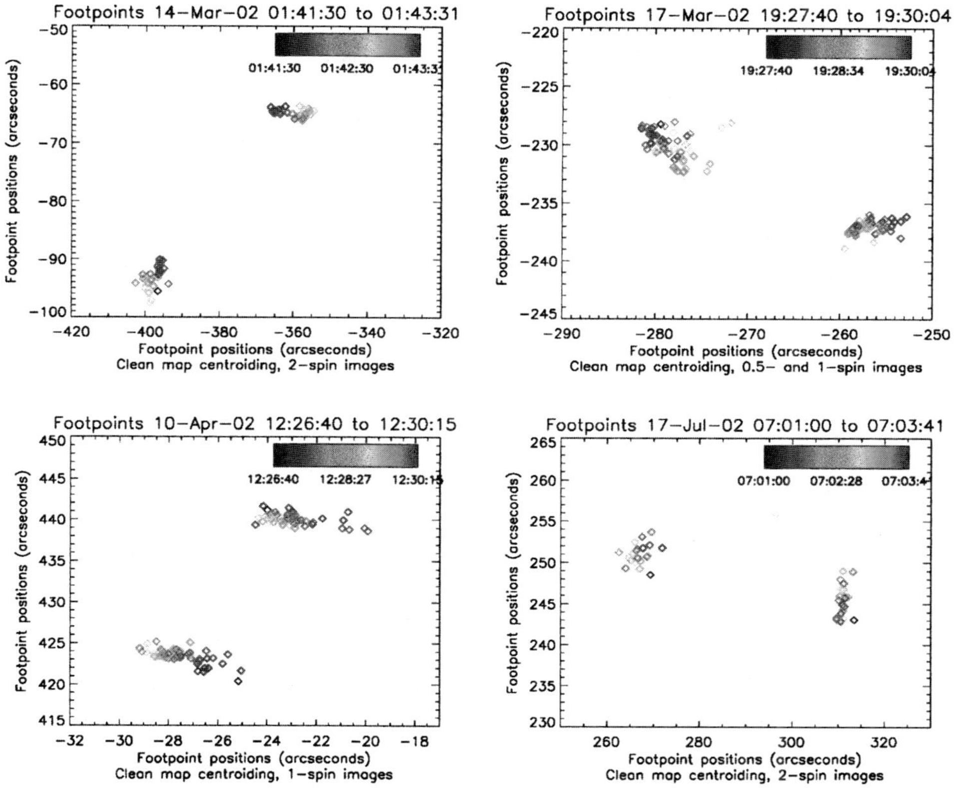


*Figure 3.* Footpoint locations derived from the CLEAN maps, color-coded by time in the original (see CD-ROM enclosure).

TABLE II

Mean speeds during periods of rapid variation in footpoint separation.

| Date | Time (UT) | Average separation rate (km s$^{-1}$) | Flux ratios/ fluxes |
|---|---|---|---|
| 14 March 2002 | 01:42:20–01:42:40 | $+390 \pm 30$ | increasing |
| | 01:43:15–01:43:30 | $-370 \pm 30$ | increasing |
| 17 March 2002 | 19:28:30–19:28:45 | $-480 \pm 80$ | increasing |
| | 19:28:45–19:29:00 | $+360 \pm 60$ | decreasing |
| | 19:29:00–19:29:15 | $-360 \pm 60$ | decreasing |
| 10 April 2002 | 12:27:00–12:27:30 | $-90 \pm 15$ | increasing |
| | 12:27:30–12:27:50 | $+95 \pm 15$ | increasing |
| 17 July 2002 | 07:01:20–07:02:30 | $+55 \pm 5$ | increasing |

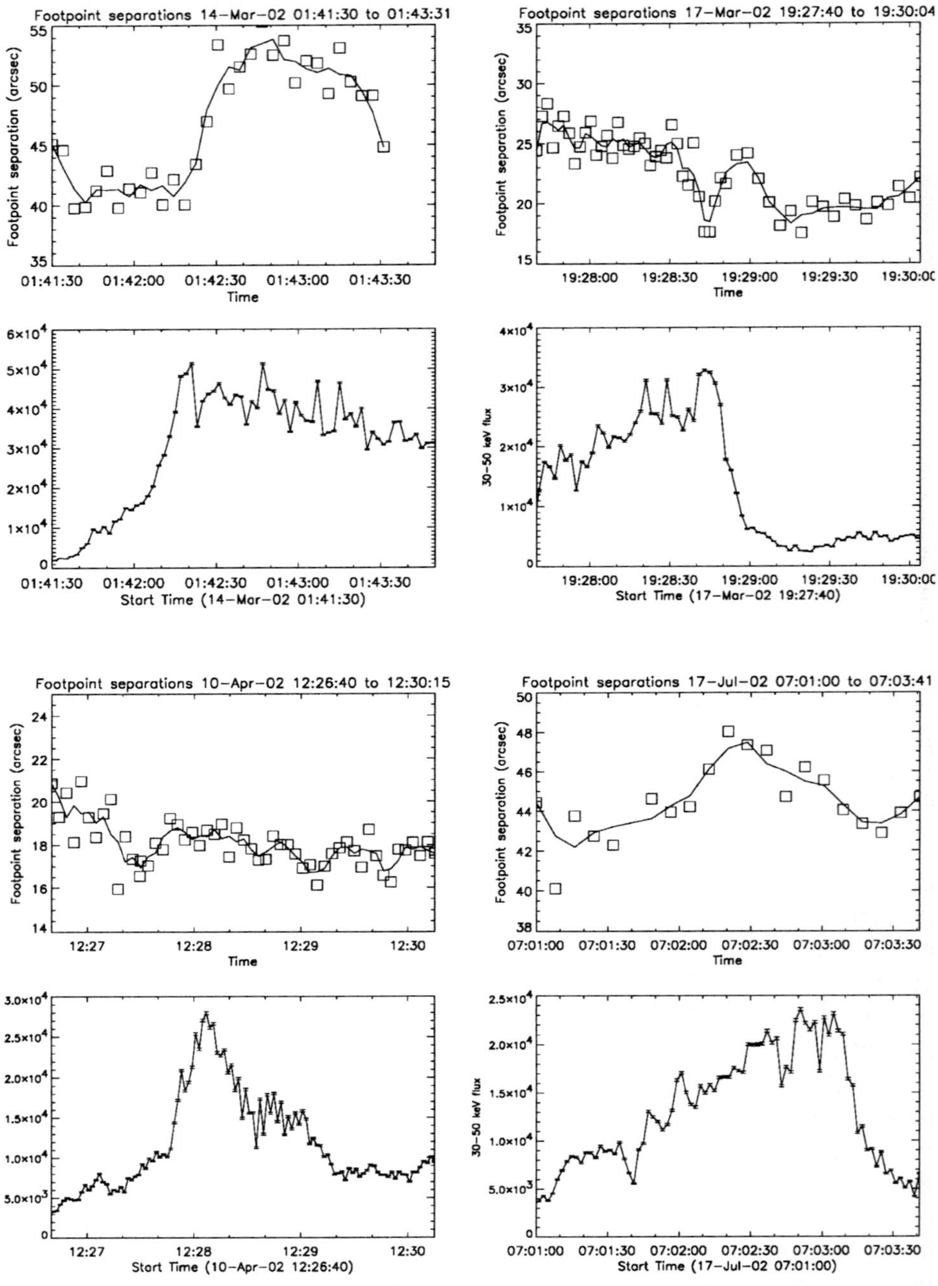


*Figure 4.* Apparent separation of footpoint sources, and the corresponding 30–50 keV flux. A 3-point running mean is drawn through the source positions. In the 14 March 2002 flare the separation varies smoothly, with an r.m.s. fluctuation around its smooth trajectory on the order of one arc sec. In the 17 March 2002 event the variation is not so systematic, but changes well above the level of the centroiding noise are clear. The 10 April 2002 event shows strong fluctuations in footpoint separation, and the 17 July 2002 shows a fairly systematic increase and decrease of separation throughout the impulsive phase. (All times in UT.)

TABLE III
Steady separation intervals.

| Date | Time (UT) | Flux ratios |
|---|---|---|
| 14 March 2002 | 01:41:30–01:42:20 | ratios dip |
| | 01:42:30–01:43:40 | decreasing slightly |
| 17 March 2002 | 19:27:50–19:28:30 | decreasing |
| | 19:29:20–19:30:00 | ratios dip |
| 10 April 2002 | 12:27:50–12:28:30 | peak then decrease |

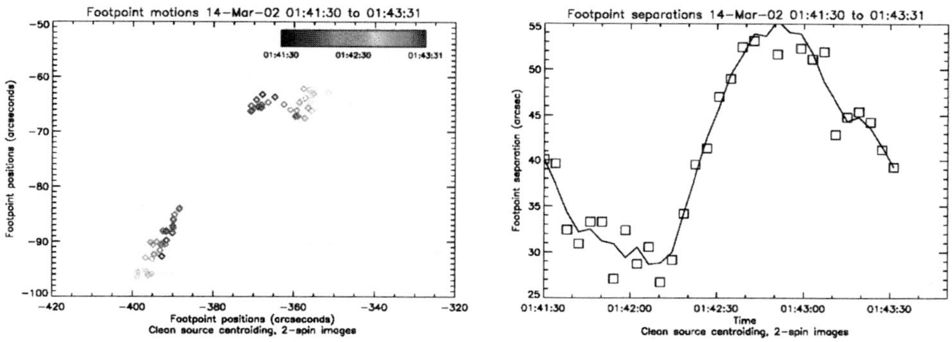


*Figure 5. Left:* CLEAN source centroids (as opposed to CLEAN map centroids) during the 14 March 2002 flare, showing systematic motion along a ribbon. *Right:* the footpoint separations derived from the CLEAN sources (see CD-ROM enclosure). (Times in UT.)

the evolution of the 14 March 2002 event, that the CLEAN sources alone contain valuable position information. This event was well-covered by Transition Region and Coronal Explorer (TRACE) observations in the 171 Å band (Handy *et al.*, 1999) primarily imaging plasma emitting in the Fe IX/X complex. Using these images allows us to form a better idea of the geometry of this event, and confirms the ribbon interpretation.

The TRACE images were de-spiked and corrected for the known offset between the pointing information (from the white-light channel) and the true pointing of the EUV channel. The cadence of the TRACE images is 12 s, and at this cadence the flare shows rapid evolution in the EUV, with a number of bright sources appearing to move along two ribbons, suggestive of initial rapid changes in the core of a highly sheared arcade. Post-flare images confirm bright, still sheared, loops joining these ribbons. A further correction, due to systematic TRACE pointing drifts as the telescope flexes thermally through its orbital motion, is needed to align the brightest TRACE and RHESSI sources; this is done by eye but the same offset of [−6 arc sec, 12 arc sec] is adequate to give good correspondence through all of the impulsive phase. Figure 6 shows the flare ribbons and footpoints through the

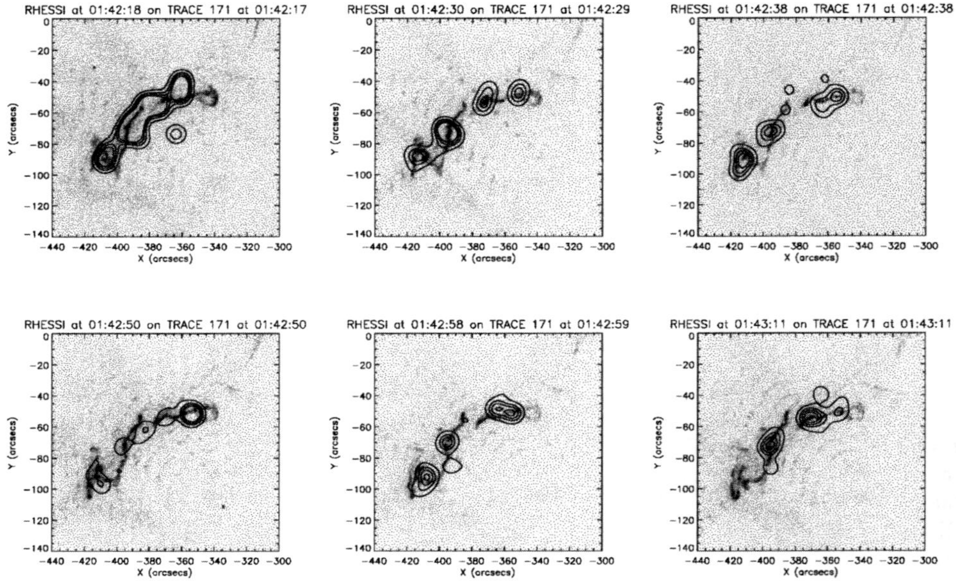

*Figure 6.* The impulsive peak of the flare of 14 March 2002. RHESSI contours, at 25%, 50%, 70%, and 90% of the maximum in each frame are overlaid on the nearest TRACE image. At the relatively low spatial resolution afforded by grid 3, there is a good correspondence between HXR sources and the brightest 171 Å flare kernels. The coordinates shown here are those given by the TRACE pointing, which, during this event, is offset from that of RHESSI by (+6 arc sec, −12 arc sec). (Times in UT.)

peak of the impulsive phase. The brightest HXR sources tend to align well with the brightest TRACE 171 Å sources. This has been noted previously by Fletcher and Hudson (2001).

Following these rapid changes at the beginning of the flare, the field settles down to a slower evolution, with HXRs emitted by two main sources (the two strongest RHESSI sources visible in the bottom right panel of Figure 6). We have made reconstructions in the 20–30 keV energy band throughout the first 12 min of this flare. There were inadequate counts in the 30–50 keV band to use this for the whole interval. Following our analysis of the contribution of thermal or gradual phase fluxes to the various energy bands, we can be certain that the majority of 20–30 keV emission is non-thermal and thus reflects footpoint locations. The footpoint separation rate as a function of time is shown in Figure 7. We see that following the main impulsive peak there is a drop in the separation, as the source at (−395 arc sec, −70 arc sec; see caption for Figure 6) becomes dominant, then a slow rise over the next five minutes or so, as these two strong footpoints separate. Fitting a smooth line between 01:45 and 01:50 UT, we find an approximately linear separation increase of about 8.9 km $s^{-1}$, suggesting a phase of slow reconnection. After an impulsive phase which was rather complicated, the gradual phase field evolution appears much simpler.

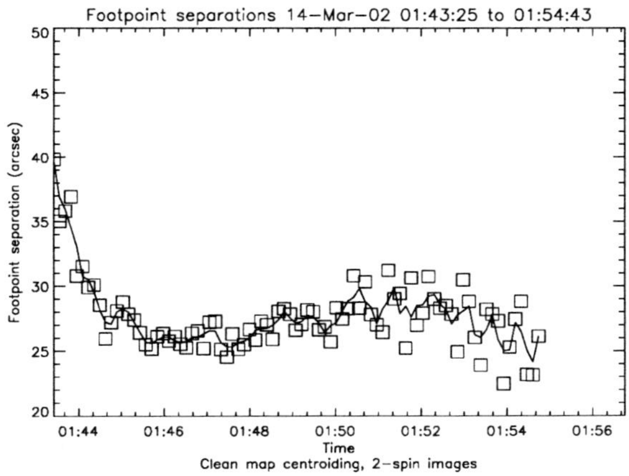


*Figure 7.* The footpoint separation in the 14 March 2002 event as measured in the 20–30 keV band, for the remainder of the full event. For the interval 01:45–01:50 UT the separation increases almost linearly at an apparent speed of 8.9 km $s^{-1}$ with a residual scatter of 0.7 arc sec, a measure of the RHESSI centroiding precision for 8-s integrations in a flare of this brightness.

## 6. Discussion

In the extraction of coronal magnetic energy, the magnetic mapping between the energy source and the footpoint brightenings should lead to correlations between the rate of dissipation of stored coronal energy and the footpoint motions. We have investigated one possible correlation: that between the spectral hardness (defined by the ratio of two broad energy bands) and the centroided footpoint motion.

Our small number of events makes it impossible to draw any general conclusions about the time behavior of flare footpoints, but we have noted the following patterns with some frequency: (1) systematic HXR footpoint motions along flare ribbons; (2) the occurrence of cases of approach of footpoints, and (3) episodes of rapid change followed by relative stability. The first of these is consistent with what has been seen previously by Qiu *et al.* (2002) in H$\alpha$, and the separation–time curves for these events do show a smooth evolution in time, suggestive of evolution through an arcade structure, albeit a highly sheared one. However, it is clear in some cases that the impulsive-phase evolution is not particularly smooth, and may result from the lighting-up of different flux elements in a patchy way as time proceeds, rather than a smooth evolution through a continuous field. We find an overall tendency for episodes of rapid footpoint motion to correspond to times when the flux ratio is increasing, whereas when the footpoint separation is not varying much the flux often decreases or goes through a dip. However, a pattern of high flux occurring with rapid footpoint motion is not readily visible.

The event of 14 March 2002 (and to a lesser extent 17 March 2002) is an example which suggests a link between flare energy production and apparent footpoint motions, as expected from the magnetic nature of a flare (Priest and Forbes,

2002). Figure 3 shows a sudden increase of footpoint separation in the time range 01:42:20–01:42:40 UT, just the time of maximum HXR luminosity (see Figure 1 and Table II). The rapid separation speed during this interval is approximately 390 km/s, but as can be seen in Figures 3 and 5 the motion may be a combination of a continuous motion of one footpoint, and a discrete sideways 'jump' in the location of the other. However, until more precise theoretical work can be done (see Section 1), and until we include the measured magnetic field strength in our calculations, we cannot be sure that the observed speed has even the right order of magnitude to explain the energy release in the context of the coronal magnetic field at the time of the flare. It is worth noting that a speed of this magnitude exceeds the Alfvén or sound speeds below the transition region, the assumed footpoint height, and would thus have to be the result of a coronal process.

## 7. Conclusions

RHESSI confirms, with high spectral and temporal resolution, the soft–hard–soft pattern in solar flares. This paper has focused on an effort to relate this spectral pattern to the flare geometry, and at the same time to explore the RHESSI capability for such studies. We find that the RHESSI data can be used to delineate footpoint motions in flares as small as a higher M class on time scales of a few seconds, and with a precision on the order of 1 arc sec (r.m.s.) even omitting data from the two finest collimators. The observations show systematic patterns of apparent footpoint motion, but these vary from flare to flare and do not resemble the simple increase of footpoint separation expected from 2D reconnection models and from H$\alpha$ ribbon behavior. The observations suggest a pattern of behavior in which the initial hardening of the spectrum is accompanied by rapid source movements or new sources appearing, whereas during episodes of negligible change in source position, the spectrum softens again. This is consistent with the spectral hardness being linked to the instantaneous conditions in the reconnection region. Further studies, particularly on larger X-class flares, combined with magnetic field reconstructions, may be able to open this link to study; for example, is there a direct correlation with global magnetic conditions in a reconnecting current sheet, as described for example by Litvinenko (1996)? Or a less direct correlation might suggest that the micro-physical conditions – turbulence level, temperature, etc. are important as would be the case for a stochastic accelerator. Benz (1977) has shown that the soft–hard–soft pattern is consistent with a stochastic acceleration process, and geometrical studies based on hard X-ray energy release may be able to locate the regions where this happens. Following the discussion in the Introduction, we would expect these regions to correspond to the sources of energy made available for particle acceleration by the coronal restructuring.

The soft–hard–soft spectral pattern dominates the bursty impulsive phase of a flare, but some events show a soft–hard–harder pattern (e.g., Frost and Dennis, 1971) on longer time scales; this may be correlated with geo-effectiveness (Kiplinger, 1995). We would therefore like to extend the present study, taking advantage of RHESSI's capability for imaging spectroscopy to relate spectral and spatial morphologies, to study this different kind of behavior as well.

## Acknowledgements

NASA supported this work via NAS 5-9833 at UC Berkeley (HSH). L.F. would like to acknowledge travel support from the Nuffield Foundation and from PPARC. We thank many RHESSI team members, especially S. Krucker on image technique, for help of various kinds.

## References

Aschwanden, M. J. Fletcher, L., Sakao, T., Kosugi, T., and Hudson, H.: 1999, *Astrophys. J.* **517**, 977.
Benz, A. O.: 1977, *Astrophys. J.* **211**, 270.
Christiansen, W. N. and Högbom, J. A.: 1969, *Radiotelescopes*, Cambridge, University Press, Cambridge.
Dennis, B. R.: 1985, *Solar Phys.* **100**, 465.
Elcan, M. J.: 1978, *Astrophys. J.* **226**, L99.
Fletcher, L. and Hudson, H.: 2001, *Solar Phys.* **204**, 69.
Handy, B. N. *et al.*: 1987, *Solar Phys.* **187**, 229.
Kiplinger, A. L.: 1995, *Astrophys. J.* **453**, 973.
Kosugi, T., Masuda, S., Makishima, K., Inda, M., Murakami, T., Dotani, T., Ogawara, Y., Sakao, T., Kai, K., and Nakajima, H.: 1991, *Solar Phys.* **136**, 17.
Lin, R. P. and Schwartz, R. A.: 1987, *Astrophys. J.* **312**, 462.
Lin, R. P. *et al.*: 2002, *Solar Phys.*, this issue.
Litvinenko, Y. E.: 1996, *Astrophys. J.* **462**, 997.
Masuda, S., Kosugi, T., and Hudson, H. S.: 2001, *Solar Phys.* **204**, 55.
Parks, G. K. and Winckler, J. R.: 1969, *Astrophys. J.* **155**, L117.
Poletto, G. and Kopp, R. A.: 1986, in D. F. Neidig (ed.), *The Lower Atmosphere of Solar Flares; Proceedings of the Solar Maximum Mission Symposium*, Sunspot, New Mexico, National Solar Observatory, p. 453.
Priest, E. R. and Forbes, T. G.: 2002, *Astron. Astrophys. Rev.* **10**, 313.
Qiu, J., Lee, J., Gary, D. E., and Wang, H.: 2002, *Astrophys. J.* **565**, 1335.
Saba, J. L. R., Gaeng, T., and Tarbell, T. D.: 2001, in P. C. H. Martens and D. Caufman (ed.), *Multi-Wavelength Observations of Coronal Structure and Dynamics – Yohkoh 10th Anniversary Meeting*, Kailua-Kona, Hawaii, Elsevier Science, p. 96S.
Sakao, T., Kosugi, T., and Masuda, S.: 1998, in T. Watanabe, T. Kosugi, and A. C. Sterling (ed.), *ASSL Vol. 229: Observational Plasma Astrophysics: Five Years of YOHKOH and Beyond*, Tokyo, p. 273.

# ENERGY DEPENDENCE OF ELECTRON TRAPPING IN A SOLAR FLARE

DAVID ALEXANDER and THOMAS R. METCALF
*Lockheed Martin Solar and Astrophysics Laboratory, 3251 Hanover Street, Palo Alto, CA 94304, U.S.A.*

(Received 8 August 2002; accepted 18 August 2002)

**Abstract.** Observations of an energy-dependent asymmetry in footpoint hard X-ray emission by RHESSI for the M4.0 solar flare of 17 March 2002 allows us to probe the dynamics of particle transport with energy and time. The presence of such an asymmetry is most readily explained by the effects of a converging magnetic field with different rates of convergence at the different footpoints, as would be expected from realistic surface field distributions. Such a geometry has been discussed in the context of a trap-plus-precipitation model where the transport of energetic particles in the flare is governed by the precipitation out of the coronal trap via collisions, wave-particle interactions or some other scattering process, into the high-density chromosphere. Comparison of RHESSI observations with a trap-plus-precipitation model allows us to use the energy dependence of the asymmetry and the observed ratio of footpoint to coronal emission at the different energies to assess the role of the trapping in the transport of energetic electrons and to probe the nature of the particle precipitation process inside the loss cone.

## 1. Introduction

Hard X-ray observations of solar flares provide important diagnostic information on the acceleration and subsequent transport of energetic particles in the solar corona. While it is now generally accepted that the bulk of the observed hard X-ray emission from flares is a result of the interaction of energetic electrons, accelerated in the corona, with the ambient medium via non-thermal electron bremsstrahlung, the details of the particle transport through the flaring solar atmosphere and the effects of the magnetic field geometry on this transport are still uncertain. Instrumentation on board a variety of spacecraft over the last two decades has generated a wealth of spectroscopic and imaging data which has significantly improved our understanding of the acceleration and transport processes in solar flares (see Aschwanden, 2002, for a recent and comprehensive review). The launch of RHESSI provides us with an unprecedented opportunity to explore the energy-dependence of the hard X-ray emission in space and time enabling new insight into the particle acceleration and transport processes.

In recent years much attention has been focused on the effects of coronal trapping, a natural consequence of the expected convergence of magnetic field between the corona, where the acceleration is presumed to occur, and the chromosphere where the bulk of the hard X-ray production is assumed to take place

*Solar Physics* **210:** 323–340, 2002.

(Brown, 1971). The idea of the magnetic trap in a solar context was originally proposed by Takakura and Kai (1966) and subsequently developed into the trap-plus-precipitation model by Melrose and Brown (1976) which was further refined by a series of authors (e.g., MacKinnon, 1988, 1991; Alexander, 1990; McClements, 1990a, b). One of the principal theoretical attractions of the trap model is that it circumvents the electrodynamic problems associated with the very large electron beam fluxes required to generate the observed blue shifts during the impulsive phases of solar flares (LaRosa and Emslie, 1988).

As observations have improved, considerable evidence in support of the trap-plus-precipitation scenario for energetic electrons in solar flares has accrued (Bai and Ramaty, 1979; Bai *et al.*, 1983; Bai and Dennis, 1985; Ramaty *et al.*, 1994; Alexander and Metcalf, 1997; Aschwanden *et al.*, 1999; Metcalf and Alexander, 1999). In particular, imaging observations at low energies ($< 40$ keV) have often shown hard X-ray emission located purely in the corona (Tsuneta *et al.*, 1984; Takakura *et al.*, 1987) with higher energy emissions concentrated at low altitudes (Kane, 1983; Aschwanden *et al.*, 1999). Less direct evidence for the need of a coronal trap lies in presence of a strong stationary component in soft X-ray emission lines (McClements and Alexander, 1989) contrary to what is expected from the thick-target electron beam interpretation of hard X-ray observations (e.g., Emslie and Alexander, 1987) which predicts large systematic blue shifts in these lines.

In this paper we are particularly interested in the exploration of the particle transport through the analysis of asymmetric trap geometries: a situation necessitated by the spatially resolved observations of asymmetric hard X-ray footpoints in a large number of solar flares. An asymmetric trap model was developed by Aschwanden *et al.* (1999) who used combined measurements from CGRO/BATSE (Fishman *et al.*, 1992) and the *Yohkoh*/HXT (Kosugi *et al.*, 1991) to quantify the relative fractions of the directly precipitating and trap-precipitating electrons at both magnetically conjugate footpoints. The presence of an asymmetry allows us to examine in detail the energy and time dependence of the particle transport.

When coupled to the increased sensitivity and improved energy resolution of RHESSI (Lin *et al.*, 2002), the characteristics of the observed asymmetry and its behavior with energy and time yield important clues to the physics governing the particle transport in solar flares. In this paper we address this issue via RHESSI observations of a solar flare which exhibited a well-defined and energy-dependent footpoint asymmetry and put these observations in context with the trap-plus-precipitation models. In Section 2 we present the observations and analysis while our results and interpretation are discussed in Section 3. Conclusions are subsequently presented in Section 4.

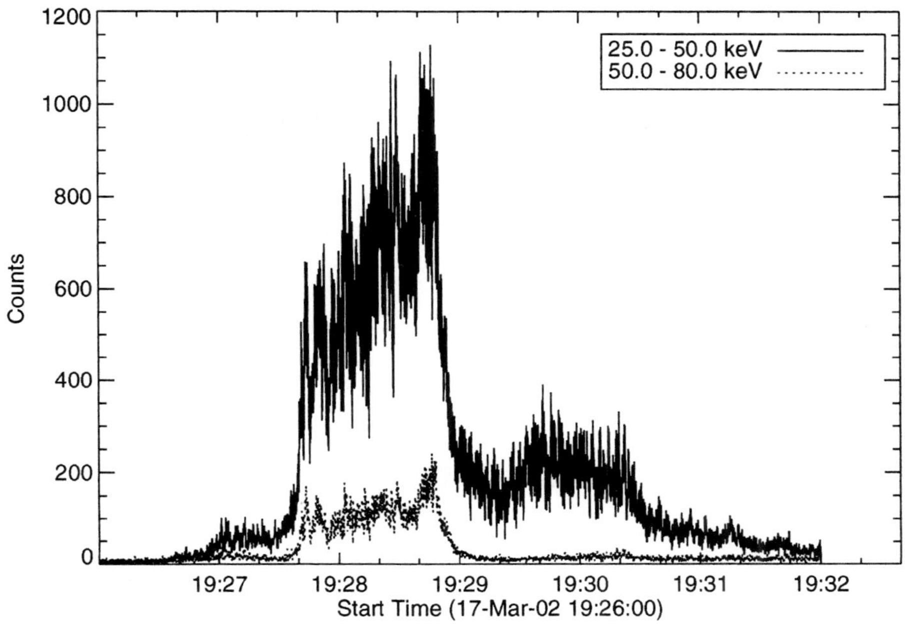


*Figure 1.* The light curve for the 17 March 2002 flare in two energy bands: 25–50 keV (*solid*) and 50–80 kev (*dotted*). The counts are summed into 0.1 s bins.

## 2. Observations and Analysis

RHESSI observed an M4.0 solar flare at 19:28 UT on 17 March 2002. The impulsive flare was in NOAA active region number 9871 at S21 E17. Figure 1 shows the hard X-ray light curve for this event. The analysis presented here is restricted to the flare impulsive phase from 19:27:30 through 19:29:00 UT. The inferred spectral index for this event is $\delta \approx 3.5$.

Figure 2 displays images in a subset of the energy bands considered. Images were reconstructed using the Pixon algorithm which gives the best suppression of spurious sources and hence the best photometry (Metcalf *et al.*, 1996). Energies above 30 keV are not shown since they are similar in appearance to the 28–30 keV image. The images shown in Figure 2 were reconstructed using the entire impulsive phase of the flare from 19:27:30 to 19:29:00 UT. To examine the time variability of the hard X-ray sources images were also reconstructed in time intervals from 19:27:30–19:28:00, 19:28:00–19:28:30, and 19:28:30–19:29:00 UT. All images were reconstructed using RHESSI detectors 3–9. The images show two distinct footpoint regions (east and west) and a coronal source (e.g., Masuda *et al.*, 1994) between the two footpoint regions and provide the basic data for the analysis described below.

For our investigation of the asymmetry in the hard X-ray emission from the two footpoint regions it is important to understand the character of the magnetic

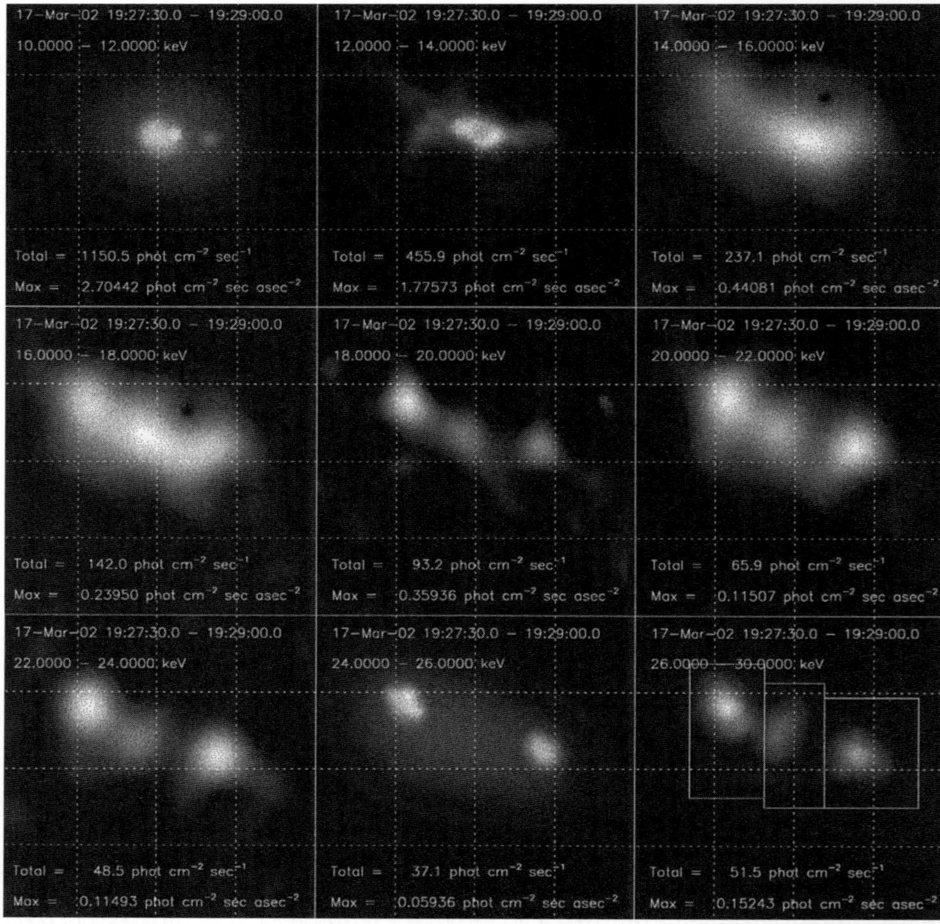


*Figure 2.* RHESSI images reconstructed using the Pixon algorithm in energy bands from 12 to 30 keV. At low energies the image is dominated by a coronal source. At high energies the images are dominated by two footpoint regions. In the *lower right* the boxes used to separate emission from the footpoints and the corona are shown. The images are each 64 arc sec $\times$ 64 arc sec with 1 arc sec pixels and are independently scaled. The total and maximum brightness for each image is given.

field at the footpoints. Figure 3 shows an overlay of the RHESSI hard X-rays on the SOHO/MDI (Scherrer *et al.*, 1995) photospheric vertical magnetic field. The vertical field was derived from the observed line-of-sight field using a simple potential field extrapolation. The MDI and RHESSI data were co-aligned using the known pointing of each instrument after correcting the MDI data for the parallax between the L1 view and the Earth view. The magnetic field in the vicinity of the west footpont is 650 G while the magnetic field in the vicinity of the east footpoint is quite variable since the footpoint sits at the edge of a strong spatial gradient in the magnetic field. The field in the vicinity of the east footpoint varies from $-450$ G

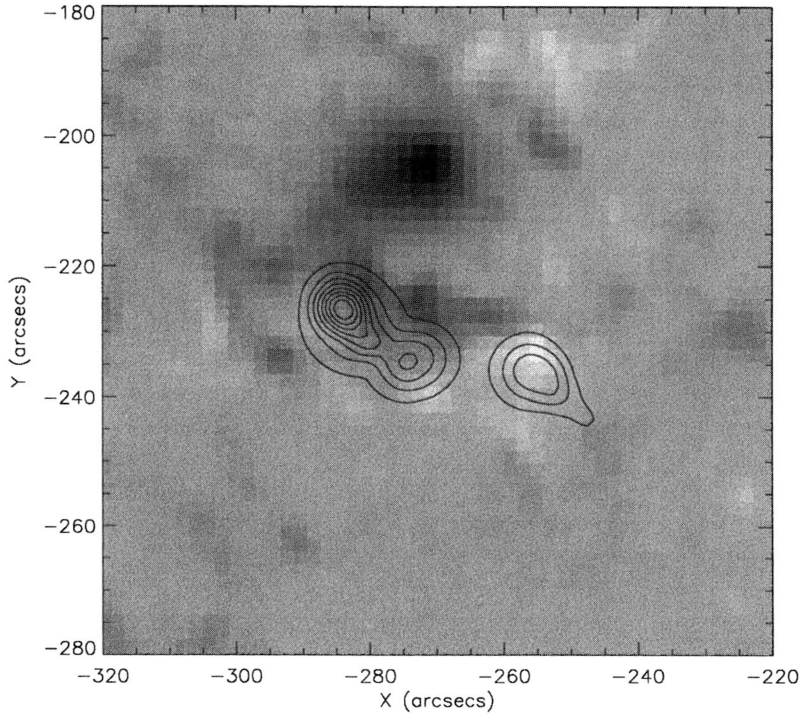


*Figure 3.* Contours of RHESSI hard X-ray emission (19:27:30–19:28:30 UT, 12–25 keV) overlayed on the MDI vertical magnetic field at 19:46:30 UT (*dark*: downward field, *white*: upward field).

to −1000 G depending on where the true flare footpoint is located. The spatial resolution achieved in the RHESSI images, and the uncertainty in the RHESSI pointing, make it difficult to determine the exact ratio of the magnetic field at the two footpoints. However, the range of possible values we derive is from 0.7 to 1.5 (west/east).

The asymmetry in the hard X-ray footpoint emission, $A$, is derived by summing the X-rays in boxes (Figure 2) surrounding the footpoints and computing the difference in brightness between the two footpoints divided by the sum of the brightness in the two footpoints, viz.,

$$A = \frac{I_2 - I_1}{I_2 + I_1}, \tag{1}$$

where $I_i$ denotes the count rate at footpoint $i$. For perfect symmetry, $A = 0.0$ and for perfect asymmetry, $A = \pm 1.0$. Figure 4 shows the footpoint asymmetry for the early impulsive phase of the flare, 19:27:30–19:28:00 UT. Although there is a strong degree of scatter, there is evidence of a reduction in the asymmetry with increasing photon energy above 20 keV. A fit to the data above 20 keV, and excluding the last point at 70–80 keV where the count rates are small, yields an

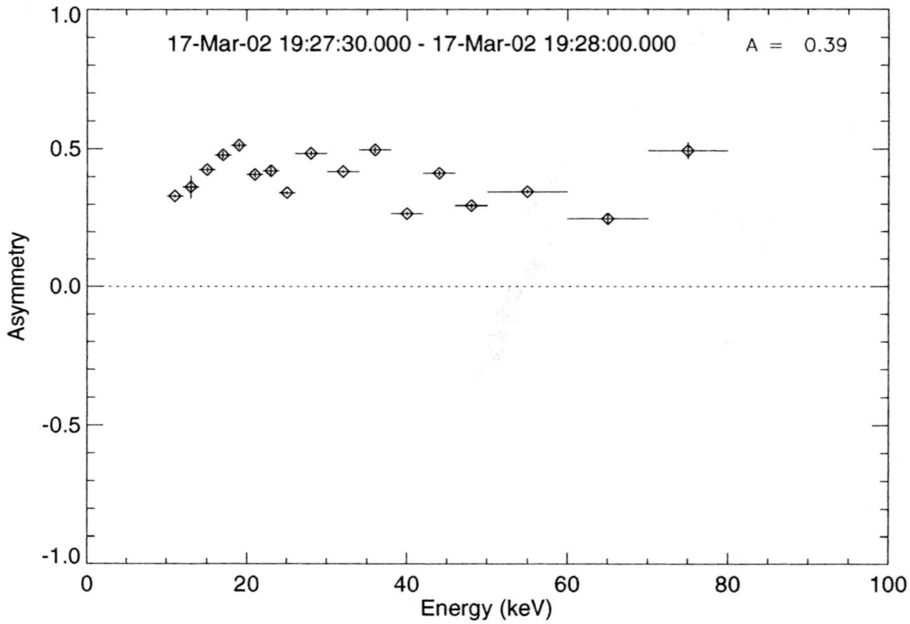


*Figure 4.* The asymmetry in the footpoint emission in the early impulsive phase from 19:27:30 to 19:28:00 UT, as a function of energy. The *horizontal bars* show the energy bands used and the *vertical bars* show the statistical errors. The energy integrated asymmetry is 0.39.

energy dependence of $4.1 \pm 1.4 \times 10^{-3}$ (keV)$^{-1}$, which is significant at the 3-sigma level. The quoted uncertainty was calculated from the scatter about the linear fit rather than from the statistical error bars which are considerably smaller than the scatter. The energy-integrated asymmetry is found to be $A_I = 0.39$. There is also some evidence for an increase in the asymmetry with energy at low photon energies ($\varepsilon < 20$ keV).

Late in the impulsive phase the character of the asymmetry clearly changes (Figure 5). From 19:28:00 to 19:28:30 UT, the asymmetry is reduced and from 19:28:30 to 19:29:00 UT the asymmetry has changed sign and shows a complex variation with energy. We will discuss this more fully in the following section.

Figure 6 shows the ratio, $R = I_{fp}/I_{\rm cor}$, of the footpoint emission to the emission from the corona, the latter being dominated by the Masuda-like source. Similar to the energy variation in the asymmetry, $R$ shows a general decay with energy in the early impulsive phase as would be expected from the early development of the particle transport when $R$ is dominated by directly precipitated electrons: higher energies stay in the trap longer. In the late impulsive phase, $R$ is progressively reduced with time at low energies as the corona becomes bright via heating and chromospheric evaporation and as the once-trapped particle population grows to dominate the directly-precipitated particles in the loss cone. At higher energies, $R$ is consistently of the order of 2.5, i.e., the combined hard X-ray emission from the

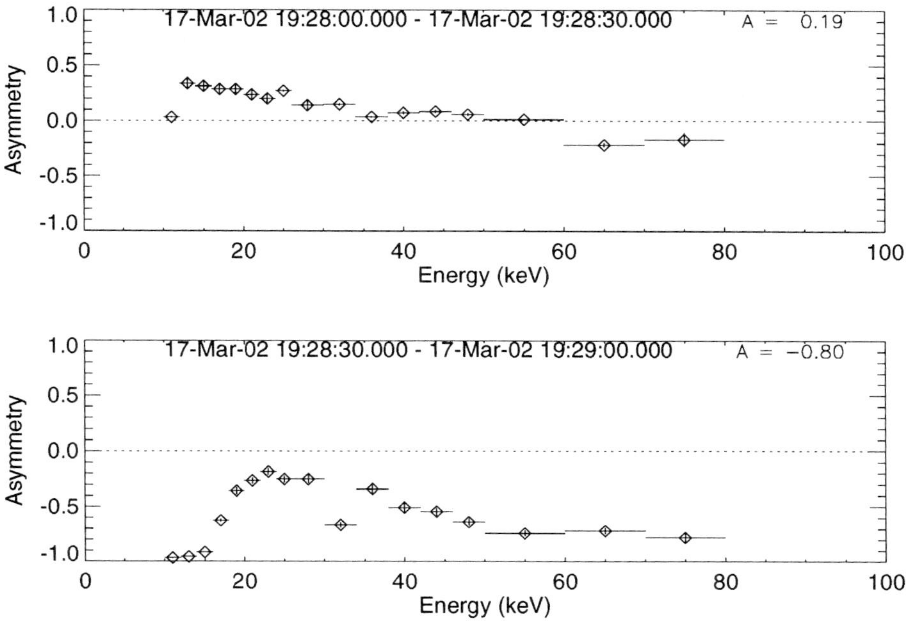


*Figure 5.* The asymmetry in the footpoint emission late in the impulsive phase from 19:28:00 to 19:28:30 UT (*upper plot*) and 19:28:30 to 19:29:00 UT (*lower plot*), as a function of energy. The *horizontal bars* show the energy bands used and the small *vertical bars* show the statistical errors.

two footpoint sources is about 2.5 times the emission from the corona. This demonstrates that, while the coronal emission is strongly influenced by the presence of a trap, the bulk of the emission at early times is dominated by directly precipitated particles.

The vertical bars in Figures 4–7 show the statistical errors on the asymmetry. These errors are derived from the uncertainties provided by the Pixon reconstruction algorithm and represent the statistical errors but do not represent any systematic errors which might be caused by the reconstruction itself.

Two things to bear in mind when interpreting the data shown in Figures 4–7 are the necessity of separating the coronal and footpoint emission and the requirement of potentially long ($\sim 30$ s) image accumulation times (cf., Metcalf and Alexander, 1999). The selection of boxes invariably results in some coronal emission being defined as footpoint emission with the impacts that the observed asymmetry could be lower than the true value and that $R$ is too high. This leads to lower inferred loss-cone angles and larger asymmetries implying a larger ratio of footpoint fields. Additionally, a 30 s accumulation time could be comparable to the trapping time for low energy electrons, depending upon the ambient density, resulting in a significant impact on the inferred asymmetry and emission ratio at these energies (see below). For stronger events we will be able to reduce the image accumulation times significantly.

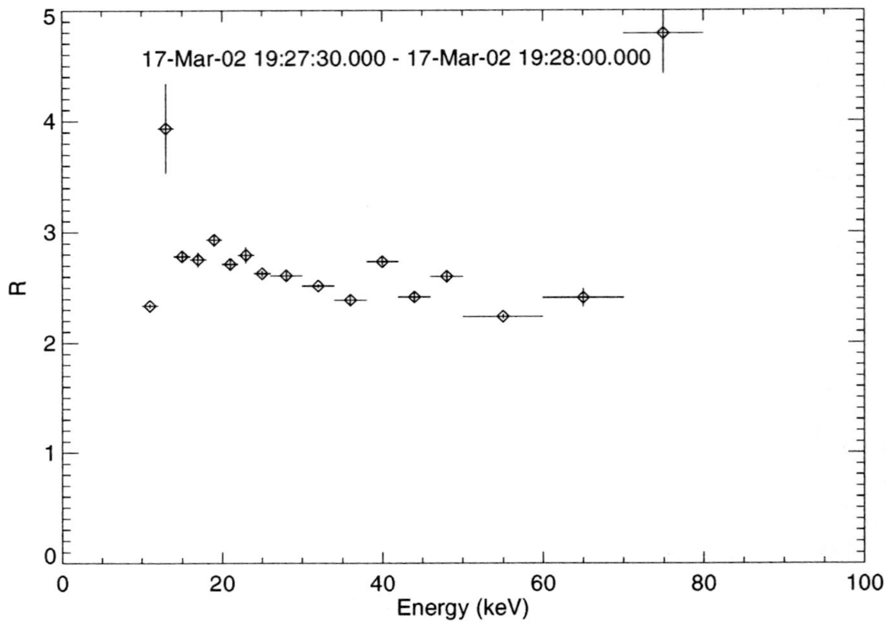


*Figure 6.* The ratio of the footpoint emission to the coronal emission as a function of energy early in the impulsive phase. The *horizontal bars* show the energy bands used and the small *vertical bars* show the statistical errors.

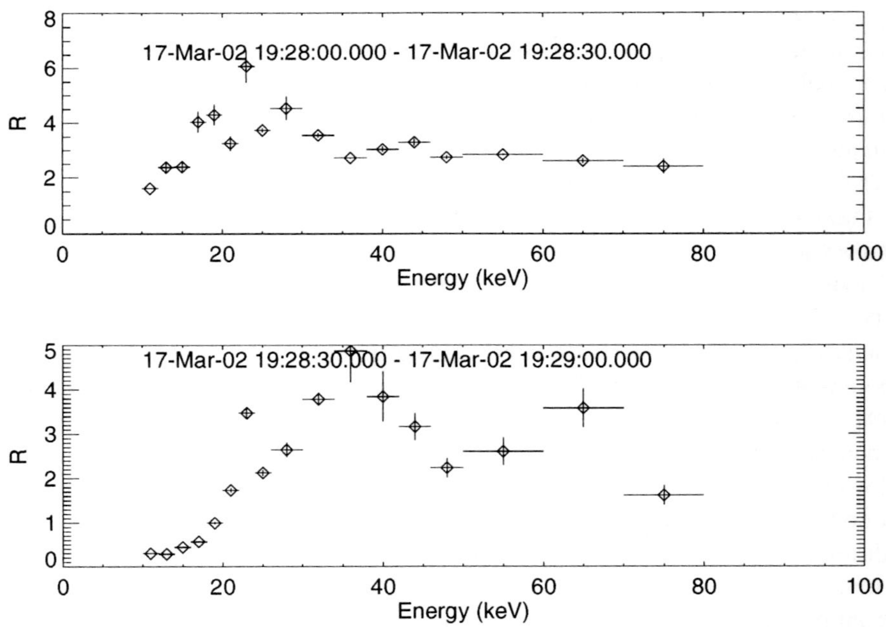


*Figure 7.* The ratio of the footpoint emission to the coronal emission as a function of energy late in the impulsive phase. The *horizontal bars* show the energy bands used and the small *vertical bars* show the statistical errors.

## 3. Results and Discussion

The analysis discussed in the preceding section demonstrates a clear asymmetry in the hard X-ray emission emanating from the footpoints of the flaring loop. Below we will discuss the energy and time dependence of this asymmetry, and the subsequent spatial distribution of the hard X-ray emission, but it should be noted that the mere presence of an asymmetry alone provides us with a powerful diagnostic tool for understanding the particle transport in this flare (cf., Aschwanden *et al.*, 1999).

As we will show, comparison of the observed asymmetry and hard X-ray spatial distribution with existing trap-plus-precipitation models (e.g., McClements, 1990a, b; MacKinnon, 1991) for non-thermal electrons allows us to use the observed asymmetry to constrain the loss-cone angles responsible for the hard X-ray emission in the conjugate footpoints. Moreover, the observed ratio of coronal ('trapped') to footpoint ('precipitated') hard X-ray emission provides an additional derivation of the loss-cone angles which can be readily compared with those derived from the asymmetry.

### 3.1. Modeling the Asymmetry

In this paper we assume that the unspecified acceleration process generates an energetic particle population which is isotropic in pitch angle space (cf., Alexander, 1990; McClements, 1990a) and is symmetrically injected into either side of the loop. Any accelerated particle distribution with a highly anisotropic injection profile, at either large, e.g., MacKinnon (1988), or small, e.g., electron beam model of Brown (1971), pitch angles would likely be rapidly isotropized by plasma instabilities (e.g., Holman, 1985). McClements (1990a) demonstrated clearly that the spatial distribution of the hard X-ray emission predicted by the trap-plus-precipitation model is strongly dependent on the pitch-angle distribution of the accelerated electrons and so it is important to represent the accelerated population as realistically as possible.

Figure 4 shows that early in the impulsive phase there is a strong asymmetry in the emission from the conjugate footpoints of the flaring loop. There is clear evidence for an energy dependence in this asymmetry (not found by Aschwanden *et al.*, 1999, using *Yohkoh*/HXT data). However, in this section, we will concentrate on the energy-integrated asymmetry which was found to be $A_I = 0.39$.

We define the asymmetry as in Equation (1) with $I_1$ and $I_2$ being defined here as the directly precipitated fluxes at each of the two footpoints; we are implicitly assuming that early in the impulsive phase the footpoint hard X-ray production is dominated by directly precipitating electrons. This is a slightly different definition of the asymmetry than used by Aschwanden *et al.* (1999): $A_{\mathrm{Asch}} = I_2/(I_1 + I_2)$. The asymmetry defined in Equation (1) is simply $A = 2A_{\mathrm{Asch}} - 1$.

The magnetic field is assumed to have a uniform value $B_{\mathrm{cor}}$ in the corona, increasing rapidly to a value $B_{\mathrm{chrom},1}$, $B_{\mathrm{chrom},2}$ at each of the loop footpoints respectively (cf., MacKinnon, 1988). With this definition the loss-cone angle takes the form

$$\theta_{0i} = \sin^{-1}\left(\frac{B_{\rm cor}}{B_{{\rm chrom},i}}\right)^{1/2}, \tag{2}$$

where $i \in (1, 2)$ denotes the particular footpoint of interest, and is uniquely defined throughout the coronal part of the loop. Assuming isotropic injection with a source function given by

$$S(E, \sigma) = s_0 E^{-\delta} \frac{\sigma}{(1-\sigma^2)^{1/2}} \equiv s_0 E^{-\delta} f(\sigma), \tag{3}$$

where $\sigma = \sin\theta$ is the pitch-angle sine (see Alexander, 1990), we can obtain the asymmetry by only considering electrons with pitch angles $\theta < \theta_{02}$. The source function $S$ defined in Equation (3) denotes the injection rate per unit $\sigma$. This results in an injection rate per unit solid angle which is independent of the pitch angle, $\theta$. Particles injected towards footpoint 1 with pitch angles $\theta_{01} < \theta < \theta_{02}$ will precipitate at footpoint 2 after a single bounce at mirror point 1. This is the source of the observed asymmetry at early times.

Ignoring the details of the photon production, the photon emission at each footpoint is given by

$$I_1 = \text{const.} \times \int_0^{\sigma_{01}} f(\sigma)\,{\rm d}\sigma = \int_0^{\theta_{01}} \sin\theta\,{\rm d}\theta = 1 - \mu_{01}, \tag{4}$$

$$I_2 = \text{const.} \times \int_0^{\sigma_{02}} f(\sigma)\,{\rm d}\sigma + \int_{\sigma_{01}}^{\sigma_{02}} f(\sigma)\,{\rm d}\sigma = 1 + \mu_{01} - 2\mu_{02}, \tag{5}$$

yielding an asymmetry

$$A = \frac{I_2 - I_1}{I_2 + I_1} = \frac{\mu_{01} - \mu_{02}}{1 - \mu_{02}}, \tag{6}$$

where $\sigma_{0i} = \sin\theta_{0i}$, and $\mu_{0i} = \cos\theta_{0i}$. We have further assumed that the distribution of those particles with pitch angles $\theta_{01} < \theta < \theta_{02}$ does not undergo significant modification during their single 'bounce' from the mirror point at $B_{{\rm chrom},1}$ to the loss cone defined by $\theta_{02}$.

Our definition of the theoretical asymmetry, $A$, assumes no effects due to Coulomb collisions in the loss cone. This is a good assumption, except at very small energies, since the loss cone is effectively empty for energies greater than $\sim$6 keV in the corona; the energy at which the loop transit time equals the collisional pitch angle scattering time. Therefore, assuming identical injection profiles into both sides of the flaring loop, our asymmetry is defined to first order by the form above (see MacKinnon, 1991). Moreover, our theoretical asymmetry is really only valid

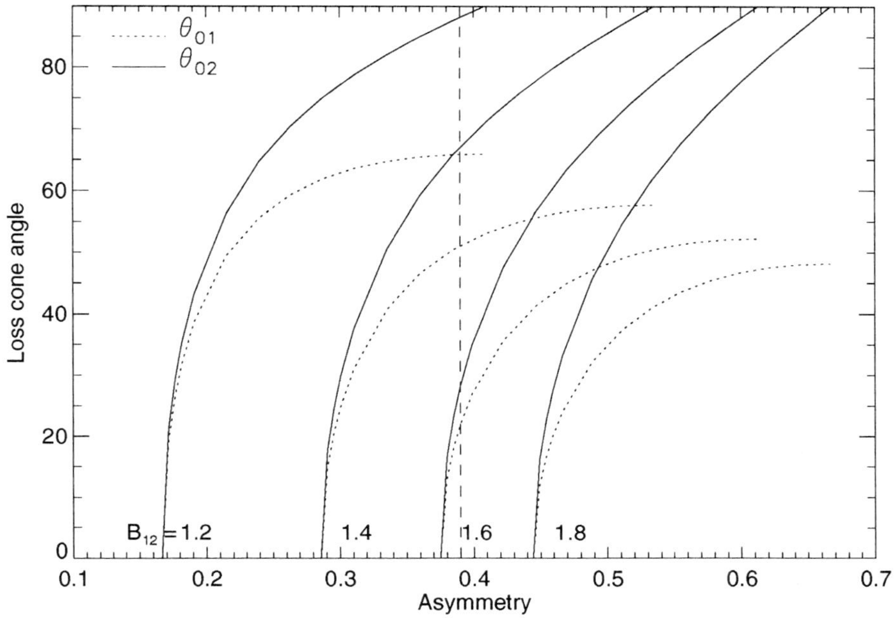


*Figure 8.* The relationship between the loop loss-cone angles and the asymmetry for different values of the footpoint magnetic field ratio. The *dashed line* indicates the observed value of the energy-integrated asymmetry, $A_I = 0.39$.

for short times such that the footpoint emission is dominated by the flux of directly precipitated electrons, before significant scattering from the trap has occurred, and assuming no pitch-angle dependence with time of the injected particle distribution.

From Equation (2) we obtain the relationship between the asymmetric loss-cone angles and the ratio of the magnetic field at the conjugate footpoints:

$$\left(\frac{\sigma_{02}}{\sigma_{01}}\right)^2 = \frac{B_{\mathrm{chrom},1}}{B_{\mathrm{chrom},2}} \equiv B_{12}. \tag{7}$$

Figure 8 illustrates the relationship between the loss-cone angles and the asymmetry for a given footpoint field ratio, $B_{12}$. The measured range of $0.7 < B_{12} < 1.5$ limits the asymmetry to $A < 0.56$ with $I_2 > I_1$ and for asymmetries as high as 0.53 if the footpoint asymmetry switches such that $I_1 > I_2$. The observed energy-integrated asymmetries at the beginning of the impulsive phase ($A_I = 0.39$) would suggest that $1.2 \leq B_{12} \leq 1.6$, since higher values do not accommodate asymmetries as low as 0.39 in this pre-scattering limit. A more certain measurement of the magnetic field at the footpoints would have enabled us to determine the loss-cone angles directly from the measured asymmetry, thereby yielding the field convergence on each side of the loop and providing an estimate of the magnetic field in the corona.

In the absence of a well-defined footpoint field, particularly at the eastern footpoint (see Section 2), we can use the measurement of the spatial distribution of the hard X-ray emission to provide an independent inference of the loss-cone angles. Comparison with the dispersive calculation of McClements (1990a), shows that the energy integrated value of $R_I = 2.5$ would correspond to a loss-cone angle of $\theta_0 \sim 53°$ (McClements, 1990a, defines $R = I_{cor}/I_{fp}$ as the ratio of the coronal to the footpoint emission). Taking $\theta_0$ as an average loss-cone angle, the observed asymmetry indicates that $B_{12} \simeq 1.44$ with $\theta_{01} = 46°$ and $\theta_{02} = 61°$, assuming that we are only observing effects from directly precipitating electrons in the early phase of the flare. Using these values for the loss cone and our determination of $B_{12}$ we can estimate that for this flaring structure the magnetic field in the corona is of order 170 G, where we assume that the chromospheric field where the particles thermalize is 50% of the measured photospheric field. This value is high but reflects the large loss-cone angles inferred from the large asymmetries in the RHESSI footpoint fluxes.

The high values of $R$ seen in Figure 6 imply a weak convergence of the magnetic field in the coronal loop (McClements, 1990a). It is interesting to note that Melrose and Brown (1976) obtained the result

$$\frac{4}{\delta + 2} < R < 2, \tag{8}$$

where $\delta$ is the electron energy spectral index. The high observed values, $R > 2$, indicate that the energy loss and pitch-angle scattering due to Coulomb collisions, neglected by Melrose and Brown (1976), are important for the particle transport in the trapped portion of the loop.

The asymmetric trap analysis of Aschwanden *et al.* (1999) obtains statistical averages for the loss cone angles from a data set of some 54 flares. Similar to the values quoted in the present work, they find $\theta_{01} = 42° \pm 11°$ and $\theta_{02} = 52° \pm 10°$, where subscript '1' denotes the less bright footpoint, yielding mirror ratios $1.6 < B_{\mathrm{chrom},1}/B_{\mathrm{cor}} < 4.0$ and $1.3 < B_{\mathrm{chrom},2}/B_{\mathrm{cor}} < 2.5$, respectively. The ratio of footpoint field by Aschwanden *et al.* (1999), $B_1/B_2 \sim 1.2$ with a range of 1.1–1.8, inferred from the measured asymmetry, is consistent with measured ratio here (see Figure 3).

### 3.2. Energy Dependence of Asymmetry and Flux Ratio

Although the loss-cone angle is completely specified by the field convergence as per Equation (2) and the mirror point is strictly a function of the particle's pitch angle and not its energy, the presence of collisional pitch angle scattering and energy loss introduce an energy dependence in the precipitation rate, particularly at later times. The isotropic injection solution for the ratio of footpoint to coronal emission, derived from the fully-dispersive calculation of McClements (1990a), exhibited a weak dependence of $R$ on energy, with $R$ varying by $\leq 20\%$ between photon energies, $\varepsilon = 20$ keV and $\varepsilon = 200$ keV. From Figure 6 we find a stronger dependence with $R$ varying by $\sim 25\%$ between 20 keV and 60 keV. Aschwanden *et al.* (1999) found little variation in the asymmetry, averaged over 29 flares with well-separated footpoints, at different energies using BATSE and *Yohkoh*/HXT: $A_{23-33}/A_{14-23} = 0.96 \pm 0.07$ and $A_{33-53}/A_{23-33} = 1.06 \pm 0.18$.

Figure 4 shows that during the impulsive phase there is an energy dependence in the asymmetry with $A$ decreasing as we move to higher energies: from 0.5 at $\varepsilon = 20$ keV to 0.35 at $\varepsilon = 60$ keV. This trend of larger asymmetry at lower energies would be expected from the energy dependence of the Coulomb collisional scattering, with preferential scattering into the larger loss cone.

The footpoint-to-coronal-flux ratio shows a weak energy dependence, with $R$ decreasing from $R \sim 2.9$ to $R \sim 2.3$ as $\varepsilon$ increases from 20 keV to 60 keV. A simple application of the energy integrated results of McClements (1990a), assuming the loss cone angles derived from the energy dependence of the asymmetry, would suggest a variation of $R$ much greater than that observed ($R_{69^\circ} > 5$; $R_{42^\circ} \sim 1$). Clearly, a more targeted investigation combining the RHESSI observations with theoretical calculations is required before we can utilize the observed energy dependence of the spatial distribution of the hard X-ray emission to fully understand the particle transport in solar flares. The results presented here provide a strong basis from which to build such an investigation.

At low electron energies for which the loop transit time is comparable to the collisional energy loss time, $E \sim 6$ keV, the weak diffusion limit does not apply and we must consider the energy losses and pitch-angle scattering which occur as the electrons traverse the corona within the loss cone. The consequence is that we might expect to see a fall off of $R$ with decreasing energy. There is some slight evidence for this during the start of the impulsive phase (Figure 6) which gets stronger as time proceeds (Figure 7) and as the trapped particles start to enter the loss cone. In the dispersive calculation of McClements (1990a) there are a significant number of low energy electrons for which the loss-cone escape time is sufficiently long to be comparable to the electron collisional loss time and, consequently, these electrons may be ejected from the loss cone, thereby reducing the precipitation rate at these energies. Since the larger loss-cone will suffer from proportionately more degradation as low energy electrons are scattered out of the loss-cone, the effect of this will be to lower the asymmetry and the footpoint to corona emission ratio as is evident in Figures 4 and 6, particularly at later times (Figures 5 and 7).

A complication at late times is the dynamic development of the coronal plasma as the energy deposition heats and evaporates the chromosphere. The evolution of the source region as the heating develops and the density in the loop increases, skews the determination of $R$, especially at low energies. Thus, if we are truly to determine $R$ from observations we should observe as early in the impulsive phase as count rates allow. As a result, the complex behavior in Figure 7 is less amenable to the analysis presented here.

On timescales shorter than the trapping times for electrons with $E > 20$ keV (cf., Metcalf and Alexander, 1999) no dependence on the energy spectrum is expected in the ratio of footpoint to coronal emission when the loss cone is subject to the weak diffusion limit (Kennel and Petschek, 1966; MacKinnon, 1988), defined by the loop transit time, $t_e$, being much shorter than the pitch-angle scattering time, $t_s$. However, in the strong limit, $t_s \le t_e$, $R \sim \epsilon^2$, which is not observed except perhaps at low energies near the peak of the impulsive phase (19:28:00–19:29:00 UT, see Figure 7).

Finally, we note that in this paper we have ignored the spectral properties of the injected electron distribution (cf., Metcalf and Alexander, 1999) but calculations of the spatial distribution of hard X-ray emission in most of the extant models have been shown to only exhibit a very weak dependence on the spectral index.

### 3.3. Time Dependence of Asymmetry and Flux Ratio

In order to get good photometry when reconstructing hard X-ray images from the RHESSI data one has to integrate over times, typically of order 10–30 s depending upon the strength of the flare. For the event discussed here our image accumulation time is 30 s. This has a direct impact on the interpretation of the observed energy dependence in the asymmetry and flux ratio parameters (Alexander and Metcalf, 1997; Metcalf and Alexander, 1999). The time spent in a coronal trap where the dominant scattering mechanism is Coulomb collisions is energy dependent with the form

$$t_{\text{trap}} = 0.95 \times 10^8 \left(\frac{E^{3/2}}{n_e}\right)\left(\frac{20}{\ln \Lambda}\right) \text{ s}, \tag{9}$$

where $E$ is the electron energy in keV, $n_e$ is the ambient electron density (cm$^{-3}$), and $\ln \Lambda$ is the Coulomb logarithm. Thus, a 100 keV electron will remain in the trap for $\sim 11$ times longer than a 20 keV electron and the trapping time for a 20 keV electron is $\sim 85(10^8/n_e)$ s. Assuming the ambient density early in the flare is low $\sim 2 \times 10^8$ cm$^{-3}$, then low energy electrons should leave the trap in about 45 s. We, therefore, assume that the results depicted in Figure 4 are uninfluenced by the trapping and the arguments in the preceding sections apply. However, precipitation of particles from the trap will have a marked effect on the inferred asymmetries and flux ratios at later times (Figures 5 and 7).

Metcalf and Alexander (1999) discussed a range of possible consequences related to the comparison of the various timescales in the trap-plus-precipitation problem. They found that if the image accumulation time, required to recover a high-confidence image reconstruction, was large relative to the trapping time then the average spectrum in the trap is harder than the injected spectrum by a power of $\sim 1.5$. Conversely, if the accumulation time was short compared to the trapping time the inferred spectrum in the trap would be identical to the injected spectrum but would harden with time as lower energy electrons escape the trap before the higher-energy electrons.

Precipitation from the trap over the relatively long integration times mean that we would measure an $R$ 'contaminated' by pre-trapped particles. Since collisional deflection is an accumulation of small-angle scattering with the escape probability higher on the side of the loop with the larger loss-cone angle/weaker field (Aschwanden *et al.*, 1999), electrons scattered from the trap will tend to increase the asymmetry, $A$, and the footpoint/corona emission ratio, $R$, particularly at low energies. Consequently, the loss-cone angles inferred from these quantities are higher than would be derived from a distribution uninfluenced by significant pre-trapped electrons (Figure 8).

Figures 5 and 7 show the behavior of the asymmetry and flux ratio with energy as we progress to later times during the flare impulsive phase. Over the time range from 19:28:00 UT to 19:29:00 UT the asymmetry shows a transition from the clear dominance of one footpoint to the clear dominance of the other. The most likely explanation for this is that, as the flare develops, more loops begin to participate in the flaring. This is confirmed by the study of this event by Krucker and Lin (2002) in this volume, where there is a clear transition to different footpoints with an opposite sense of asymmetry. As was pointed out in Section 2 the eastern footpoint lies in a region of strong magnetic gradient (Figure 3) and, given that both footpoints lie in comparably strong magnetic fields, it is not unreasonable to suppose that a newly flaring loop in the same region might have dramatically different footpoint field ratios. The effect on the observed energy dependence of the asymmetry is quite marked as can be seen in Figure 5 and would require a much more sophisticated modeling than is currently available. Consequently we will forego discussion of the asymmetry at late times here and concentrate on the footpoint to coronal flux ratio.

As can be seen in Figure 7, the ratio of footpoint to coronal emission demonstrates some interesting behavior as we progress through the impulsive phase. Although we believe that there is a transition to different loop structures with different conjugate field properties, this should not have as strong an effect on the footpoint to coronal emission as it did for the asymmetry. This statement is based on the assumption that, whatever the field connectivity, the acceleration process and, consequently, the properties of the injected energetic particle distribution remain unchanged throughout this transition. Essentially, this constitutes a series of traps with the same particle injection profile, which significantly affects the observed asymmetries but not the flux ratios. We, thus, assume that the observed behavior of $R$ with time is a direct consequence of the escape of particles from the trap into the loss cone.

The first thing to notice is that, compared to the early impulsive phase, $R$ increases significantly at all but the lowest and highest energies, signifying an increase in the footpoint emission relative to that in the corona. The decrease observed at the lowest energies is a direct consequence of the increased density in the trap as chromospheric evaporation proceeds and the temperature increases (MacKinnon, 1986; Antonucci *et al.*, 1998). This results in more of the electrons thermalizing in the corona, regardless of their pitch angles. As one would expect, the effect of this process with time would be to continue lowering $R$ and having the degradation of $R$ spread to higher energies, as is seen in a comparison between the top and bottom panels of Figure 7. Ultimately, the coronal emission comes to dominate the footpoint emission at low energies, $R < 1$, as is frequently observed (Antonucci *et al.*, 1998).

The enhancement in the flux ratio at energies 20–30 keV (19:28:00–19:28:30 UT) and 30–45 keV (19:28:30–19:29:00 UT), prior to the increasing density affecting these energies, can be understood in terms of the escape of particles from the trap into the loss cone, enhancing the footpoint emission at the expense of the coronal emission. Electrons with energies in the 20–30 keV range have trapping times of order 45–83 s for our assumed density of $2 \times 10^8$ cm$^{-3}$, comparable to the 60 s time from the beginning of the impulsive phase considered in the top panel of Figure 7 (slightly higher density at this time would lower these escape times). Electrons in the 30–45 keV energy range would have trapping times as long as $\sim 150(2 \times 10^8/n_e(t = 19{:}29))$ s, which compares with the 90 s time into the event considered in the bottom panel of Figure 7.

At the highest energies, $\varepsilon \gtrsim 65$ keV there is little change in $R$ with time indicating that the trapped particles at these energies have not yet escaped the trap, having trapping times in excess of 260 s. Overall, the variation of $R$ with time and energy would seem to match qualitatively the expected evolution of energetic particles in a trap-plus-precipitation scenario.

MacKinnon (1991) calculates the ratio of footpoint to trapped emission as a function of energy and time with the assumption that the source function for the injected accelerated particles is time dependent, effectively shutting off after a time $t = 2t_0$. He finds that at long times $t > 2t_0$ the ratio of footpoint to coronal hard X-ray emission decreases with increasing energy. Physically, this behavior is a result of the change in dominance of directly-precipitated particles to trap-precipitated particles as the injection rate falls to zero. While we have no direct indication of the time dependence of the injected profile from our data, the presence of a strong asymmetry over the whole of the impulsive phase suggests that the injection continues for $t_0 \gtrsim 30$ s. Over the energy range covered in our RHESSI analysis (12–60 keV) we expect $R$ to fall from a value $\sim 4$ to $\sim 2$–3, extrapolating the curves of MacKinnon (1991). This is close to the observed behavior shown in Figures 6 and 7, indicating that trapping played a strong role in the hard X-ray source evolution for this event at late times.

## 4. Conclusions

The trap-plus-precipitation model has been with us for some time now. The new RHESSI solar flare observations enable us to study observationally the predictions and implications of this model with unprecedented energy and spatial resolution.

Our observations of the energy-dependent coronal emission and the asymmetry in the footpoint hard X-ray emission for the 17 March 2002 flare provide a new look at how electrons are trapped in the solar corona during flares and show that trapping played a significant role in the hard X-ray source evolution for this event. Observations of the coronal trap may be intimately connected with the particle acceleration process. For example, Tsuneta and Naito (1998) demonstrate how electrons can be accelerated through a first-order Fermi process at the fast shock: the region between fast- and slow-mode shocks trap the electrons.

The observations presented above are consistent with the trap-plus-precipitation model in which a portion of the coronal electron population is trapped as a result of the converging magnetic field in the transition region and chromosphere. The asymmetry in the footpoint emission early in the impulsive phase, combined with the ratio of the footpoint to coronal emission, predicts loss cone angles at the two footpoints of about 46 and 61 deg for the western and eastern footpoints, respectively. The small 15-deg difference in the loss cone angles well describes the observations early in the impulsive phase.

Later in the impulsive phase, the asymmetry changes sign. This is likely to be the result of the flare lighting up new magnetic field lines as time progresses so that the new field lines sample an ever shifting magnetic environment at the footpoints. For the observed flare, we surmise that the magnetic field strength at the west footpoint increases as the footpoint moves until the field at the western footpoint becomes larger than the field at the eastern footpoint. This is fully consistent with the MDI magnetic field observations which show a strong spatial gradient in the magnetic field at the western footpoint.

This study has pointed out some aspects of the data analysis to expand upon as we continue using RHESSI data to study the transport of flare particles. First, a more intense hard X-ray flare than the M4.0 described here will enable us to extend this work to shorter time integrations without sacrificing energy resolution. With these shorter time integrations, we will be able to study the dynamics of the trapped particles more directly if the time resolution can be set below the trapping time for the trapped particles. Further, we have not yet fully exploited the spectral characteristics of this event, a study which will also yield new insight into the trapped particle population.

## Acknowledgements

We would like to thank the RHESSI team for their excellent work in providing a great scientific instrument and the software tools to access and analyze the data and to Säm Krucker for useful discussions. We are also particularly grateful to the referee for a rapid and insightful review. This work was supported in part by NASA grant NAS5-02048 (DA) and a direct grant from UC Berkeley (TRM).

## References


Alexander, D.: 1990, *Astron. Astrophys.* **235**, 431.

Alexander, D. and Metcalf, T. R.: 1997, *Astrophys. J.* **489**, 442.

Antonucci, E. *et al.*: 1998, in J. L. R. Saba, B. M. Haisch, and J. T. Schmelz (eds.), 'Flare Dynamics', *The Many Faces of the Sun*, pp. 331–393.

Aschwanden, M. J.: 2002, *Space Sci. Reviews* **101**, 1.

Aschwanden, M. J., Fletcher, L., Sakao, T., Kosugi, T., and Hudson, H.: 1999, *Astrophys. J.* **517**, 977.

Aschwanden, M. J., Hudson, H. S., Kosugi, T., and Schwartz, R. A.: 1996, *Astrophys. J.* **464**, 985.

Bai, T. and Dennis, B. R.: 1985, *Astrophys. J.* **292**, 699.

Bai, T. and Ramaty, R.: 1979, *Astrophys. J.* **227**, 1072.

Bai, T., Hudson, H. S., Pelling, R. M., Lin, R. P., Schwartz, R. A., and von Rosenvinge, T. T.: 1983, *Astrophys. J.* **267**, 433.

Brown, J. C.: 1971, *Solar Phys.* **18**, 489.

Emslie, A. G. and Alexander, D.: 1987, *Solar Phys.* **110**, 295.

Holman, G. D.: 1985, *Astrophys. J.* **293**, 584.

Fishman, G. J., Meegan, C. A., Wilson, R. B., Paciesas, W. S., and Pendleton, G. N.: 1992,'The BATSE Experiment on the Compton Gamma Ray Observatory: Status and some Early Results', *The Compton Observatory Science Workshop*, pp. 26–34.

Kane, S.: 1983, *Solar Phys.* **86**, 355.

Kennel, C. F. and Petschek, H. E.: 1966, *J. Geophys. Res.* **71**, 1.

Kosugi T. *et al.*: 1991, *Solar Phys.* **136**, 17.

Krucker, S. and Lin, R. P.: 2002, *Solar Phys.* (this volume).

LaRosa, T. N. and Emslie, A. G.: 1988, *Astrophys. J* **326**, 997.

Lin R. P. *et al.*: 2002, *Solar Phys.*, this volume.

MacKinnon, A. L.: 1986, *Astron. Astrophys.* **163**, 2391.

MacKinnon, A. L.: 1988, *Astron. Astrophys.* **194**, 279.

MacKinnon, A. L.: 1991, *Astron. Astrophys.* **242**, 256.

Masuda, S., Kosugi, T., Hara, H., Tsuneta, S., and Ogawara, Y.: 1994, *Nature* **371**, 495.

McClements, K. G.: 1990a, *Astron. Astrophys.* **230**, 213.

McClements, K. G.: 1990b, *Astron. Astrophys.* **234**, 487.

McClements, K. G.: 1992, *Astron. Astrophys.* **253**, 261.

McClements, K. G. and Alexander, D.: 1989, *Solar Phys.* **123**, 161.

Melrose, D. B. and Brown, J. C.: 1976, *Monthly Notices Royal Astron. Soc.* **176**, 15.

Metcalf, T. R. and Alexander, D..: 1999, *Astrophys. J.* **522**, 1108.

Metcalf, T. R., Hudson, H. S., Kosugi, T., Puetter, R. C., and Piña, R. K.: 1996, *Astrophys. J.* **466**, 585.

Ramaty, R., Schwartz, R. A., Enome, S., and Nakajima, H.: 1985, *Astrophys. J.* **436**, 941.

Scherrer, P. *et al.*: 1995, *Solar Phys.* **162**, 129.

Takakura, T. and Kai, K.: 1966, *Publ. Astron. Soc. Japan* **18**, 57.

Takakura, T., Tanaka, K., Nitta, N., Kai, K., and Ohki, K.: 1987, *Solar Phys.* **107**, 109.

Tsuneta, S. and Naito, T.: 1998, *Astrophys. J.* **495**, L67.

Tsuneta, S. *et al.*: 1984, *Astrophys. J.* **280**, 887.

# RHESSI AND TRACE OBSERVATIONS OF THE 21 APRIL 2002 X1.5 FLARE

PETER T. GALLAGHER[1,2], BRIAN R. DENNIS[1], SÄM KRUCKER[3], RICHARD A. SCHWARTZ[1,4] and A. KIMBERLEY TOLBERT[1,4]

[1]*Laboratory for Astronomy and Solar Physics, NASA Goddard Space Flight Center, Greenbelt, MD 20771, U.S.A.*

[2]*L-3 Com Analytics Corp., 1801 McCormick Drive, Largo, MD 20774, U.S.A.*

[3]*Space Sciences Laboratory, University of California, Berkeley, CA 94720-7450, U.S.A.*

[4]*Science Systems and Applications, Inc., 10210 Greenbelt Road, Lanham, MD 20706, U.S.A.*

(Received 7 August 2002; accepted 23 August 2002)

**Abstract.** Observations of the X1.5 flare on 21 April 2002 are reviewed using the Reuven Ramaty High-Energy Solar Spectroscopic Imager (RHESSI) and the Transition Region and Coronal Explorer (TRACE). The major findings are as follows: (1) The 3–25 keV X-rays started < 4 min before the EUV (195 Å) emission suggesting that the initial energy release heated plasma directly to $\gtrsim$20 MK, well above the 1.6 MK needed to produce the Fe XII (195 Å) line. (2) Using coaligned 12–25 keV RHESSI and TRACE images, further evidence is found for the existence of hot (15–20 MK) plasma in the 195 Å passband. This hot, diffuse emission is attributed to the presence of the Fe XXIV (192 Å) line within the TRACE 195 Å passband. (3) The 12–25 keV source centroid moves away from the limb with an apparent velocity of $\sim$9.9 km s$^{-1}$, slowing to $\sim$1.7 km s$^{-1}$ after 3 hours, its final altitude being $\sim$ 140 Mm after $\sim$ 12 hours. This suggests that the energy release site moves to higher altitudes in agreement with classical flare models. (4) The 50–100 keV emission correlates well with EUV flare ribbons, suggesting thick-target interactions at the footpoints of the magnetic arcade. The 50–100 keV time profile matches the time derivative of the GOES light curve (Neupert effect), which suggests that the same electrons that produced the thick-target hard X-ray emission also heat the plasma seen in soft X-rays. (5) X-ray footpoint emission has an E$^{-3}$ spectrum down to $\sim$ 10 keV suggesting a lower electron cutoff energy than previously thought. (6) The hard X-ray (25–200 keV) peaks have FWHM durations of $\sim$ 1 min suggesting a more gradual energy release process than expected. (7) The TRACE images reveal a bright symmetric front propagating away from the main flare site at speeds of $\geq$ 120 km s$^{-1}$. This may be associated with the fast CME observed several minutes later by LASCO. (8) Dark sinuous lanes are observed in the TRACE images that extend almost radially from the post-flare loop system. This 'fan of spines' becomes visible well into the decay phase of the flare and shows evidence for both lateral and downward motions.

## 1. Introduction

On 21 April 2002, RHESSI (Lin *et al.*, 2002) observed its first X-class flare, an X1.5/1F at S14 W84. The complete rise-phase of the flare was observed by RHESSI during the daytime part of the first orbit. X-ray emission was detected from 3 keV to about 200 keV during this time. The fluxes were sufficiently high that imaging spectroscopy was possible with angular resolutions as fine as 2″ and spectral resolution of $\sim$ 1 keV. The decay phase was clearly visible above background

*Solar Physics* **210:** 341–356, 2002.

to energies of $\gtrsim 10$ keV on subsequent orbits for the next 12 hours. This flare was also imaged by TRACE (Handy *et al.*, 1999) in its 195 Å band with $1''$ resolution and 20 s cadence, providing information on the 1.5 MK plasma from the Fe XII line emission and on the 15–20 MK plasma from the Fe XXIV emission line in the same band. A TRACE movie of this flare with RHESSI contours overlaid is available on the accompanying CD-ROM. A companion paper discusses the detailed time history of the soft and hard X-ray emission and the extent to which they show the Neupert effect (Young *et al.*, 2002).

This flare is an example of an eruptive two-ribbon long-duration event, a type C flare in the classification scheme first proposed by Tanaka (1983) and expanded by Tsuneta *et al.* (1984) and Tanaka (1987) for grouping flares according to their combined properties. It is relatively modest in peak intensity and in the maximum X-ray energies but it warrants detailed study because of the excellent coverage by RHESSI, TRACE, and other instruments. Its near-limb location allows the loops to be seen edge on and the altitude profile of the emissions to be determined into the high corona to over 140 Mm. During the rise phase, both the footpoint and the coronal X-ray emission are imaged and independent spectra can be obtained. During the long decay, the X-ray source is followed to higher and higher altitudes through the RHESSI imaging observations. The subsequent cooling of the hot loops is revealed as they appear in the TRACE passband. A fast ($\sim 2500$ km s$^{-1}$) coronal mass ejection accompanied this event, which can be seen in the TRACE movie as a faint feature moving through the field of view with a velocity of $\sim 120$ km s$^{-1}$. This will be discussed further in a follow-on publication currently in preparation (Gallagher and Dennis, 2002).

Long-duration flares similar to the one reported here were studied in the 1980s with X-ray and UV instruments on the Solar Maximum Mission (e.g., Švestka *et al.*, 1982) and on Hinotori (Tsuneta *et al.*, 1984). However, they proved difficult to study with the Hard X-ray Telescope on *Yohkoh*, particularly in the decay phase, because of their large-scale structures and relatively soft spectra (Masuda, Kosugi, and Hudson, 2001). Also, the energy range of $\sim 2$–15 keV was not covered, thus making it difficult to image plasma in the critical temperature range of $\sim 10$–20 MK.

This earlier work has resulted in an understanding of such events based on the Kopp and Pneuman (1976) model of magnetic reconnection at an X-point or neutral sheet in the corona. Particles are accelerated and plasma is heated as a result of the reconnection. The accelerated electrons stream down to the footpoints and produce hard X-ray bremsstrahlung as they are thermalized by Coulomb collisions in the higher density regions of the lower corona and chromosphere. The combination of directly heated plasma already in the loop and evaporated chromospheric material (Czaykowska *et al.*, 1999) produces hot loops below the reconnection site with temperatures that can be 20 MK or higher and densities as high as $10^{13}$ cm$^{-3}$. These loops are visible in soft X-rays. They subsequently cool and eventually become visible as post-flare loops in the lower temperature EUV lines and in H$\alpha$. The reconnection site gradually moves upwards and continues to release energy, even as the total X-ray flux decreases during the decay phase of the event. This is evidenced by the appearance of higher and higher hot loops and the lower altitudes of the cooler loops. The multithermal observations reported by Švestka *et al.* (1987) show this scenario for a flare on 6 November 1980.

This paper contains the results of a preliminary analysis of the RHESSI and TRACE observations of the X1.5 flare on 21 April 2002. With the current state of our understanding of the instrument response and the capabilities of the analysis software, it is possible to obtain images at all energies where the number of counts from the flare is $\gtrsim 1000$. The highest dynamic range in any one image is currently limited to $\sim 20:1$, although image reconstruction artifacts can be present at somewhat higher levels. Currently, the shortest time resolution for obtaining an image is also limited to one spacecraft rotation period of $\sim 4$ s. Spectral information is available down to $\sim 3$ keV when all shutters are out of the field of view. However, detailed knowledge of the instrument sensitivity is still uncertain, especially at energies below 10 keV. Thus, absolute fluxes have relatively large error bars at these low energies but the images are reliable.

The following section describes the overall time-line of the flare as seen with RHESSI and TRACE. The preflare conditions and possibly related earlier flares are described. The rise phase of the flare is discussed with the appearance of hard X-ray emission and TRACE brightenings during the first RHESSI orbit. The decay phase of the flare is then discussed as it appears in hard X-rays on 7 subsequent RHESSI orbits and in the lower-temperature lines seen in the TRACE passband. In the last section, the observations are discussed in relation to possible models of energy release and particle acceleration throughout the over 12-hour duration of this event.

## 2. Observations

### 2.1. Pre-flare Activity

The period 14–21 April 2002 was a time of moderate to high solar activity, due mainly to the transit across the disk of three large regions NOAA 9901, 9906 and 9907. The X1.5 event studied here occurred in NOAA 9906 on 21 April 2002 at $\sim$00:43 UT. This region was classified as a large complex $\beta\gamma\delta$ region consisting of an asymmetric leader and follower spot with a compact area of mixed magnetic polarities in between. This can be seen clearly in Figure 1 which shows an MDI continuum image and magnetogram from the Active Region Monitor (`http://www.bbso.njit.edu/arm`; Gallagher, Moon, and Wang, 2002). Between 18 and 19 April, there was some decay evident in the interior of the group, which appears to have continued into 20–21 April, although this is difficult to confirm due to the region's proximity to the limb. When the flare started, the leader spot and

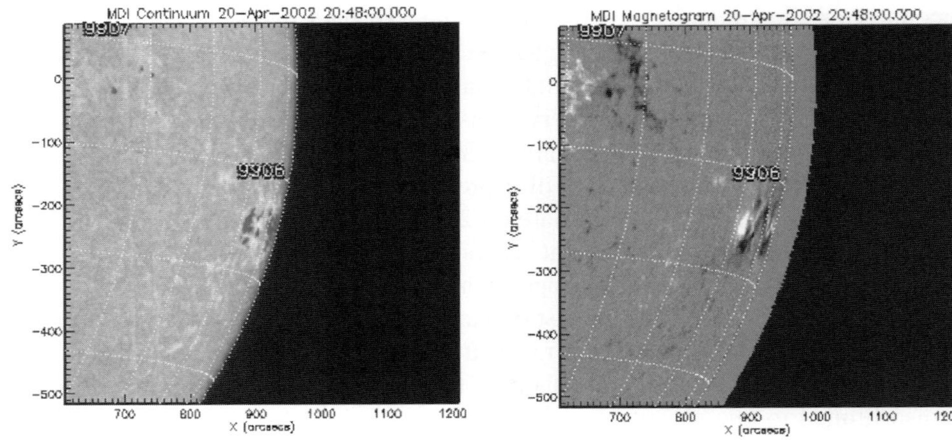


*Figure 1.* NOAA 9906 at 20:48:00 UT on 20 April 2002. The *left-hand panel* gives the MDI continuum image while the corresponding MDI magnetogram is given at *right*. A color version of this and other figures is on the accompanying CD-ROM.

almost 50% of the region's center had rotated off the solar disk possibly resulting in some of the X-ray emission being occulted.

The flare time-lines from GOES, TRACE, and RHESSI are shown in Figure 2. The initial activity seen in NOAA 9906 started as early as 23:00 UT on 20 April 2002 with an extremely impulsive soft X-ray event peaking at 23:27 UT. This was imaged with RHESSI as a compact source from the same location as the start of the X-flare over an hour later. This source is one of the few flares that does not show any extent beyond the finest RHESSI resolution of 2″. The small flare in the GOES time profile starting at 00:20 UT was identified as being from the opposite limb. RHESSI observed the decay phase of this event from the night-to-day transition at 00:32 UT until a second small increase was seen from NOAA 9906 starting at 00:33 UT. This decayed away until the start of the main flare.

### 2.2. Rise Phase

The first increase from the main flare was seen in the GOES 1–8 Å band at 00:42 UT. RHESSI saw an increase in the 3–12 keV flux starting some 2 min earlier, thus demonstrating its greater sensitivity when all shutters are out of the detector fields of view as they were at that time. The flux in this energy band rose gradually until 00:42 UT, when it became visible in the 12–25 keV band and started to rise more rapidly. The images at this time show a single source on the disc centered at heliocentric coordinates of (930″, −230″ ). In the top left-hand panel of Figure 3, the contours of the RHESSI image are overlaid on the TRACE image taken at the same time to within 10 s. Note that no brightening is seen in the TRACE image at this time. In fact the first increase in the 195 Å intensity was not evident until some 3 min later, at ∼00:45 UT. The TRACE image with the

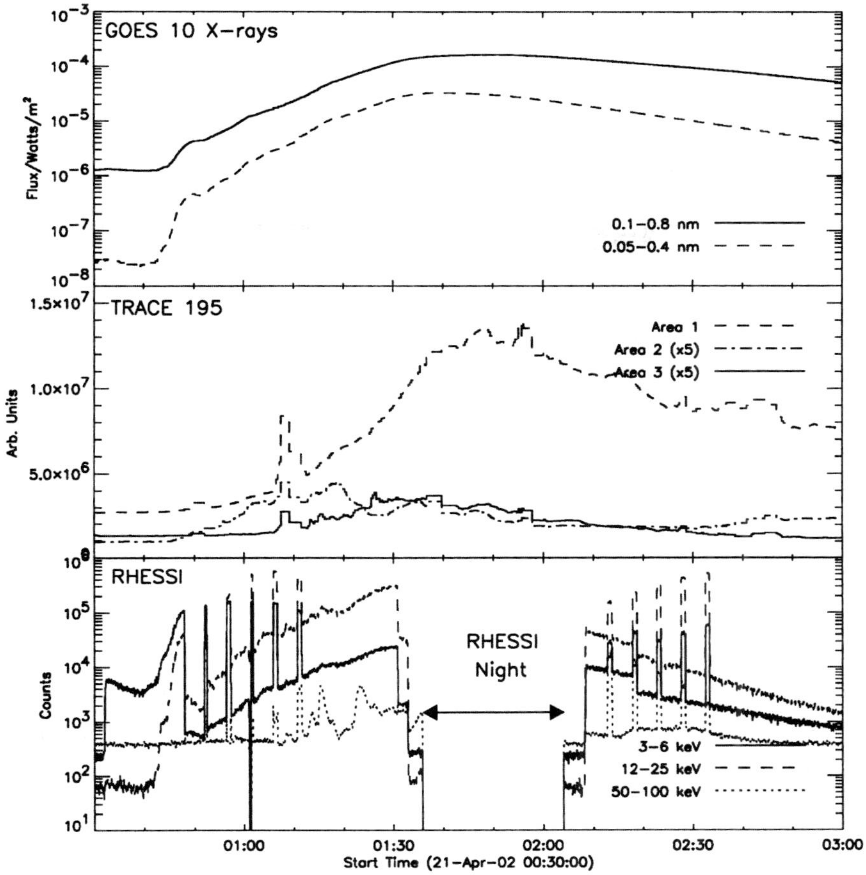


*Figure 2.* GOES 10, TRACE 195 Å and RHESSI lightcurves for 00:30–03:00 UT. The systematic spikes visible in the bottom panel are due to RHESSI shutter movements.

RHESSI contours overlaid at 00:48:54 UT is shown in the top right-hand panel of the same figure. As there is a small error in TRACE's pointing, the 195 Å passband images were cross-correlated with the EIT 195 Å image taken at the same time and then shifted by $\sim 2.5''$ in solar x and $\sim 4.3''$ in solar y before being coaligned with RHESSI. Further information on this technique can be found at `http://hesperia.gsfc.nasa.gov/~ptg/trace-align/`.

Once the total count rate in each of the RHESSI detector front segments reached about 20 000 counts $s^{-1}$, the thin shutters moved automatically into the detector fields of view. This effect can be seen clearly in the bottom panel of Figure 2. The shutters are thin discs of aluminum that absorb the lower energy X-rays and reduce the total detector count rate by a large factor. They are put in place when the counting rate becomes so high that pulse pile-up takes place causing significant spectral distortion. The effect of putting the thin shutters into the field of view can

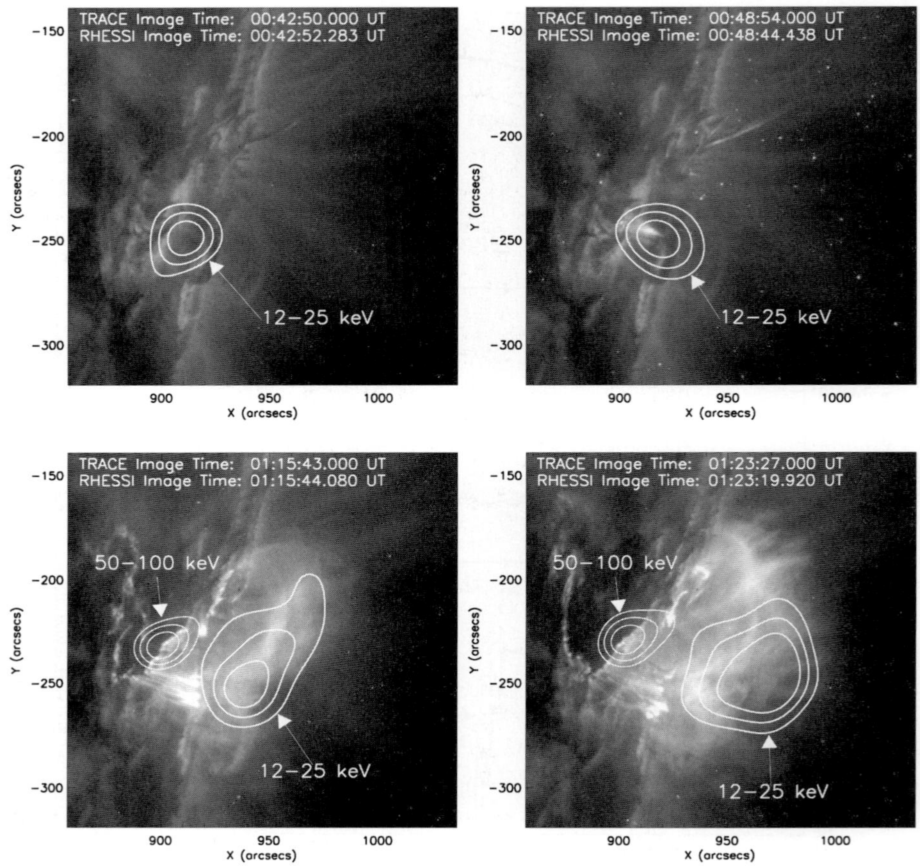


*Figure 3.* TRACE 195 Å and RHESSI 12–25 and 50–100 keV images for four distinctive times during the flare. The RHESSI contour levels are at 25, 50, and 75% of the peak flux for each energy band.

be seen at 00:48 UT, when the counting rate is reduced by a large factor at the lower energies. Unfortunately, the on-board algorithm that controlled the shutter movement used the now lower counting rate and determined that the shutters could be removed. Thus, after a built-in delay of 5 min the shutters were moved out. But, of course, the rate was now too high again and, after another built-in 1-min delay, the shutters were moved back in. This cycle was repeated 5 times, the maximum number allowed by the algorithm, and then finally at 01:12 UT, the shutters were put in place for the remainder of the orbit. Near the end of the daytime part of the orbit, at 01:31 UT, the total counting rate had again risen to exceed the threshold for putting the thick shutters in place, and they were automatically moved into the fields of view at that time. The same cycling of the thin shutters occurred on the next two orbits but on subsequent orbits the counting rates were sufficiently low that all shutters were removed allowing for the most sensitive observations down

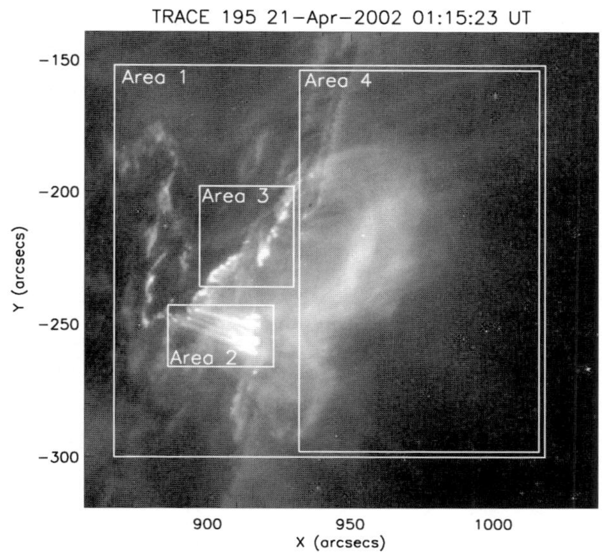


*Figure 4.* TRACE 195 Å image at 01:15:23 UT showing the four areas from which lightcurves were extracted. The hot, diffuse emission attributed to Fe XXIV (192 Å) is clearly visible in Area 4.

to as low as 3 keV. This sequence of shutter movements explains the large steps and peaks in the time profiles at different energies. Since the thicknesses of the shutters are accurately known from optical and X-ray measurements made before launch, the incident photon flux can still be accurately determined at all times during the event except for the times lasting for $\lesssim 1$ s when the shutters are actually moving or when pile-up becomes an issue.

The two bottom panels of Figure 3 show the RHESSI contours in the 12–25 and 50–100 keV energy bands overlaid on the TRACE images taken at the same time within 10 s. The 12–25 keV emission now extends to higher coronal altitudes and appears to show a loop-like feature connecting the initial energy release site to a point just over the limb. In the TRACE 195 Å images, this hot X-ray feature is found to be coincident with a broad area of diffuse emission which first becomes visible at ~01:02 UT in the TRACE 195 Å images. This diffuse coronal brightening is believed to be from 15–20 MK plasma emitted by the Fe XXIV (192 Å) line which lies within the 195 Å band and can be seen quite clearly in Figure 4, Area 4. A similar high-temperature region was also noted by Warren and Reeves (2001) in the TRACE 195 Å passband, also near the top of a flare arcade. In this case, they derived electron temperatures of $\lesssim 20$ MK using the TRACE 195 Å/171 Å ratio for an X1.8 flare on 24 March 2000. In the bottom right-hand panel of Figure 3 at 01:23 UT, the 12–25 keV flux is now from a considerably higher altitude in the corona than seen earlier and is well above the diffuse Fe XXIV source.

The 50–100 keV emission, on the other hand, is emitted exclusively from low-altitude regions, presumably at the footpoints of the main flare arcade. When

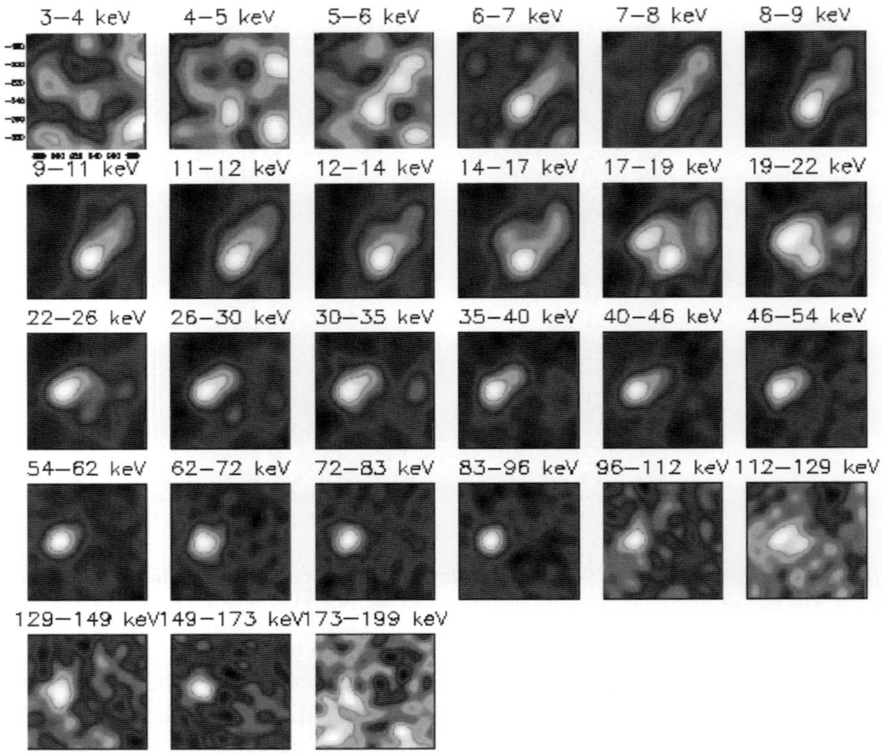


*Figure 5.* RHESSI images obtained in the 27 indicated energy bands using the CLEAN algorithm with collimators 3–9 for a period of 25 s (6 rotations) at the time of the peak in the hard X-ray emission near 01:15 UT. (See a color version on the CD-ROM.)

coaligned with TRACE 195 Å images, this harder emission correlates both spatially and temporally with bright EUV flare ribbons. This is consistent with models in which flare-accelerated electrons stream down magnetic field lines and lose their energy in the higher densities of the chromosphere. This then results in heating of the chromospheric plasma to coronal temperatures and the formation of bright flare ribbons, visible in numerous lines, including H$\alpha$ (Qiu *et al.*, 2001). In the images on the bottom panels of Figure 3, the hard X-ray emission can be seen overlying the EUV flare ribbons, which are presumably heated by the same flare electrons.

Various peaks in the hard X-ray emission were seen up to the 100–300 keV energy band, most notably at 01:15 UT and again at 01:23 UT. The time profile for the complete rising phase of the flare was characterized, however, by gradual variations; the most impulsive emission, even at the highest energies, had timescales not much shorter than 1 min. We have looked in detail at the first of these peaks. The RHESSI images in Figure 5 for this peak show that the X-ray source can be located with $\lesssim 5''$ accuracy in all energy bands from 6–173 keV. At the higher energies above $\sim 50$ keV, the source is unresolved at the $7''$ resolution of these

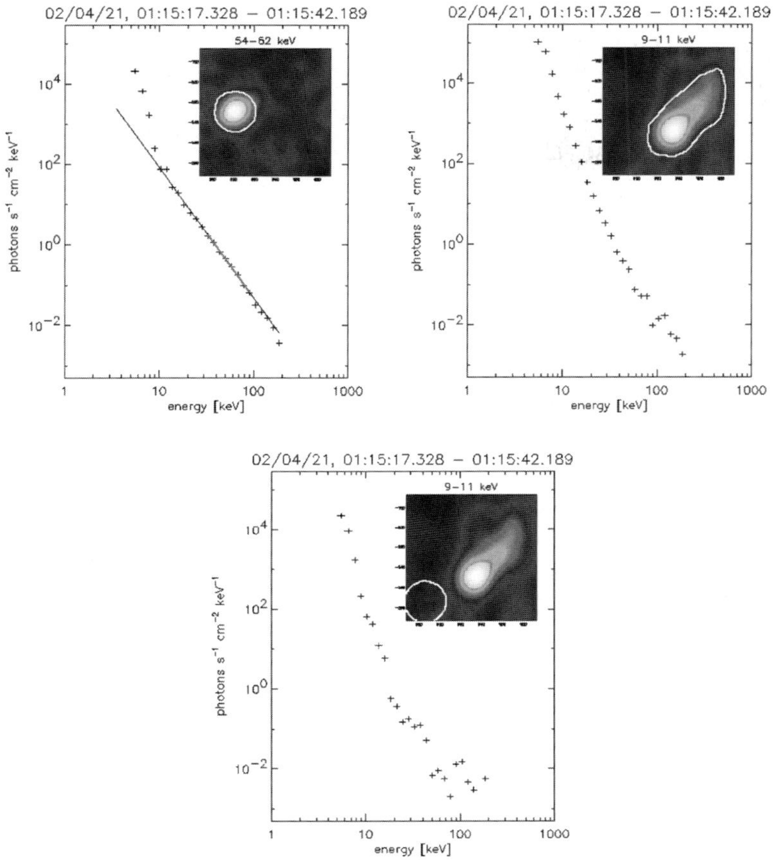


*Figure 6.* Spectra of the two bright sources seen in the images of Figure 5. *Top left*: the source on the solar disk with an $E^{-3.2}$ power-law fit between 30 and 100 keV. *Top right*: the coronal source with an $E^{-7}$ power-law fit from 8–30 keV. *Bottom*: the spectrum from the indicated region well away from either of the two bright sources. (A color version on the CD-ROM.)

images. It is located on the solar disk coincident with the EUV ribbons seen in the TRACE images, presumably at the foot points of the arcade of magnetic loops. At lower energies, below about 14 keV, the source is at a different location above the limb. Both sources are evident in the transition that takes place in the images between $\sim 14$ and $\sim 46$ keV.

The spectra of these two different sources in this same time interval are shown in Figure 6. The spectrum of the source on the disk is shown in the top left of Figure 6. Between 30 and 100 keV, a power-law function with an index of $-3.2$ provides a good fit to the data points. The spectrum of the coronal source is shown to the right. It can be fitted with an $E^{-7}$ power law between 8 and 30 keV. Note that the spectrum of the disk source steepens to about this same $E^{-7}$ form below about 15 keV whereas the spectrum of the coronal source flattens to about an $E^{-3}$

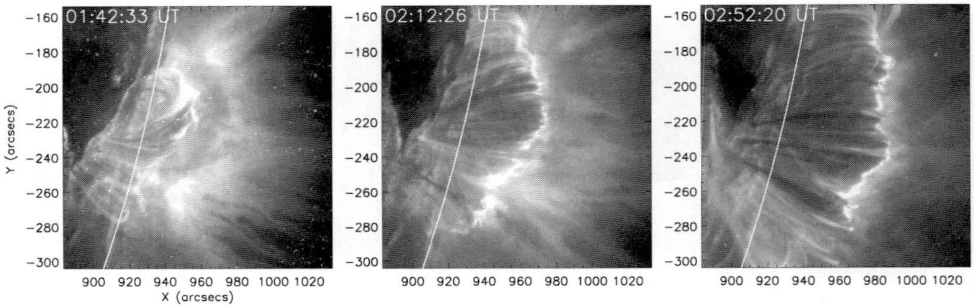


*Figure 7.* Rising post-flare loop system seen in TRACE 195 Å from 01:42 UT to 02:52 UT.

form above about 50 keV. This effect seems to be caused by contamination of the spectrum of one bright source by a background from the other bright source. That such a background is present in the images can be seen from the spectrum shown in the lower panel of Figure 6. This is from the indicated region that appears free from either of the two bright sources and yet it has a similar steep spectrum at lower energies that becomes flatter at higher energies.

We conclude from these spectra that it is not possible to determine, at this stage in the evolution of the data analysis procedures, if the coronal source is a thermal exponential or a nonthermal power law. It is not yet clear exactly what background must be subtracted from the spectrum determined from the location of the coronal source at the higher energies. Detailed modeling is probably the only way that this question can be answered.

It seems clear, however, that the footpoint source spectrum extends down to as low as 10 keV with the same $E^{-3.2}$ spectrum as seen to energies at least as high as 100 keV. If this is correct, then it would indicate that the spectrum of the precipitating electrons must extend down to at least this energy without a turnover. This is a factor of at least two lower in energy than that often assumed for the cutoff in the electron spectrum, thus leading to a factor of 5 greater energy in electrons than would have been calculated with the higher cutoff energy.

The softer X-ray emission continued to rise monotonically for the remainder of the daytime part of the RHESSI orbit that ended at 00:01:37 UT. The GOES light curves show that the peak was reached at 01:50 UT. Thus, the rise phase of the flare lasted for 70 min. The temperature at this time derived from the ratio of the fluxes in the two GOES channels was 17 MK and the emission measure was $2 \times 10^{50}$ cm$^{-3}$.

### 2.3. Decay Phase

The decay phase of the flare lasted for a further 14 h or at least until 16:00 UT. There is a notable lack of smaller events during this period compared with the earlier and later average rates of such events. Only two or three small events are

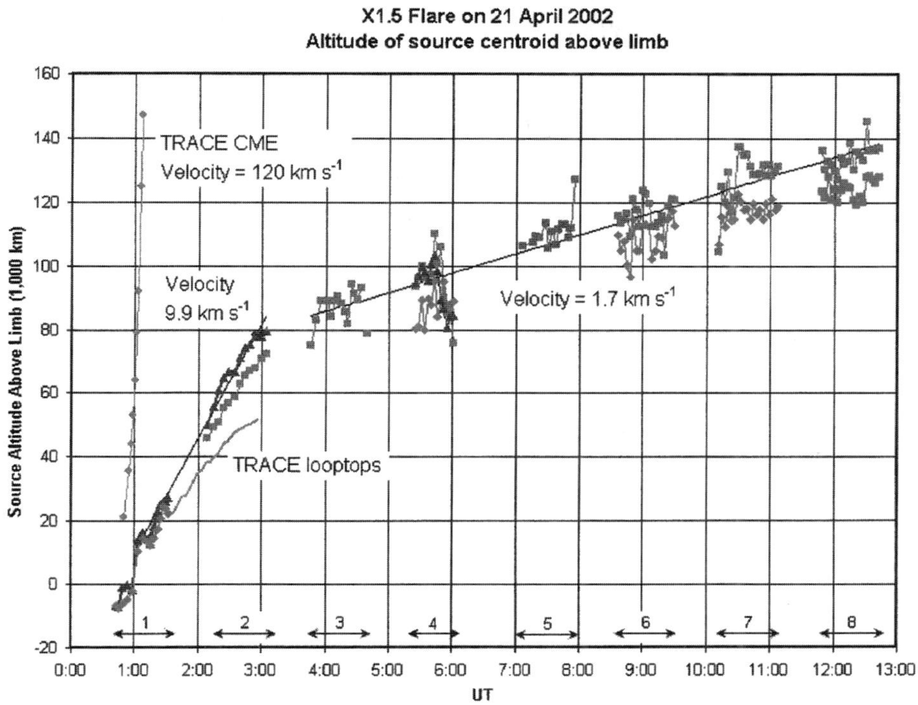


*Figure 8.* Apparent altitude above the solar limb as a function of time for several features seen in RHESSI and TRACE images. The RHESSI 3–6, 6–12, and 12–25 keV source centroids are indicated by *diamonds*, *squares*, and *triangles*, respectively.

seen in the GOES plot above the decaying flux from the main event, suggesting that additional activity was suppressed during this 16-hour interval.

RHESSI observations are available up until almost 13:00 UT after which the data are missing. The thin shutters were in the detector fields of view for the first two orbits of the decay (apart from five 1-min intervals when they were moved out as discussed above). On the five subsequent orbits, all the shutters were out of the field of view and the flare emission is seen above background below 10 keV for the whole time.

Of particular interest during the decay phase is the rise of the sources of emission, both in X-rays as seen with RHESSI and in the EUV as seen with TRACE. Figure 7 shows the rising post-flare loop system as seen by TRACE at approximately 1, 1.5, and 2 h after the flare begins. We have made images with RHESSI throughout the decay phase in the 3–6, 6–12, and 12–25 keV energy bands. For each image, we have determined the location of the centroid of the emission inside the contour level at 50% of the peak value in each image. Since the source is about 20″ in extent during this whole period with no evidence of any finer structure, we have used only the detectors behind collimators 6–9 to make the images. We used the back-projection image reconstruction technique for speed since we were not concerned with the side lobes that were always below the 50% level.

Figure 8 shows the computed height of the centroid above the limb in the plane of the sky. A similar analysis has been done for the TRACE images for both the post-flare loops and a faint CME-like feature observed propagating away from the main flare site. The resultant altitudes above the limb are plotted on the same figure for comparison with the X-ray information. Detailed analysis of the X-ray centroid locations and their movements with time shows immediately that they all rise to higher and higher altitudes above the limb, reaching over 140 000 km. The X-ray centroids rise initially at a rate of about 10 km $s^{-1}$ until about 03:00 UT and then slow down to about 1.7 km $s^{-1}$ for the rest of the time. Note that the lower-energy X-ray centroids are consistently at a lower altitude than the higher-energy centroids. The centroids of the 12–25 keV X-rays are about 10 000 km lower than the 6–12 keV centroids during the second orbit from 02:00 to 03:00 UT. The 3–6 keV centroids are about 10 000 km below the 6–12 keV centroids after 05:15 UT. Similar motions of the TRACE features are also seen in the figure. The post-flare loop tops rise rapidly between approximately 01:00 and 03:00 UT although at a slightly slower rate than the X-ray features seen in RHESSI. The Fe XII loop-tops are always well below the centroids of the X-ray emission. This pattern of higher loops being hotter than the loops below them is consistent with the flare scenario described above.

Also plotted in Figure 8 are heights as a function of time of a faint feature ejected from the main flare site at the flare onset time. Due to the morphology, timing and direction of this fast-moving feature, it is almost certainly associated with a fast CME observed to have velocities of $\sim$2500 km $s^{-1}$. In TRACE the measured velocity was $\sim$ 120 km $s^{-1}$ (Gallagher and Dennis, 2002).

At any given time, the energy release site is at or above the highest sources that are observed. This site is rising continuously and the released energy results in higher and higher loops becoming filled with hot plasma. The newly heated loops appear first in X-rays. The lower energy X-ray source and the TRACE features are loops that have cooled to temperatures at which these band passes are sensitive. An estimate can be made of the loop cooling times from the altitudes and times at which they are seen in the different temperature regimes.

## 3. Summary and Conclusions

It is clear that RHESSI's imaging and spectroscopic capabilities are already providing new insights into high-energy processes in solar flares, such as particle acceleration, propagation, and precipitation. In combination with observations from ground- and space-based observatories such as SOHO and TRACE, it is now possible to gain an unprecedented understanding of flaring plasmas from the corona, through the transition region, to the chromosphere. In this paper, soft and hard X-ray images and spectra from RHESSI together with EUV images from TRACE were used to study the X1.5 of 21 April 2002. The major findings are as follows:

(1) The 3–25 keV X-rays started <4 min before the EUV (195 Å) emission at the same location. This suggests that the initial energy release heated plasma directly to $\sim$20 MK, well above the 1.5 MK needed to produce the Fe XII emission line at 195 Å. Presumably it took several minutes for a sufficient amount of plasma to cool down to be visible in Fe XII. The Fe XXIV resonance line at 192 Å from 15 MK plasma must have been too weak to be detectable with TRACE in the 195 Å band although it is not clear what line produced the first detectable EUV emission.

(2) Some parts of the flare may have been occulted by the solar limb. There is some evidence for this in EUV brightenings at and above the limb earlier than, and at a different location to that of the initial X-ray brightening. Also, the appearance of the high-altitude Fe XXIV and 12–25 keV X-ray emissions later in the flare suggests a large loop-like structure extending beyond the limb. It is not clear what effects the over-the-limb parts of the flare could have had on the appearance of the flare as seen with RHESSI and TRACE, although the indications are that they were probably not major. Based on the earlier longitudinal extent ($\sim 200''$) of this active region (NOAA 9906) when it was all visible on the disk 1 day earlier, it is unlikely that any part of the flare was more than about $50''$ over the limb.

(3) The 3–25 keV X-rays came from higher and higher altitudes as the flare progressed. The 12–25 keV source centroid moved away from the limb at an apparent velocity of 9.9 km s$^{-1}$ from 01:00 to 03:00 UT. The velocity decreased at about this time to 1.7 km s$^{-1}$ and continued at this rate until at least 13:00 UT, when the source of 6–12 keV X-rays was at $\sim$140 000 km above the limb. This suggests that the energy release site moved to higher altitudes also.

(4) During the rise and early decay phase of the flare, the 12–25 keV source was at the same or greater altitude above the limb as the source of the Fe XXIV emission, and both were well above the tops of the Fe XII loops. By 03:00 UT, the 12–25 keV source was at 80 000 km above the limb while the Fe XII source was at 50 000 km. Later in the decay phase, the 3–6 keV source was consistently about 10 000 km lower than the 6–12 keV source. All of these observations are consistent with the generally held picture of a rising energy release site that heats the magnetic structures below it. These then cool and become visible at the lower temperatures. It is not clear if there is any relation between the velocity of the energy release site and the velocity of the associated CME or why the velocity changed quite abruptly at about 03:00 UT from 10 to 1.7 km s$^{-1}$. Also, why should the energy release on the decay be so smooth compared to the relatively impulsive release on the rise?

(5) The $\geq$25 keV X-ray emission later in the flare came from the location of the ribbons seen in the TRACE images suggesting electron beams and thick-target interactions at the footpoints of the magnetic arcade of loops. The 50–100 keV time profile matches the time derivative of the GOES light curve (Neupert effect; see Young *et al.*, 2002). This suggests that the same electrons that produced the thick-target hard X-ray emission also heat the plasma that is seen in the GOES soft X-ray

emission. The peak temperature derived from the GOES 2-channel measurements is 17 MK and a peak emission measure of $2 \times 10^{50}$ cm$^{-3}$.

(6) X-ray footpoint emission has an $E^{-3.2}$ spectrum all the way down to $\sim$ 10 keV suggesting a lower electron cutoff energy than previously thought. This results in a factor of $(20/10)^{2.2} \approx 5$ greater energy in the electrons than if a 20 keV cutoff were assumed as has generally been used in the past. However, this increase in total energy in electrons is offset by the steep spectrum ($E^{-7}$) of the emission from high in the corona that would dominate the spatially integrated spectrum below 25 keV. It should be possible to estimate the total time-integrated energy in the electrons that produced this emission assuming a 10 keV cutoff energy.

(7) There was no impulsive X-ray emission with time constants of much less than 1 min. The main hard X-ray (25–200 keV) peaks have FWHM durations of $\sim$ 1 min. This suggests a more gradual energy release process, possibly reflecting weaker than usual magnetic fields and the increasing altitude of the energy release site.

(8) The TRACE images reveal a 'feature' propagating outward into the corona at a speed of $\gtrsim 120$ km s$^{-1}$ following the initial brightening. The time and place of origin of this feature is not clear, nor is its physical interpretation. This may be connected to the coronal mass ejection seen with LASCO following the flare with a very high velocity of 2500 km s$^{-1}$. Extrapolating the LASCO height-time profile to the solar surface reveals a start time of 01:13 UT, some 30 min after the start of the X-ray event. From the TRACE height-time profile, it is clear that height $\sim$ time$^{\alpha}$, where $\alpha > 1$. This therefore places the CME launch time closer to the flare peak (Gallagher and Dennis, 2002).

(9) Dark sinuous lanes that seem to be propagating downwards from the high corona appear in the TRACE images after the peak of the soft X-ray event. These features are similar to those observed previously in *Yohkoh*/SXT images, which McKenzie and Hudson (1999) attributed to downward propagating voids created by evacuated flux tubes. Švestka *et al.* (1998), on the other hand, speculate that this 'fan of spines' represents a set of mini-streamers or extensions similar to polar plumes. Our preliminary analysis indicate that theses features are most likely due to the presence of downward propagating voids or cool material. An interesting possibility is that these alternating dark and bright spines are generated by an MHD Kelvin–Helmholtz type shear instability similar to that described by Andries, Tirry, and Goossens (2000) and Andries and Goossens (2001) for plumes.

## 4. Future Work

This paper has provided an overview of the new RHESSI and TRACE observations of the X1.5 flare on 21 April 2002. Many conclusions are obvious from the preliminary analysis of these two data sets that we have been able to do so far. Thus, we feel it is useful to publish these early results. Clearly, further work can be done once more refined analysis procedures are perfected and the RHESSI instrument responses are more accurately and reliably understood. In particular, further analysis of the rise phase offers new insights into the electron acceleration and heating processes going on at this time. The association between these traditional flare phenomena and the apparently near simultaneity of the CME onset as seen in the TRACE movie is particularly intriguing.

Many advanced analysis tools are now becoming available for RHESSI. Recently, it has been possible to include collimators 1 and 2 in the image reconstruction process, thus allowing the possibility of obtaining images for this flare with up to 2″ resolution, at least a factor of 3 better than the images presented here. Also, knowledge of the instrument response function is being improved, particularly at energies below 10 keV, where most of the emission was recorded in the decay phase of this event. Modeling of sources similar to those observed here is ongoing to determine what features are real and what are artifacts of the image reconstruction algorithm. This should also allow for a better understanding of the backgrounds underlying the sources in the images and how best to obtain the spectrum of each source region free from contamination from other sources in the image. As knowledge of the grid modulation parameters improves, it should be possible to increase the image dynamic range towards the design goal of 100:1, thus making source contamination less of a problem. Other image reconstruction algorithms can also be tried such as maximum entropy, forward fitting (Aschwanden *et al.*, 2002), and pixons. These may well be better suited to providing more accurate photometry than the CLEAN algorithm used for the images shown in this paper. For TRACE, we will determine the sensitivity of the 195 Å bandpass and compare the emission measure from its flux to the temperature and emission measure of the near cospatial source seen with RHESSI.

## Acknowledgements

The authors would like to thank the referee, František Fárník, and RHESSI team-member, Hugh Hudson, for their comments and discussion during the preparation of this work.

## References

Andries, J. and Goossens, M.: 2001, *Astron. Astrophys.* **368**, 1083.
Andries, J., Tirry, W. J., and Goossens, M.: 2000, *Astrophys. J.* **531**, 561.
Aschwanden, M. J., Schmahl, E., and the RHESSI Team: 2002, *Solar Phys.*, this volume.
Czaykowska, A., De Pontieu, B., Alexander, D., and Rank, G.: 1999, *Astrophys. J.* **521**, L75.
Gallagher, P. T. and Dennis, B. R.: 2002, *Astrophys. J.*, in preparation.
Gallagher, P. T., Moon, Y.-J., and Wang, H.: 2002, *Solar Phys.* **209**, 169.
Handy, B. N. *et al.*: 1999, *Solar Phys.* **187**, 229.

Kopp, R. A. and Pneuman, G. W.: 1976, *Solar Phys.* **50**, 85.

Lin, R. P. *et al.*: 2002, *Solar Phys.*, this volume.

Masuda, S., Kosugi, T., and Hudson, H. S.: 2001, *Solar Phys.* **204**, 57.

McKenzie, D. E. and Hudson, H. S.: 1999, *Astrophys. J.* **519**, L93.

Qiu, J., Ding, M. D., Wang, H., Gallagher, P. T., Sato, J., Denker, C., Goode, P. R.: 2001, *Astrophys. J.* **554**, 445.

Švestka, Z., Stewart, R. T., Hoyng, P., van Tend, W., Acton, L. W., Gabriel, A. H., Rapley, C. G., Boelee, A., Bruner, E. C., de Jager, C., Lafleur, H., Nelson, G., Simnett, G. M., van Beek, H. F., and Wagner, W. J.: 1982, *Solar Phys.* **75**, 305.

Švestka, Z. F., Fontenla, J. M., Machado, M. E., Martin, S. F., Neidig, D. F., and Poletto, G.: 1987, *Solar Phys.* **108**, 237.

Švestka, Z. F., Fárník, F., Hudson, H. S., and Hick, P.: 1998, *Solar Phys.* **182**, 179.

Tanaka, K.: 1983, in P. B. Byrne and M. Rodono (eds.), 'Activity in Red-Dwarf Stars', *IAU Colloq.* **71**, 307.

Tanaka, K.: 1987, *Publ. Astron. Soc. Japan* **39**, 1.

Tsuneta, S.: 1997, *Astrophys. J.* **483**, 507.

Tsuneta, S., Takakura, T., Nitta, N., Ohki, K., Tanaka, K., Makishima, K., Murakami, T., Oda, M., and Ogawara, Y.: 1984, *Astrophys. J.* **280**, 887.

Warren, H. P. and Reeves, K. K.: 2001, *Astrophys. J.* **554**, L103.

Young, C. A., Dennis, B. R., Schwartz, R. A., and Tolbert, A. K.: 2002, *Solar Phys.*, in preparation.

# RHESSI OBSERVATION OF ATMOSPHERIC GAMMA RAYS FROM IMPACT OF SOLAR ENERGETIC PARTICLES ON 21 APRIL 2002

G. H. SHARE[1], R. J. MURPHY[1], B. R. DENNIS[2], R. A. SCHWARTZ[2],
A. K. TOLBERT[2], R. P. LIN[3] and D. M. SMITH[3]

[1]*E.O. Hulburt Center for Space Research, Naval Research Laboratory, Washington, D.C. 20375, U.S.A.*

[2]*NASA Goddard Space Flight Center, Greenbelt, MD 20771, U.S.A.*

[3]*Space Sciences Laboratory and Dep't. of Physics, University of California, Berkeley, CA 94720, U.S.A.*

(Received 5 August 2002; accepted 10 September 2002)

**Abstract.** The RHESSI high-resolution spectrometer detected $\gamma$-ray lines and continuum emitted by the Earth's atmosphere during impact of solar energetic particles in the south polar region from 16:00–17:00 UT on 21 April 2002. The particle intensity at the time of the observation was a factor of 10–100 weaker than previous events when gamma-rays were detected by other instruments. This is the first high-resolution observation of atmospheric gamma-ray lines produced by solar energetic particles. De-excitation lines were resolved that, in part, come from $^{14}$N at 728, 1635, 2313, 3890, and 5106 keV, and the $^{12}$C spallation product at $\sim$4439 keV. Other unresolved lines were also detected. We provide best-fit line energies and widths and compare these with moderate resolution measurements by SMM of lines from an SEP event and with high-resolution measurements made by HEAO 3 of lines excited by cosmic rays. We use line ratios to estimate the spectrum of solar energetic particles that impacted the atmosphere. The 21 April spectrum was significantly harder than that measured by SMM during the 20 October 1989 shock event; it is comparable to that measured by *Yohkoh* on 15 July 2000. This is consistent with measurements of 10–50 MeV protons made in space at the time of the $\gamma$-ray observations.

## 1. Introduction

A moderate-sized solar energetic particle (SEP) event began shortly after the X1.5 solar flare from AR 9906 (S14 W84) that commenced at 00:43 UT and peaked at 01:51 UT on 21 April 2002 in soft X-rays. The flare was well observed by RHESSI which detected radiation up to about 400 keV; there was no evidence for nuclear line emission. A CME with a launch time of $\sim$01:15 and a speed of $2200 \pm 300$ km s$^{-1}$ was observed by the LASCO instrument on SOHO (M. Andrews, private communication). Fast CME's such as this can produce shocks that accelerate the particles comprising the bulk of the SEP event. In Figure 1 we plot the integral fluxes of protons $> 10$ MeV observed by the GOES satellites from the 21 April 2002 particle event, in addition to those from the 19–20 October 1989 and 14–15 July 2000 events. The July particle event initially rose to an intensity ten times higher than that observed in October; however, the peak flux observed

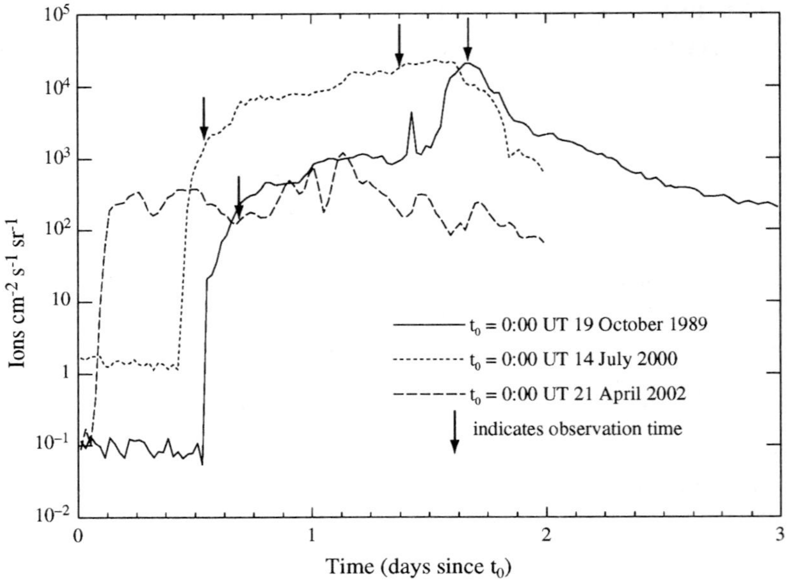


*Figure 1.* Integral fluxes of protons > 10 MeV observed by the GOES satellites during the 19 October 1989, 14 July 2000, and 21 April 2002 SEPs. The arrows depict times when atmospheric gamma rays produced by particle interactions were detected by SMM, *Yohkoh*, and RHESSI.

on 20 October from particles in the shock as it passed the Earth reached the level observed in the July event. The peak proton flux > 10 MeV from the 21 April 2002 event was at least a factor of 20 below the peak intensities observed from the earlier events.

Gamma-ray lines produced by solar energetic particle impact of the Earth's atmosphere were first detected by the moderate-resolution SMM spectrometer during the intense 20 October 1989 shock event (Share and Murphy, 2001). During this event the line intensities increased by over 2 orders of magnitude above the quiescent level. The quiescent line radiation is primarily from nuclei excited by neutrons produced in cosmic-ray collisions with atmospheric $^{14}$N and $^{16}$O (Ling, 1973). Details of the quiescent atmospheric positron-electron annihilation line at 511 keV were first provided by high-resolution measurements made by the HEAO 3 germanium spectrometer (Mahoney, Ling, and Jacobson, 1981). De-excitation line radiation from cosmic-ray impact was also observed by SMM (Letaw *et al.*, 1989) and HEAO 3 (Willett and Mahoney, 1992). Share and Murphy (2001) identified 19 resolved nuclear lines and the positron-electron annihilation line in the quiescent spectrum. Almost all of these lines appeared in the spectrum produced by the intense SEP event on 20 October. The nuclear lines were emitted from excited states of $^{14}$N, $^{16}$O, and various spallation products of high-energy interactions such as $^{11}$B and $^{12}$C. The *Yohkoh* gamma-ray spectrometer detected atmospheric radiation from the strong SEP event that reached earth beginning on 14 July 2000 following the

X5.7 'Bastille Day' flare (Share *et al.*, 2001). This event was the third most intense proton event observed $> 10$ MeV. The resolution of the *Yohkoh* spectrometer was good enough to resolve only the $^{14}N$ 2.31 MeV de-excitation line and the $^{11}B/^{12}C$ complex near 4.4 MeV.

The arrows in Figure 1 denote the times when gamma rays were measured by SMM on 20 October 1989, by *Yohkoh* on 14–15 July 2002, and by RHESSI on 21 April 2002. These observations all were made as the satellites reached high magnetic latitude and viewed the atmosphere in the polar region where the low-energy protons are not impeded by the magnetic field and impact the earth's atmosphere. On 21 April 2002 RHESSI observed the atmospheric glow of gamma-rays beginning at 16:05 UT as it approached its most southerly passage at a magnetic latitude of 48° S (magnetic cutoff $\sim 2$ GV). At that time the SEP intensity was about two orders of magnitude below the intensities at the times of the gamma-ray observations by SMM on 20 October 1989 and by *Yohkoh* on 15 July 2000. In spite of this low particle flux RHESSI has made significant new measurements of atmospheric gamma-ray lines, which we discuss below.

## 2. Instrument Response to Orbital Background and Atmospheric Lines

The scientific objectives and design of the RHESSI instrument were described by Lin *et al.* (2000) and Smith *et al.* (2000). More details of the full design and instrument performance can be found in the current volume. RHESSI was launched on 5 February 2002 and has performed nominally since then. The spectral resolution of 8 of its germanium spectrometers has not degraded significantly during its 5 months of operation, although there are some differences in detector performance, especially at low energies. In our current analysis we sum data from the rear segments of 8 detectors as we are interested in the gamma-ray spectrum above a few hundred keV; we exclude data from detector 2 which has degraded resolution (Smith *et al.*, 2002).

The radiation observed by RHESSI is made up of several components and is highly variable within each orbit and from orbit to orbit. Due to weight constraints there is no anti-coincidence shield for rejecting charged particles, thus cosmic-ray protons and electrons that deposit energies in the RHESSI spectral range will be detected. Several passages per day through the intense proton belts irradiate the detector and spacecraft producing hundreds of gamma-ray lines. Wheaton *et al.* (1989) provided a detailed listing of the lines observed in the HEAO 3 germanium spectrometer. An excellent summary of the various backgrounds encountered in space missions can be found in the compendium of articles on the subject contained in the AIP publication 'High-Energy Radiation Background in Space' in which the Wheaton *et al.* (1989) article appeared. The instrument is also activated by the cosmic-ray flux that varies with magnetic latitude. This same flux generates the quiescent atmospheric gamma rays discussed in the Introduction that are also de-

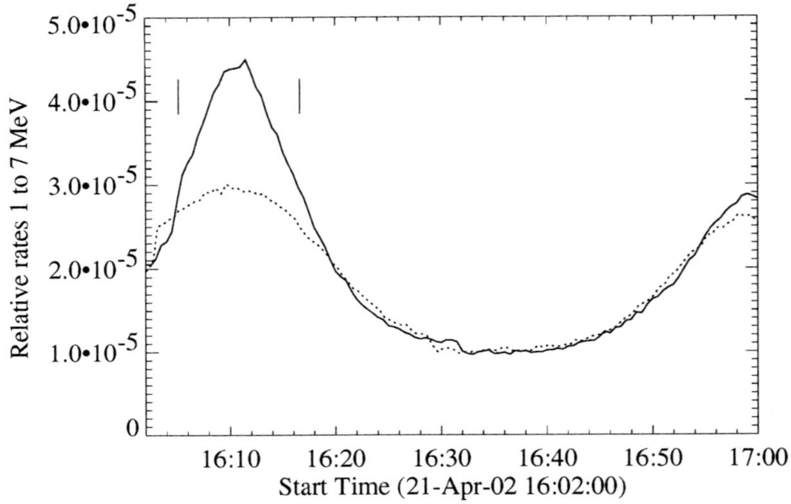


*Figure 2.* Rates observed in the 1 to 7 MeV band on 21 April 2002 (*solid line*) during the orbit when atmospheric gamma-ray lines were detected by RHESSI. Peak rates occurred when the satellite was at 48° S and 48° N magnetic latitudes. For comparison we plot the rates observed on 19 April shifted in time by + 640 s (*dotted line*). The spectrum of atmospheric gamma-rays was accumulated between the two lines.

tected. As RHESSI's orbit reaches magnetic latitudes near 50° it also can encounter precipitating electrons whose energies can reach MeV energies (Foat *et al.*, 1998).

In order to look for differences that may be related to an SEP enhancement in gamma-ray flux, we use the fact that the satellite returns to approximately the same physical location every 15 orbits. Typically we can use data taken on the previous and subsequent days shifted by an appropriate amount to estimate the background on a given day. However, there were insufficient data available at comparable times on 20 and 22 April. In Figure 2 we plot the relative rates observed in the 1 to 7 MeV band on 21 April 2002 (solid line), about 15 h after the X1.5 flare, at a time when the satellite reached high magnetic latitudes where the cosmic-ray intensity increases the rates significantly. (N.B. These and other rates presented in this paper are not absolute and have not been corrected for an energy-independent normalization error in the original data analysis software.) The rates are seen to peak near 16:10 and 17:00 UT when the satellite reached magnetic latitudes of $\sim$ 48° S and $\sim$ 48° N, respectively. For comparison, we also plot the rates (with a + 640 s shift in time) taken two days earlier on 19 April (dotted line) when the satellite reached magnetic latitudes $\sim$ 1° higher.

There is a clear excess flux in the 1 to 7 MeV band on 21 April near 16:10 UT relative to the flux on 19 April, when the satellite was in the southern hemisphere; there was only a small increase over the 19 April flux 50 min later when RHESSI was in the northern hemisphere. The excess flux observed on 21 April is not likely to be due to precipitating electrons because the geomagnetic field was relatively

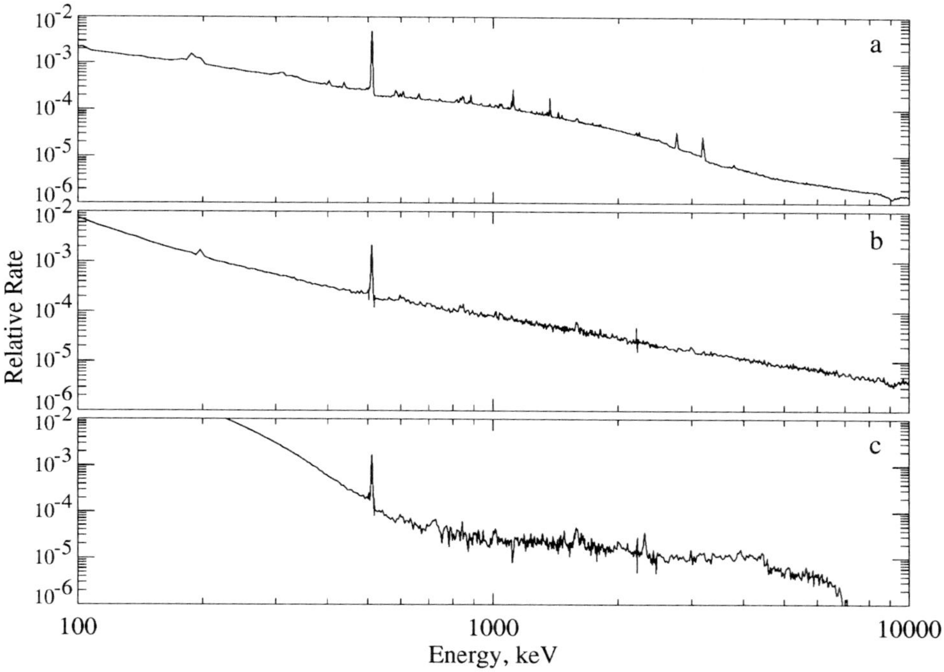


*Figure 3.* Comparison of three different RHESSI spectral accumulations. (a) Total background spectrum primarily from internal radioactivity, charged cosmic rays, and atmospheric radiation accumulated between 0–6 UT on 19 April 2002. (b) Spectrum primarily from charged cosmic rays, atmospheric radiation, and short-lived excited nuclei accumulated between 15:54:20 and 16:05:50 UT on 19 April after subtracting background at low magnetic latitude. (c) Gamma-ray spectrum from impact of SEP protons obtained by subtracting the data accumulated from 15:54:20 to 16:05:50 UT on 19 April from data accumulated from 16:05:00–16:16:30 UT on 21 April. (Note that a feature appearing in the spectra near 2.2 MeV is due to different energy binning around the neutron capture line and features near 1.6 MeV and 9 MeV are due to instrumental artifacts.)

quiet ($K_p = 1-$) at that time while it was significantly disturbed ($K_p = 6+$) on 19 April. We are interested in studying the spectrum accumulated near the peak between the two times marked by the lines in the figure, and comparing it with background spectra detected by RHESSI during a typical day. In Figure 3(a) we plot the spectrum from 200 keV to 10 MeV accumulated from 8 detectors in the first 6 hours of 19 April. The 511 keV annihilation line is the dominant feature and the power-law continuum extends up to at least 10 MeV. This continuum is primarily due to charged cosmic rays. Visible in the spectrum are over 120 narrow background lines that we show in more detail below. $\beta$-decay from radioactive species in the Ge detectors produces the curved continuum observed up to about 3 MeV.

In Figure 3(b) we show the typical spectrum of radiation accumulated at high magnetic latitudes, after subtracting the long-term radioactive background. This

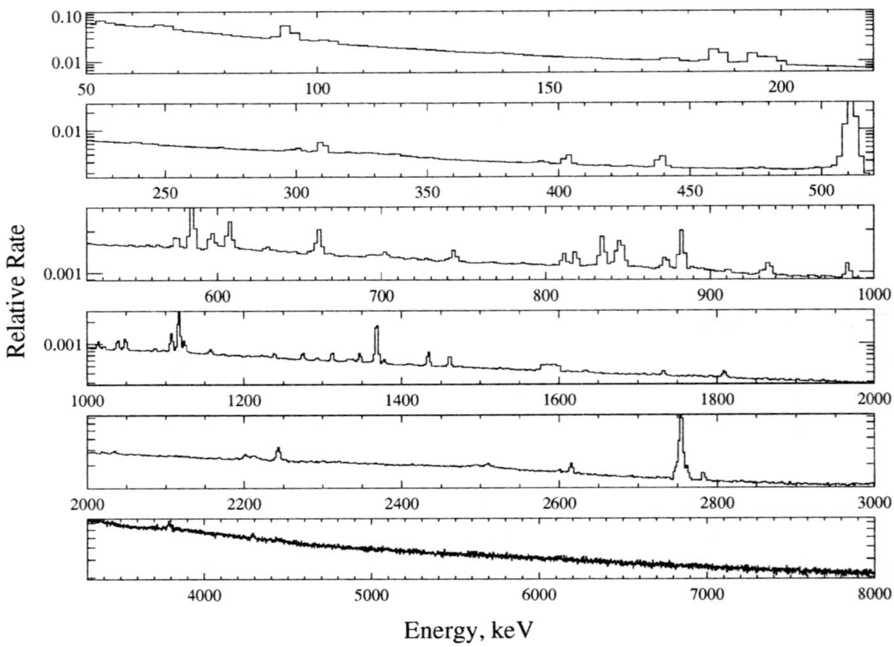


*Figure 4.* Detailed 24-hour background spectrum accumulated on 19 April with a resolution of 2.1 keV. (The broad feature below 1.6 MeV is an instrumental artifact.)

will be compared with the spectrum observed from the SEP event. The spectrum in Figure 3(b) was obtained by accumulating data taken at high latitude on 19 April, in the time interval between the two lines plotted in Figure 2, after subtracting background taken on both sides of the peak near the magnetic equator where the cosmic ray flux is lower. The annihilation line is once again the dominant spectral feature and sits above a power-law continuum from charged cosmic rays that extends to energies above 10 MeV. Most of the long-lived radioactive lines have been subtracted, but there are some residual background lines that we will discuss below.

The spectrum plotted in Figure 3(c) was obtained by subtracting the total 19 April spectrum between the times indicated by the lines in Figure 2 from the spectrum accumulated during the same relative interval on 21 April. This spectrum is markedly different from the others and is similar to the spectra observed by SMM and *Yohkoh* from the SEP events on 20 October 1989 and 15 July 2000. There is also an intense contribution at low energies that is likely to be due to precipitating electrons (the auxiliary particle detector observed an increase at that time).

In order to study the features of the three spectra plotted in Figure 3 in greater detail, we accumulated the data at higher spectral resolution. We plot a 24-hr background spectrum accumulated on 19 April 2002 at 2.1 keV resolution in Figure 4. There are over 120 gamma-ray lines in this spectrum, most of which have been

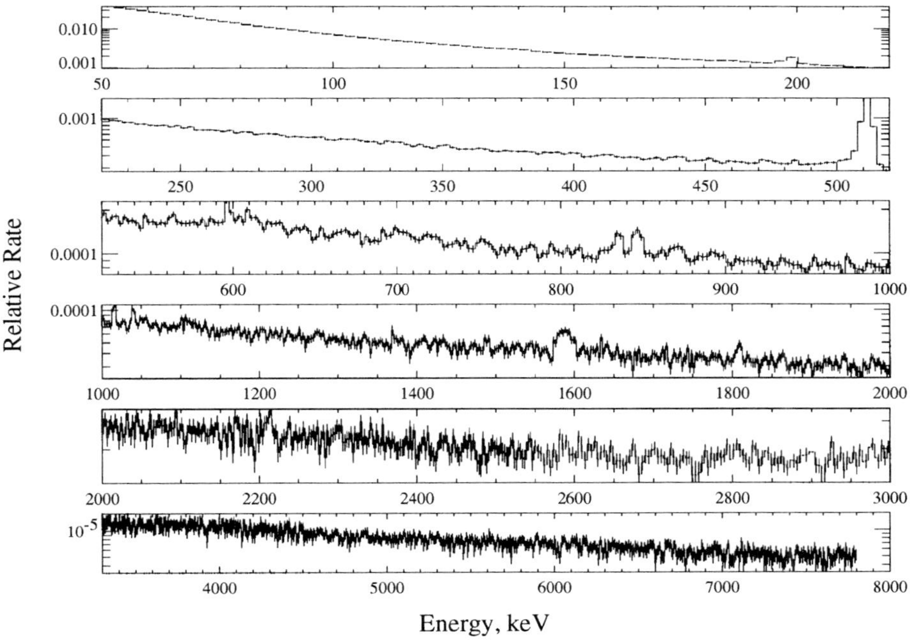


*Figure 5.* High-resolution spectrum of the background-subtracted data accumulated between 15:54:20 and 16:05:50 UT on 19 April 2002. Most of the radioactive lines have been removed by subtracting the spectrum accumulated at low magnetic latitudes. (2.5 keV bins to 2.5 MeV, 4 keV bins to 4.5 MeV, and 6 keV bins at higher energy; the broad feature below 1.6 MeV is an instrumental artifact.)

detected earlier using the study of the HEAO 3 background (Wheaton *et al.*, 1989) and an unpublished tabulation of TGRS/WIND data (Wiedenspointner, Ferguson, and Harris, private communication).

In Figure 5 we plot a high-resolution version of the background-corrected spectrum shown in Figure 3(b). The spectrum is dominated by a hard continuum due to the higher flux of charged cosmic rays $> 45°$S that deposit energy directly in the detectors and also produce bremsstrahlung from the Earth's atmosphere. Most of the long-lived lines shown in Figure 4 have been removed by the background subtraction. The strong annihilation line comes from radioactive nuclei and electromagnetic showers in the instrument, satellite, and Earth's atmosphere. There are also some residual lines evident from isotopes with short half lives that are not subtracted using data where the cosmic-ray background is lower. We have fitted these lines to identify their origins. The line at 198.3 keV is due to $^{71m}$Ge that has a half-life of 20 ms and could be produced in an $n,\gamma$ reaction on $^{70}$Ge. There is a blend of lines near 600 keV that come from several excited states of $^{74}$Ge, including lines at 595.9 and 608.4 keV from the first and second excited states, respectively. A strong line at 835 keV is likely to come from the first excited state of $^{54}$Cr. Another strong line feature near 845 keV comes from a blend of the first excited

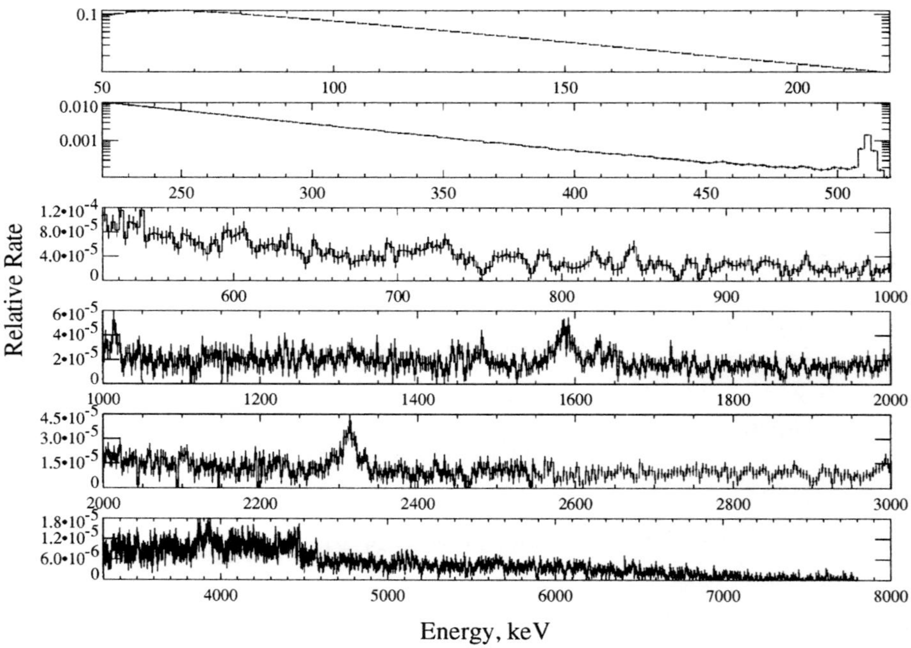


*Figure 6.* High-resolution spectrum of SEP-produced atmospheric line and continuum emission plotted in Figure 3(c). The spectrum was obtained by subtracting the total data accumulated from 15:54:20 to 16:05:50 UT on 19 April from data accumulated from 16:05:00–16:16:30 UT on 21 April. (The broad feature below 1.6 MeV is a data artifact.)

states of $^{27}$Al and $^{56}$Fe. The line near 1014 keV is from the second excited state of $^{27}$Al, while an equally strong line near 1039 keV is likely from the first excited state of $^{70}$Ge. The line near 1809 keV is consistent with the first excited state of $^{26}$Mg, produced by an *n,pn* reaction on $^{27}$Al. A broadened line feature appears centered near 2211 keV that is in part due to emission from the third excited state of $^{27}$Al.

In Figure 6 we plot at high resolution the SEP-produced atmospheric spectrum on 21 April between 16:05:00 and 16:16:30 UT after subtracting background accumulated on 19 April. We first search for the same background features that were observed in the spectrum at high latitudes shown in Figure 5. There is no evidence for the line at 198.3 keV. Weak line features appear near 597 and 608 keV that are consistent with high-latitude background lines. A narrow line feature is evident near 843 keV that could be due to $^{27}$Al (843.7 keV); however, there is no evidence for the $^{56}$Fe line (846.8 keV). A second line from $^{27}$Al may be present in the spectrum near 1015 keV but there is no evidence for the higher energy line at 2211 keV. There is also no evidence for the $^{26}$Mg line at 1809 keV. The presence of any of these instrumental lines would be puzzling as we believe that the SEP protons can not penetrate the magnetosphere at a rigidities near 2 GV, where the satellite was positioned. The geomagnetic field was also much quieter on 21 April

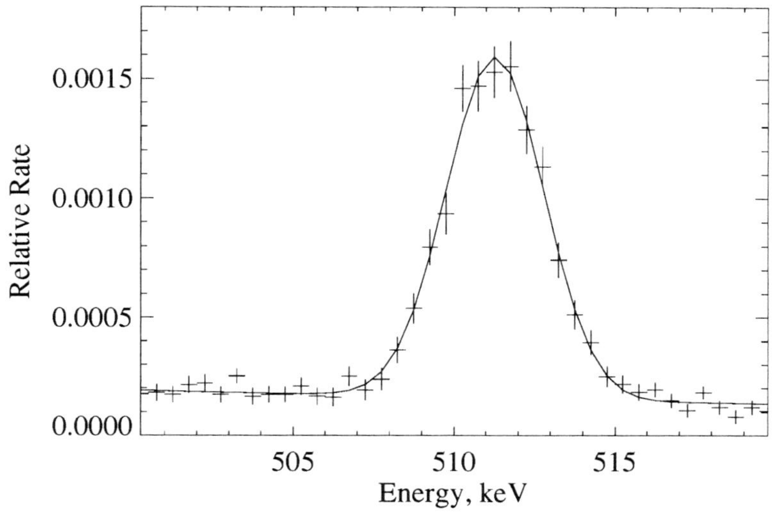


*Figure 7.* Fit to the 511 keV line observed during the SEP event.

than it was on 19 April when the background was accumulated. We also note that the possible line features discussed above are not significantly stronger than other fluctuations that appear in the spectrum between about 520 and 1000 keV.

## 3. Measurements of SEP-Produced Atmospheric Gamma-Ray Lines

In this section we specifically discuss line features associated with gamma rays produced by the SEP protons that impacted the Earth's atmosphere in the polar region and were observed remotely by RHESSI. We specifically will concentrate on the strong lines observed in the SMM measurement of the 20 October 1989 event (Share and Murphy, 2001). The annihilation line is the dominant feature in the spectrum. In Figure 7 we plot a fit to the RHESSI annihilation line. We have performed a $\chi^2$ map to obtain the best fitting values for the energy and width; they are $511.3 \pm 0.1$ keV and $2.44 \pm 0.14$ keV (FWHM).

A weak broad feature appears above 700 keV in the RHESSI spectrum that is near the position of a line observed in the 20 October 1989 SEP event by SMM and attributed to $^{10}$B (718 keV) and $^{14}$N (728 keV). We have fit this feature with both a single line and with two lines having fixed energies at 718 and 728 keV. We plot the data and the latter fit in Figure 8. We have performed a $\chi^2$ map to obtain the best fitting values for the energies and widths. For a single line, the energy is $723.0 \pm 2.0$ keV and width is $21.3 \pm 5.2$ keV (FWHM). The fit with two lines is slightly improved. The 718 keV line is about two times stronger than the 728 keV line. Its width (FWHM) is $24 \pm 8$ keV while the 728 keV line is probably less than about 15 keV wide.

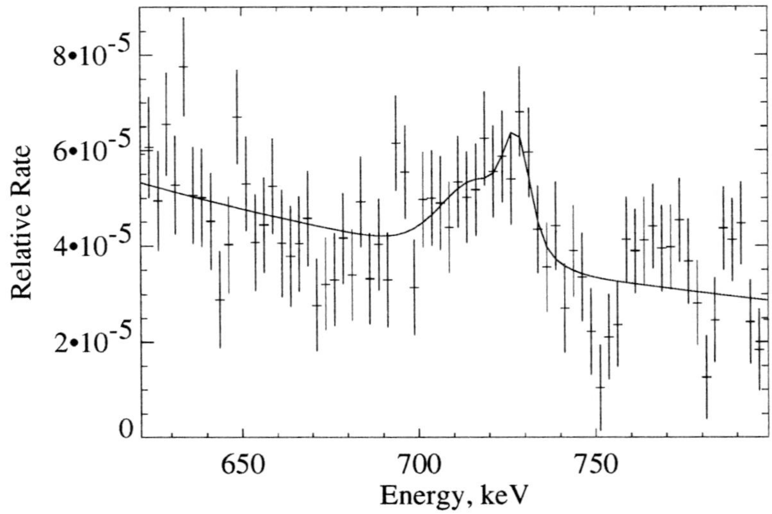


*Figure 8.* Fit to the ~ 720 keV line feature from the Earth's atmosphere produced by the SEP.

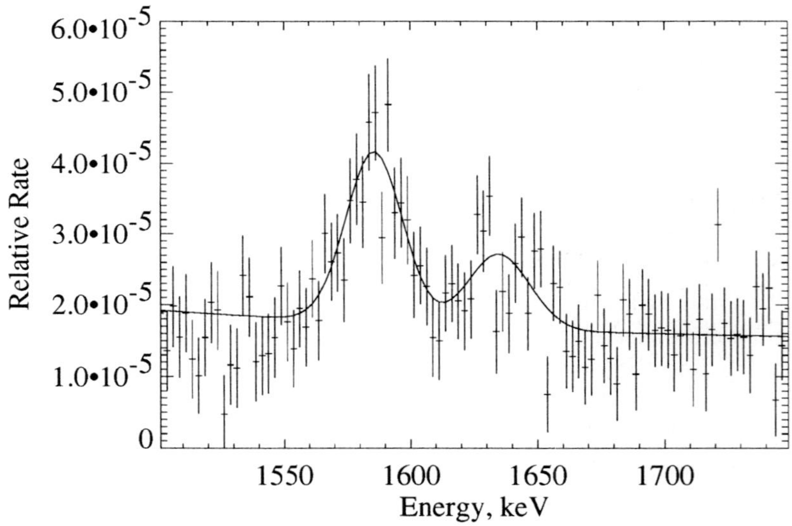


*Figure 9.* Fit to the 1630 keV $^{14}$N line from the Earth's atmosphere produced by the SEP. In order to do this we also simultaneously fit the instrumental artifact with a Gaussian shaped line.

One of the strong lines in the SMM spectrum appears at 1630 keV and is associated with de-excitation of the 3948 keV level in $^{14}$N to the first excited state. For this reason we have studied this region in more detail in the RHESSI spectrum. We plot the spectrum in this region in Figure 9. The dominant feature is an instrumental artifact that appears between 1550 and 1600 keV. There is evidence for a broadened line just above 1600 keV. We fit both the artifact and the atmospheric

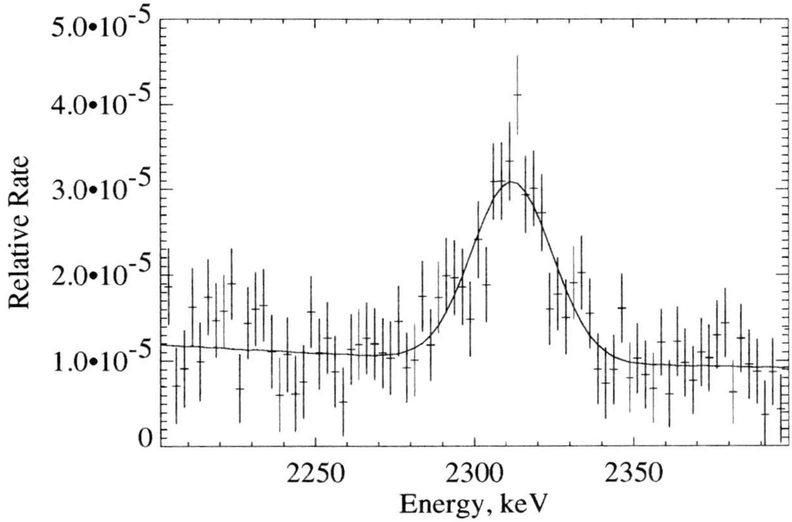


*Figure 10.* Fit to the 2313 keV $^{14}N$ line from the Earth's atmosphere produced by the SEP.

line with Gaussians and performed a $\chi^2$ analysis in order to obtain the energy and width of the $^{14}N$ feature. The best fitting energy and width of the $^{14}N$ line are $1635.0 \pm 3.0$ keV and $30.0 \pm 6.9$ keV (FWHM), respectively.

The second most striking line in the atmospheric spectrum observed during the SEP event appears near 2310 keV. This is the location of the line that arises in the transition from the first excited state of $^{14}N$ to its ground state. In Figure 10 we plot the RHESSI spectrum and best fit to the data. Our $\chi^2$ analysis produced a measured energy of $2312.5 \pm 1.2$ keV and width of $25.8 \pm 3.9$ keV (FWHM).

The energy range between 3500 and 7000 keV is quite complex. According to the SMM measurements, there may be at least 10 identifiable lines in this energy range. We plot the RHESSI spectrum in Figure 11. The features in the spectrum are generally complex and not easy to resolve. This is especially true at the highest energies where the efficiency of the instrument is rather low. There are three rather well-defined features in the spectrum, near 3900, 4420, and 5115 keV. These features have been identified in the SMM spectrum as primarily being due to transition from the 6204 to 2313 keV level (3890 keV) in $^{14}N$, from $^{11}B$ (4444 keV) and $^{12}C$ (4439 keV) spallation products, and from transition from the 5106 keV level to the ground state in $^{14}N$. In fitting the spectrum we included lines at 3900, 4180, 4420, 5115, 5830, 6133, 6350, and 7000 keV and used the full instrument response for an incident angle of 60°, that includes escape peaks and Compton continuum. We fixed the line energies and widths of some of the weaker features at the values found in the SMM observation. We show the best fit to the spectrum in the figure. We then performed a $\chi^2$ analysis of lines near 3900, 4420, and 5115 keV to obtain measurements of their energies and widths. The derived energies of the lines are

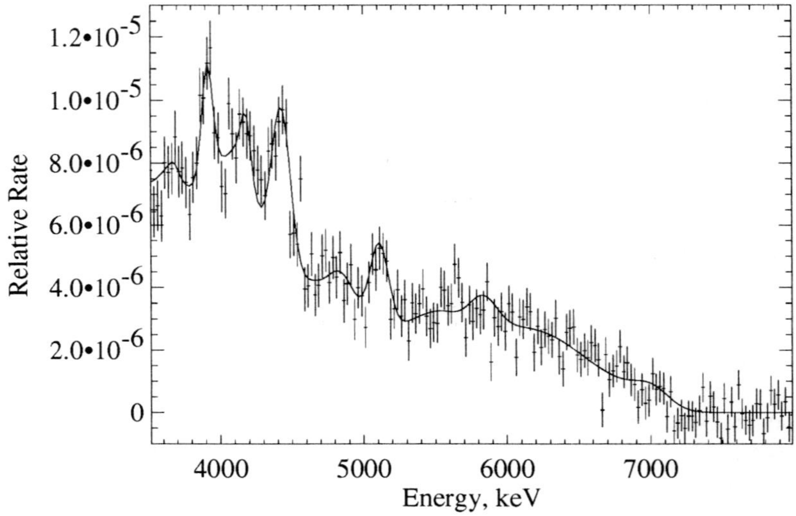


*Figure 11.* Fit to the RHESSI spectrum between 3500 and 8000 keV, including Gaussian lines near 3900, 4180, 4420, 5115, 5830, 6133, 6350, and 7000 keV.

3911.5 ± 8.5, 4422.0 ± 8.0 keV, and 5116.0 ± 12.0 keV, and their widths are 80.5 ± 13.8, 200 ± 16 keV and 157.6 ± 26.5 keV, respectively.

## 4. Summary and Discussion

A moderate-sized solar energetic particle event followed an X1.5 solar flare on 21 April 2002. RHESSI observed hard X-rays up to about 400 keV from the flare. About 16 hours after the flare RHESSI also detected gamma-ray lines emanating from the polar region in the southern hemisphere produced by the solar energetic particles impacting the Earth's atmosphere. In Table I we compare the integral fluxes of protons $> 10$ MeV and flux ratios at the times of the detection of atmospheric gamma-rays for this and two other events. The SMM spectrometer (Share and Murphy, 2001) observed the October 1989 event and the *Yohkoh* GRS detector the event in July 2000 (Share *et al.*, 2001). At the time of the RHESSI observation, the proton flux was one to two orders of magnitude weaker than fluxes at the times of the previously detected atmospheric gamma-ray events.

In order to validate the RHESSI atmospheric-line detection during this relatively weak SEP event, we accumulated a 24-hr background spectrum two days before the flare. There are over 120 narrow lines and a $\beta$-decay continuum visible in the spectrum over a power-law continuum from charged cosmic rays that extends to at least 10 MeV (Figures 3(a) and 4). Most of the narrow lines have been observed previously by the HEAO 3 germanium spectrometer (Wheaton *et al.*, 1989) and are produced by SAA and cosmic-ray irradiation of the instrument and spacecraft.

TABLE I
GOES proton measurements.

| Date/time | Flux > 10 MeV $p$ (cm$^2$ s sr)$^{-1}$ | Flux ratios >10/>30 | >30/>50 | >50/>100 |
|---|---|---|---|---|
| 20 Oct. 1989/16 UT | $4 \times 10^4$ | 6 | 3 | 5.5 |
| 14 July 2000/13 UT | $2 \times 10^3$ | 1.5 | 1.5 | 3 |
| 15 July 2000/09 UT | $2 \times 10^4$ | 4 | 3.5 | 14 |
| 21 Apr. 2002/16 UT | $1.4 \times 10^2$ | 1.6 | 1.9 | 4.9 |

The $\beta$-decay continuum is produced by irradiation of the germanium detectors. Long-lived background lines can be removed from the high-latitude spectrum by subtracting data taken near the equator. Residual narrow lines appear in the spectrum at high latitudes from proton irradiation of the detector and surrounding material, such as Fe and Al, that produce lines with half-lives shorter than a few minutes (Figures 3(b) and 5). The positron-electron annihilation line is the most intense line observed in both the total background and high-latitude spectra. The power law continuum is due to cosmic rays and atmospheric bremsstrahlung.

We used data accumulated on 19 April, at roughly the same geographic location as the time when the SEP-produced atmospheric gamma rays were observed on 21 April, in order to remove most of the background-line and continuum features. There is evidence for some weak residual background features from excited states of nuclei such as $^{27}$Al. As there are also several sources of short-lived $\beta^+$-unstable nuclei in the spacecraft, the intense SEP-produced atmospheric annihilation line may also be contaminated by this background. RHESSI observed several atmospheric lines from excitation of $^{14}$N and $^{16}$O and their spallation products. Resolved line features were detected near 720, 1630, 2310, 3900, 4420, and 5115 keV. There is also evidence for many of the other lines detected by the SMM spectrometer during the intense 20 October 1989 event (100 times more intense than the 21 April 2002 event).

In Table II we compare the line energies and widths of some of the most intense lines in the atmospheric spectrum in different instruments. The lines measured by SMM and RHESSI were of the SEP irradiated atmosphere while the lines measured by HEAO 3 (Mahoney, Ling, and Jacobsen, 1981; Willett and Mahoney, 1992) were made during quiescent times and were produced by cosmic-ray irradiation. (The *Yohkoh* spectral resolution was not adequate for these line studies.) The RHESSI and HEAO 3 annihilation line energies and widths are consistent with one another in spite of the fact that the RHESSI observation may be partly contaminated by 511 keV photons from the instrument and spacecraft. Mahoney (1981) showed that the 511 keV instrumental background line was a few tenths of

TABLE II

Comparison of RHESSI, SMM, and HEAO 3 line observations.

| Source | Energy, keV | | | Width, keV (FWHM) | | |
|---|---|---|---|---|---|---|
| | SMM | RHESSI | HEAO 3 | SMM | RHESSI | HEAO 3 |
| $e^+e^-$ | 512.1 ± 3.0 | 511.3 ± 0.1 | 511.07 ± 0.1 | – | 2.44 ± 0.14 | 2.29 ± 0.3 |
| $^{10}B$, $^{14}N$ | 719.7 ± 3.0 | 723.0 ± 2.0 | – | 30.5 ± 8.0 | 21.3 ± 5.2 | – |
| $^{10}B$ | – | 718.0 | – | – | 24.0 ± 8.0 | – |
| $^{14}N$ | – | 728.0 | – | – | ≤15 | – |
| $^{14}N$ | 1629.9 ± 3.0 | 1635.0 ± 3.0 | 1634.8 ± 1.4 | 32.0 ± 7.0 | 30.0 ± 6.9 | 20.2 ± 5.7 |
| $^{14}N$ | 2298.0 ± 15.1 | 2312.5 ± 1.2 | 2309.4 ± 1.9 | 43.0 ± 10.0 | 25.8 ± 3.9 | 24.0 ± 5.4 |
| $^{14}N$ | 3890.9 ± 20.2 | 3911.5 ± 8.5 | – | <270 | 80.5 ± 13.8 | – |
| $^{11}B$, $^{12}C$ | 4435.0 ± 5.0 | 4422.0 ± 8.0 | 4428.5 ± 7.5 | 159.8 ± 14.0 | 200.0 ± 16.0 | 135.0 ± 12.0 |
| $^{14}N$ | 5115.0 ± 10.1 | 5116.0 ± 12.0 | – | 152.8 ± 23.5 | 157.6 ± 26.5 | – |

TABLE III
Relative gamma-ray line fluxes from SEP impact on Earth's atmosphere.

| Date/time | 1.63 MeV | Relative line flux | | |
|---|---|---|---|---|
| | | 2.31 MeV | 4.44 MeV | 0.511 MeV |
| 20 October 1989/16 UT | $0.41 \pm 0.04$ | $0.79 \pm 0.09$ | 1.00 | $1.29 \pm 0.10$ |
| 14 July 2000/13 UT | – | <0.3 | 1.00 | $1.20 \pm 0.42$ |
| 15 July 2000/09 UT | $0.23 \pm 0.03$ | $0.48 \pm 0.03$ | 1.00 | $1.1 \pm 0.1$ |
| 21 April 2002/16 UT | $0.19 \pm 0.05$ | $0.44 \pm 0.06$ | 1.00 | $1.8 \pm 0.1$ |

a keV broader than the atmospheric line. Within statistics, the RHESSI measured instrumental and SEP annihilation lines are consistent with each other.

The RHESSI observations appear to confirm the suggestion that two lines contribute to a feature observed near 720 keV by SMM (Share and Murphy, 2001). A single line fit to the RHESSI spectrum has a center energy ($723 \pm 2.0$ keV) that lies between the 718 keV $^{10}$B line produced in the $p,p\alpha$ reaction on $^{14}$N and the 728 keV $^{14}$N de-excitation line. With two lines fixed at 718 and 728 keV, we get a slightly improved fit, and find that the $^{10}$B line is likely to be broader than the $^{14}$N line (see Table II). The $^{10}$B spallation line also appears to be about twice as intense as the $^{14}$N de-excitation line, suggesting that the particle spectrum may be rather hard.

There is excellent agreement between the RHESSI and HEAO 3 line energies and widths of the strong 2313 keV line emitted in transition from the first excited state to the ground state in $^{14}$N. This is true even though the line measured by HEAO 3 was produced by secondary neutrons from cosmic-ray interactions. The 1635 keV line emitted in the transition from the 3948 keV level to the first-excited state in $^{14}$N, was not as well detected by RHESSI, as it is about half as strong as the 2313 keV line and lies near an instrumental artifact. This may account for the width and uncertainty being larger than that measured by HEAO 3. The energy and width of the complex at 4440 keV from the $^{11}$B and $^{12}$C spallation products was not as well measured by RHESSI as it was by SMM because of the weakness of the 21 April 2002 SEP event and RHESSI's lower efficiency at high energy. The measured energy of the line is about $2\sigma$ below the rest energies of $^{11}$B and $^{12}$C (4444 and 4439 keV). If the measured line was due to $^{11}$B then we would expect to see a comparably strong line at 2124 keV from its first excited state. We searched for this line and set a $2\sigma$ limit of 8% of the 4440 keV line. This demonstrates that the measured 4440 keV line is dominated by emission from the first excited state of $^{12}$C. The RHESSI measured energy for the $^{14}$N line at 3890 keV is about $2\sigma$ high while the measured energy for 5105 keV line is within the statistical uncertainties.

It is possible to estimate the spectrum of SEP particles that interact in the Earth's atmosphere by comparing the relative flux in the spallation line at 4440 keV with those in the $^{14}$N de-excitation lines at 1635 and 2313 keV (Share and Murphy, 2001). We list relative gamma-ray line fluxes (for an assumed incident angle of 60°) normalized to the 4440 keV line for the SEP events in Table III. The $^{14}$N lines measured by RHESSI appear to be weaker relative to the 4440 keV spallation lines than those in the 20 October 1989 and roughly comparable to the 15 July 2002 measurements. Using these ratios, we have used the revised solar gamma-ray production code (Kozlovsky, Murphy, and Ramaty, 2002) to estimate the spectrum of SEP between about that interact with the atmosphere. For an assumed power-law, the SEP spectrum on 20 October 1989 has a spectral index that is steeper by $\sim 0.5$ than those obtained on 15 July 2000 and 21 April 2002. The softer spectrum in the October event is consistent with proton flux ratios between 10 and 50 MeV listed in Table I. The relatively high 511 keV annihilation flux observed by RHESSI may reflect this harder spectrum; however, there is also evidence that there may be a contribution from an instrumental line due to an imprecise background subtraction (The instrument response includes production of 511 keV photons produced by high-energy photons impacting the satellite).

## Acknowledgements

This work was supported by NASA DPR's W-10049, W-19746, and W-19916 and by the Office of Naval Research. We wish to thank the entire RHESSI team for the success of this mission and for the tools to analyze these excellent data. We also thank Benz Kozlovsky for discussions concerning nuclear cross sections.

## References

Foat, J. E. *et al.*: 1998, *Geophys. Res. Lett.* **25**, 4109.

Kozlovsky, B., Murphy, R. J., and Ramaty, R.: 2002, *Astrophys. J. Suppl.* **141**, 523.

Lampton, M., Margon, B., and Bowyer, S.: 1976, *Astrophys. J.* **208**, 177.

Letaw, J. R., Share, G. H., Kinzer, R. L., Silberberg, R., Chupp, E. L., Forrest, D. J., and Rieger, E: 1989, *J. Geophys. Res.* **94**, 1211.

Lin, R. P. *et al.*: 2000, in R. Ramaty and N. Mandzhavidze (eds.), *High Energy Solar Physics – Anticipating HESSI*, American Soc. of the Pacific Conf. Series 206, San Francisco, p. 1.

Ling, J. C.: 1973, *J. Geophys. Res.* **80**, 3241.

Mahoney, W. A., Ling, J. C., and Jacobson, A. S.: 1981, *J. Geophys. Res.* **86**, 11 098.

Share, G. H. and Murphy, R. J.: 2001, *J. Geophys. Res.* **106**, 77.

Share, G. H., Murphy, R. J., Tylka, A. J., and Schwartz, R. A.: 2001, *Solar Phys.* **204**, 43.

Smith, D. M. *et al.*: 2000, in R. Ramaty and N. Mandzhavidze (eds.), *High Energy Solar Physics–Anticipating HESSI*, American Soc. of the Pacific Conf. Series 206, San Francisco, p. 92.

Smith, D. M. *et al.*: 2002, *Solar Phys.*, this volume.

Wheaton, W. A., Ling, J. C., Mahoney, W. A., Varnell, L. S., and Jacobson, A. S.: 1989, in A. C. Rester and J. I. Trombka (eds.), *High-Energy Radiation Background in Space*, AIP Conf. Proc. 186, AIP, New York, p. 304.

Willett, J. B. and Mahoney, W. A.: 1992, *J. Geophys. Res.* **97**, 131.

# CHROMOSPHERIC HEIGHT AND DENSITY MEASUREMENTS IN A SOLAR FLARE OBSERVED WITH RHESSI

## I. *Theory*

JOHN C. BROWN[1], MARKUS J. ASCHWANDEN[2] and EDUARD P. KONTAR[1]

[1]*Astronomy and Astrophysics Group, Dept. of Physics and Astronomy, University of Glasgow, Glasgow G12 8QQ, U.K. (e-mail: john@astro.gla.ac.uk)*

[2]*Lockheed Martin Advanced Technology Center, Solar & Astrophysics Laboratory, Dept. L9-41, Bldg.252, 3251 Hanover St., Palo Alto, CA 94304, U.S.A. (e-mail: aschwanden@lmsal.com)*

(Received 19 July 2002; accepted 15 August 2002)

**Abstract.** We obtain a theoretical description of the height ($z$) distribution of flare hard X-rays in the collisional thick-target model as a function of photon energy $\varepsilon$. This depends on the target atmosphere density structure $n(z)$ and on the beam spectral index $\delta$. We show that by representing the data in terms of the 1-D function $z_{\max}(\varepsilon)$ defining where the emission peaks as a function of $\varepsilon$ it is possible to derive $n(z)$ from data on $z_{\max}(\varepsilon)$. This is done first on the basis of a simple stopping depth argument then refined to allow for the dependence on spectral index $\delta$. The latter is worked out in detail for the case of a parameterization $n(z) = n_0\ (z/z_0)^{-b}$ which yields numerical results for $z_{\max}(\varepsilon)$ well fit by $z_{\max}(\varepsilon) \sim \varepsilon^{-\alpha}$, with $\alpha$ dependent on $\delta$, which is also found to fit well to actual observations. This enables derivation of flare loop $n(z)$ in terms of $n_0$, $b$ from RHESSI data in an entirely novel way, independent of other density diagnostic methods, and also of how $n(z)$ varies with time in flares such as by evaporation, as detailed in companion Paper II.

## 1. Introduction

The spatial distribution of energy loss in solar flares can provide us with vital information on the origin and propagation of accelerated nonthermal particles, which carry a significant amount of the energy released in flares. Today we assume that in most flares the particle acceleration site is located in coronal magnetic reconnection sites, where the plasma is nearly collisionless. The accelerated particles propagate along the newly-reconnected field lines, either downward toward the chromosphere, or upward along large-scale loops or open field lines. The downward propagating electrons reach the chromosphere either directly, if they have sufficiently small pitch angles, or after some intermediate trapping with subsequent precipitation, in the case of large pitch angles. The overall scenario seems to hold for most flare observations (for a review see Aschwanden, 2002). A major new aspect of the present study is use of the high-energy resolution of RHESSI to measure the energy loss as function of height with high precision, which can be used for reconstruction of the chromospheric density profile, if we assume collisional losses only. Since the observed nonthermal hard X-ray emission is a direct result of energetic beam electrons, we can probe the density locally in regions where fast electrons interact with a dense target. Most other chromospheric density modeling is based on line-of-sight integrated EUV line emission that contain contributions from a very long path through an inhomogeneous solar atmosphere, and thus are subject to unknown filling factors. Moreover, X-rays are least affected by propagation effects, unlike radio emission that is generated by the same electrons and can be strongly affected by the plasma where electromagnetic waves propagate.

In this paper we derive the relevant theory in terms of the thick-target model, while the data analysis for the case of the 20 February 2002 flare is presented in a companion paper (Aschwanden, Brown, and Kontar, 2002, Paper II).

## 2. Theory

The aim of this study is to show that one can infer a source plasma density structure $n(z)$ from hard X-ray source height data obtained by RHESSI, assuming a purely collisional thick-target model (Brown, 1971), and to see whether this $n(z)$ is consistent with the results from other data. The underlying theory has been previously developed by calculations of the thick-target model height structure (Brown and McClymont, 1976; Emslie and Vlahos, 1980) based on treatments of the collisional transport (Brown, 1972, 1973; Emslie, 1978; Brown *et al.*, 1983) but we re-derive it here to suit our specific application.

### 2.1. Simplifying assumptions

In previous work, the following simplifying assumptions have been made as a first approximation of the problem: (1) full target ionization , (2) one-dimensional Coulomb collisional transport, neglecting pitch-angle changes (pitch angle $\alpha = 0$ or $\mu = \cos\alpha = 1$), (3) no mirroring of particles, and (4) power-law function for electron injection flux energy spectrum. As in Brown (1971), $F_0(E_0)$ is the flux spectrum of injected electrons, while $F(E, N)$ denotes the instantaneous electron flux spectrum at a column depth $N(z)$, and $n(z)$ denotes the local number density of the ambient plasma. The height $z$ refers to the altitude above some reference level, such as the photosphere. The acceleration site is assumed to be localized, somewhere in the corona where collisional energy loss is negligible (in the framework of the thick-target model). To avoid complicating geometric projection effects, we concentrate here only on limb flares, where flare loops are viewed side-on, without (appreciable) projection effects. The collisional thick-target region of interest is located close to the flare loop footpoints, so that this loop segment can be approximated with a vertical flux tube.

### 2.2. Collisional Transport

Collisional transport gives, in the limit of zero pitch angle ($\mu = 1$) and neglecting pitch angle scattering, the following dependence of the kinetic electron energy $E$ as function of the column depth $N(z) = \int_z^{z_{\max}} n(z')\, dz'$ as in Brown (1972):

$$E(E_0, N) = (E_0^2 - 2KN)^{1/2}, \tag{1}$$

which defines the following stopping depth $N_s$ for energy $E_0$,

$$N_s(E_0) = \frac{E_0^2}{2K}, \tag{2}$$

where $K = 2\pi e^4 \Lambda$ is a constant with $\Lambda$ the Coulomb logarithm (see Spitzer, 1962).

The continuity condition then yields (using Equation (1))

$$F(E, N) = F_0(E_0)\frac{dE_0}{dE} = F_0([E^2 + 2KN]^{1/2})\frac{E}{(E^2 + 2KN)^{1/2}} \tag{3}$$

for the local electron spectrum $F(E, N)$ as a function of $N$. Note that we neglect here the fact that $\Lambda$ may vary somewhat in lower regions where the ionization falls (e.g., Emslie, 1978).

### 2.3. Simple Treatment

For a steep electron spectrum, bremsstrahlung photons with energy $\varepsilon$ are mainly produced by electrons of energy $E \approx \varepsilon$. Such electrons typically stop at a depth with column density $N_s$ given by Equation (2),

$$N_s(\varepsilon) \approx \frac{\varepsilon^2}{2K}. \tag{4}$$

So, ignoring the spectral dependence for now, we roughly expect to see photons of energy $\varepsilon$ from height $z$ around

$$z(\varepsilon) = z\left(N = N_s(\varepsilon) = \frac{\varepsilon^2}{2K}\right) \tag{5}$$

(but see Section 2.4 below for more accurate analysis). If the centroid of flare loop footpoint sources at energies $\varepsilon$ is observed with RHESSI to occur at height $z(\varepsilon)$, then we can estimate the source height dependence of the column depth $N(z)$ by equating Equation (5) to the data values, i.e.,

$$N(z) = \frac{\varepsilon^2(z)}{2K}. \tag{6}$$

This $N(z)$ defines the atmospheric structure required for the data to be consistent with the collisional thick-target model. To obtain the chromospheric density $n(z)$ we take the derivative d/d$z$ of Equation (6),

$$n(z) = -\frac{\mathrm{d}N(z)}{\mathrm{d}z} = -\frac{\varepsilon}{K}\frac{\mathrm{d}\varepsilon}{\mathrm{d}z}\,. \tag{7}$$

For real measurements at discrete energy values $\varepsilon_1$ and $\varepsilon_2$ we can express the derivative by

$$n(z) \approx \frac{(\varepsilon_1+\varepsilon_2)}{2K}\frac{\varepsilon_2-\varepsilon_1}{z_1-z_2} = \frac{1}{2K}\frac{\varepsilon_2^2-\varepsilon_1^2}{(z_1-z_2)}. \tag{8}$$

In terms of numerical values we use a Coulomb logarithm of $\Lambda \approx 20$, so that $K = 2\pi e^4\Lambda \approx 6.7 \times 10^{-36}$ (erg cm)$^2$, yielding

$$N(z) \approx 10^{20}\ \mathrm{cm}^{-2}\left(\frac{\varepsilon_2}{20\ \mathrm{keV}}\right)^2, \tag{9}$$

$$n_e \approx 10^{12}\ \mathrm{cm}^{-3}\,\frac{(\varepsilon_2/20\ \mathrm{keV})^2-(\varepsilon_1/20\ \mathrm{keV})^2}{(z_1/1000\ \mathrm{km})-(z_2/1000\ \mathrm{km})}\,. \tag{10}$$

### 2.4. Exact Expression with Spectral Dependence

The photon flux per unit energy $\varepsilon$ per unit source height range $dz$ at Earth distance $D$ is, for a bremsstrahlung cross-section $Q(\varepsilon, E)$ and beam area $A$,

$$\frac{\mathrm{d}I}{\mathrm{d}z}(\varepsilon, z) = \frac{A}{4\pi D^2}n(z)\int_{\varepsilon}^{\infty} F(E,N)Q(\varepsilon,E)\,\mathrm{d}E \qquad (\mathrm{cm}^{-2}\ \mathrm{s}^{-1}\ \mathrm{erg}^{-1}\ \mathrm{cm}^{-1})\,. \tag{11}$$

We use Kramer's cross-section

$$Q(\varepsilon, E) = \frac{Q_0 m_e c^2}{\varepsilon E}, \tag{12}$$

where $Q_0$ is as in Brown and Emslie (1988) and approximate the injection spectrum $F_0(E_0)$ with a power-law function,

$$F_0(E_0) = (\delta-1)\frac{F_1}{E_1}\left(\frac{E_0}{E_1}\right)^{-\delta}\,. \tag{13}$$

Using Equations (3) and (1), Equation (11) becomes with Equation (12)

$$\frac{\mathrm{d}I}{\mathrm{d}z}(\varepsilon, z) = \frac{AQ_0}{4\pi D^2}\frac{m_e c^2}{\varepsilon}n(z)\int_{\varepsilon}^{\infty}\frac{F_0((E^2+2KN)^{1/2})}{(E^2+2KN)^{1/2}}\,\mathrm{d}E \tag{14}$$

and for a power-law injection function (14) is

$$\frac{\mathrm{d}I_\varepsilon}{\mathrm{d}z} = (\delta-1)\frac{AF_1}{E_1^{\delta+1}}\frac{Q_0}{4\pi D^2}\frac{m_e c^2}{\varepsilon}n(z)\int_{\varepsilon}^{\infty}\frac{\mathrm{d}E}{(E^2+2KN)^{(\delta+1)/2}}. \tag{15}$$

The product $AF_1 = \mathcal{F}_1$ is the total flux of electrons $s^{-1}$ at $E_0 \geq E_1$ over the area $A$. Introducing $u(\varepsilon, z) = \varepsilon^2/2KN(z)$, one immediately obtains

$$\frac{\mathrm{d}I}{\mathrm{d}z}(\varepsilon, z) = (\delta - 1)\frac{\mathcal{F}_1}{E_1}\frac{Q_0}{8\pi D^2}\frac{m_e c^2}{\varepsilon}n(z)\left(\frac{2KN}{E_1^2}\right)^{-\delta/2} B\left(\frac{1}{1+u}, \frac{\delta}{2}, \frac{1}{2}\right), \tag{16}$$

where $B$ is an incomplete Beta function,

$$B\left(\frac{1}{1+u}, \frac{\delta}{2}, \frac{1}{2}\right) \equiv \int_0^{1/(1+u)} x^{\delta/2-1}(1-x)^{-1/2}\,\mathrm{d}x. \tag{17}$$

Note that the incomplete beta function $B$ (Equation (17)) depends on $\varepsilon$ only in the combination $u = \varepsilon^2/2KN$, which is similar to the expression considered in Section 2.3 above. On the other hand, $n(z)[2KN/E_1^2]^{-\delta/2}$ depends only on $z$ for a given $\delta$, and is independent of $\varepsilon$.

### 2.5. Peak of $\mathrm{d}I/\mathrm{d}z(\varepsilon, z)$ for Power-Law Density Model $n(z)$

Based on the argument of the stopping depth $N(\varepsilon)$,

$$N(\varepsilon) \approx \frac{\varepsilon^2(z)}{2K}, \tag{18}$$

we found from data fitting that the energy dependency $\varepsilon(z)$ can be approximated by a power-law function (see Paper II), that immediately brings us the density power law via Equation (18). Therefore, we assume that

$$n(z) = n_0\left(\frac{z}{z_0}\right)^{-b}, \tag{19}$$

so that the column depth $N(z)$ becomes

$$N(z) = \int_z^{\infty} n(z)\,\mathrm{d}z = N_0\left(\frac{z}{z_0}\right)^{1-b}, \tag{20}$$

where

$$N_0 = \frac{n_0 z_0}{b-1}. \tag{21}$$

The upper limit of the column depth integral is approximated with $z_{\mathrm{max}} \approx \infty$, which is saying essentially that the beam encounters little material before it reaches the chromosphere, so that $z \ll z_{\mathrm{max}}$.

The proposal is now to see how well we can fit the data on $z(\varepsilon)$ from gaussian models of RHESSI data to the predictions of $\mathrm{d}I/\mathrm{d}z$ $(\varepsilon, z)$ in the previous theory framework (Equation (16)), but using a chromospheric model of $n(z)$ and $N(z)$ defined in terms of power laws like (Equations (19) and (20)), estimated from the simple model approach but now adjusting the parameters $(b, n_0)$ to take account of the spectral dependence of the peak emission heights.

Inserting the chromospheric density model (Equations (19) and (20)) into Equation (16) gives

$$\frac{dI}{dz}(\varepsilon, z) = I_0 \frac{\sqrt{2KN_0}}{\varepsilon} \left(\frac{z}{z_0}\right)^{-b+(b-1)\delta/2} B\left(\frac{1}{1+\frac{\varepsilon^2}{2KN_0}\left(\frac{z}{z_0}\right)^{b-1}}, \frac{\delta}{2}, \frac{1}{2}\right), \tag{22}$$

where $I_0$ is just a flux scale factor,

$$I_0 = (\delta - 1)\frac{\mathcal{F}_1}{E_1} \frac{Q_0}{8\pi D^2} \frac{m_e c^2}{\sqrt{2KN_0}} n_0 \left(\frac{2KN_0}{E_1^2}\right)^{-\delta/2} \tag{23}$$

independent of the altitude $z$ and the photon energy $\varepsilon$.

What we want to find out is where Equation (22) peaks in $z$ for each $\varepsilon$. This will depend on the spectral slope $\delta$, the reference height $z_0$, and the parameters $b$, $N_0$ defining the atmosphere model. We can consider the shape of Equation (22) in terms of the dimensionless parameters $\tilde{\varepsilon} = (\varepsilon/\varepsilon_0)$ and $\tilde{z} = (z/z_0)$, by defining the reference energy $\varepsilon_0$ via Equation (4),

$$\varepsilon_0 = \sqrt{2KN_0}\,, \tag{24}$$

which yields the dimensionless form of Equation (22),

$$\frac{dI}{dz}(\tilde{\varepsilon}, \tilde{z}) = I_0\, \tilde{\varepsilon}^{-1} \tilde{z}^{-b+(b-1)\delta/2}\, B\left(\frac{1}{1+\tilde{\varepsilon}^2 \tilde{z}^{b-1}}, \frac{\delta}{2}, \frac{1}{2}\right). \tag{25}$$

For a given chromospheric model (with free parameters $n_0, b$, with $N_0 = n_0 z_0/(b-1)$), an arbitrary reference height $z_0$, and an electron spectrum slope $\delta$, we can numerically determine the surface $dI/dz\ (\tilde{\varepsilon}, \tilde{z})$ in dimensionless parameter space $(\tilde{\varepsilon}, \tilde{z})$ and find the maximum $\tilde{z}_{\max}(\tilde{\varepsilon})$ as a function of photon energy $\tilde{\varepsilon}$. The independent parameters $(n_0, b)$ can then be varied until a best fit of the model (Equation (25)) to the observed height dependence $\tilde{z}_{\max}^{\text{obs}}(\tilde{\varepsilon})$ is obtained.

The spatial distribution of the flux $dI/dz\ (\varepsilon, z)$ is shown in Figures 1 and 2. We calculated examples of this function for a density power-law parameter of $b = 2.5$ in the chromospheric model, a spectral slope of $\gamma = 4.0$ of the photon spectrum (appropriate to Paper II), and by varying the energy $\varepsilon = 10, \ldots, 60$ keV (Figure 1). The spectral dependence of the peak height is shown for the same density power-law parameter $b = 2.5$, a photon energy of $\varepsilon = 20$ keV, and by varying the spectral slope in the range of $\delta = 2.5, \ldots, 5.0$ (Figure 2). Note the strong spectral dependence of the peak emission height $z_{\max}(\varepsilon)/z_0$. The flux profiles in Figure 2 show clearly that the radiation is spread mainly along the upper loop for soft spectra ($\delta = 6.0$), because the majority of low-energy electrons are not able to penetrate to the dense chromosphere before they lose their energy, while harder

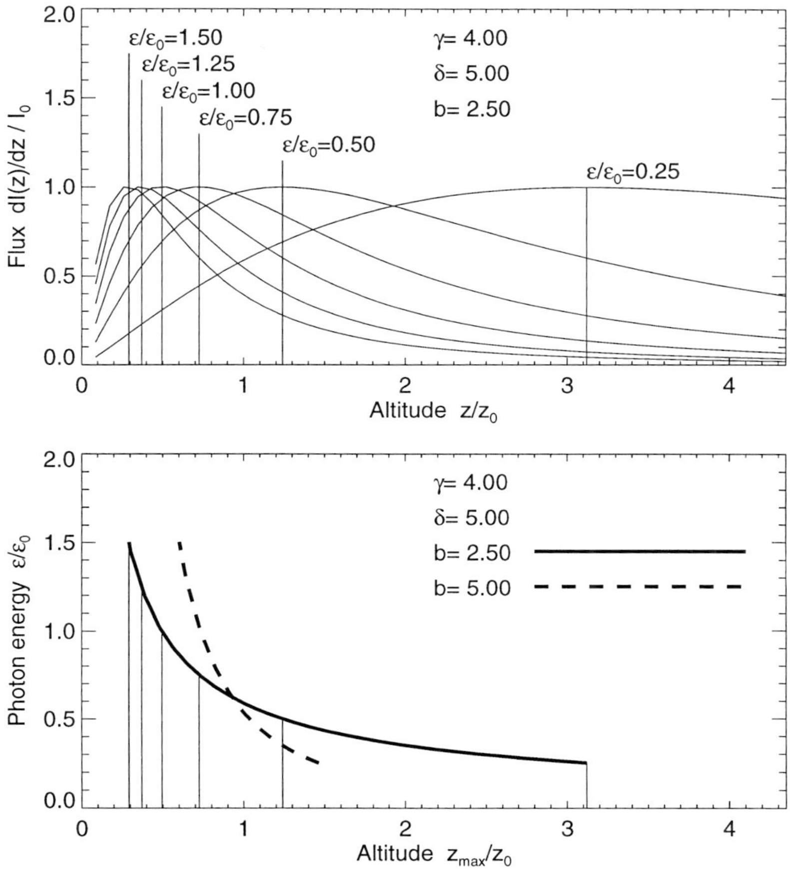


*Figure 1.* The analytical functions of the energy flux per height range, d$I$/d$z$ $(\varepsilon, z)$ (see Equation (25)), for six different energies, $\varepsilon = 10, 20, \ldots, 60$ keV, as function of the dimensionless altitude $z/z_0$ (*top panel*). The functions are calculated for $b = 2.5$, $\gamma = 4.0$, and are normalized to unity. The locations of the maximum of these functions, $z_{\max}(\varepsilon)/z_0$, are found numerically (*bottom panel*).

spectra ($\delta = 3.5$) contain a larger fraction of high-energy electrons that propagate all the way to the dense chromosphere, yielding maximum radiation at heights of $z_{\max}/z_0 \lesssim 1$ (Figure 2, top).

## 3. Discussion and Conclusions

We see from Equation (10) how data on $z_{\max}(\varepsilon)$ can yield estimates of the atmospheric density structure $n(z)$ from hard X-ray spectral images. Using the more refined treatment allowing for spectral index $\delta$ dependence shows (Figure 1) how

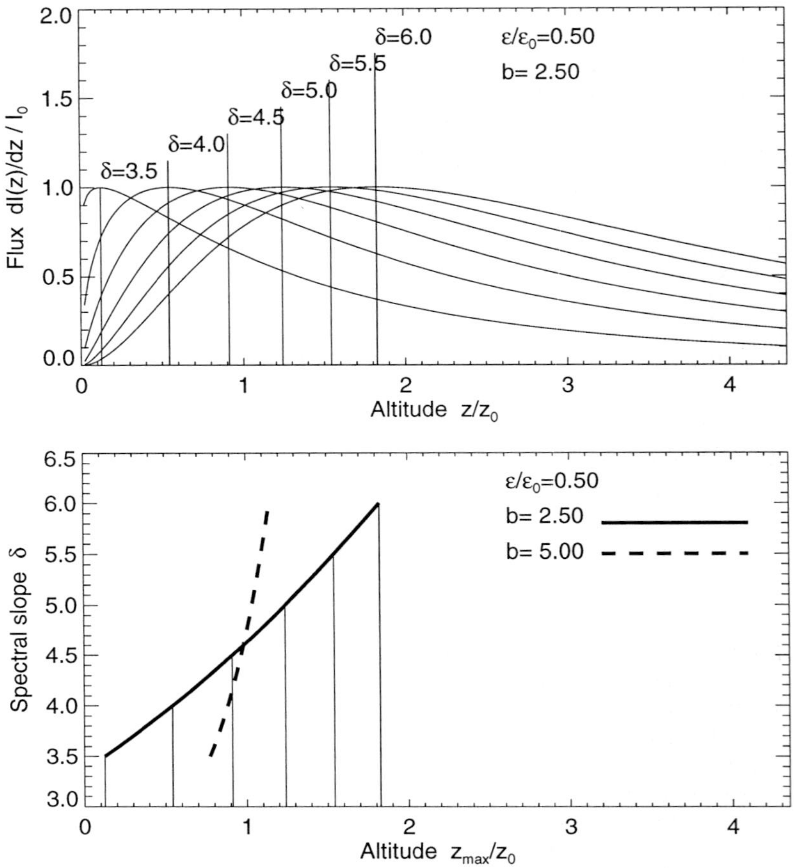


*Figure 2.* The spectral dependence of the analytical function $\mathrm{d}I/\mathrm{d}z\ (\varepsilon, z)$ (see Equation (25)) is shown for six different spectral slopes $\gamma = 2.5, \ldots, 5.0$ of the photon spectrum (i.e., $\delta = 3.5, \ldots, 6.0$). The spatial distribution of the fluxes are shown as function of the the dimensionless altitude $z/z_0$ (*top panel*), and the maxima $z_{\max}(\varepsilon)/z_0$ have been calculated numerically (*bottom panel*).

the peak emission altitude $z_{\max}(\varepsilon)$ depends monotonically, for given $\delta$, on the values of the parameters $n_0, b$ in our (power-law) parameterized model of $n(z)$ and hence enables their inference from data as shown for RHESSI observations in Paper II. There we also discuss the effect of relaxing simplifying assumptions in our theoretical treatment of the thick-target model and how our resulting $n(z)$ compares with results from other density diagnostic methods.

Some specific simplifying assumptions we have made can also affect the results. In particular we used a 1-D treatment of the transport, ignoring pitch angle increase, and assumed zero injection pitch angle. These assumptions were relaxed in Brown (1972) from which it is easy to see that their inclusion would have the qualitative effect of an increase by a factor $\geq 1.5$ in the collisional transport $K$ value and hence a decrease in the inferred values of $N(z), n(z)$ for given height structure data. Secondly in $K$ we used $\Lambda = \Lambda_{ee}$ for an ionized target rather than $\Lambda = \Lambda_{eH}$ for a neutral one which is smaller by a factor of about 0.4 and hence will make inferred $N(z), n(z)$ values larger by a factor of about 2.5. Ignoring return current losses will also tend to make estimated $N(z), n(z)$ too large. Clearly if the fullest use is to be made of the hard X-ray method of measuring densities and studying energy loss process the more fully these effects can be included the better, though our results seem to prove that the basic ionized target collisional thick-target model is good within factors of a few and so an excellent starting point.

## Acknowledgements

Support for this work was provided by the NASA SMEX grant NAS5-98033 through University of California, Berkeley (subcontract SA2241-26308PG) and by UK PPARC Standard and Visitor Grants.

## References

Aschwanden, M. J.: 2002, *Space Sci. Rev.* **101**, 240.
Aschwanden, M. J., Brown, J. C., and Kontar, E. P.: 2002, *Solar Phys.*, this volume, Paper II.
Brown, J. C.: 1971, *Solar Phys.* **18**, 489.
Brown, J. C.: 1972, *Solar Phys.* **26**, 441.
Brown, J. C.: 1973, *Solar Phys.* **32**, 227.
Brown, J. C. and MacKinnon, A. L.: 1985, *Astrophys. J.* **292**, L31.
Brown, J. C. and Emslie, A. G.: 1988, *Astrophys. J.* **331**, 554.
Brown, J. C. and McClymont, A. N.: 1976, *Solar Phys.* **49**, 329.
Brown, J. C., Carlaw, V. A., Cromwell, D., and Kane, S. R.: 1983, *Solar Phys.* **88**, 281.
Emslie, G. A.: 1978, *Astrophys. J.* **224**, 241.
Emslie, G. A. and Vlahos, L.: 1980, *Astrophys. J.* **242**, 359.
Spitzer, L. W.: 1962, *Physics of Fully Ionized Gases*, 2nd ed., Interscience, New York.

# CHROMOSPHERIC HEIGHT AND DENSITY MEASUREMENTS IN A SOLAR FLARE OBSERVED WITH RHESSI

## II. *Data Analysis*

MARKUS J. ASCHWANDEN[1], JOHN C. BROWN[2] and EDUARD P. KONTAR[2]

[1] *Lockheed Martin Advanced Technology Center, Solar & Astrophysics Laboratory, Department L9-41, Bldg. 252, 3251 Hanover St., Palo Alto, CA 94304, U.S.A. (e-mail: aschwanden@lmsal.com)*

[2] *Astronomy and Astrophysics Group, Department of Physics and Astronomy, University of Glasgow, Glasgow G12 8QQ, U.K. (e-mail: john@astro.gla.ac.uk)*

(Received 19 July 2002; accepted 15 August 2002)

**Abstract.** We present an analysis of hard X-ray imaging observations from one of the first solar flares observed with the Reuven Ramaty High-Energy Solar Spectroscopic Imager (RHESSI) spacecraft, launched on 5 February 2002. The data were obtained from the 22 February 2002, 11:06 UT flare, which occurred close to the northwest limb. Thanks to the high energy resolution of the germanium-cooled hard X-ray detectors on RHESSI we can measure the flare source positions with a high accuracy as a function of energy. Using a forward-fitting algorithm for image reconstruction, we find a systematic decrease in the altitudes of the source centroids $z(\varepsilon)$ as a function of increasing hard X-ray energy $\varepsilon$, as expected in the thick-target bremsstrahlung model of Brown. The altitude of hard X-ray emission as a function of photon energy $\varepsilon$ can be characterized by a power-law function in the $\varepsilon = 15-50$ keV energy range, viz., $z(\varepsilon) \approx 2.3(\varepsilon/20 \text{ keV})^{-1.3}$ Mm. Based on a purely collisional 1-D thick-target model, this height dependence can be inverted into a chromospheric density model $n(z)$, as derived in Paper I, which follows the power-law function $n_e(z) = 1.25 \times 10^{13}(z/1 \text{ Mm})^{-2.5}$ cm$^{-3}$. This density is comparable with models based on optical/UV spectrometry in the chromospheric height range of $h \lesssim 1000$ km, suggesting that the collisional thick-target model is a reasonable first approximation to hard X-ray footpoint sources. At $h \approx 1000-2500$ km, the hard X-ray based density model, however, is more consistent with the '*spicular extended-chromosphere model*' inferred from radio sub-mm observations, than with standard models based on hydrostatic equilibrium. At coronal heights, $h \approx 2.5-12.4$ Mm, the average flare loop density inferred from RHESSI is comparable with values from hydrodynamic simulations of flare chromospheric evaporation, soft X-ray, and radio-based measurements, but below the upper limits set by filling-factor insensitive iron line pairs.

## 1. Introduction

In this paper we measure for the first time hard X-ray source altitudes with the RHESSI spacecraft, which has an unprecedented energy resolution thanks to the germanium-cooled detectors, and thus is expected to provide the most accurate spectral height measurements. The reason is because the observed height distribution of hard X-ray emission, which is a convolution of the electron injection spectrum, the bremsstrahlung cross-section, and the instrumental spatial/energy resolution, can most accurately be modeled for a high energy resolution. We model the spatial distribution of hard X-ray sources observed with RHESSI during the 20

*Solar Physics* **210:** 383–405, 2002.

February 2002 flare according to the theoretical model of thick-target bremsstrahlung derived in Paper I (Brown, Aschwanden, and Kontar, 2002).

Hard X-ray height limits on the regions of particle acceleration and energy deposition have first been obtained using stereoscopic observations with multiple spacecraft. Kane (1983) found that 95% of the $\approx 150$ keV X-ray emission in impulsive flares originates at attitudes $\lesssim 2.5$ Mm, and also 75% of the 100–150 keV emission of gradual flares was found at altitudes of $\lesssim 2.5$ Mm. Brown *et al.* (1983) modeled the height structure of stereoscopically observed flares and found that those with heights of $\lesssim 2.5$ Mm could satisfactorily be explained by the collisional thick-target model (Brown, 1971; Brown and McClymont, 1976) with direct precipitation (no trapping) while hard X-ray emission from occulted flares with height levels of $\approx 25.0$ Mm could not be explained with the basic thick-target model. Height measurements of hard X-ray sources have also been obtained by measuring the displacement of hard X-ray source centroids from the reported H$\alpha$ flare positions. Takakura *et al.* (1986) measured a hard X-ray source height of $h = 7.0 \pm 3.5$ Mm above the neutral line seen in H$\alpha$ in the 20–40 keV range with *Hinotori*. Matsushita *et al.* (1992) measured the statistical displacement of hard X-ray source centroids to reported H$\alpha$ flare positions and found heights of $h_{14} = 9.7 \pm 2.0$ Mm for 14–23 keV (*Yohkoh* Lo band), and height differences of $h_{23} - h_{14} = -1.0 \pm 0.3$ Mm for 23–33 keV (M1 band), $h_{33} - h_{14} = -2.0 \pm 0.5$ Mm for 33–33 keV (M2 band), and $h_{53} - h_{14} = -3.2 \pm 0.7$ Mm for 53–96 keV (Hi band). Fletcher (1996) modeled the theoretical energy dependence of the hard X-ray source heights with a numerical Fokker–Planck collisional transport code and found that Matsushita's measurements can be reproduced with loops that have coronal densities between $n_e = 2 \times 10^{10}$ and $3 \times 10^{11}$ cm$^{-3}$ and half lengths of $L \approx 13$–$27$ Mm.

In Section 2 we present the data analysis for the case of the 20 February 2002 flare, in Section 3 we discuss comparisons with previous measurements and alternative models, and in Section 4 a summary and conclusions are given.

## 2. Data Analysis

### 2.1. Observations

One of the first prominent flares observed with RHESSI occurred on 20 February 2002, 11:06 UT. It was identified as GOES C7-class flare, occurring in NOAA/USAF active region AR 9825. Big Bear Observatory pinpointed the flare location at N16 W80, i.e., 919″ west and 285″ north from solar disk center, and classified the sunspots as $\beta/\gamma$ Hale configuration. A light curve of the flare observed by RHESSI in the energy band of $\varepsilon = 12$–$50$ keV is shown in Figure 1 (bottom), binned in time intervals of 4.0 s, which roughly corresponds to the RHESSI spin rate. The actual spin rate of RHESSI was determined by the Photo-Multiplier-Tube Roll Aspect System (PMTRAS) to be $R = 4.3329$ s at the time of the flare

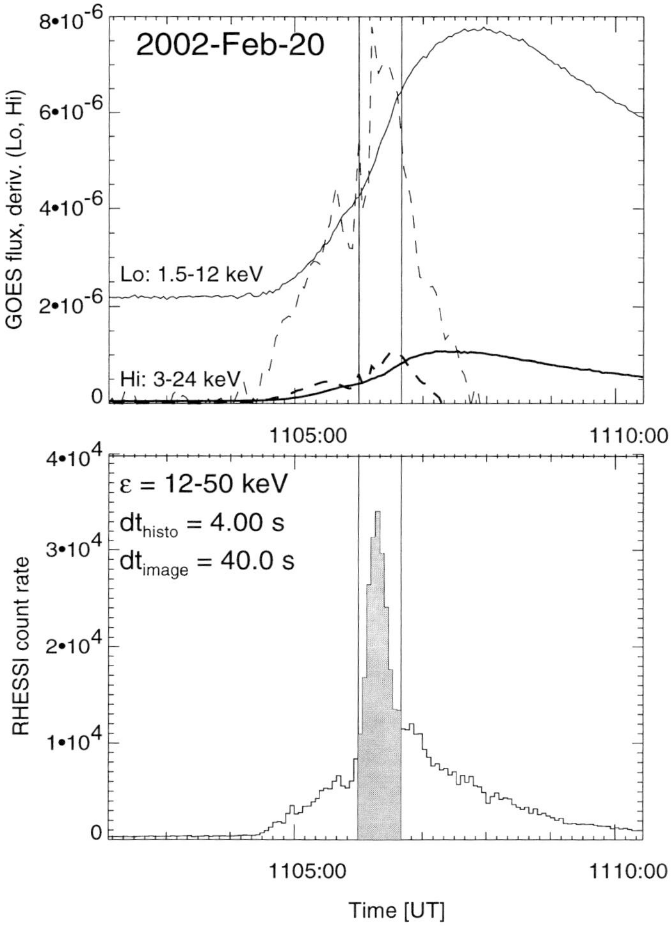


*Figure 1.* RHESSI (*bottom frame*) and GOES light curves (*top frame*). The GOES light curves are shown for the low-energy (*thin line*) and high-energy channel (*thick line*), along with the time derivative of the GOES flux (*dashed lines*). The RHESSI light curve is binned in steps of $dt_{\text{histo}} = 4.0$ s. The time interval $dt_{\text{image}} = 40.0$ s used in the images, 16:06:00–16:06:40 UT, is marked with *grey color*.

peak, at 11:06:16 UT. The rotation rate needs to be known with an accuracy of at least $10^{-4}$ to calibrate the phase of the rotational modulation time profiles properly. Two soft X-ray light curves of the flare observed by GOES are also shown in Figure 1 (top) from the two channels, 1–8 Å (1.5–12 keV) and 0.5–4 Å (3–24 keV), respectively. The time derivative of the soft X-ray light curves (dashed lines in Figure 1, top) matches closely the hard X-ray light curve observed by RHESSI (Figure 1, bottom), as expected from the Neupert effect (e.g., Dennis and Zarro, 1993). For imaging, in different energy ranges we will use the time interval

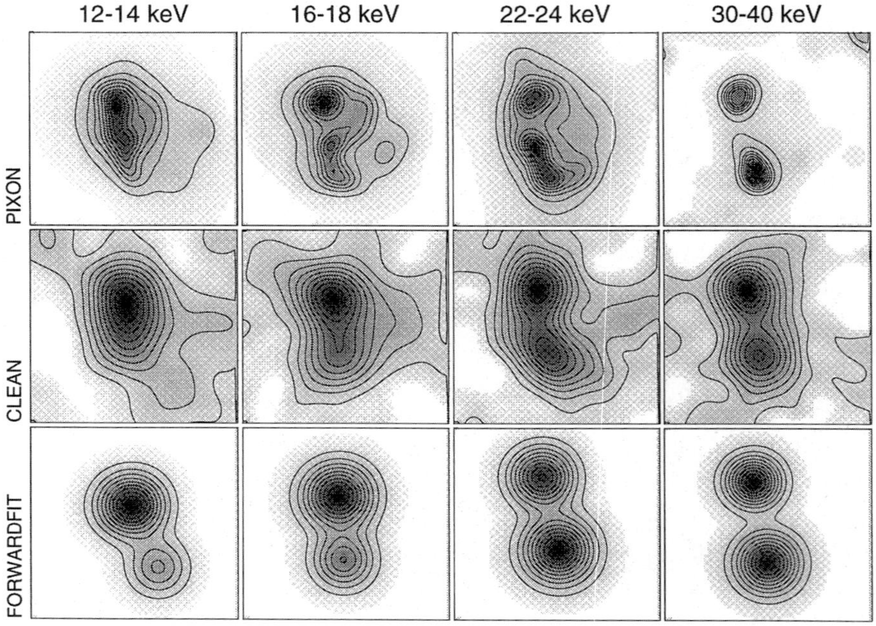


*Figure 2.* Image reconstructions of the 20 February 2002, 11:06:00–11:06:40 UT flare, observed with RHESSI, using grid Nos. 3–9. The images are reconstructed in four different energy ranges, the lowest showing mainly thermal looptop emission (12–14 keV), while the higher energy ranges (16–18 keV, 22–24 keV, 30–40 keV) display double footpoint sources. Three different imaging algorithms were used: pixon method (*top row*), clean (*middle row*), and forward-fitting of two gaussian sources (*bottom row*). The field of view is in all cases 64″ or 46 Mm. All maps are normalized to the same peak brightness, with contour levels given in increments of 10% of the peak brightness.

of 11:06:00–11:06:40 UT, which extends roughly over 10 RHESSI rotations and encompasses the flare peak (see time interval marked in Figure 1, bottom).

### 2.2. Flare Loop Geometry

In Figure 2 we show RHESSI image reconstructions of the flare during the time interval 20 February 2002 11:06:00–11:06:40 UT in four different energy ranges (12–14, 16–18, 22–24, and 30–40 keV), obtained with three different imaging algorithms, i.e., with the 'pixon', 'clean', and 'forward-fit' algorithms. These image reconstructions have been performed using the RHESSI grids Nos. 3–9, which contain information on spatial scales of 6.8″–183″ (see Table I in Aschwanden *et al.*, 2002b).

Here we discuss the imaging only for RHESSI grids Nos. 3–9 and not for the finest grids Nos. 1–2. This is because the finest grids Nos. 1–2 have not been fully calibrated at the time of writing which, together with the limited photon statistics makes their use rather unreliable for now. Nevertheless we obtain angular measurements of a precision of order that of the finest grids because we are fitting the centroids of parametric models rather than recovering the full image structure. At the time of image reconstruction (May 2002), the RHESSI software had already implemented (1) the roll-angle aspect solution (PMTRAS), (2) the corrected grid No. 5 calibration, and (3) corrections for small ($\lesssim 100$ ms) data drop-outs due to cosmic-ray hits. A comparison of the three imaging algorithms shows both similarities and significant differences. The clean algorithm iteratively deconvolves a backprojection map with a calculated pointspread function, which has roughly a FWHM of the finest grid with significant modulation ($\approx 7''$ here). The clean images (Figure 2, middle row) contain also a residual map which provides a measure of the data noise. The pixon images were produced, unlike the clean algorithm, by fitting the theoretically calculated modulation time profiles from a pixonized model map to the observed time profiles. In this comparison in Figure 2 they tend to show somewhat sharper source structures and thus reveal some more detailed structures than the clean maps. The forward-fitting maps were produced by fitting the modulation time profiles of a parameterized model map with two gaussian source structures to the observed time profiles. This forwardfit method places all the flux into two locations that represent the best possible approximation to the real flux distribution, under the constraint of the particular parameterizations of 2 circular gaussian sources. The feature that is commonly reproduced by all three algorithms is the double-source structure for energies above 15 keV, while a single-source structure is seen at all lower energies (as imaged in the range of 3–15 keV).

The contours at the half peak flux of the forward-fitting maps are overlaid in Figure 3, for energies of $\varepsilon = 15, 20, \ldots, 45$ keV. Because the centroids of the double gaussians are nearly co-spatial at higher energies, we interpret the two locations as conjugate footpoints of a flare loop (or a compact loop arcade). With forward-fitting, the centroids of the gaussian source structures can be localized more precisely than the angular resolution. This super-resolution, however, hinges on the assumption of a gaussian (or some other few-parameter) source model. For asymmetric sources there would be a possible bias depending on the skewness (3rd-order moment) of the true source distribution. The locations of the two footpoints, as determined from the averaged centroids of the forward-fitted gaussians at high energies ($\gtrsim 40$ keV), are found at the coordinates $X_1, Y_1 = [906.6'', 252.3'']$ and $X_2, Y_2 = [900.9'', 277.0'']$ west and north of Sun center. This yields a midpoint position of $[903.7'', 264.6'']$, a loop baseline of $l = 24.8$ Mm, and an azimuth of $\alpha = 102°$ of the baseline to the east–west direction. If we assume a semi-circular geometry for the flare loop (indicated with dashed line style in Figure 2) we have a loop radius of $r_{\rm loop} = l/2 = 12.4$ Mm. The midpoint between the hard X-ray footpoints corresponds to a location of (N16 W75), close to the Big Bear reported flare position (N16 W80). Because this location is close to the limb, we see essentially the loop plane (assumed vertical) unforeshortened, with $\sin(75°) = 0.96$. Radial distances to Sun center correspond therefore to altitude differences in good approximation.

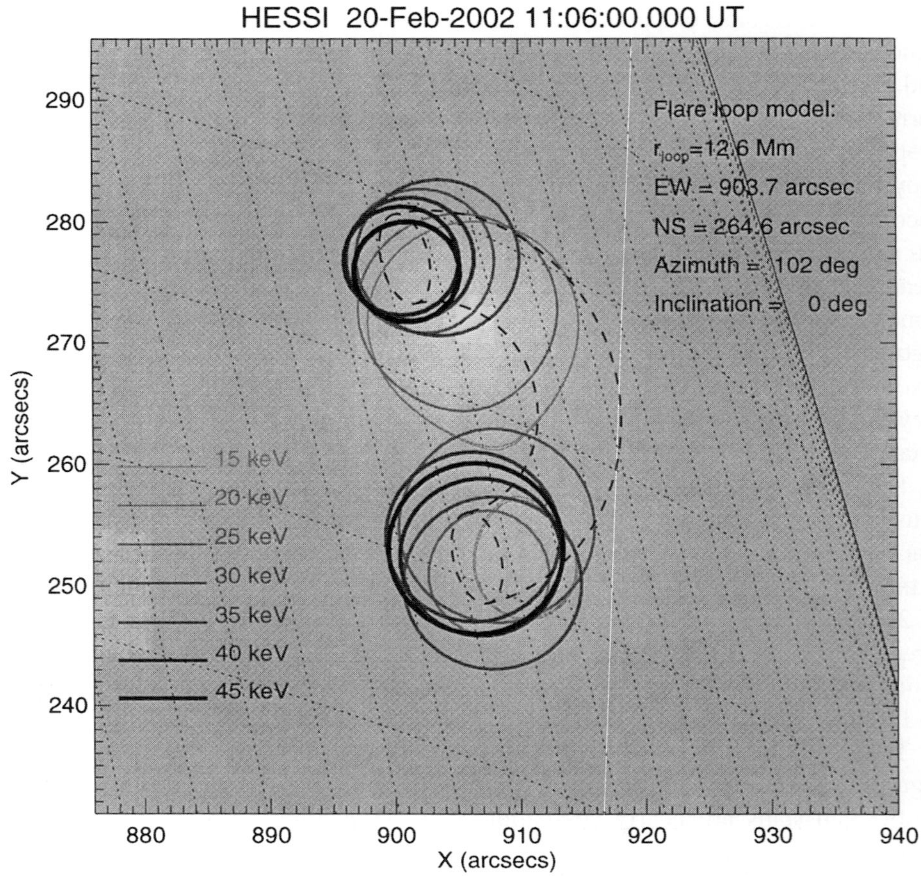


*Figure 3.* FWHM contours of forward-fit maps overlaid at different energies (15, 20, ..., 45 keV). The greyscale map represents the 15 keV map. A heliographic grid with 1° spacing is overlaid. The *dashed lines* demarcate a flare loop with semi-circular geometry and vertical loop plane. Note that the centroids of the hard X-ray sources systematically increase in altitude with lower energies.

## 2.3. Altitude measurements of hard X-ray sources

The next step is to measure the radial distances $r(\varepsilon)$ of the hard X-ray footpoint sources from Sun center as a function of energy $\varepsilon$. We reconstructed 68 images with the forward-fitting method in the energy range of $\varepsilon = 3, \ldots, 7$ keV, in steps of 1 keV. The energy widths $\Delta\varepsilon$ of the images were chosen to be wide enough to ensure acceptable image quality, i.e. $\Delta\varepsilon = 2$ keV for energies of $\varepsilon \leq 25$ keV, $\Delta\varepsilon = 4$ keV for energies of 25 keV $< \varepsilon <$ 35 keV, and $\Delta\varepsilon = 10$ keV for energies of $\varepsilon \geq 35$ keV, respectively. From the forwardfit maps we obtain the image coordinates $x_i$, $y_i$ of the gaussian source centroids (with respect to Sun center). We sort the coordinates of the two gaussian sources into northern (N) and southern (S) sources and determine their distances to Sun center, e.g.,

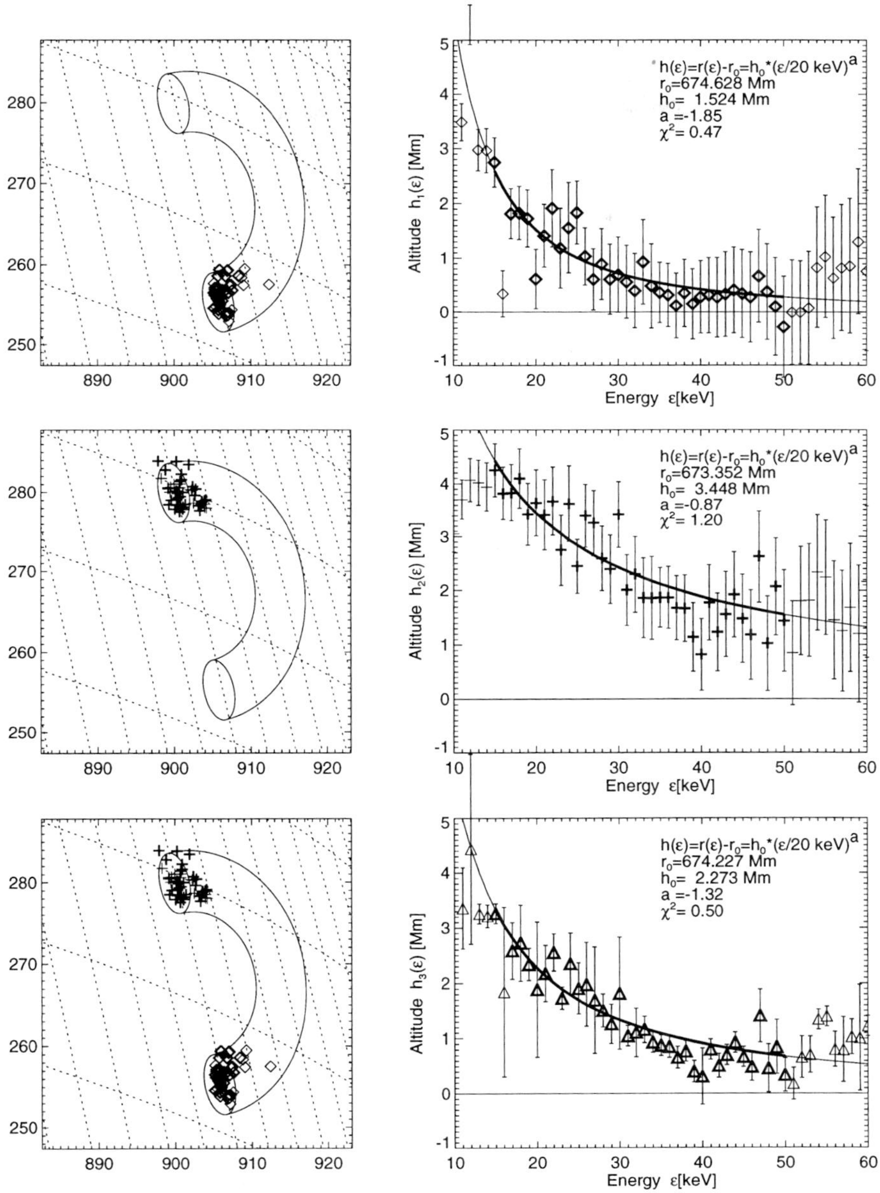


*Figure 4.* Height measurements of the centroids of the gaussian fits to the hard X-ray sources. The spatial location of the source centroids are shown in the *left side*, the altitudes as functions of energy in the *right panels*. Measurements at the southern footpoint are shown in the *top panels*, at the northern footpoint in the *middle panels*, and combined measurements in the *bottom panel*. The error bars of the height measurements were estimated from Poisson statistics in the cases of single footpoints, or from the differences in the case of combined footpoints. The curves indicate power-law fits, marked with thick linestyle in the energy range of the fit (15–50 keV). The best-fit parameters are also indicated.

$$r(\varepsilon_i) = \sqrt{x_i^2 + y_i^2}. \tag{1}$$

The positions of the source centroids are shown in Figure 4 (with cross symbols for the northern sources in the middle left panel and with diamond symbols for the southern sources in the top left panel in Figure 4). The location of the centroid positions shown in Figure 4 have been rotated by an empirical (energy-dependent) roll angle of $7°/\varepsilon_{\rm keV}$ to correct for a spin rate drift of $\approx 14''/40$ s that is presently not fully accounted for in the imaging software. This roll-angle drift is energy-dependent because the image reconstruction is flux-weighted and the flux at lower energies peaks later than at higher energies (Neupert effect). This roll angle correction places all source locations within a loop diameter of $\approx 5''$) (Figure 4, left panels), but in no way affects the height measurements $r(\varepsilon_i)$ here, because it is strictly in the orthogonal direction.

The measured source altitudes $z(\varepsilon_i) = r(\varepsilon_i) - r_0$ are shown in Figure 4 (right panels), defined by a height difference of the radial distance $r$ to Sun center to a reference height $r_0$. We choose a power-law function to fit the observed source altitudes $z(\varepsilon)$ as a function of the energy $\varepsilon$,

$$z(\varepsilon) = r(\varepsilon) - r_0 = z_0 \left(\frac{\varepsilon}{20 \text{ keV}}\right)^{-a}. \tag{2}$$

In the high-energy limit ($\varepsilon \mapsto \infty$) this power-law function yields as asymptotic limit a zero height ($z \mapsto 0$), which we can associate with the solar surface. This power-law function has three free parameters, the source height $z_0$ at 20 keV, the reference level $r_0$, and the power-law slope $a$. We determine these three parameters from a least-square fit (with the Powell minimization method) to the observed values $r(\varepsilon_i)$. We fit only over the energy range 15 keV $< \varepsilon_i <$ 50 keV, which is the energy range that exhibits clear double-footpoint morphology. For the northern footpoint we find the parameters $r_0^N = 674.6$ Mm, $z_0 = 1.5$ Mm, and $a = 1.85$, and for the southern footpoint $r_0^N = 673.4$ Mm, $z_0 = 3.4$ Mm, and $a = 0.87$, respectively. If we average the altitudes from both footpoints (and adjust them to the same reference height) we find $r_0 = 674.2$ Mm, $z_2 = 2.3$ Mm, and $a = 1.32$ (Figure 4, bottom right).

We attempted to estimate the uncertainties $\sigma_r$ of the height measurements $r$ from the simulations of double footpoint sources in Aschwanden *et al.* (2002b), which scale as a function of the gaussian source width $w$ (in arcsec) and photon count statistics as

$$\sigma_r = w'' \left(\frac{696 \text{ Mm}}{R''_\odot}\right) 10^{-1.27} \left(\frac{10^4}{R\,\Delta t}\right)^{1/2} \text{(Mm)}, \tag{3}$$

where $R_\odot = 970.5''$ is the solar radius, $R$ the count rate, and $\Delta t$ the imaging time interval. These formal error bars are shown in Figure 4 (right top and middle panel), yielding a $\chi^2 = 0.47$ and 1.20, for the power-law height model fit of Equation (2). Another estimate of the systematic uncertainty of height measurements can be obtained from the difference of heights at the northern and southern location, assuming that the same chromospheric density model holds at both places. Using the half height difference $\sigma_z = |z_N - z_S|/2$ as an estimate of the uncertainty, we find a $\chi^2 = 0.50$ for a model fit to the combined heights $z(\varepsilon) = (z_N + z_S)/2$ (Figure 4, bottom right).

The height dependence $z(\varepsilon)$ can also be fitted with an exponential function or a logarithmic function over this limited range of energy $\varepsilon = 15$–$50$ keV. We fitted these alternative functions and obtained comparable $\chi^2$-values in the order of unity, so that the data cannot distinguish between these different models. However, a power-law function $z(\varepsilon)$ is analytically (Equation (2)) most convenient and seems to provide reasonable physical limits even beyond the observed parameter range.

### 2.4. CHROMOSPHERIC DENSITY MODEL

The power-law dependence $z(\varepsilon)$ (Equation (2)) deduced from the observations can be inverted to give,

$$\varepsilon(z) = 20 \left( \frac{z}{z_0} \right)^{-1/a} \quad \text{(keV)}. \tag{4}$$

This expression can first be inserted into the simple form solution (Equation (7) in Paper I) for the chromospheric density model $n(z)$, after calculating the derivative $d\varepsilon/dz$. This yields also a power-law function for the density model $n(z)$,

$$n(z) = n_0 \left( \frac{z}{z_0} \right)^{-b}, \tag{5}$$

with the density constant $n_0$,

$$\begin{aligned} n_0 &\approx \frac{(20 \text{ keV} \times 1.610^{-9} \text{ erg keV}^{-1})^2}{\text{K}} \left( \frac{1}{a} \right) \left( \frac{1 \text{ Mm}}{z_0} \right) \\ &\approx 1.5 \times 10^{12} \left( \frac{1}{a} \right) \left( \frac{1 \text{ Mm}}{z_0} \right) \quad (\text{cm}^{-3}), \end{aligned} \tag{6}$$

and the power-law slope $b$,

$$b = 1 + \frac{2}{a}. \tag{7}$$

From the best-fit parameters ($a = 1.32$, $b = 2.52$, $n_0 = 1.14 \times 10^{12}$ cm$^{-3}$, $z_0 = 2.3$ Mm) to the combined footpoint sources we obtain a chromospheric density model of

$$n(z) \approx 1.14 \times 10^{12} \left( \frac{z}{z_0} \right)^{-2.5} = 9.2 \times 10^{12} \left( \frac{z}{1 \text{ Mm}} \right)^{-2.5} \quad (\text{cm}^{-3}). \tag{8}$$

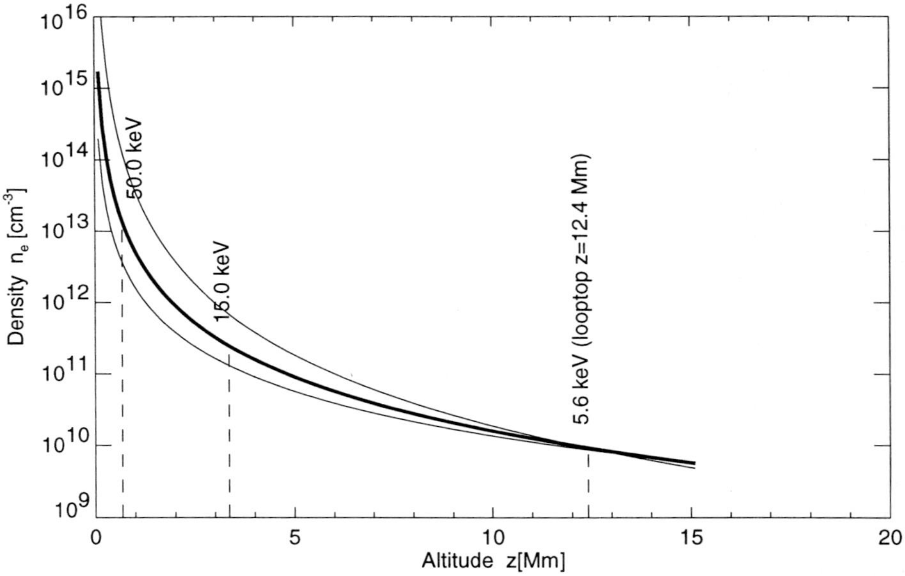


*Figure 5.* The chromospheric density model $n_e(z)$ resulting from the best fit of the accurate model to the data. The accurate model includes the spectral dependence in terms of Kramer's cross-section (Paper I, Sections 2.4 and 2.5). The densities are a factor of 1.36 lower in the simple model (Paper I, Section 2.3). The thin curve represents the model at the northern and southern loop side, while the thick curve represents the averaged positions. The power-law model is constrained in the height range of $h = 0.7$–$3.3$ Mm, corresponding to the energy range (15–50 keV) of the fit.

Figure 5 shows the chromospheric density model resulting from the accurate model (including the spectral dependence of $\varepsilon \leq E$ in terms of Kramer's cross-section), which differs only by a density factor of 1.36 from the simple model (with the simplified assumption $z = z(N(\varepsilon)) = z(N = \varepsilon^2/2K)$). In this model, electrons that produce 15–50 keV hard X-rays are stopped at a heights of $z(\varepsilon = 15 \text{ keV}) =$ 3.4 Mm and $z(\varepsilon = 50 \text{ keV}) = 0.7$ Mm, respectively.

### 2.5. HEIGHT FITTING

In Figure 6 we show the observed photon spectrum at the flare peak time, averaged over the time interval of 40 s over which the images were accumulated. This photon spectrum covers the same time interval and energy range of the 68 images reconstructed with forward-fitting. To find $\gamma, \delta$ the spectral fit was performed for the whole source volume with the standard SPEX software adapted for RHESSI in the SSW software. The spectrum can be fitted with a thermal plus a power-law function, as shown in Figure 6. The thermal spectrum that dominates at energies of $\lesssim 15$ keV corresponds to a temperature of $T = 15.3$ MK. Above an energy of $\gtrsim 15$ keV, the nonthermal spectrum dominates, which was fitted with a power-law slope of $\gamma = 3.89$, corresponding to a power-law slope of $\delta = \gamma + 1 = 4.89$

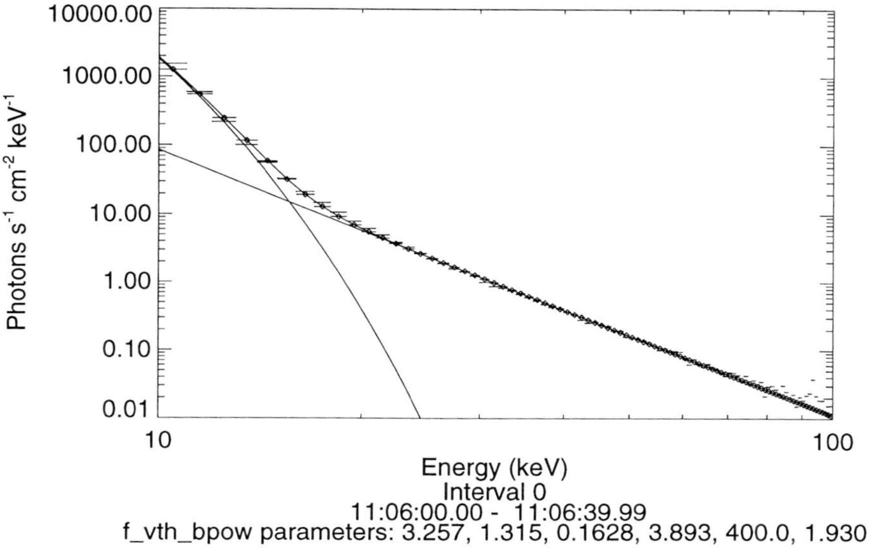


*Figure 6.* Photon spectrum (photons $s^{-1}$ $cm^{-2}$ $keV^{-1}$) and spectral fits of a combined thermal and single-power-law spectrum in the energy range of 10–100 keV and time interval 11:06:00–11:06:40 UT. In the range of 20–100 keV, the spectrum can be approximated by a power law with a slope of $\gamma = 3.89$. The thermal spectrum that dominates $\lesssim 15$ keV corresponds to a temperature of $T = 15.3$ MK (or 1.315 keV).

for the electron injection spectrum in terms of the thick-target model. The accurate value of the spectral slope depends on various effects, i.e., (1) the temporal variation during the imaging time interval, (2) the energy range selected for fitting, (3) the spectral model chosen for fitting (e.g., single-power law vs. thermal-plus-power-law function), and (4) the chosen spatial volume (whole source versus single footpoints). By varying all these effects using different image reconstruction methods (CLEAN, MEM-Sato, MEM-Vis) we found a range of $\gamma \approx 3.1$–$3.9$ for this particular flare time interval, which translates into an uncertainty range of 2.6 for the density scale factor $n_0$ in the inferred chromospheric density model. This is less than the uncertainty due to the simplifications made in the thick-target model in Paper I.

In order to derive the heights $z(\varepsilon)$ that correspond to the flux maxima of the gaussian fits we have to determine the location $z_{max}$ of the peaks in the theoretical height distributions $dI/dz(\varepsilon, z)$ expressed in terms of incomplete beta functions (Equation (25) and Figures 1 and 2 in Paper I). The flux maxima $z_{max}(\varepsilon)$ can be determined numerically. We calculate the flux distributions $dI/dz(\varepsilon, z)$ for the observationally determined spectral slope $\delta = 4.89$ and the approximate values obtained from the simple model, i.e., $a = 1.32$, $b = 1 + 2/a = 2.5$, $n_0 = 1.14 \times 10^{12}$ cm$^{-3}$, $N_0 = n_0 z_0/(b-1) = 7.6 \times 10^{20}$ cm$^{-2}$, and $z_0 = 2.3 \times 10^8$ cm. We then vary the free parameters $b$ and $n_0$ until we find the best match to the

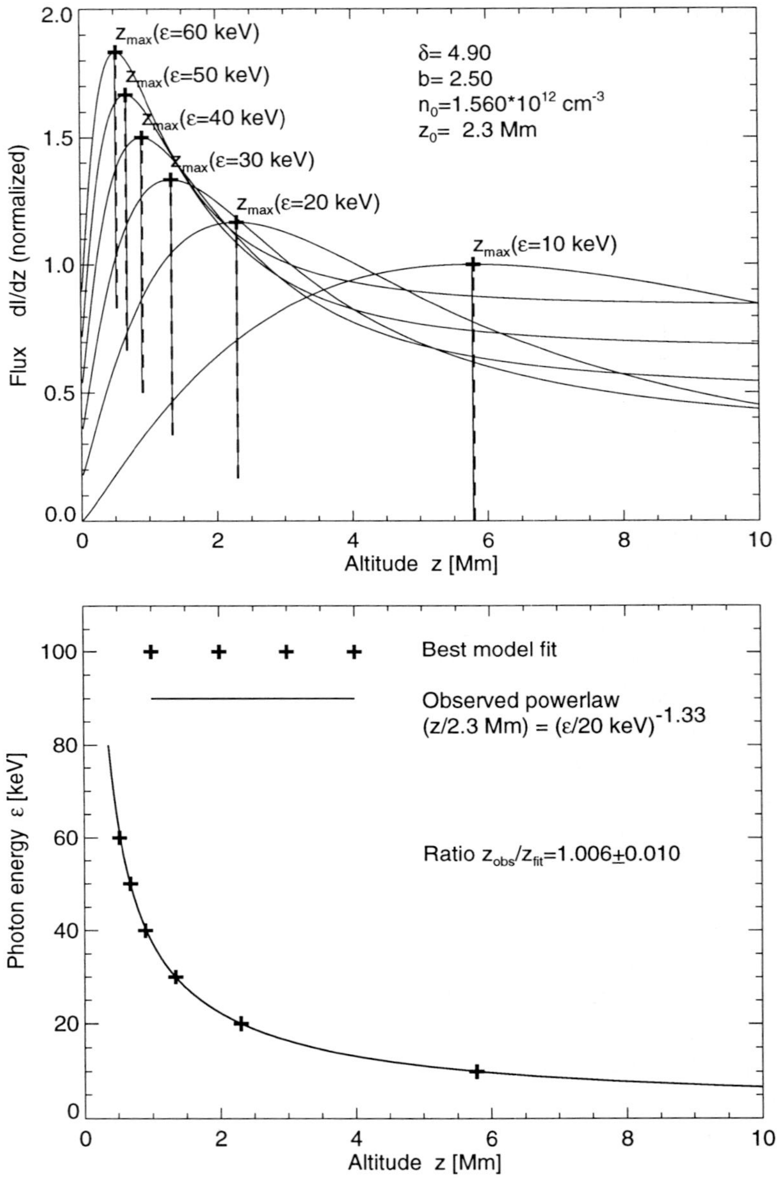


*Figure 7.* The analytical functions of the energy flux per height range, $dI/dz(\varepsilon, z)$ (see Equation (25) in Paper I), for six different energies, $\varepsilon = 10, 20, \ldots, 60$ keV, as a function of the altitude $z$ (*top panel*). The functions are normalized to unity and shifted by 0.1 per energy value to make them visible. The locations $z_{\max}(\varepsilon)$ of these functions are found numerically (marked with a *cross* and a *vertical solid line*). The locations of the observed heights of the gaussian sources are indicated with *vertical dashed lines*, which coincide with the function maxima positions for this best fit. The parameters $n_0$ and $b$ of the theoretical function have been optimized until the maxima $z_{\max}$ of the theoretical model coincided with the observed heights. The location of the same maxima are also shown as function of energy in the lower panel. The agreement between the model and the best fit amounts to $z_{\rm obs}/z_{\rm fit} = 1.006 \pm 0.010$.

observed heights of the gaussian source centroids, which follow approximately the power-law function given in Equation (2), $z_{obs}(\varepsilon) = z_0(\varepsilon/20 \text{ keV})^{-1.32}$. We find a best fit for the parameters $b = 2.5$ and a density constant $n_0 = 1.56 \times 10^{12}$ cm$^{-3}$. Thus our best-fit density model is

$$n(z) \approx 1.56 \times 10^{12} \left(\frac{z}{z_0}\right)^{-2.5} = 1.25 \times 10^{13} \left(\frac{z}{1 \text{ Mm}}\right)^{-2.5} \quad (\text{cm}^{-3}). \tag{9}$$

The display of $\varepsilon(z_{max})$ shown in Figure 7 (bottom panel) illustrates that the best model fit (marked with crosses in Figure 7, determined from the maxima of the theoretical function with the incomplete beta function (see Equation (25) in Paper I), matches the observed power-law function $(z/2.3 \text{ Mm})=(\varepsilon/20 \text{ keV})^{-1.32}$ with an accuracy of $z^{obs}/z^{fit} = 1.006 \pm 0.010$, averaged over the energy range of 10–60 keV. The height distribution $I(z, \varepsilon)$ is shown in graphical form in Figure 7, and as a brightness map in Figure 8. Note that the low-energy emission at $\varepsilon \lesssim 15$ keV is spread over a considerable fraction of the flare loop, while high-energy emission at $\varepsilon \gtrsim 20$ keV is fairly concentrated in chromospheric heights $h \lesssim 2.5$ Mm.

For the best model fit used above with $\gamma = 3.89$ the mono-energetic approximation $\varepsilon \approx E$ (made in the approximate density model of Section 2.3 in Paper I) leads to a density scale value $n_0$ which is a factor of 1.36 smaller than that for the spectrally corrected value with the (Kramer's) bremsstrahlung cross-section $Q(\varepsilon, E)$. If one uses a higher energy range to estimate $\gamma$ one gets a larger differential factor between the two methods because for flatter electron spectra more photons are emitted at well below the short-wavelength limit.

## 3. Discussion

Here we discuss three topics that are relevant to our results: comparison with previous hard X-ray altitude measurements (Section 4.1), comparison of our inferred chromospheric density model with previous chromospheric and coronal density models (Section 4.2), and comparison of our inferred average flare loop density with previous measurements and numerical simulations of the density in the chromospheric evaporation phase (Section 4.3).

### 3.1. Hard X-ray altitude measurements

Previous altitude measurements of hard X-ray sources have been performed by stereoscopic observations of occulted flares (Kane, 1983), by relative displacements of hard X-ray source centroids from H$\alpha$ flare positions (Takakura *et al.*, 1986; Matsushita *et al.*, 1992), by displacements above the solar limb (Aschwanden *et al.*, 1999), and in this study by fitting a power-law density model $h(\varepsilon) = r(\varepsilon) - r_0 \sim \varepsilon^{-a}$ to the observed distances $r(\varepsilon)$ from Sun center. Each method

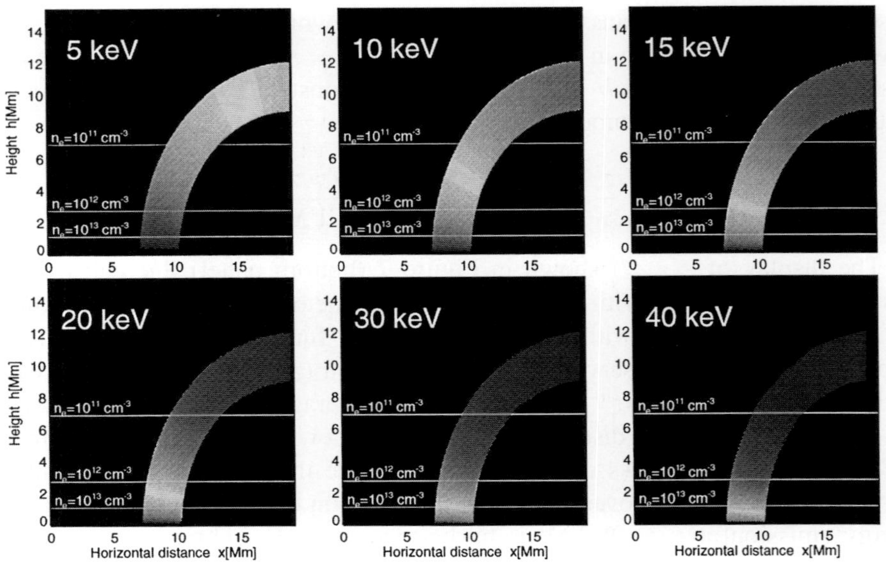


*Figure 8.* The height distributions $I(z, \varepsilon)$ are computed for $\varepsilon = 5, 10, \ldots, 40$ keV and are represented by brightness maps of the flare loops in different energy bands. The vertical axis represents the height scale $h$, for a flare loop with a loop radius of 12.4 Mm. The corresponding chromospheric densities $n_e$ are indicated on the *right side*. Note that the brightness distributions of all energies $\varepsilon \gtrsim 20$ keV are fairly concentrated inside the chromosphere (at $h \lesssim 2.0$ Mm), while the $\varepsilon \lesssim 15$ keV emission is spread along the entire flare loop.

has its own systematic errors, which have never been compared in detail. We show a comparison with the previous results in Figure 9. If we plot the reported absolute heights (Figure 9, top), we find that all previously measured altitudes are higher than those measured with RHESSI in this study. The reason for this systematic trend is not obvious, although we expect arbitrary offsets between H$\alpha$ and hard X-ray flare positions, because the brightest part of the often quite long H$\alpha$ flare ribbons do not necessarily coincide with the rather concentrated hard X-ray footpoints of flare loops. Also we know that H$\alpha$ flare positions are reported with typical accuracies of 1 heliographic degree, which corresponds to 12 Mm. In the statistical dataset of Matsushita *et al.* (1992), which contains 92 flare events, the absolute height error was estimated to $\sigma_h = 2.0$ Mm. Even in the highest energy channel (53–93 keV) analyzed by Matsushita *et al.* (1992), a source height of $h_{53} = 6.5 \pm 2.0$ Mm was deduced, which requires a thick-target density of $n_e \approx 10^{13}$ cm$^{-3}$, which is unlikely to exist at this height (Figure 5), according to standard chromospheric models (Figure 10). Thus, we conclude that most of the previous absolute height measurements have large uncertainties that were not properly accounted for.

A much better agreement is achieved in the relative height measurements of hard X-ray sources at different energies. In Figure 9 (bottom panel) we show the

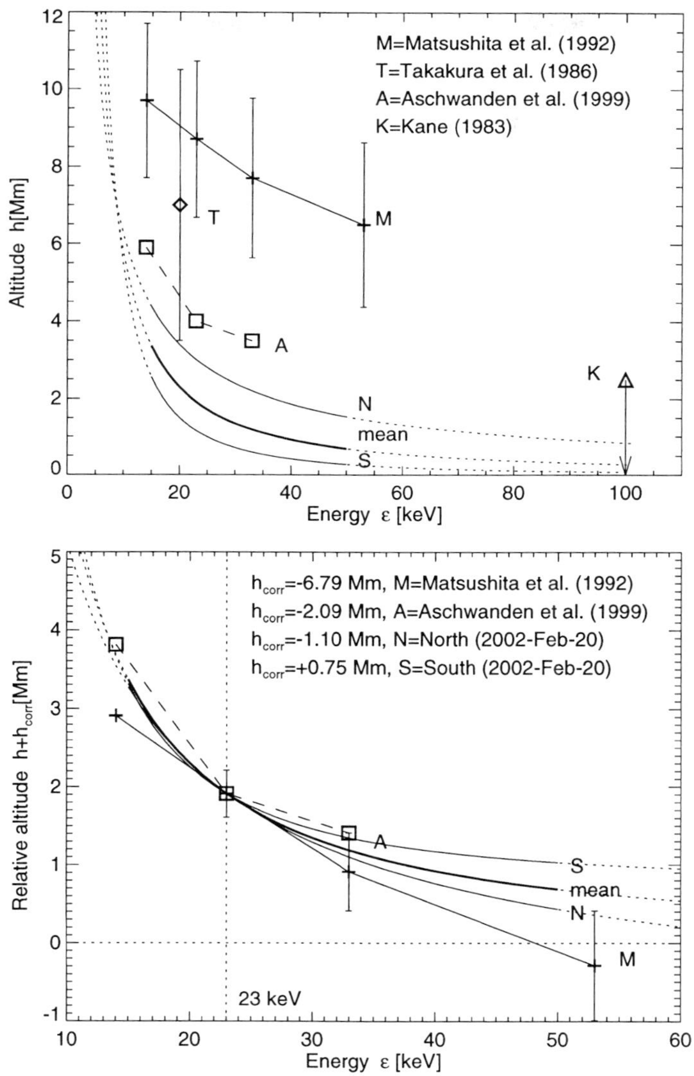


*Figure 9.* A compilation of reported hard X-ray source heights is shown in the *top panel*, from Kane (1983), Takakura *et al.* (1986), Matsushita *et al.* (1992), and Aschwanden *et al.* (1999), and from this work (N, S, mean). In the *lower panel* we shift the absolute height to a common reference level of $h = 1.9$ Mm at $\varepsilon = 23$ keV. Note that the agreement of relative source heights as function of energy (*lower panel*) agrees much better than the absolute heights (*top panel*).

relative height differences for a common reference height of $h(\varepsilon = 23$ keV$)=$ 1.9 Mm. The required corrections of the reported absolute altitudes are: $h_{corr} = -6.8$ Mm (Matsushita *et al.*, 1992), $h_{corr} - 2.1$ Mm (Aschwanden *et al.*, 1999), and $h_{corr} - 1.1$ Mm and $+ 0.8$ Mm for the north and south footpoints of the 20 February 2002 flare. The relative heights at different energies agree then within $\Delta h \approx \pm 0.5$ Mm over the entire energy range from 10 to 50 keV, all being consistent with a thick-target model with plausible chromospheric densities, even if we allow for the larger values inferred using smaller values of $\gamma$. Note, however, that $\Delta h \approx \pm 0.5$ corresponds to almost 2 density scale heights at low chromospheric temperatures, or a factor of 10 uncertainty in $n$.

### 3.2. Chromospheric Density Models

The main new result from this study is that we determine for the first time a chromospheric density model $n_e(h)$ from hard X-ray measurements, enabled by the new capabilities of RHESSI. It is therefore instructive to put our measurements into context with previous chromospheric models. A compilation of a number of chromospheric density models available in literature is shown in Figure 10. Chromospheric and coronal density models are sharply divided by a thin transition region where the temperature rises from the temperature minimum region ($\approx$ 10 000–20 000 K) to coronal temperatures ($T_e \gtrsim 1$ MK). Below the transition region, the plasma is only partially ionized, and thus the electron density $n_e$ is lower than the hydrogen density $n_H$. At the transition region, where complete ionization sets in, the neutral hydrogen density $n_{H^0}$ drops then to a very small value, at an altitude of approximately $h \approx 2.0$ Mm (Figure 10). Above the transition region, i.e., at $h \gtrsim 2.0$ Mm, the plasma is fully ionized and the electron density is almost equal to the ion density, i.e., $n_e \approx n_i$. Coronal density models therefore assume $n_e(h) \approx n_i(h)$.

Chromospheric density models have been calculated in great detail based on ion abundance measurements from a larger number of EUV lines, constrained by hydrostatic equilibrium and radiation transfer assumptions (e.g., Vernazza, Avrett, and Loeser, 1973, 1976, 1981; VAL models, see Figure 10), and ambipolar diffusion (Fontenla, Avrett, and Loeser, 1990, 1991; FAL models, see Figure 10). Newer developments include sunspot umbral models (Maltby *et al.*, 1986; Obridko and Staude, 1988), sunspot penumbral models (Ding and Fang, 1989), or stochastic multi-component models with hot flux tubes randomly embedded in a cool medium (Gu *et al.*, 1997), all shown in Figure 10. At coronal heights, i.e., at $h \gtrsim 2.0$ Mm, an average quiet sun density model was computed by Gabriel (1976), based on the expansion of the magnetic field of coronal flux tubes that line up with the boundaries of the supergranule convection cells. The geometric expansion factor and the densities at the lower boundary computed from chromospheric models then constrains coronal densities, which fall off exponentially with height, in isothermal fluxtubes with hydrostatic equilibrium. This yields electron densities of $n_e \approx 10^9$ cm$^{-3}$ at the

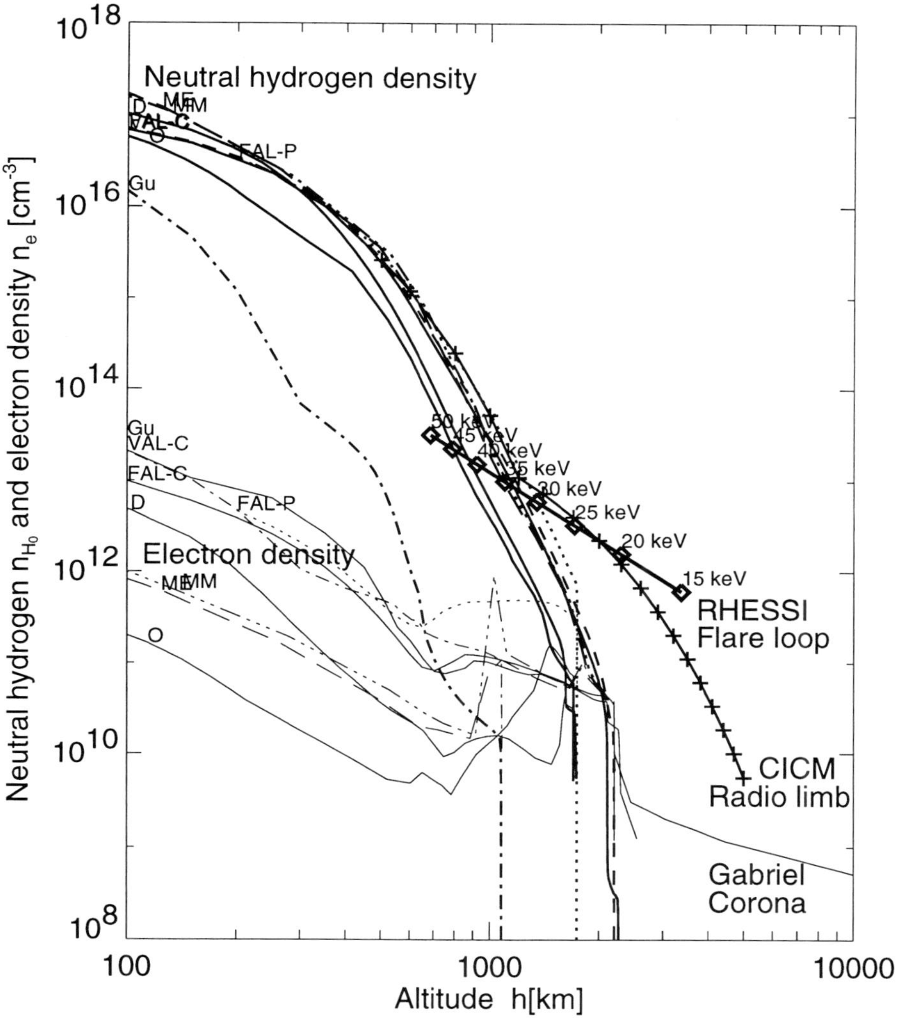


*Figure 10.* A compilation of chromospheric and coronal density models: VAL-C = Vernazza, Avrett, and Loeser (1981), model C; FAL-C = Fontenla, Avrett, and Loeser (1990), model C; FAL-P = Fontenla, Avrett, and Loeser (1990), model P; Gu = Gu *et al.* (1997); MM = Maltby *et al.*, 1986), model M; ME = Maltby *et al.* (1986), model E; D = Ding and Fang (1989); O = Obridko and Staude (1988); Gabriel = Gabriel (1976), coronal model; CICM = Caltech Irreference Chromospheric Model, radio sub-millimeter limb observations (Ewell *et al.*, 1993), RHESSI flare loop (this work).

coronal base in quiet-Sun regions (Gabriel, 1976), as shown in Figure 10. These hydrostatic models in the lower corona, however, have been criticized because of the presence of dynamic phenomena such as spicules, which contribute in the statistical average to an extended chromosphere. The spicular extension of this dynamic chromosphere has been probed with high-resolution measurements of the Normal

Incidence X-Ray Telescope (NIXT) (Daw, DeLuca, and Golub, 1995) as well as with radio submillimeter observations during a total eclipse (Ewell *et al.*, 1993). At radio wavelengths, the opacity of the chromosphere is due to free-free emission, which can be modeled straightforwardly from radio brightness temperature measurements at the limb. Using the radio limb height measurements at various mm and sub-mm wavelengths in the range of 200–3000 $\mu$m (Roellig *et al.*, 1991; Horne *et al.*, 1981; Wannier, Hurford, and Seielstad, 1983; Belkora *et al.*, 1992; Ewell *et al.*, 1993), an empirical Caltech Irreference Chromospheric Model (CICM) was established, which fits the observed limb heights between 500 km and 5000 km in a temperature regime of $T = 4410$ K to $T = 7500$ K (Ewell *et al.*, 1993), shown in Figure 10. We see that these radio limb measurements yield electron densities that are 1–2 orders of magnitude higher in the height range of 500–5000 km than predicted by hydrostatic models (VAL, FAL, Gabriel, 1976), which was interpreted in terms of the dynamic nature of spiculae (Ewell *et al.*, 1993). Comparing now our RHESSI measurements, which mainly probe the total neutral and ionised hydrogen density that governs the bremsstrahlung and the total bound and free electron density in the collisional energy losses, with the free electron density $n_e(h)$ inferred from the radio-based measurements of free-free emission, we see a remarkably good agreement, between the two density profiles in the height range of $h \approx 1000$–3000 km. The RHESSI-inferred densities in the hot parts of the flare loop ($\gtrsim$ 3000 km) exceed the radio-inferred densities, probably due to the process of chromospheric evaporation, which we discuss in the next subsection.

### 3.3. Chromospheric Evaporation

The collisional degradation of high-energy electrons precipitating during the impulsive flare phase heats up the chromospheric plasma and drives an upflow of heated plasma that fills the flare loops (Brown, 1973), most visible in the emission measure increase in soft X-rays, a scenario dubbed *chromospheric evaporation* (e.g., see reviews by Bornman, 1999 and Antonucci *et al.*, 1999). Hydrodynamic simulations of this particle-driven chromospheric heating process yield typical densities of $n_e \approx 10^{11}$ cm$^{-3}$ for the upflowing plasma (e.g., Mariska, Emslie, and Li, 1989; Mariska, 1995; Hori *et al.*, 1997, 1998). Slightly higher densities can be obtained by increasing the energy flux (of precipitating electrons and ions), but radiative loss becomes so efficient that not much more density can be pumped into the flare loops (Mariska, private communication).

We can compare the RHESSI-inferred densities with the values obtained from chromospheric evaporation models by averaging the density model $n(z)$ over a volume that corresponds to the hot flare loop segment, as seen in soft X-rays. Because the density is rapidly decreasing with height, we have to be careful where we assume the lower height limit of the hot flare plasma. According to the chromospheric models shown in Figure 10, the lower limit of the hot flare loop segment could be anywhere between $h \gtrsim 1000$ km and $h \lesssim 5000$ km, where dense cool spicular

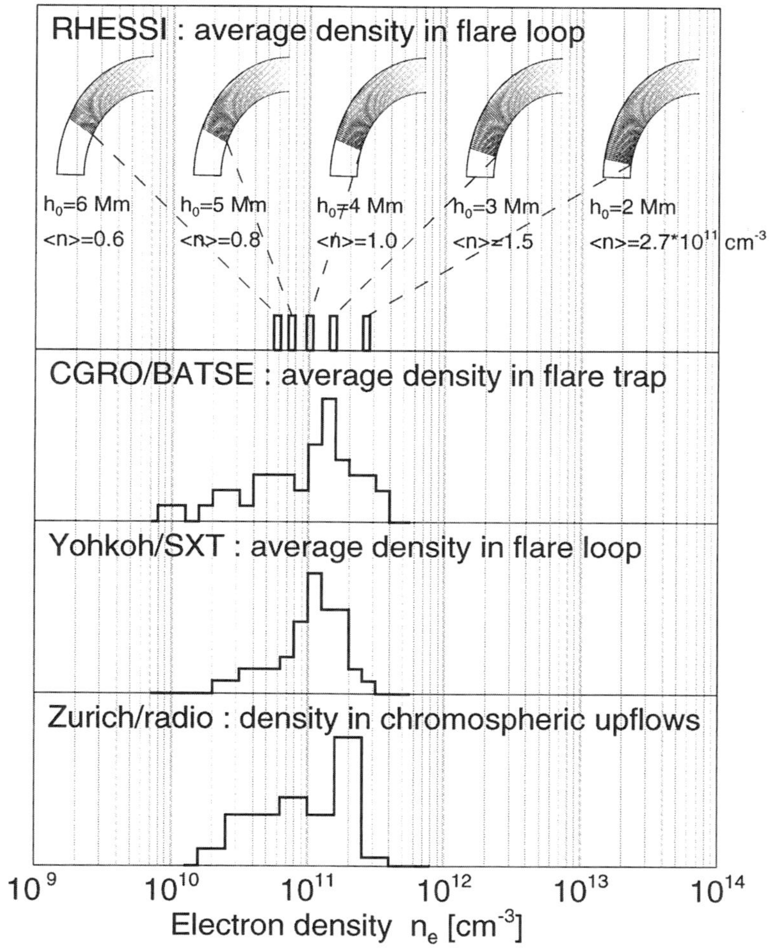


*Figure 11.* A comparison of average flare loop densities, obtained from RHESSI, CGRO/BATSE, *Yohkoh*/SXT, and Zürich radio measurements. The flare loop densities are obtained from RHESSI by averaging the thick-target density model over a loop segment between $h_0$ and the top, calculated for $h_0 = 2, 3, 4, 5, 6$ Mm, with a loop height of $r_0 = 12.4$ Mm (indicated as shaded loop segments at the top). The resulting densities vary from $\langle n \rangle = 0.6 \times 10^{11}$ cm$^{-3}$ to $\langle n \rangle = 2.7 \times 10^{11}$ cm$^{-3}$. Densities in the flare loop trap are inferred from CGRO/BATSE (*second panel*) in terms of trapping time delays interpreted as collisional deflection times (Aschwanden *et al.*, 1997). Densities in soft X-ray-bright flare loops have been determined from *Yohkoh*/SXT (*third panel*) by measuring the emission measure and loop diameters (Aschwanden *et al.*, 1997). Densities in chromospheric upflows were measured from Zürich radio measurements of slowly-drifting cutoffs at decimetric frequencies interpreted in terms of free-free opacity changes at the upflow fronts (Aschwanden and Benz, 1997).

material resides (CICM model). The average density is therefore, if we integrate the power-law function of our chromospheric density model (Equation (5)) over a loop segment of $h_0 < h < r_0$ (neglecting the curvature near the flare loop top, where the contributions to the average density are lowest anyway),

$$\langle n \rangle \approx \frac{\int_{h_0}^{r_0} n(s)\, \mathrm{d}s}{\int_{h_0}^{r_0} \mathrm{d}s} \approx n_0 \frac{z_0}{1.5(r_0 - h_0)} \left[ \left( \frac{h_0}{z_0} \right)^{-1.5} - \left( \frac{r_0}{z_0} \right)^{-1.5} \right]. \tag{10}$$

Our flare loop of the 20 February 2002 flare has a height of $r_0 = 12.4$ Mm. If we vary the lower height of the hot flare loop segment between $h_0 = 2000$ km and $h_0 = 6000$ km, we obtain average densities between $\langle n \rangle = 0.6 \times 10^{11}$ cm$^{-3}$ and $\langle n \rangle = 2.7 \times 10^{11}$ cm$^{-3}$ (illustrated in Figure 11, top panel). Thus, these average densities of the hot flare loop plasma are fully compatible with values obtained from hydrodynamic simulations, which typically yield $\langle n \rangle \approx 10^{11}$ cm$^{-3}$. This is of course in line with the simpler estimate that one needs a stopping depth of $N \approx 10^{20}$ to stop a typical beam electron which, divided by loop length of $10^9$ cm, yields $\langle n \rangle \approx 10^{11}$ cm$^{-3}$.

It is also interesting to compare these values with other methods. In Figure 11 we show comparisons with three different methods. From CGRO/BATSE data one can measure energy-dependent time delays that are associated with electron time-of-flight propagation as well as with trapping times. Interpreting the trapping times in terms of collisional deflection times, $\tau^{\mathrm{defl}}(E) = 0.95 \times 10^8 (E_{\mathrm{kev}}^{3/2}/n_e)(20/\ln \Lambda)$, one obtains average trap densities of $10^{10} \lesssim n_e \lesssim 4 \times 10^{11}$ cm$^{-3}$ (Figure 11, second panel), with an average of $n_e^{\mathrm{trap}} = 10^{11.1 \pm 0.3}$ (Aschwanden *et al.*, 1997). A second method is to measure the soft X-ray emission measure $EM = \int n_e^2(z)\mathrm{d}z \approx n_e^2 w$ and the flare loop widths $w$. This method yields density values in a range of $2 \times 10^{10} < n_e < 3 \times 10^{11}$ cm$^{-3}$ in a sample of 78 flares simultaneously observed with *Yohkoh*/SXT and CGRO (Aschwanden *et al.*, 1997). A third method can be performed with radio observations of slowly-drifting frequency cutoffs of decimetric emission at the plasma frequency, which has been interpreted as opacity change at the interface of chromospheric evaporation fronts (Aschwanden and Benz, 1997). This method yields densities in the range of $2 \times 10^{10} < n_e < 3 \times 10^{11}$ cm$^{-3}$ (Figure 11, bottom panel). All three methods yield comparable densities with an average of $n_e \approx 10^{11}$ cm$^{-3}$, but could be higher if there are filling factors involved. In this context it is highly interesting to note what highest electron densities in solar flares have been determined from density-sensitive line pairs, i.e., Fe XXI and Fe XXII, which are not sensitive to the filling factor (but see Almleaky, Brown, and Sweet, 1990, for a critical treatment). These highest densities were found at $n_e = 2{-}3 \times 10^{12}$ cm$^{-3}$, with a maximum of $n_e \approx 10^{13}$ cm$^{-3}$

(Phillips *et al.*, 1996). Therefore, our hard X-ray-inferred average flare loop densities ($n_e \approx 2$–$8 \times 10^{11}$ cm$^{-3}$) agree very well with the three other methods based on the assumption of unity filling factor, but also agree with the upper limits set by filling factor-independent density measurements at the footpoints of our hard X-ray-inferred density model.

## 4. Summary and Conclusions

We fitted the theoretical model $I(z, \varepsilon)$ derived in Paper I to gaussian forward fits of RHESSI observations, where the height $z$ above the solar surface was measured over an energy range of $\varepsilon = 15$–$50$ keV for a limb flare (20 February 2000). We found the following:

(1) The height of hard X-ray footpoint sources has an energy dependence that can be approximated by a power-law function, i.e., $z(\varepsilon) \approx 2.3(\varepsilon/20 \text{ keV})^{-1.32}$ Mm in the $\varepsilon = 15$–$50$ keV energy range for a spectral fit with $\gamma = 3.9$, though the absolute density scale could be up to a factor of $\approx 2.6$ higher due to uncertainties ($\gamma \approx 3.1$–$3.9$) in the spectral fitting procedure.

(2) The best fit of the analytical model $I(z, \varepsilon)$ to the data is consistent with a chromospheric density model of $n_e(h) = 1.25 \times 10^{13}(z/1 \text{ Mm})^{-2.5}$ cm$^{-3}$, for a spectral fit with $\gamma = 3.9$, though the absolute scale depends somewhat on the choice of energy range in spectral fitting.

(3) The unprecedented energy resolution of RHESSI yields a value of the flare plasma temperature, i.e., $T \approx 15$ MK, which dominates the observed photon spectrum at energies of $\varepsilon \lesssim 15$ keV, while the high-energy hard X-ray emission at $\varepsilon \gtrsim 15$ keV is nonthermal (or at least non-isothermal). Hard X-ray emission in the energy range of $\varepsilon = 15$–$20$ keV is produced by thick-target bremsstrahlung from larger fractions of the flare loop (see Figure 8), at heights of $h \approx 2.5$–$12.4$ Mm and densities of $n_e \approx 10^{11}$–$10^{12}$ cm$^{-3}$, while higher energy emission of $\varepsilon = 20$–$50$ keV originates from thick-target bremsstrahlung in chromospheric heights of $h \approx 500$–$2500$ km and densities of $n_e \approx 10^{12}$–$10^{14}$ cm$^{-3}$.

(4) Previous altitude measurements of hard X-ray sources are systematically higher than the RHESSI-inferred altitudes at the same energies, probably suffering from insufficient energy and spatial resolution of previous hard X-ray imagers. The relative altitude differences as a function of energy, however, agree well with the RHESSI measurements.

(5) The chromospheric density model inferred from RHESSI roughly agrees with the ion densities of previous chromospheric density models based on optical/EUV spectroscopy, at altitudes of $h \approx 500$–$1000$ km. At coronal heights, the RHESSI density model comes closest to the '*spicular extended chromospheric model*' inferred from radio sub-mm limb observations. Such spicules are indicative of a highly dynamic chromosphere, and have densities lying several orders of magnitude above quiet-Sun chromosphere/coronal models based on hydrosta-

tic equilibrium. The evaporative upflow in flares is a more extreme form of such deviation from hydrostatic conditions.

(6) The average flare loop densities obtained from the RHESSI measurements, $\langle n \rangle \approx 0.6–3 \times 10^{11}$ cm$^{-3}$, are fully comparable with other flare loop density measurements (assuming unity filling factor), $\langle n \rangle \approx 0.2–3 \times 10^{11}$ cm$^{-3}$, though substantially lower than the estimates of the highest densities present as measured by density-sensitive line ratios (cf., Almleaky, Brown, and Sweet, 1989) which reach $\langle n \rangle \approx 10^{12}–10^{13}$ cm$^{-3}$.

Our measurements and quantitative modeling have demonstrated that the Brown (1971) collisional thick-target model seems to be a pretty good description of the hard X-ray height structure for reasonable target density models. If one accepts the purely collisional thick-target model then RHESSI provides useful data to constrain the density model in the chromospheric and coronal parts of flare loops. The hard X-ray inferred density models have moreover the advantage that they determine spatial mean densities averaged over any highly localized regions, independent of filling factors. Also, no assumption on hydrodynamic equilibrium is needed, which is a restriction in most density models based on optical/EUV spectroscopy. Accurate measurements of chromospheric density are also important to modeling coronal loop oscillations regarding Alfvénic wave propagation, wave leakage, and associated damping mechanisms (e.g., Aschwanden *et al.*, 2002a). Future energy-dependent height measurements with RHESSI hopefully will also help constrain particle propagation all the way back to the acceleration regions. Some final caveats that add additional uncertainties to our density inversion method are discussed in Paper I, concerning non-collisional energy losses, pitch-angle distributions, partial target ionization, and return current losses.

## Acknowledgements

We thank Tom Metcalf for providing the pixon maps in Figure 2, Bart DePontieu for the digitized chromospheric data shown in Figure 10, and John Mariska for helpful information on hydrodynamic simulations of chromospheric evaporation. Also we thank the RHESSI team for extensive software support, useful tutorials and data analysis training workshops, and very helpful personal advice by many software team members. Support for this work was provided by the NASA SMEX grant NAS5-98033 through University of California, Berkeley (sub contract SA2241-26308PG) and by UK PPARC Standard and Visitor Grants.

## References

Almleaky, Y., Brown, J. C., and Sweet, P. A.: 1989, *Astron. Astrophys.* **224**, 328.

Antonucci, E., Alexander,D., Culhane, J. L., DeJager, C., MacNeice, P., Somov, B. V., and Zarro, D. M.: 1999, in K. T. Strong, J. L. R. Saba, B. M. Haisch, and T. Schmelz (eds.), *The Many Faces*

*of the Sun, A Summary of the Results from NASA's Solar Maximum Mission*, Springer-Verlag, Berlin, p. 331.
Aschwanden, M. J.: 2002, *Space Sci. Rev.* **101**, 1.
Aschwanden, M. J. and Benz, A. O.: 1997, *Astrophys. J.* **480**, 825.
Aschwanden, M. J., Bynum, R. M., Kosugi, T., Hudson, H. S., and Schwartz, R. A.: 1997, *Astrophys. J.* **487**, 936.
Aschwanden, M. J., Fletcher, L., Sakao, T., Kosugi, and Hudson, H.: 1999, *Astrophys. J.* **517**, 977.
Aschwanden, M. J., DePontieu, B., Schrijver, C. J., and Title, A.: 2002a, *Solar Phys.* **206**, 99.
Aschwanden, M. J., Schmahl, E. J. and the RHESSI Team, 2002b, *Solar Phys.* (this volume).
Belkora, L., Hurford, G. J., Gary, D. E., and Woody, D.: 1992, *Astrophys. J.* **400**, 629.
Bornmann, P. L.: 1999, in K. T. Strong, J. L. R. Saba, B. M. Haisch, and J. T. Schmelz (eds.), *The Many Faces of the Sun, A Summary of the Results from NASA's Solar Maximum Mission*, Springer-Verlag, Berlin, p. 301.
Brown, J. C.: 1971, *Solar Phys.* **18**, 489.
Brown, J. C.: 1972, *Solar Phys.* **26**, 441.
Brown, J. C.: 1973, *Solar Phys.* **32**, 227.
Brown, J. C. and MacKinnon, A. L.: 1985, *Astrophys. J.* **292**, L31.
Brown, J. C. and Emslie, A. G.: 1988, *Astrophys. J.* **331**, 554.
Brown, J. C., and McClymont, A. N.: 1976, *Solar Phys.* **49**, 329.
Brown, J. C., Carlaw, V. A., Cromwell, D., and Kane, S. R.: 1983, *Solar Phys.* **88**, 281.
Brown, J. C., Aschwanden, M. J., and Kontar, E. P.: 2002, *Solar Phys.*, this volume (Paper I).
Daw, A., DeLuca, E. E., and Golub, L.: 1995, *Astrophys. J.* **453**, 929.
Dennis, B. R. and Zarro, D. M.: 1993, *Solar Phys.* **146**, 177.
Ding, M. D. and Fang, C.: 1989, *Astron. Astrophys.* **225**, 204.
Ewell, M. W. Jr., Zirin, H., Jensen, J. B., and Bastian, T. S.: 1993, *Astrophys. J.* **403**, 426.
Fletcher, L.: 1996, *Astron. Astrophys.* **310**, 661.
Fontenla, J. M., Avrett, E. H., and Loeser, R.: 1990, *Astrophys. J.* **355**, 700.
Fontenla, J. M., Avrett, E. H., and Loeser, R.: 1991, *Astrophys. J.* **377**, 712.
Gabriel, A. H.: 1976, *Phil. Trans. Roy. Soc. (London)* **A281**, 339.
Gu, Y., Jefferies, J. T., Lindsey, C., and Avrett, E. H.: 1997, *Astrophys. J.* **484**, 960.
Hori, K., Yokoyama, T., Kosugi, T., and Shibata, K.: 1997, *Astrophys. J.* **489**, 426.
Hori, K., Yokoyama, T., Kosugi, T., and Shibata, K.: 1998, *Astrophys. J.* **500**, 492.
Horne, K., Hurford, G. J., Zirin, H., and DeGraauw, Th.: 1981, *Astrophys. J.* **244**, 340.
Kane, S. R.: 1983, *Solar Phys.* **86**, 355.
Maltby, P., Avrett, E. H., Carlsson, M., Kjeldseth-Moe, O., Kurucz, R. L., and Loeser, R.: 1986, *Astrophys. J.* **306**, 284.
Mariska, J. T.: 1995, *Astrophys. J.* **444**, 478.
Mariska, J. T., Emslie, A. G., and Li, P.: 1989, *Astrophys. J.* **341**, 1067.
Matsushita, K., Masuda, S., Kosugi, T., Inda, M., and Yaji, K.: 1992, *Publ. Astron. Soc. Japan* **44**, L89.
Obridko, V. N. and Staude, J.: 1988, *Astron. Astrophys.* **189**, 232.
Phillips, K. J. H., Bhatia, A. K., Mason, H. E., and Zarro, D. M.: 1996, *Astrophys. J.* **466**, 549.
Roellig, T. L., Becklin, E. E., Jefferies, J. T., Kopp, G. A., Lindsey, C. A., Orrall, F. Q., and Werner, M. W.: 1991, *Astrophys. J.* **381**, 288.
Takakura, K., Tanaka, K., Nitta, N., Kai, K., and Ohki, K.: 1986, *Solar Phys.* **107**, 109.
Vernazza, J. E., Avrett, E. H., and Loeser, R.: 1973, *Astrophys. J.* **184**, 605.
Vernazza, J. E., Avrett, E. H., and Loeser, R.: 1976, *Astrophys. J. Suppl.* **30**, 1.
Vernazza, J. E., Avrett, E. H., and Loeser, R.: 1981, *Astrophys. J. Suppl.* **45**, 635.
Wannier, P. G., Hurford, G. J., and Seielstad, G. A.: 1983, *Astrophys. J.* **264**, 660.

# EMPIRICAL CORRECTION OF RHESSI SPECTRA FOR PHOTOSPHERIC ALBEDO AND ITS EFFECT ON INFERRED ELECTRON SPECTRA

R. CALUM ALEXANDER and JOHN C. BROWN
*Department of Physics and Astronomy, University of Glasgow, Glasgow G12 8QQ, U.K.*

(Received 6 August 2002; accepted 22 August 2002)

**Abstract.** Photospheric Compton backscatter (albedo) makes a significant contribution to observed hard X-ray (HXR) spectral fluxes over the RHESSI energy range and should be allowed for in HXR spectral interpretation. The full correction problem is nonlinear and messy but we offer a simple approximate first-order correction procedure for global HXR spectra based upon empirical fits to published albedo simulations. We also illustrate the impact of this correction on inferred electron spectra for the thin- and thick-target models.

## 1. Introduction

The high HXR spectral resolution of RHESSI creates the chance for precise study of source electron spectra provided the observed spectra are well corrected for non-primary effects at the Sun including albedo, directivity, source ionisation variations and the like. General treatment of all of these is theoretically complex and computationally intensive for each source model one wants to try. Although this should be an eventual goal, for the present it is more useful to have simpler but more versatile methods for first-order correction of spectral data (after optimal removal of the already complex instrumental effects). In this paper we offer such an approach for correction of the photospheric albedo.

It was first pointed out by Tomblin (1972) and by Santangelo, Horstman, and Horstman-Moretti (1973) that some of the primary photons emitted downward around the deka-keV range in the optically thin solar atmosphere undergo Compton backscatter in the low atmosphere and add to the primary photons, emitted upwards, in the observed signal. These albedo photons come from an extended area and arrive with a spread of delay times, depending on the primary source height. It has been suggested that careful imaging (Brown, van Beek, and McClymont, 1975) or time delay (Bai, 1978) studies could help infer source heights. In this paper we restrict ourselves solely to considering the effect of the albedo contribution on the spectrum of the HXR source as a whole.

The spectral distribution of the albedo contribution was studied in detail by Bai and Ramaty (1978) via Compton scattering simulation allowing for photon absorption, and we base our empirical approach (Section 2) on their results. The

*Solar Physics* **210:** 407–418, 2002.

main need for empirical simplification is the fact that albedo correction is not just a spectral correction factor (like a 'coloured' mirror) but strictly speaking is itself a convolution. That is, the fractional albedo addition $A(\epsilon)$ at photon energy $\epsilon$ to the primary photon spectrum $I_0(\epsilon)$ itself depends on the functional form of $I_0(\epsilon)$, since Compton scattering shifts photons in energy. Thus recovery of $I_0(\epsilon)$ from the total observed spectrum $I(\epsilon)$ involves inversion of a nonlinear convolution of $I_0(\epsilon)$ with the scattering and absorption processes.

## 2. Empirical Albedo Correction to Spectra

Suppose the total primary HXR source emission rate is $I_0(\epsilon)$ (photons cm$^{-2}$ s$^{-1}$ per unit $\epsilon$) as seen at the Earth. The actual total rate $I(\epsilon)$ as seen from the Sun can be written as

$$I_{\mathrm{tot}}(\epsilon) = I_0(\epsilon)(1 + A_{I_0}(\epsilon)), \qquad (1)$$

where $A_{I_0}(\epsilon)$ is the fractional albedo contribution at energy $\epsilon$. This depends, as mentioned, on $I_0(\epsilon)$ itself as well as on $\epsilon$ explicitly. In fact $A_{I_0}$ also depends somewhat on the source geometry; specifically on the source height above the photosphere and the heliocentric-angle of the source, since Compton backscatter is angle and energy dependent. Since variation across the disk is not large and is more of a scale factor than a spectral distortion, here we ignore it and use angle-averaged results from Bai and Ramaty (1978), though our approach could readily by extended to apply to each flare heliocentric angle separately by fitting of our empirical form to the appropriate Bai and Ramaty results for that angle.

Bai and Ramaty computed $A_{I_0}(\epsilon)$ for the specific forward problem of a simple power-law $I_0(\epsilon) \sim \epsilon^{-\gamma}$ and for these cases we can write

$$A_{I_0}(\epsilon) = A_\gamma(\epsilon). \qquad (2)$$

What we would like is a simplified procedure to estimate $I_0(\epsilon)$ from $I(\epsilon)$ allowing for the influence of $A(\epsilon)$ and we propose the following:

(1) Since A, while significant, is never large or rapidly varying, obtain a first approximation to $I_0(\epsilon)$ as a power law $\sim \epsilon^{-\gamma}$ by best fitting the total data $I(\epsilon)$ to a power-law index $\gamma$. (Essentially taking $A =$ constant to zeroth order).

(2) Use the best fit $\gamma$ from step (1) with the Bai and Ramaty results on $A_\gamma(\epsilon)$ for a power law $I_0(\epsilon)$ to obtain a first-order estimate of $A_{I_0}(\epsilon)$ using Equation (2).

(3) Adopt this $A_\gamma(\epsilon)$ in Equation (1) to derive a first-order albedo-corrected $I_0(\epsilon) = I_{\mathrm{tot}}(\epsilon)/(1 + A_\gamma(\epsilon))$.

To make this easy to do in practice we have explored convenient parameterized forms of $A_\gamma(\epsilon)$ and best fit the ($\gamma$ dependent) parameters to the Bai and Ramaty results for each of the four different $\gamma$ values which they simulated. We found a convenient form was

TABLE I

Best-fit parameters to the Bai and Ramaty data.

| $\gamma$ | $A_0$ | $a$ | $b$ |
|---|---|---|---|
| 2 | 0.0077 | 1.53 | 0.31 |
| 3 | 0.0088 | 1.46 | 0.34 |
| 4 | 0.0098 | 1.41 | 0.37 |
| 5 | 0.0111 | 1.34 | 0.38 |

$$A_\gamma(\epsilon) = A_0(\gamma)\left(\frac{\epsilon}{10}\right)^{a(\gamma)} e^{-b(\gamma)(\epsilon/10)}, \tag{3}$$

where $\epsilon$ is in keV and the best fit values for of $A_0$, $a$, and $b$ are given in Table I. In Figures 1(a–d) we show the Bai and Ramaty $A_\gamma(\epsilon)$ results with our best fit superimposed.

Figures 1(a–d) show the plots of the best-fit parameters given in Table I for $A_0(\gamma)$, $a(\gamma)$, and $b(\gamma)$ substituted into (3) (full line) compared with the Bai and Ramaty data (symbols) for the four values of $\gamma$ studied.

Bai and Ramaty only computed four cases of $\gamma$. To extend these fits empirically to general $\gamma$ we show in Figures 2(a–c) plots of $A_0(\gamma)$, $a(\gamma)$, and $b(\gamma)$ versus $\gamma$ with linear fits through these four points.

Figures 2(a–c) show the plots of the best-fit parameters given in Table I for $A_0(\gamma)$, $a(\gamma)$, and $b(\gamma)$. A linear fit straight line through the data points is shown for each parameter.

## 3. Effect of Albedo Correction on Inferred Electron Spectra

### 3.1. Formulation

Given $I_0(\epsilon)$, one can derive the source mean electron spectrum $\bar{F}(E)$ ('thin-target') and the collisional thick-target electron injection spectrum $\mathcal{F}_0(E_0)$ by inversion of the bremstrahlung spectral integral. This was shown analytically by Brown (1971) for the Bethe–Heitler cross section and by Brown and Emslie (1988) for the Kramer's cross-section. Here we use the Kramer's analytic case to demonstrate how large the effect of $A(\epsilon)$ can be on inference of source electron spectra. Specifically we: (a) consider cases where the primary electron spectra are in fact power law ($E^{-\delta}$ or $E_0^{-\delta}$), resulting in power law $I_0(\epsilon)$ in both cases (with $\gamma = \delta+1, \delta-1$, respectively), (b) generate the total $I_{\text{tot}}(\epsilon) = I_0(\epsilon)(1 + A(\epsilon))$ that would be observed using Equation (3), (c) use the analytic inversion formulae to find out what

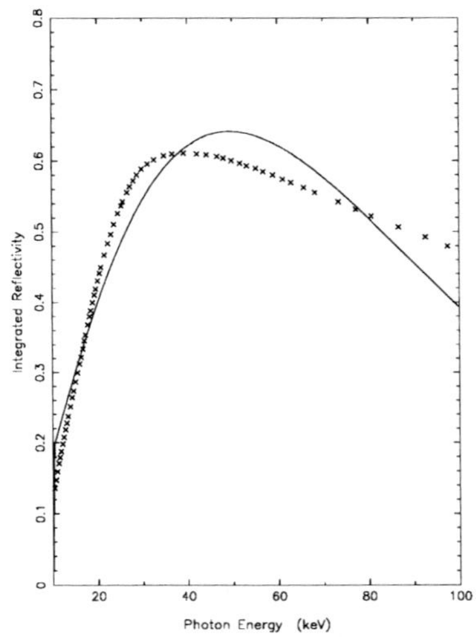


a. $\gamma = 2$

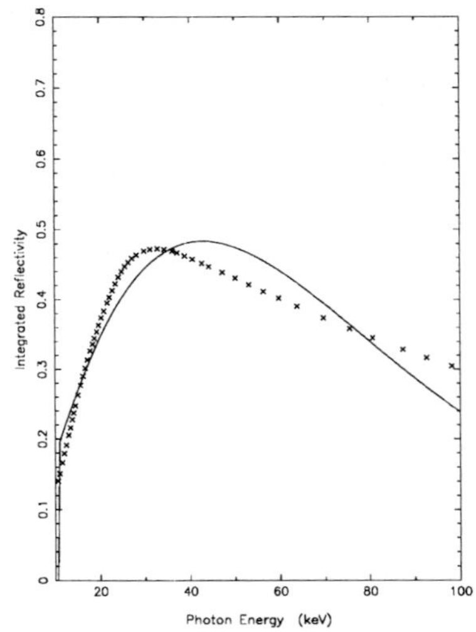


b. $\gamma = 3$

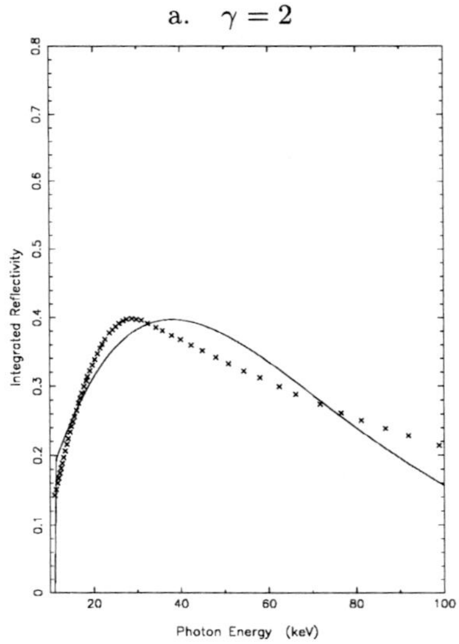


a. $\gamma = 4$

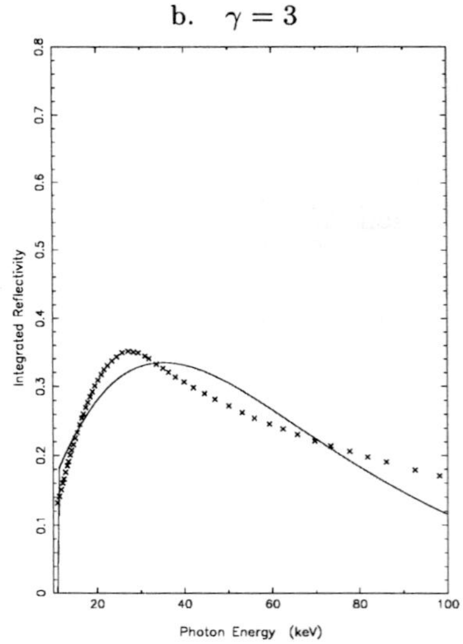


b. $\gamma = 5$

*Figure 1.* Plots of best-fit parameters (*solid line*), along with the Bai and Ramaty data for the 10–100 keV photon energy range.

$\bar{F}(E)$, $\mathcal{F}_0(E_0)$ would be derived from that $I_{\text{tot}}(\epsilon)$ if albedo were ignored, i.e., if it were assumed that $I_0 = I_{\text{tot}}$.

The *normalized* Kramer's approximation to the bremstrahlung cross-section is

$$Q(\epsilon, E) = \frac{Q_0}{\epsilon E} q(\epsilon, E),$$

where $q = 1$ and $Q_0$ is a constant.

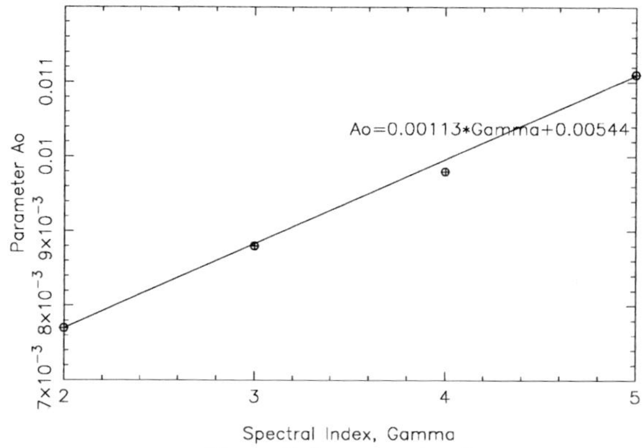


a. Linear fit for parameter $A_o$

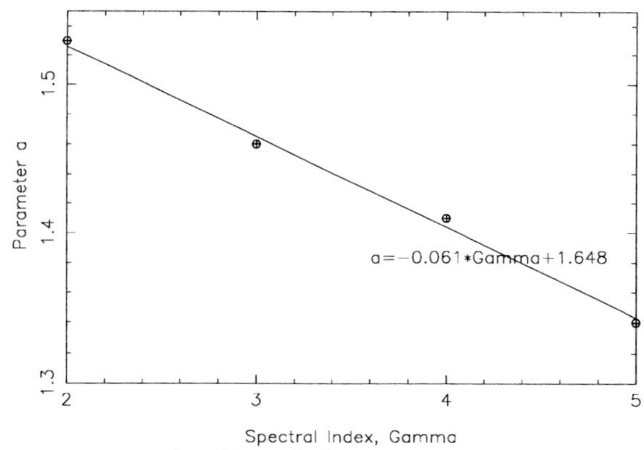


b. Linear fit for parameter a

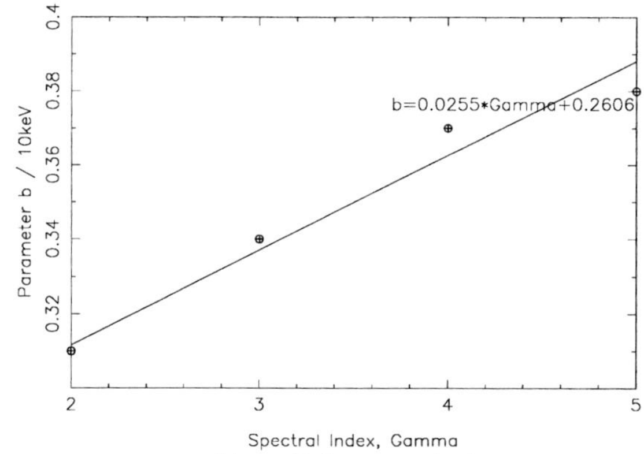


c. Linear fit for parameter b

*Figure 2.* Plots of linear fits through the best-fit parameters $A_0(\gamma)$, $a(\gamma)$, and $b(\gamma)$.

### 3.2. Thin Target

For this case the solution for the mean (thin-target) electron spectrum $\bar{F}(E)$ in a source of mean density $\bar{n}_p$, volume $V$ is (Brown and Emslie, 1988)

$$\bar{F}(E) = \frac{1}{\bar{n}_p V Q_0} E \left[ -\frac{\mathrm{d}}{\mathrm{d}\epsilon}(\epsilon I(\epsilon) \right]_{\epsilon=E} . \tag{4}$$

In the case of a power law

$$I_0(\epsilon) = C\epsilon^{-\gamma}, \tag{5}$$

this yields

$$\bar{F}(E) \equiv \bar{F}_0(E) = \frac{(\gamma - 1)C}{\bar{n}_p V Q_0} E^{-\gamma+1}. \tag{6}$$

However, in reality we observe $I_{\text{tot}} = I_0(1 + A(\epsilon))$ rather than $I_0$ and if we ignore the albedo correction by misidentifying $I_{\text{tot}}$ (Equations (3) and (5)) with $I_0$ then by (4) – we would infer

$$\bar{F}(E) = \frac{(\gamma - 1)C}{\bar{n}_p V Q_0} E^{-\gamma+1} \left[ 1 - A_0 E^a e^{-bE} \frac{(1 + a - (\gamma + bE))}{\gamma - 1} \right], \tag{7}$$

which is shown in Figure 3(a–d), or a mean electron spectrum wrong by a fractional error

$$f_{\text{thin}}(\gamma) = \frac{\Delta\bar{F}(E)}{\bar{F}_0(E)} = \frac{\bar{F}(E) - \bar{F}_0(E)}{\bar{F}_0(E)} = \frac{A_0 E^a e^{-bE} \left[ (\gamma + bE) - (a + 1) \right]}{\gamma - 1}, \tag{8}$$

which is shown in Figure 4.

### 3.3. Thick Target

Here the relevant electron spectrum is the total electron injection rate $\mathcal{F}_0(E_0)$ electrons per s per unit $E_0$ and is given from the photon spectrum by Brown and Emslie (1988) (with $K = 2\pi e^4 \Lambda$):

$$\mathcal{F}_0(E_0) = \frac{K}{Q_0} \left[ \frac{\mathrm{d}^2}{\mathrm{d}\epsilon^2}(\epsilon I_0(\epsilon)) \right]_{\epsilon=E_0} . \tag{9}$$

For a primary power-law $I_0(\epsilon)$, (5), this leads to

$$\mathcal{F}_0(E_0) = \frac{K}{Q_0} C(\gamma - 1) E_0^{-\gamma-1}, \tag{10}$$

while if the albedo correction is added to $I_0$ and $I_{\text{tot}}$ is misinterpreted as $I_0$ then one obtains

$$\mathcal{F}(E_0) = \frac{K}{Q_0} C E_0^{-\gamma} \times \\ \times \left[ \frac{\gamma}{E_0}(\gamma - 1) + A_0 \epsilon^a e^{-b\epsilon} \left( (\gamma - 1)(\frac{\gamma}{E_0} - 2(\frac{a}{E_0} - b)) + \right. \right. \\ \left. \left. + E_0 \left( (\frac{a}{E_0} - b)^2 - \frac{a}{E_0^2} \right) \right) \right], \tag{11}$$

which is incorrect by a fractional amount

$$f_{\text{thick}}(\gamma) = \frac{\mathcal{F}(E_0) - \mathcal{F}_0(E_0)}{\mathcal{F}_0(E_0)} = \frac{A_0 \epsilon^a e^{-b\epsilon} E_0}{\gamma(\gamma - 1)} \times \\ \times \left[ (\gamma - 1)(\frac{\gamma}{E_0} - 2(\frac{a}{E_0} - b)) + E_0 \left( (\frac{a}{E_0} - b)^2 - \frac{a}{E_0^2} \right) \right],$$

which is shown in Figure 6.

### 3.4. Resulting Correction for the Kramer's Cross Section

Figures 3(a–d) show the plots of the recovered thin-target Kramer electron spectrum (solid lines) along with their respective primary electron spectrum (broken lines) for the four values of $\gamma$ studied. The error in the recovered spectrum can be observed as a 'bump' in the spectrum.

Figure 4 shows the fractional difference decreasing as $\gamma$ increases. It also shows that the inferred electron flux at 'X' keV will be *greater* by 'Y' percent when the effect of an albedo has been included. This percentage also varies with energy. For example, at 100 keV, an error of approximately 35% is obtained for $\gamma = 2$.

Figures 5(a–d) show the log plots of the recovered thick-target Kramer electron spectrum $\mathcal{F}(E_0)$ (solid lines) along with their respective primary electron spectrum $\mathcal{F}_0(E_0)$ (broken lines) for the four values of $\gamma$ studied.

Figure 6 shows the fractional difference decreasing as $\gamma$ increases. As previously mentioned for the thin-target case, there is an energy dependent difference in the inferred electron flux where the effect of an albedo has been included. It is also apparent from Figure 6 that the error in the thick-target is significantly greater than in the thin-target case.

## 4. Discussion

RHESSI spectra have not been reported as exhibiting evident spectral bulges like those shown in Figures 3(a–d) and 5(a–d). In order to redress this apparent discrepancy the following explanations might be offered:

(1) The primary spectrum $I_0(\epsilon)$ had a dip where $A(\epsilon)$ has a bulge, the two offsetting one another. This seems too much of a coincidence to be plausible.

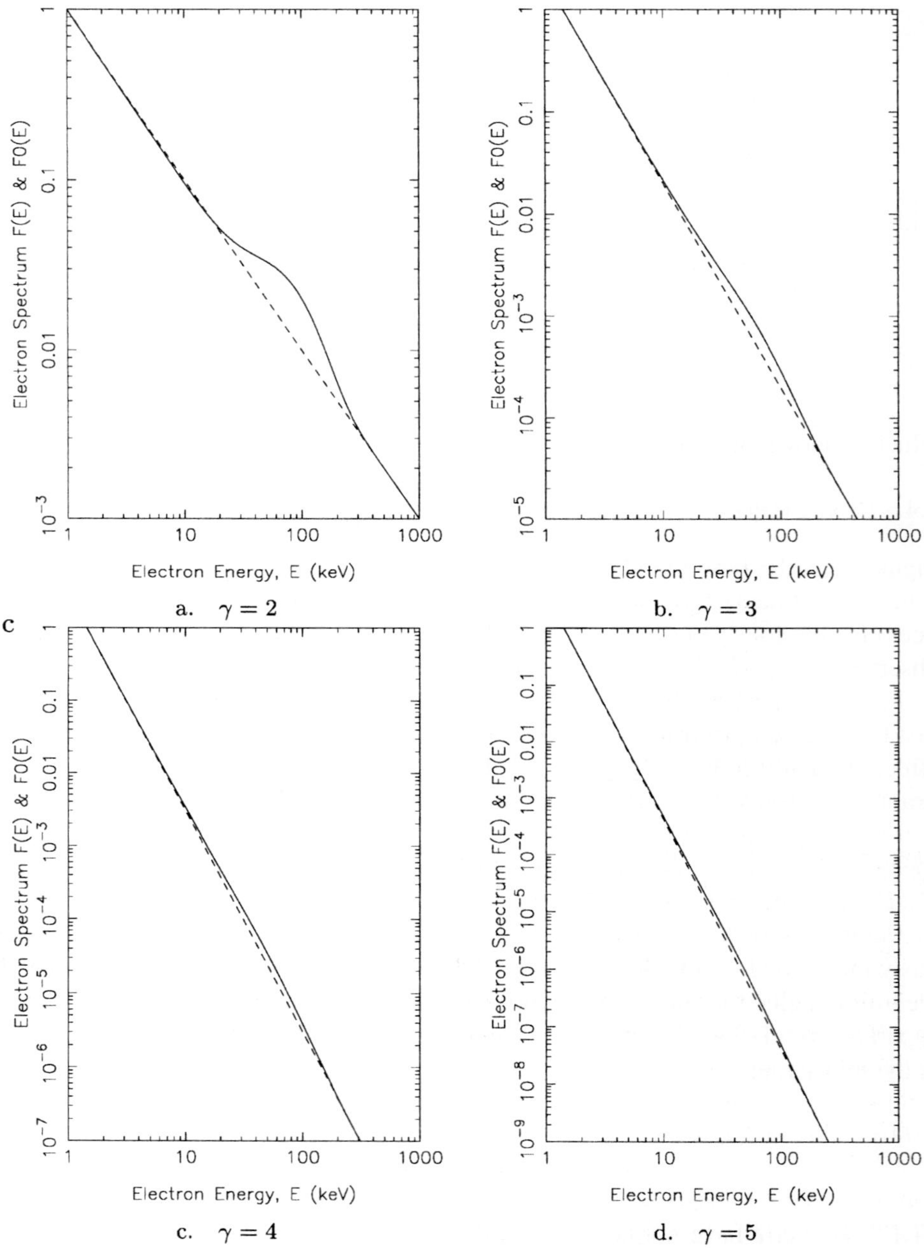


*Figure 3.* Plots of observed electron spectrum $\bar{F}(E)$ (*solid line*) and primary electron spectrum $\bar{F}_0(E)$ (*dashed line*).

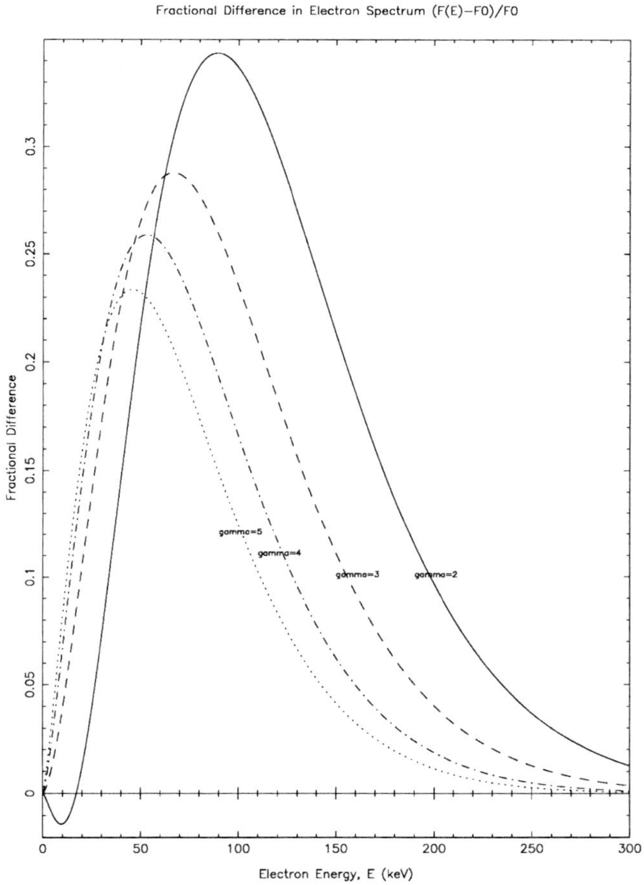


*Figure 4.* Fractional difference between the observed and primary thin target (Kramer's) electron spectrum showing the fraction difference decreasing with increasing $\gamma$.

(2) The bulges are present but have not been specifically noticed or reported as such. In fact, the lower energy end of the bulge is down around 10 keV which may be lost in the thermal emission component. The middle and upper end energy range of the bulge looks somewhat like a downward knee in the deka-keV range. Such features are regularly seen in data, cf., discussion in Kontar, Brown, and McArthur (2002). Albedo may be a partial explanation of these.

(3) There are other corrections – especially that for non-uniform target ionisation in the case of thick-target primary sources discussed by Kontar, Brown, and McArthur (2002) which have been ignored in this paper. Depending on the depth ('energy') of the transition region, this correction might tend to either augment or hide the effect of albedo on the spectrum.

(4) The assumption of an isotropic, point source (Bai and Ramaty, 1978) which provided our source data for $A(\epsilon)$ may require modification.

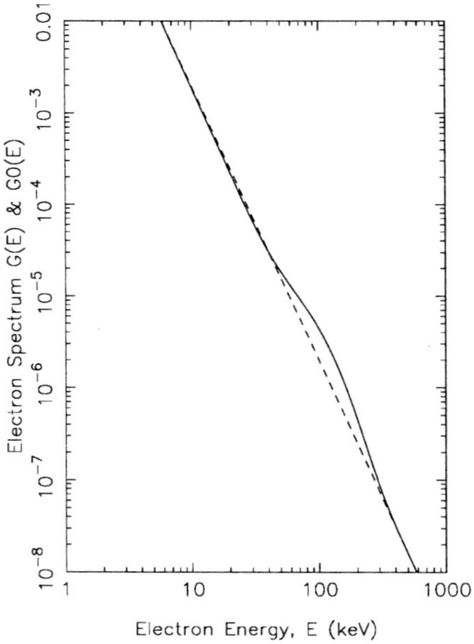


a. $\gamma = 2$

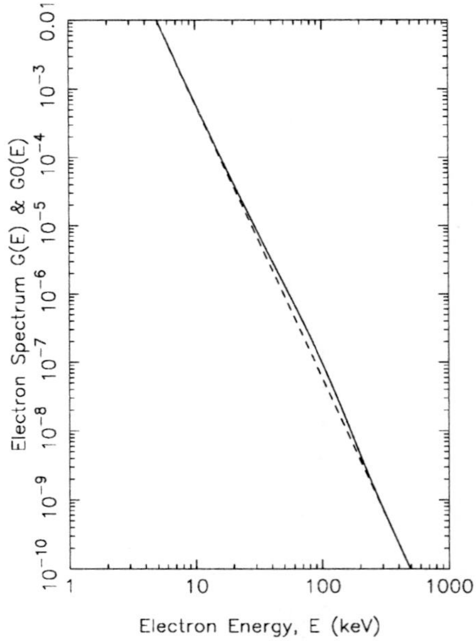


b. $\gamma = 3$

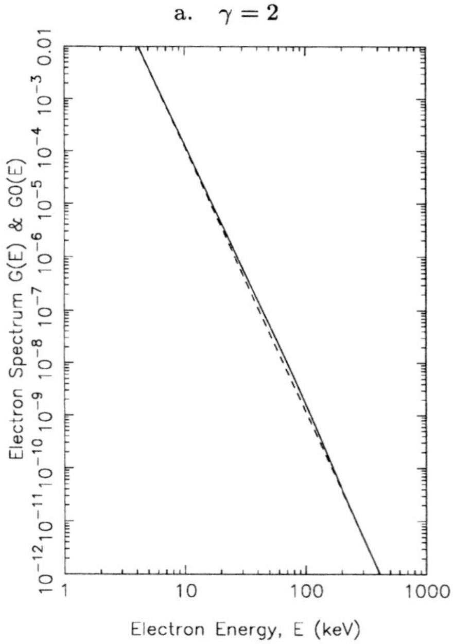


c. $\gamma = 4$

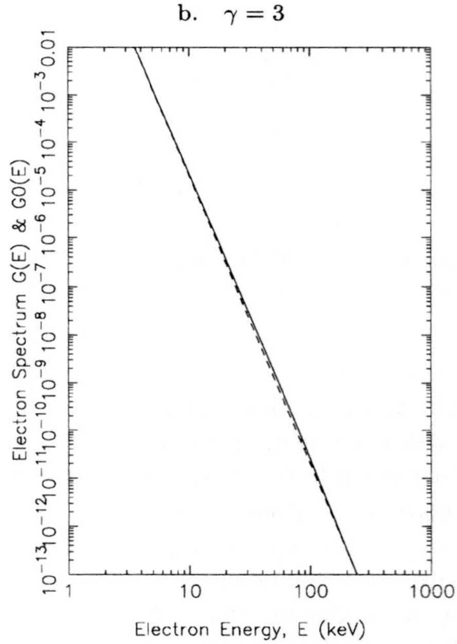


d. $\gamma = 5$

*Figure 5.* Plots of cold thick-target electron spectrum $\mathcal{F}(E_0)$ (*solid line*) and primary electron spectrum $\mathcal{F}_0(E_0)$ (*dashed line*).

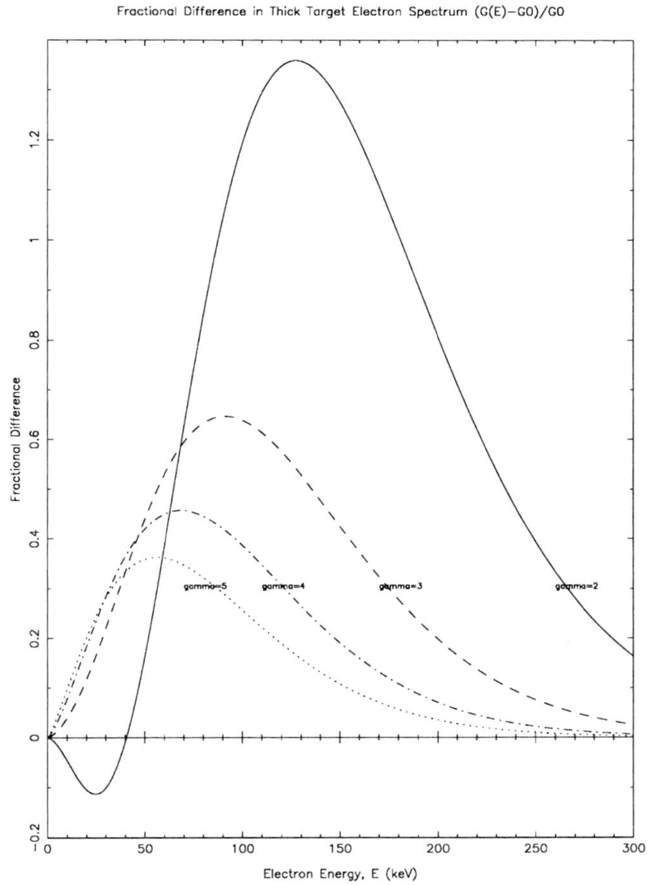


*Figure 6.* Fractional difference between the observed and primary thick-target electron spectrum.

## 5. Conclusion

We have explored the effect of photospheric albedo on observations of global flare hard X-ray spectra and derived an expression to allow approximate correction for this in the case of primary power-law photon spectra. We have also examined, for the Kramer's cross-section, the consequences of ignoring the albedo correction in using observed spectra to infer flare source electron spectra for thin and thick target interpretations and shown that the effects are very significant in terms of inferred spectral shape, especially for hard spectra. We have not extended the analysis to other cross sections but we note that the effects of albedo on deriving electron spectra will be even larger for more realistic smoother cross-section approximations, such as the Bethe–Heitler, than for Kramer's because they filter the electron spectral features even more. This is confirmed by our preliminary results for the Bethe–Heitler case, to be presented in a future paper. We also note that the effects of albedo should be considered alongside other corrections such as that of nonuniform target ionisation in the case of the thick target beam model as discussed by Kontar, Brown and McArthur (2002).

## Acknowledgements

This research was in part funded by PPARC. We would also like to acknowledge the input from useful discussions with A. Conway, M. Hendry, and L. Fletcher.

## References

Bai, T.: 1978, *Solar Phys.* **59**, 141.
Bai, T. and Ramaty, R.: 1978, *Astrophys. J.* **219**, 705.
Brown, J. C.: 1971, *Solar Phys.* **18**, 489.
Brown, J. C. and Emslie, A. G.: 1988, *Astrophys. J.* **331**, 554.
Brown, J. C., Van Beek, H. F., and McClymont, A. N.: 1975, *Astron. Astrophys.* **41**, 395.
Kontar, E. P., Brown, J. C., and McArthur, G., 2002, *Solar Phys.*, this volume.
Santangelo, N., Horstman, H., and Horstman-Moretti, E.: 1973, *Solar Phys.* **29**, 143.
Tomblin, F. F.: 1972, *Astrophys. J.* **171**, 377.

# NONUNIFORM TARGET IONIZATION AND FITTING THICK TARGET ELECTRON INJECTION SPECTRA TO RHESSI DATA

EDUARD P. KONTAR[1], JOHN C. BROWN[1] and GUILLIAN K. McARTHUR[1,2]
[1]*Astronomy and Astrophysics Group, Dept. of Physics and Astronomy, University of Glasgow, Glasgow G12 8QQ, U.K. (e-mail: john@astro.gla.ac.uk)*
[2]*Catholic University of America, Goddard Space Flight Center*

(Received 1 August 2002; accepted 31 August 2002)

**Abstract.** Past analyses of flare hard X-ray (HXR) spectra have largely ignored the effect of nonuniform ionization along the electron paths in the thick-target model, though it is very significant for well-resolved spectra. The inverse problem (photon spectrum to electron injection spectrum $\mathcal{F}_0(E_0)$) is disturbingly non-unique. However, we show that it is relatively simple to allow for the effect in forward fitting of parametric models of $\mathcal{F}_0(E_0)$ and provide an expression to evaluate it for the usual single power-law form of $\mathcal{F}_0(E_0)$. The expression involves the column depth $N_*$ of the transition region in the flare loop as one of the parameters so data fitting can enable derivation of $N_*$ (and its evaporative evolution) as part of the fitting procedure. The fit to RHESSI data on four flares for a single power law $\mathcal{F}_0(E_0)$ is much improved when ionization structure is included compared to when the usual fully ionized approximation is used. This removes the need, in these events at least, to invoke broken power laws, or other forms, of the acceleration spectrum $\mathcal{F}_0(E_0)$ to explain the observed photon spectrum

## 1. Introduction

To utilize fully the high spectral resolution of RHESSI in diagnosing flare photons and particles, it is essential to correct the observed radiation spectra (already instrument-corrected) for any effects at the Sun which contaminate or distort them. For example, correction for photospheric albedo is discussed elsewhere in this volume (Alexander and Brown, 2002). In the case of thick target model interpretation of hard X-ray (HXR) spectra, one effect, which has been largely ignored in data modeling till now, is that of varying ionization along the thick target beam electron paths. As first discussed by Brown (1973a) the fall of ionization with depth in the atmosphere reduces long-range collisional energy losses and so enhances the HXR bremsstrahlung efficiency there, elevating the high energy end of the HXR spectrum by factors of up to 2.8 above that for an ionized target. The net result is that a power-law electron spectrum of index $\delta$ produces a photon spectrum of index $\gamma = \delta - 1$ at low and high energies but with $\gamma < \delta - 1$ in between. The upward knee occurs at fairly low energies, probably masked in data by the tail of the thermal component, while the downward knee occurs in the few deka-keV range, depending on the column depth of the transition zone.

*Solar Physics* **210:** 419–429, 2002.

The inverse problem of determining the electron 'injection' spectrum $\mathcal{F}_0(E_0)$ from the HXR spectrum was solved analytically by Brown (1971) for the Bethe-Heitler cross-section, and by Brown and Emslie (1988) for the Kramer's cross-section, and numerically by a variety of approaches (see references in Emslie, Barrett, and Brown, 2001) but only for a uniformly ionized target. Brown *et al.* (1998) finally addressed the inverse problem (using the Kramer's cross-section) for the case of non-uniform ionization and showed the inversion to be not only unstable but non-unique (cf., also Piana *et al.*, 2000). As yet no systematic way to deal with the non-uniqueness has been found, though normal regularisation techniques to deal with the usual instability to noise (Craig and Brown, 1986) do help to suppress it (McArthur, 2000). The mathematical issues involved make this inverse problem rather inaccessible to many users of RHESSI spectral data, despite the fact that the important effect of ionization variation should be included in their modeling to get dependable results. However, many such modelers will be happy to proceed by forward fitting of models of $\mathcal{F}_0(E_0)$ to the HXR spectral data $I(\varepsilon)$ rather than trying to invert the latter to derive $\mathcal{F}_0(E_0)$. This forward approach is satisfactory in that a model of $\mathcal{F}_0(E_0)$ predicts a definite $I(\varepsilon)$ and can be parametrically adjusted to best fit the data (though other $\mathcal{F}_0(E_0)$ may also fit it within the data noise). In this paper we therefore revisit the work of Brown *et al.* (1998), expressing it purely in terms of the forward problem and obtaining expressions we feel will be very useful for thick target modelers of RHESSI data. We also examine whether its inclusion improves the fit to data of photon spectra predicted for single power-law electron injection spectra $\mathcal{F}_0(E_0)$.

## 2. Thick-Target Spectrum for Nonuniform Target Ionisation

We will denote the electron injection spectrum by $\mathcal{F}_0(E_0)$ (electrons $s^{-1}$ per unit injection energy $E_0$) and the emitted bremsstrahlung HXR spectrum at the Sun by $I(\varepsilon)$ (photons $s^{-1}$ per unit photon energy $\varepsilon$). Following Brown *et al.* (1998) we treat the electron propagation as 1-D and ignore trapping, relativistic, and directivity effects. The bremsstrahlung cross-section differential in $\varepsilon$,

$$Q(\varepsilon, E) = Q_0 \frac{q(\varepsilon, E)}{\varepsilon E}, \tag{1}$$

yields for the Kramer's approximation $q = 1$

$$J(\varepsilon) = \frac{Q_0}{K'\varepsilon} \int_{\varepsilon}^{\infty} \int_{E}^{\infty} \frac{\mathcal{F}_0(E_0)\, \mathrm{d}E_0\, \mathrm{d}E}{\lambda + x((E_0^2 - E^2)/2K')}, \tag{2}$$

where $K' = 2\pi e^4 \Lambda$, $\Lambda = \Lambda_{ee} - \Lambda_{eH}$ and $\lambda = \Lambda_{eH}/\Lambda$ (numerically $\Lambda_{ee} = 20$, $\Lambda_{eH} = 7.1$ so $\Lambda = 12.9$ and $\lambda = \lambda_{eH}/\Lambda \simeq 0.55$). $x(M)$ is the atmospheric ionization at the ionization weighted target column density variable $M$ ($cm^{-2}$) defined in Section 2 of Brown *et al.* (1998).

$x$ changes from 1 to near 0 over a small spatial range on the Sun and the change is even more abrupt in column density range. So, following Brown *et al.* (1998), we approximate $x(M)$ as a step function at depth $M_*$

$$x(M) = \begin{cases} 1, & M < M_*, \\ 0, & M > M_*, \end{cases} \tag{3}$$

for which Equation (2) becomes, with 'nonuni' denoting non-uniform target,

$$J(\varepsilon)_{\text{nonuni}} = \frac{Q_0}{(\lambda+1)K'\varepsilon} \int_{\varepsilon}^{\infty} \left[ \int_{E}^{(E^2+2K'M_*)^{1/2}} \mathcal{F}_0(E_0)\,\mathrm{d}E_0 + \right. \left. + \frac{\lambda+1}{\lambda} \int_{(E^2+2K'M_*)^{1/2}}^{\infty} \mathcal{F}_0(E_0)\,\mathrm{d}E_0 \right] \mathrm{d}E. \tag{4}$$

The limiting case of a uniformly ionized target corresponds to letting $M_* \to \infty$ here or, with "uni" denoting uniform target,

$$J_{\text{uni}}(\varepsilon) = \frac{Q_0}{K\varepsilon} \int_{\varepsilon}^{\infty} \int_{E}^{\infty} \mathcal{F}_0(E_0)\,\mathrm{d}E_0\,\mathrm{d}E, \tag{5}$$

where $K = 2\pi e^4 \Lambda_{ee}$.

For any model of $\mathcal{F}_0(E_0)$ one wants to test against data, Equation (4) can be used to compute $J(\varepsilon)$ allowing for the nonuniform ionization effect (as compared to Equation (5)) and the parameters of $\mathcal{F}_0(E_0)$ adjusted to find the best fit. Note that in doing so one must either specify the depth $M_*$ of the ionization step (transition region) or regard it as a free parameter to be found as a part of the fitting procedure as was done for *Yokhoh* data by McArthur (2000). This can provide a method of studying changes in $M_*$ due to evaporation as the flare proceeds and, for example, seeing if evolution of $M_*$ is in line with the evaporation rate expected from electron beam heating as first estimated by Brown (1973b) and more precisely studied using various hydrodynamic codes (see, e.g., Aschwanden, 2002).

Note that in the analysis here we have ignored the contribution to the observed $J(\varepsilon)$ of the photospheric albedo $A(\varepsilon)$. The 'reflection' of HXRs in the lower atmosphere leads to a modified observed flux $J_{\text{obs}}(\varepsilon) = J_{\text{flare}}(\varepsilon)(1 + A(\varepsilon))$, where $A(\varepsilon)$ is the integrated reflection coefficient. Note that $A(\varepsilon)$ is somewhat dependent on flare position and as a result should demonstrate center to limb variation, while we are interested in general properties of $J(\varepsilon)$ common to all the flares selected. Moreover, the value of the coefficient $A$ never exceeds about 0.5 so the correction factor is small compared with the ionization effect on energy losses and hence on $J(\varepsilon)$. Therefore, while the effect of albedo should be included in future work, its effect on forward model fitting is less important than ionization for the thick-target situation. On the other hand, in the problem of inversion of $J(\varepsilon)$ to infer $\mathcal{F}_0(E_0)$ the solution involves derivatives of the albedo $A(\varepsilon)$ as a function of energy and can also be important (Alexander and Brown, 2002).

## 3. Effect of Ionisation Structure on $J(\varepsilon)$ for Single Power-Law $\mathcal{F}_0(E_0)$

Here we consider the form of $J_{\rm nonuni}(\varepsilon)$ given by Equation (4) (and also $J_{\rm uni}(\varepsilon)$ by Equation (5) for comparison) for the commonly used single power-law form of electron injection spectrum $\mathcal{F}_0(E_0)$. In doing so we replace $M_*$ by $E_* = (2K'M_*)^{1/2}$, the transition zone 'stopping energy'.

### 3.1. Power-Law $\mathcal{F}_0(E_0)$

Writing

$$\mathcal{F}_0(E_0) = (\delta - 1)\frac{\mathcal{F}_1}{E_1}(E_0/E_1)^{-\delta} \tag{6}$$

where $\mathcal{F}_1$ is the total injection rate ($\mathrm{s}^{-1}$) above $E_0 = E_1$, results, after one integration, in

$$\begin{aligned} J(\varepsilon)_{\rm nonuni} &= \frac{Q_0\mathcal{F}_1}{(\lambda+1)K'\varepsilon}\int_\varepsilon^\infty \left[\left(\frac{E}{E_1}\right)^{-\delta+1} + \frac{1}{\lambda}\left(\frac{E^2+E_*^2}{E_1^2}\right)^{(-\delta+1)/2}\right]\mathrm{d}E \\ &= \frac{Q_0\mathcal{F}_1E_1}{(\lambda+1)K'\varepsilon}\left[\frac{1}{\delta-2}\left(\frac{\varepsilon}{E_1}\right)^{-\delta+2} + \right. \\ &\left. + \frac{1}{2\lambda}\left(\frac{E_*}{E_1}\right)^{(-\delta+2)} B\left(\frac{1}{1+(\varepsilon/E_*)^2}, \frac{\delta}{2}-1, \frac{1}{2}\right)\right], \end{aligned} \tag{7}$$

where $B$ is the incomplete beta function,

$$B\left(\frac{1}{1+(\varepsilon/E_*)^2}, \frac{\delta}{2}-1, \frac{1}{2}\right) \equiv \int_0^{\frac{1}{1+(\varepsilon/E_*)^2}} x^{\delta/2-2}(1-x)^{-1/2}\,\mathrm{d}x, \tag{8}$$

and is a well-tabulated function.

For comparison the thick-target photon power law (with $\gamma = \delta - 1$) for the fully ionized target is immediately obtained from Equation (5) (Equation (7) as $E_* \to \infty$),

$$J(\varepsilon)_{\rm uni} = \frac{Q_0\mathcal{F}_1}{K(\delta-2)}\left(\frac{\varepsilon}{E_1}\right)^{-\delta+1}. \tag{9}$$

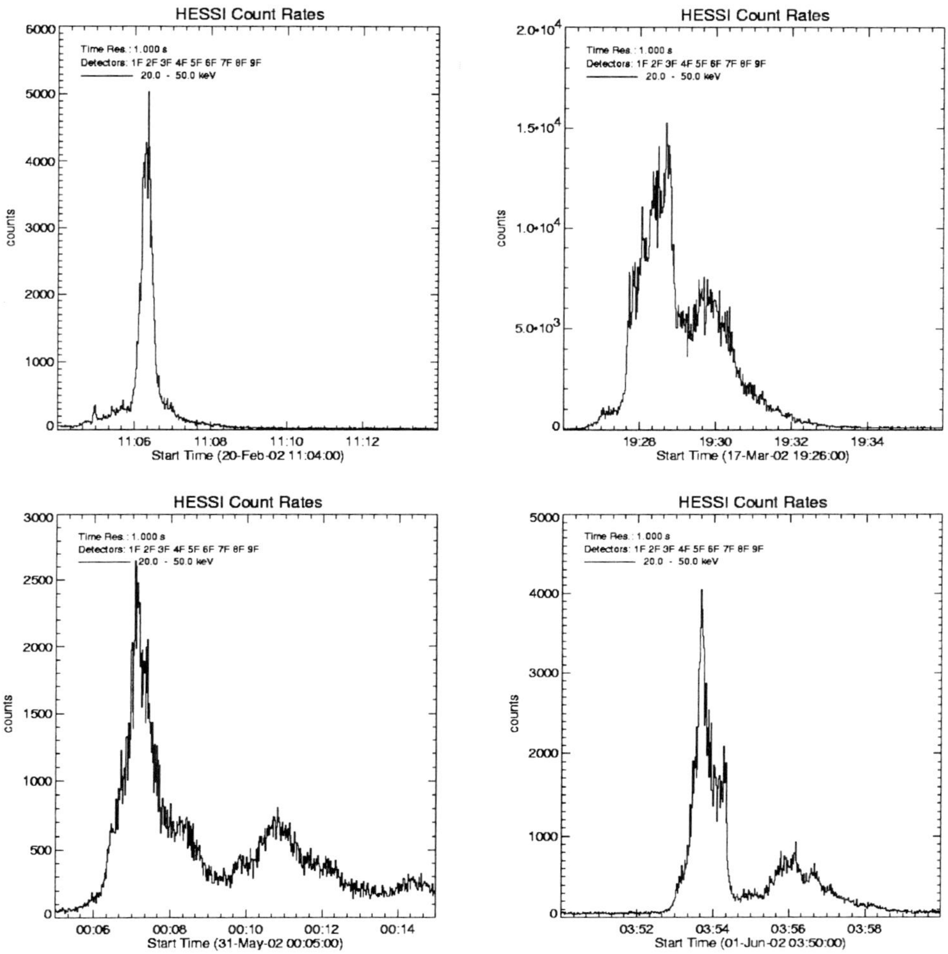


*Figure 1.* The RHESSI light curves are shown for the low-energy range 20–50 keV, binned in steps of $dt_{\text{histo}} = 1.0$ s. The time intervals taken for the spectral fits are given in Table II.

## 4. Applications to RHESSI Data

We have chosen four examples of RHESSI flare spectral data sets. These are the events of 20 February at 11:06 UT, 17 March at 19:26 UT, 31 May at 23:52 UT, and 1 June at 00:06 UT. The corresponding light curves are presented in Figure 1 while the locations and classes of the flares are given in Table I. Note that the event of 31 May is very near the limb and if this involves any footpoint occultation our analysis based on the whole thick target source volume may be inappropriate.

TABLE I

Location and flare classification of selected events.

| Date | Class | Location | AR |
|---|---|---|---|
| 20 Feb. 2002 | C7.5 | N15 W77 | 9825 |
| 17 Mar. 2002 | M4.0 | S21 E18 | 9871 |
| 31 May 2002 | M2.4 | S30 E90 | 9973 |
| 1 June 2002 | M1.5 | S19 E29 | 9973 |

### 4.1. Fully ionized target with single power-law $\mathcal{F}_0(E_0)$

After allowance for all instrumental effects (as in Aschwanden, Brown, and Kontar, 2002) the spatially integrated RHESSI spectra in these events can be modeled well as the sum of an isothermal component and a non-thermal component. The energy range below around 10–20 keV (depending on the flare) can be best fit with an exponential function while the range well above this is approximately a power law, though with some steepening at higher energies. The best fit parameter ($\delta = \gamma + 1, kT$) values for a single power law are presented in Table I while the spectral fits, made with the help of SPEX, are shown in Figure 2. The figures clearly show that the greatest deviation from a single power law occurs in the range 30–60 keV (see Figure 2) where the spectra show a downward 'knee'. To improve the fit we need to take a more complex function than the simple power law of Equation (9). This spectral 'knee' in the deka-keV range of $\varepsilon$ is often attributed to some sharp feature in $\mathcal{F}_0(E_0)$ such as a broken power law or spectral bump (e.g., Lin and Johns, 1993). However these interpretations are based on the fully ionized target approximation. Here we consider how well the photon data are fit by reverting to a single power law in $\mathcal{F}_0(E_0)$ but allowing for the ionization structure which is certainly present – i.e, by the form predicted in Equation (7).

### 4.2. Non-uniformly ionized target with power-law $\mathcal{F}_0(E_0)$

We therefore fitted the data with the following model prediction (Equation (7))

$$J(\varepsilon) = \frac{I_0}{(\lambda+1)\varepsilon}\left[\frac{\varepsilon^{-\delta+2}}{\delta-2} + \frac{E_*^{-\delta+2}}{2\lambda} B\left(\frac{1}{1+(\varepsilon/E_*)^2}, \frac{\delta}{2}-1, \frac{1}{2}\right)\right], \quad (10)$$

where $\lambda = 0.55$ and $I_0$, $E_*$ and $\delta$ are adjustable parameters. Note that $I_0$ is just a scale factor depending on $\mathcal{F}_0$, $E_1$ – see Equation (7).

The best-fit results for $E_*$ and $\delta$ are presented in Table III, along with the column depth $N_*$ equivalent to $E_*$ and $\chi^2_{\mathrm{nonuni}}/\chi^2_{\mathrm{uni}}$ is the ratio between total $\chi^2$ deviations of the model prediction from the data,

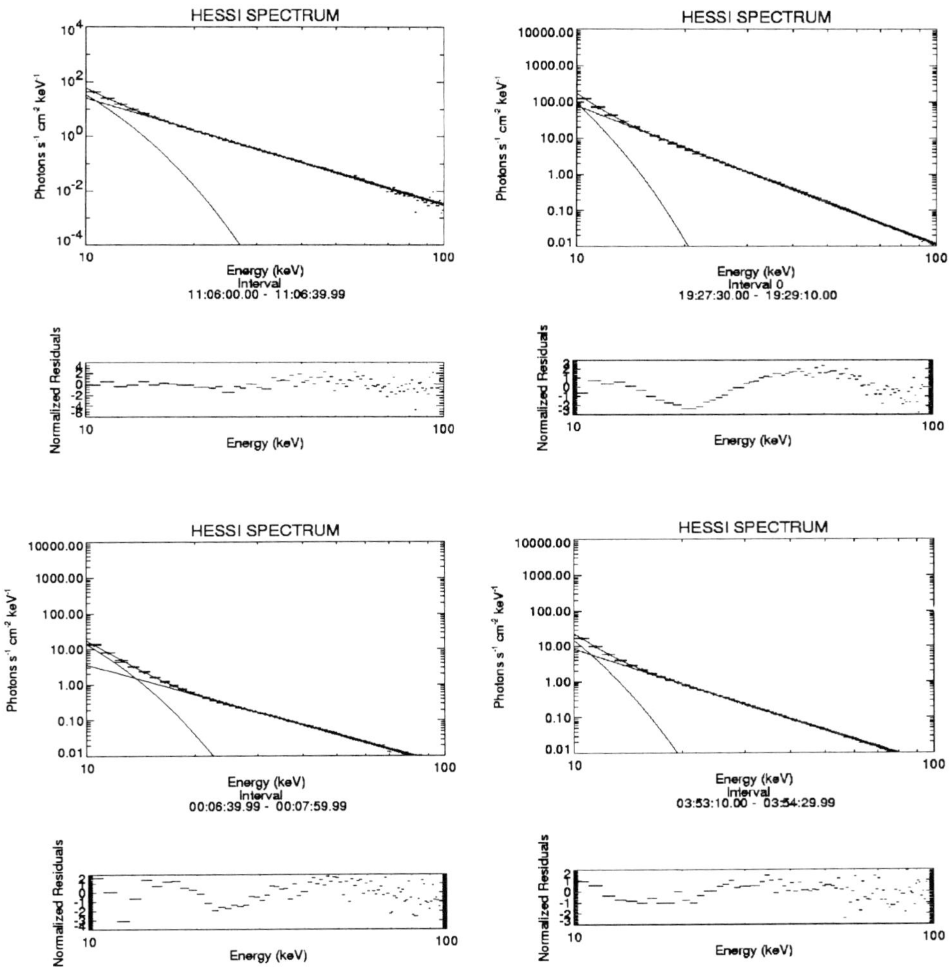


*Figure 2.* The photon spectra and spectral fits with photon power law plus thermal component in the energy range 10–100 keV (*top panel*). The fit parameters are given in Table II. Normalized residuals are presented in the bottom panel showing the deviation of real data from the power law, $(J_{\rm model}(\varepsilon) - J_{\rm RHESSI}(\varepsilon))/J_{\rm RHESSI}(\varepsilon)$.

$$\chi^2_{\rm model} = \sum_{i=0}^{N} (J_{\rm model}(\varepsilon) - J_{\rm RHESSI}(\varepsilon))^2, \tag{11}$$

for the two models – non-uniformly ($_{\rm nonuni}$) and uniformly ($_{\rm uni}$) ionized targets. In the four panels of Figure 3 we show the best-fit spectra superimposed on the data for each of the four events. The overall $\chi^2$ value summed over all data points (11) is much lower for the model allowing for the ionization structure than for the fully ionized case (see Table I) and the individual residuals are also mostly much smaller (note the log scale) in the range 20–60 keV. Above 60 keV the data are pro-

TABLE II

Uniformly ionized target model with single power-law $\mathcal{F}_0(E_0)$ plus thermal component fit for the events – results produced with SPEX.

| Date | Time (UT) | $\delta$ | $kT$ (keV) |
|---|---|---|---|
| 20 Feb. 2002 | 11:06:00–11:06:40 | 4.89 | 1.47 |
| 17 Mar. 2002 | 19:27:30–19:29:10 | 4.84 | 1.27 |
| 31 May 2002 | 00:06:40–00:08:00 | 3.79 | 2.02 |
| 1 June 2002 | 03:53:10–03:54:30 | 4.26 | 1.45 |

TABLE III

Best fit non-uniformly ionized target model parameters for single power-law $\mathcal{F}_0(E_0)$, and equivalent $N_*$ (energy range 20–100 keV), and the ratio of $\chi^2_{\mathrm{nonuni}}/\chi^2_{\mathrm{uni}}$.

| Date | $\delta$ | $E_*$ (keV) | $N_*$ (cm$^2$) | $\chi^2_{\mathrm{nonuni}}/\chi^2_{\mathrm{uni}}$, |
|---|---|---|---|---|
| 20 Feb. 2002 | 5.29 | 37.4 | $2.7 \times 10^{20}$ | 0.032 |
| 17 Mar. 2002 | 4.99 | 24.4 | $1.1 \times 10^{20}$ | 0.047 |
| 31 May 2002 | 4.15 | 56.2 | $6.1 \times 10^{20}$ | 0.041 |
| 1 June 2002 | 4.46 | 21.0 | $8.4 \times 10^{19}$ | 0.055 |

gressively noise-dominated and therefore both models give similar residuals. Thus, inclusion of the ionization effect allows a much better fit to these RHESSI spectra with a single power-law $\mathcal{F}_0(E_0)$ than does the usual ionized approximation and removes the need to invoke features in the acceleration process producing $\mathcal{F}_0(E_0)$. The inferred best fit $\delta$ values are slightly higher in the more realistic ionization model (compare Tables II and III).

### 4.3. Location of the Transition Region

By assuming that the main spectral feature observed in the HXR spectra is caused by the increased bremsstrahlung efficiency of the chromosphere and not an original feature in the electron spectra, then for a pure power-law $\mathcal{F}_0(E_0)$ the non-thermal part of the HXR spectra can be expressed analytically as the combination of a power law and an incomplete beta function (10). The HXR spectra can be fitted with (10) to determine the three parameters of this model i.e., photon flux constant $I_0$, electron spectral index $\delta$ and the transition zone electron stopping energy $E_*$.

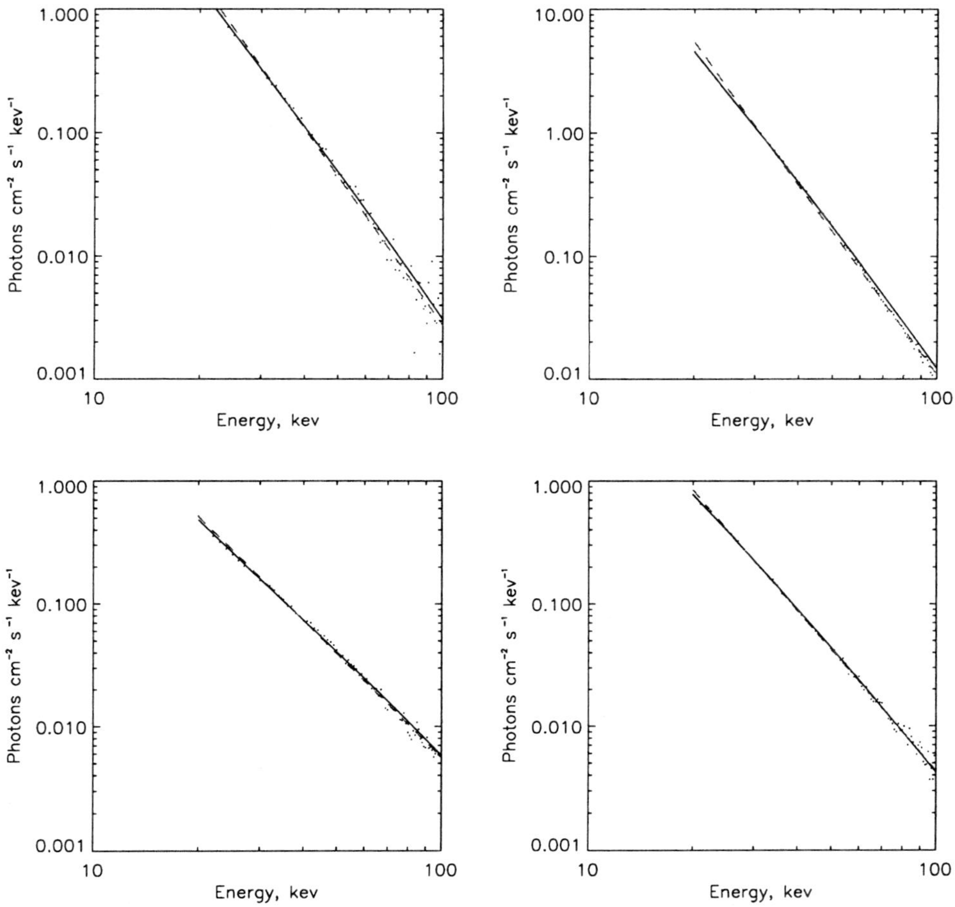


*Figure 3.* The RHESSI spectra (*dots*) and the spectral fit (cf., Table III) for a non-uniformly ionized target for the 4 selected events. The *solid line* represents the model for the nonuniform fit, *dashed lines* correspond to the single power-law fit.

For these flares, the best fit values of the transition region energy parameter $E_*$ lie in the 20–60 keV range. The stopping energy can be readily converted into a column density of the total material the electron passed through to reach the chromosphere, so this fitting procedure allows estimation of the coronal column density from $E_* = (2K'M_*)^{1/2}$. This corresponds for our events to $M_* = (1.2 - 11) \times 10^{20}$ cm$^{-2}$ with the equivalent $N_*$ in the range $(0.8 - 7) \times 10^{20}$ cm$^{-2}$. Thus the spectral power of RHESSI enables us to derive the transition region column depth. By extending the work of Aschwanden, Brown, and Kontar (2002) using RHESSI spectral images to derive hard X-ray source heights for limb events and combining this with the spectral estimate of $N_*$ it should be possible to determine the geometric height of the transition region inside the thick target flare loop. This is the subject of a future paper where we will also attempt to follow the evaporative evolution of $N_*$ as the flare progresses, using our spectral fitting technique.

## 5. Discussion and Conclusions

Starting from the basic principles of collisional propagation of a thick-target beam in the non-uniformly ionized solar plasma we derived a simple model for its bremsstrahlung spectrum for a general electron injection spectrum $\mathcal{F}(E_0)$. For a pure power-law $\mathcal{F}(E_0)$, with two adjustable parameters $\mathcal{F}_1$, $\delta$ plus the stopping energy $E_*$, we can physically explain the observed deviation from a power-law photon spectrum, agreeing much better with RHESSI data than the pure power law obtained in the usual uniformly ionized target analysis. The method is also transparent and easy to use for general $\mathcal{F}(E_0)$ . Finally we note that there are known additional effects which should be included in refined spectral modeling which might lead to deviations from a single photon power law. In particular there is the effect of solar HXR albedo (Alexander and Brown, 2002). In summary, the results show that

– detailed RHESSI spectral data show a deviation from a simple power law in the range 20–100 keV often attributed to a feature in $\mathcal{F}_0(E_0)$ , i.e., in the acceleration process.

– inclusion of the effect of target ionization change across the transition region removes the need for any such feature in $\mathcal{F}_0(E_0)$ which can then be a pure power law, favouring a stochastic process.

– using the technique presented in the paper one can determine the energy $E_*$ corresponding to the transition zone depth and hence the value of that column depth.

## Acknowledgement

This work was supported by a PPARC Rolling Grant.

## References

Alexander, C. R. and Brown, J. C.: 2002, *Solar Phys.*, this volume.
Aschwanden, M. J.: 2002, *Space Sci. Rev.* (in press).
Aschwanden, M. J., Brown, J. C., and Kontar, E. P.: 2002, *Solar Phys.*, this volume.
Brown, J. C.: 1971, *Solar Phys.* **18**, 489.
Brown, J. C.: 1973a, *Solar Phys.* **28**, 151.
Brown, J. C.: 1973b, *Solar Phys.* **31**, 143.
Brown, J. C. and Emslie, A. G.: 1988, *Astrophys. J.* **331**, 554.
Brown, J. C., McArthur, G. K., Barrett, R. K., McIntosh, S. W., and Emslie, A. G.: 1998, *Solar Phys.* **179**, 379.
Craig, I. J. D. and Brown, J. C.: 1986, *Inverse Problems in Astronomy*, Hilger, Bristol.

Emslie, A.G., Barrett, R.K., and Brown, J.C.: 2001, *Astrophys. J.* **557**, 921.
Lin, R. P. and Johns, C. M.: 1993, *Astrophys. J. Lett.* **417**, L53.
McArthur, G. K.: 2000, PhD Thesis, University of Glasgow.
Piana, M., Barrett, R. K., Brown, J. C., and McIntosh, S. W.: 2000, *Inverse Problems* **15**, 1469.

# MICROFLARES AND HOT COMPONENT IN SOLAR ACTIVE REGIONS

ARNOLD O. BENZ and PAOLO C. GRIGIS
*Institute of Astronomy, ETH Zürich, CH-8092 Zurich, Switzerland*

(Received 31 July 2002; accepted 4 September 2002)

**Abstract.** Open-shutter RHESSI observations of 3–15 keV X-rays are found to exhibit active-region transient brightenings and microflares at a rate of a least 10 per hour occurring even during the periods of lowest solar activity so far in the mission. A thermal component fitted by temperatures of 6–14 MK dominates from 3 keV to about 9 keV, but can be traced up to 14 keV in some cases, and has an average duration of 131(±103) s at 7–8 keV. The duration increases with decreasing photon energy. The peak count rate defined by cross-correlation is delayed at low energies. The temperature peaks early in the event and then decreases, whereas the emission measure increases throughout the event. The properties are consistent with thermal conduction dominating the evolution. In some of the bigger events, a second component was found in the 11–14 keV range extending down to 8 keV in some cases. The duration is typically 3 times shorter and ends near the peak time of the thermal component consistent with the Neupert effect of regular flares. Therefore the second component is suggested to be of non-thermal origin, presumably causing the beam-driven evaporation of the first component. The two components can be separated and analyzed in detail for the first time. Low-keV measurements allow a reliable estimate of the energy input by microflares necessary to assess their relevance for coronal heating.

## 1. Introduction

Plasma of many million degrees and unknown origin has been reported from solar active regions (e.g., Brosius *et al.*, 1996). Observers of the full Sun find a power-law distribution for the differential emission measure (DEM) decreasing with temperature. The higher the level of activity, the larger is the coronal emission measure and the flatter its temperature distribution. In the presence of large active regions, it has been observed to extend to temperatures higher that 10 MK (e.g., Peres *et al.*, 2000, and references therein), but is not well-observed beyond 10 MK. On top of the power-law distribution, flares produce transient peaks. The DEM of flares has a peak at 6 (±1) MK for A2 flares increasing up to 23 (±3) MK for X1 flares (Feldman *et al.*, 1996). At any phase, the flare DEM is quite narrow and covers less than a decade in temperature (Reale, Peres, and Orlando, 2001). From observations unresolved in time and space thus emerges the question whether the high temperature plasma apparent outside flares is just the remnant of previous larger flares, is heated by a stationary process or results from a superposition of unresolved small flares.

*Solar Physics* **210:** 431–444, 2002.

Small flares in the 3–8 keV range lasting a few minutes have first been observed by HXIS on SMM (Simnett *et al.*, 1989). Shimizu (1995) found microflares (active region transient brightenings) in *Yohkoh*/SXT observations sensitive to 4–7 MK. He reported events having energies as low as a few $10^{26}$ ergs and a correlation of peak flux with source size (loop length) and density. Watanabe *et al.* (1995) have observed the full Sun in the high-temperature sulfur XV lines using the BCS instrument on board *Yohkoh*, indicating brightenings in excess of 10 MK. Note in contrast that the thermal energy content of micro-events in the quiet corona is more than two orders of magnitude smaller, and temperatures of only 1.0–1.6 MK are reported (Krucker *et al.*, 1997, and reviewed by Benz and Krucker, 2002).

Lin *et al.* (1984, and references therein) have discovered small HXR flares produced by >20 keV electrons. Comparing the >25 keV channel of LAD/CGRO with microflares observed in the 8–13 keV channel of SPEC/CGRO, Lin, Feffer, and Schwartz (2001) reported many low-energy events that have no counterpart at high energies. If interpreted as emission of the same non-thermal electron population, it would suggest that the spectrum steepens or cuts off between 10 and 25 keV. Thus the nature of the low-energy events and their relation to the non-thermal events at high energy remain questionable.

In this paper, we present microflare observations by the RHESSI satellite (Lin *et al.*, 2002) at unprecedented spectral and temporal resolution of photons above 3 keV, highly sensitive to the highest temperatures reported for active regions as well as to the lowest energies of non-thermal photons expected from microflares. First RHESSI results on microflare imaging are presented by Krucker *et al.* (2002). Here we explore the spectral and temporal characteristics of microflares and concentrate on the question whether the emissions are thermal or non-thermal.

## 2. Observations and Results

The Reuven Ramaty High-Energy Solar Spectroscopic Imager (RHESSI) was launched into orbit on 5 February 2002. The germanium detectors register photons in the energy range from 3 keV to 17 MeV with 1 keV resolution at low energies (Smith *et al.*, 2002). Imaging is achieved by nine absorbing grids modulating by satellite rotation. It allows reconstruction of the full-Sun image with a resolution of $2''$ at low energies (Hurford *et al.*, 2002). At low count rates, the spatial resolution is degraded (Saint-Hilaire and Benz, 2002). RHESSI observes the Sun full time interrupted only by satellite night and the South Atlantic Anomaly. Energetic electrons sometimes cause spurious counts when the satellite is at high geographical latitudes. We use standard RHESSI software and calibrations updated to mid-August 2002.

As the energy, time and detector information of each photon is registered, the large number of low-energy photons during large flares causes pile-up in the detector. To bring it down to a manageable level, one or two shutters are usually put in front of the detectors to reduce the count rate at low energies. Early in the mission, the shutters were open only in relatively quiet times, when no large flares were expected.

Here we have selected RHESSI observing times when no shutter was absorbing the low-energy photons. During these times the full flux of low-energy photons is available. Thus these times yield maximum sensitivity for low-energy studies. We have found same 50 open-shutter orbits in the months of March, April and May 2002. Uninterrupted intervals at low solar activity and low background were selected for further analysis. In particular, we required (i) a count rate less than 600 cts per second and per detector in the 3–12 keV range and (ii) a constant background in the 100–300 keV range. A total of seven intervals satisfied these requirements. The GOES level in all of them is below C, and all flares are below GOES class A9 after subtracting the background.

### 2.1. Light curves

Figure 1 displays RHESSI full-Sun observations at low energies during one-orbit intervals. The low peak count rates justify neglecting the pile-up effect of photons in the detectors. The photons have been integrated in one keV bins that we will refer to as channels. The 10–11 and 11–12 keV channels have a lower signal-to-noise ratio because of an instrumental background line feature. Also displayed in Figure 1 are the GOES 8 observations in the low-energy band, extending the energy coverage below the RHESSI range.

Visual inspection of Figure 1 yields many events per orbit. There is no time of stationary flux. During the total selected observing time of 373 min, we found 64 events, thus 10.3 per hour on average in one solar hemisphere. Lin *et al.* (1984) reported 10 $hr^{-1}$, Lin *et al.* (2001) 5.5 $hr^{-1}$, and Shimizu (1995) 26 $hr^{-1}$, all of them observing during higher levels of solar activity.

Some general properties follow immediately from Figure 1:

– The largest variability occurs in the channels between 5–9 keV .

– Some events in the 6–9 keV channels are not visible in the GOES light curve, being dominated by photons $< 3$ keV (see, e.g., event after B in Figure 1(d). GOES shows no event larger than A0.1 above background).

– The emission above about 9 keV peaks generally before the emission at lower energies.

– The event duration decreases continuously from 1.6 keV (GOES) to 15 keV.

– In most events there is no emission perceivable beyond about 12 keV.

### 2.2. Photon energy spectra

Two spectra are shown in Figure 2. The data were calibrated using the full detector response matrix. The first spectrum was integrated during event A shown in Figure 1(d) from 01:46:40 to 01:47:40 UT, the maximum phase at the higher energies. The total background, taken at the time of minimum flux before the event, was

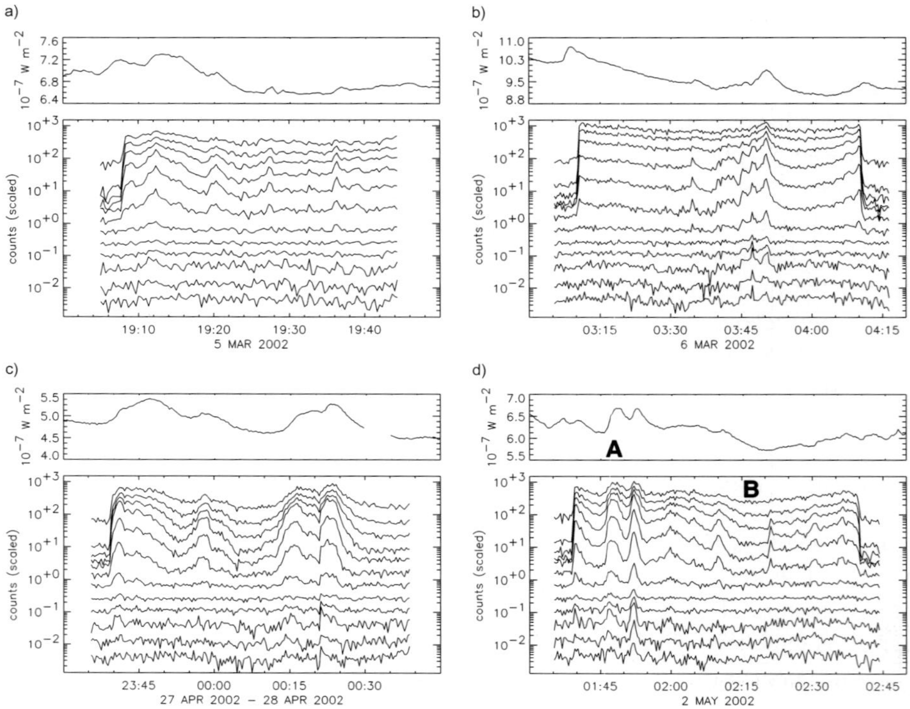


*Figure 1. Top*: light curve observed by GOES 8 in the 1–8 Å band (1.6–12.4 keV) of soft X-rays. *Bottom*: RHESSI light curves in one keV channels from 3–4 keV (*top*) to 14–15 keV (*bottom*). Each channel is multiplied such that it does not overlap with others. The time resolution is 20 s. (a) Sun light starts at 19:08 UT. (b) Sun light starts at 03:11 UT and ends at 04:10 UT. (c) Sun light starts at 23:40 UT. (d) Sun light starts at 01:40 UT and ends at 02:40 UT.

subtracted. Only the front detectors 1, 3, 5, and 9 have been used, having the lowest dropout rates in the time interval. Data dropouts are suspected to be caused by heavy cosmic rays interfering with the RHESSI electronics. Although the software corrects for the deadtime due to these data gaps, they introduce additional noise.

At low energies the spectrum fits well with an isothermal plasma at a temperature of 12.1 ($\pm$0.4) MK and an emission measure of 7 ($\pm$2) $\times$ $10^{45}$ cm$^{-3}$. The best fit is shown in Figure 2, where all normalized residuals are below 0.4. The errors originate mainly from the choices of the background and the low-energy break point for non-thermal emission. As the power-law distribution of the non-thermal component must turn over at low energies, it was approximated by a constant below some energy. The quality of the fit is not sensitive the break energy; best fits are between 6 and 9 keV. In the case of Figure 2 different background choices changed temperatures in the range of $\pm$ 0.8 MK, and different choices for the low-energy break for non-thermal emission produced a range of $\pm$ 0.7 MK. The standard error was estimated by taking a third of the rms sum of the two. Note that the error in temperature and emission measure are anti-correlated.

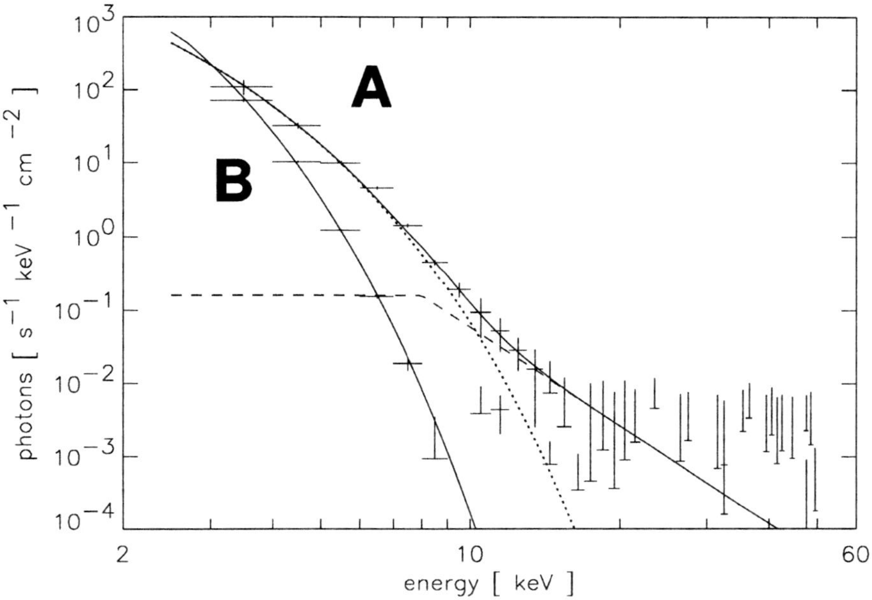


*Figure 2.* Spectrum of event A and background B on 2 May 2002 (Figure 1(d)) observed with RHESSI. The event is fitted with a thermal contribution (*dotted*) and a non-thermal component (*dashed*) having a break point at 8 keV. The background is fitted with only a thermal contribution.

At the time of writing, the energies below 5 keV are not yet fully calibrated. This can be noted in the fact that single collimators give a range of temperature values of $\pm 1.7$ MK if using only the lowest 3 keV. The scatter decreases by a factor of 3 for taking a larger range. At 6–7 keV, and possibly also at 7–8 keV, the flux is often enhanced. This cannot be reduced substantially by the choice of the break point. The enhancement could be caused by the 6.7 keV iron line emissions not included in the fit. Therefore we did not use 6–8 keV for fitting. Instead the break point was put at 8 keV in accordance with other information (Figures 4 and 5).

In the energy range of 10–15 keV, the isothermal model does not fit the observations of microflare A (Figure 2). A power-law photon distribution was therefore added. Its fitted slope depends on the break point and has an exponent in the range $-4.5 \lesssim \gamma \lesssim -5.7$. For the choice of the break point at 8 keV, the best fit yields $-4.5$. The spectrum disappears in the background beyond 15 keV.

Instead of a power-law, the 10–15 keV spectrum could also be fitted by a second thermal component with higher temperature and smaller emission measure. The best fit is achieved for a temperature of 24.7 MK and an emission measure of $0.14 \times 10^{45}$ cm$^{-3}$. The contribution of the hot kernel would not be discernible below 8 keV and fit all other constraints.

The thermal energy content of an isothermal plasma is

$$E_{th} \approx 3\sqrt{\mathcal{M} V} k_B T. \quad (1)$$

It is assumed in Equation (1) that the electron and ion densities are about equal and constant in the volume $V$. $\mathcal{M}$ is the emission measure, and $k_B$ is the Boltzmann constant. Putting in the observed values, we find for the thermal content of the microflare (index A) and of the hot kernel (index k)

$$E_{th}^{\mathrm{A}} \approx 4.2 \times 10^{14} \sqrt{V_{\mathrm{A}}} \quad \text{(erg)}, \quad (2)$$

$$E_{th}^{\mathrm{k}} \approx 1.2 \times 10^{14} \sqrt{V_{\mathrm{k}}} \quad \text{(erg)}, \quad (3)$$

Thus the hot kernel would contain less energy than the main component for $V_{\mathrm{k}} < V_{\mathrm{A}}$.

The spectrum at the time of the lowest count rate in the orbit (marked B in Figure 1(d)) is also displayed in Figure 2. It is the time interval from 02:14:40–02:19:56 UT, when no microflare was observed and the most recent one had occurred more than 4 minutes before (Figure 1(d)). For this case, the background interpolated from the satellite's night time before and after the observations was subtracted. An isothermal model having a temperature of 6.4 MK fits well. No non-thermal or hot kernel component is above the background level. The emission measure of the hemisphere is measured $45 \times 10^{45}$ cm$^{-3}$, and its thermal energy content is

$$E_{th}^{\mathrm{B}} \approx 5.6 \times 10^{14} \sqrt{V_{\mathrm{B}}} \quad \text{(erg)}. \quad (4)$$

The fact that Equations (2) and (5) have similar factors does not mean that $E_{th}^{\mathrm{A}} \approx E_{th}^{\mathrm{B}}$ since probably $V_{\mathrm{B}} \gg V_{\mathrm{A}}$.

Microflare A can be imaged with RHESSI collimators 3 to 9. The microflare rook place in NOAA Region 9932 (see image and movie on CD) and had a nominal FWHM size of $8'' \times 11''$. It is close to the resolution limit, thus $V_{\mathrm{A}} \lesssim 3 \times 10^{26}$ cm$^3$. Equation (2) yields an upper limit on the thermal energy content of $7.2 \times 10^{27}$ erg. A lower limit for the electron density of $4.8 \times 10^9$ cm$^{-3}$ is derived from emission measure and volume assuming a filling factor of unity. Microflare imaging is studied in detail by Krucker *et al.* (2002).

The relation between the temperature of the thermal component and the emission measure is further explored in Figure 3. Temperature and emission measure have been determined identically to Figure 2. Using the same background and break point that produce similar systematic errors, the remaining relative error between measurements is estimated about $\pm$ 0.2 MK, as determined from different fits. The temperature maximum occurs between peak fluxes in 12–15 keV and 3–7 keV. The product $T\sqrt{\mathcal{M}}$ is proportional to the energy content if the volume remains constant. The product increases initially until about 01:48:03 s, when the 12–15 keV count rate has dropped to the half-maximum level, and then remains approximately constant throughout the microflare.

### 2.3. Low- and High-Energy Photons

The light curves integrated from 3–7 keV (low energy) and 12–15 keV (high energy) are shown in Figure 3. The low-energy band peaks at 01:48:10 (±20) UT. The high-energy count rate reaches the preflare background level of 1.8 counts $s^{-1}$ at 01:48:40 (±20) UT. Thus the two light curves exhibit the Neupert (1968) effect known in regular flares. It is generally interpreted there as the signatures of different emissions relating to a thermal and a non-thermal electron population in the scenario of electron-boom driven evaporation.

The high-energy band peaks 80 seconds before the low-energy band. The FWHP duration at 12–15 keV is only 45 s, whereas it is 148 s at 3–5 keV.

Further differences between low and high-energy photons appear in the spectrogram (Figure 4). The data are calibrated using a diagonal response matrix. The high-energy peak at 01:46:50 UT is clearly visible. It takes a straight vertical shape emerging at about 8 keV out of the thermal part. The latter has a triangular shape with an apex at about 01:47:25 UT and 12 keV.

The timing of photon fluxes at different energies was studied also by cross-correlation. The light curves at 4 s resolution were each cross-correlated relative to the most variable channel (7–8 keV). The derived cross-correlation coefficients revealed a local maximum at zero lag even at energies >20 keV, where no solar photons were detected above background. As the zero-lag cross-correlation peak also exists at satellite night, it must be an instrumental effect. Most likely it is due to data dropouts, except for the auto-correlation in the 7–8 keV channel. The effect was eliminated by interpolation between the neighboring lags and integration to 8 s.

Figure 5 depicts the result of the cross-correlation in spectrogram form (correlogram). Channels with large background, such as 10–11, and 11–12 keV, have higher noise and yield lower cross-correlation coefficients. Again, the difference between low- and high-energy components appears. Most relevant is the change in drift rate of the cross-correlation peak. At energies higher than 9 keV, the peak does not drift but forms a vertical structure at a constant lag of −25 s. Below 9 keV, the peak drifts to larger lags (from upper left to lower right), reaching about +25 s at 3–4 keV.

### 2.4. Low-Energy Component

The 25 largest microflares in the sample have been analyzed similarly to the above analysis. Their average peak flux in the 7–8 keV band above the preceding minimum is 0.7 photons $s^{-1}$ $cm^{-2}$. The average FWHM duration is 131(±103) s (standard deviation), and the median value is 114 s. We found no correlation between peak flux and duration in agreement with Shimizu's (1995) measurements at lower energies.

A relation between peak flux and absolute drift rate of the cross-correlation peak is apparent from Figure 6. The drift rate was determined from the cross-correlation

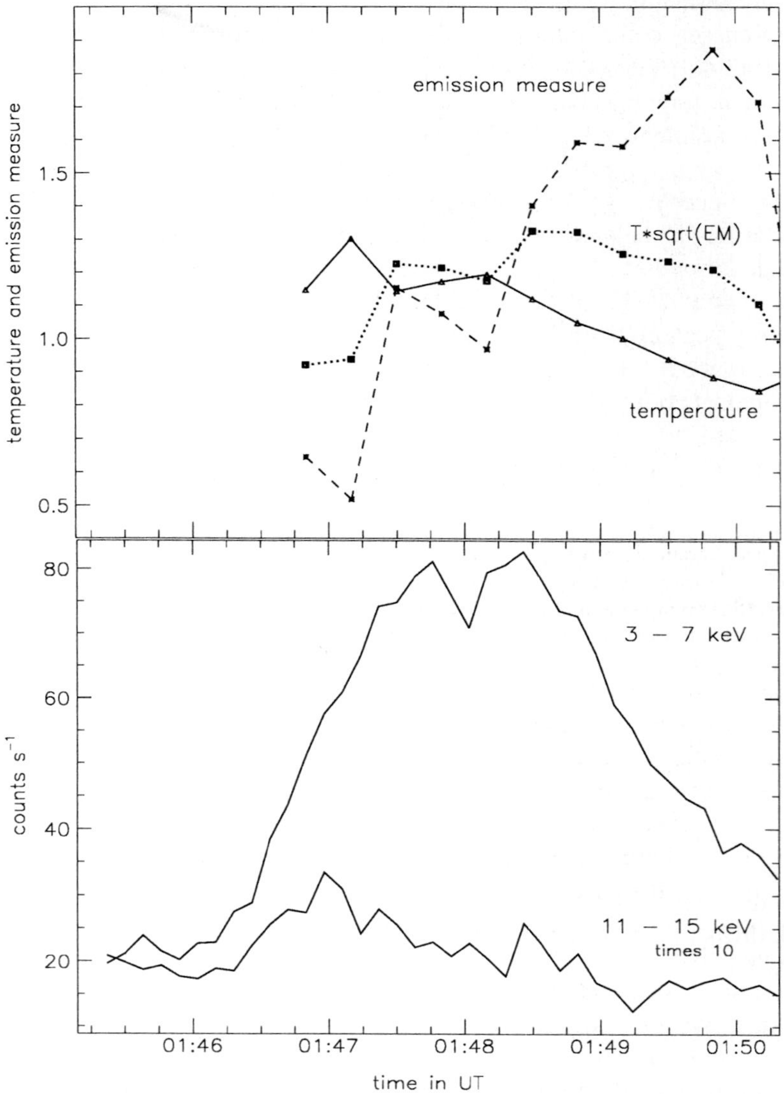


*Figure 3. Top*: Temperature (*solid*) in units of 10 Mk, emission measure (*dashed*) in $10^{-46}$ cm$^{-3}$ and the product $T\sqrt{\mathcal{M}}$ (*dotted*) of event A in Figure 1(d) (thermal component). *Bottom*: light curve of event A in the energy bands 3–8 keV and 11–15 keV (multiplied by factor of 10). The integration time is 8 s, the data are calibrated by the diagonal matrix elements.

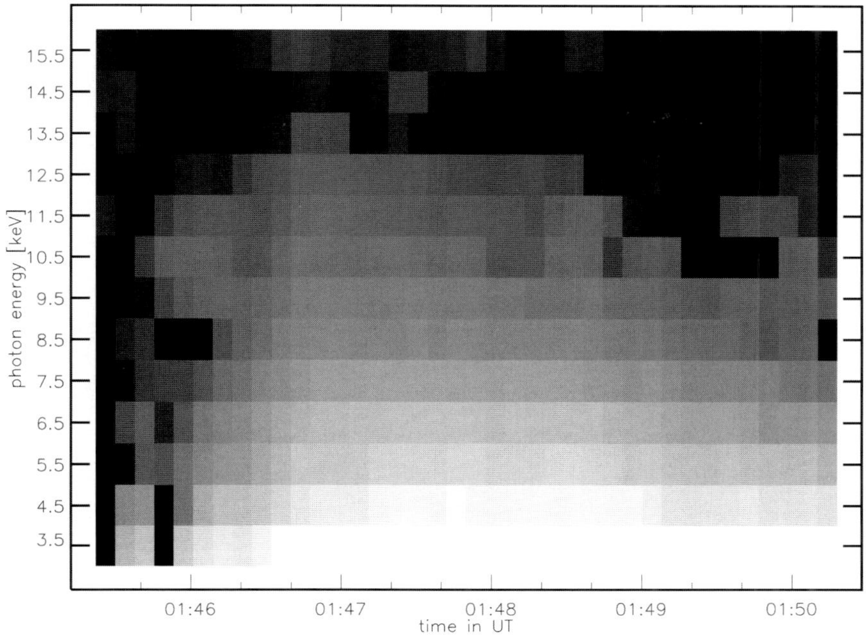


*Figure 4.* Spectrogram observed by RHESSI of event A in Figure 1(d). The energy channels (1 keV) and time bins (8 s) are presented by pixels. A logarithmic scale is applied, bright meaning enhanced photon flux. (See color version of this figure on the CD-ROM).

coefficients of the 1-keV channels in the 3–9 keV range by linear (Figure 5) of the peak values in energy and time. The peaks in each channel are determined from a boxcar smoothing of 40 s and spline interpolation. Spline and regression allow a higher resolution and the determination of larger drift rates than from the pixels in the spectrogram. Nevertheless, the error increases enormously at drift rates beyond about 2 keV s.$^{-1}$ Large microflares have slower absolute drift rate, which is usually negative (peak drifting from high to low energies). Some smaller microflares have large absolute drift. In one case out of 25 it was found positive, but the error range includes negative values. The anti-correlation between photon flux and drift rate is more prominent for events during the same orbit.

The spectrum of X-ray microflares at energies in the 3–9 keV range is well modeled by an isothermal plasma as shown in the example of the previous section. The best fitting temperatures in the 25 largest and well isolated events are in the range 10.2–14.7 MK, with an average value of 12.0 ($\pm$1.2) MK. This is considerably more than reported previously (Shimizu, 1995; Feldman *et al.*, 1996). The difference seems to originate from *Yohkoh*/SXT and GOES favoring the 4–7 MK range, but RHESSI observing the most energetic thermal photons and thus emphasizing the highest temperatures. This is supported by GOES observations of the microflare of Figure 2 (GOES class A5) where the two GOES channels having

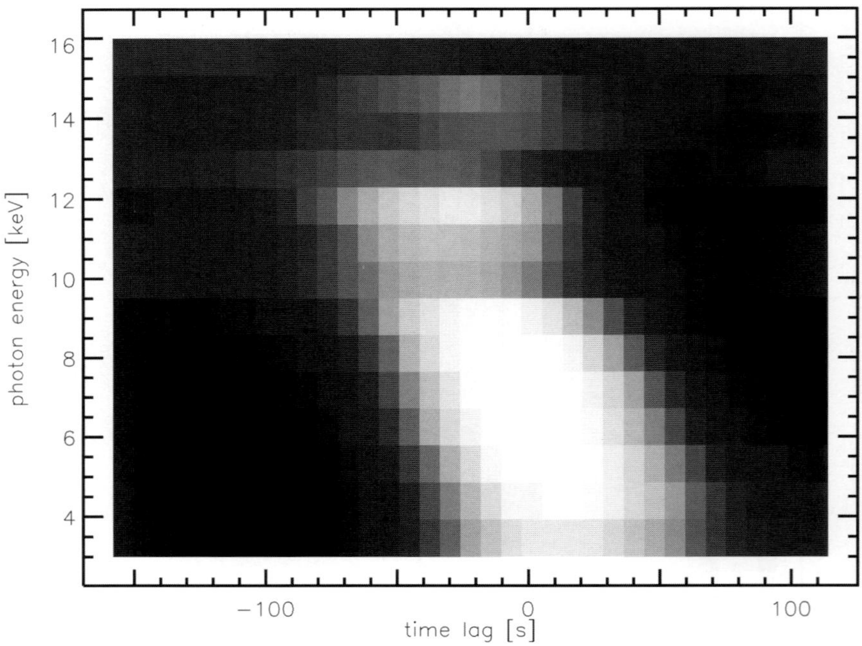


*Figure 5.* Correlogram of Figure 4 showing the cross-correlation of relative to the 7−8 keV channel. Bright pixels represent high correlation coefficients. (See color version of this figure on the CD-ROM.)

lower energy bounds of 1.6 and 3.2 keV indicate a temperature of only 7±1 MK (background subtracted).

### 2.5. High-energy component

The second component is best observable in 11−14 keV photons. As a test of its presence the <8 keV spectrum was fitted by a thermal model, and the 12−13 keV channel compared to the fit. In 6 of the 25 largest microflares this channel was clearly above the thermal, 2 more cases were doubtful. In 4 cases, all at relatively high temperature, the 12−13 keV flux was within the error bars consistent with the thermal model. In the rest, 13 cases, the 12−13 keV flux was not significantly above background and could not be classified.

## 3. Discussion

The existence of a second, non-thermal component at high energies in microflares is not obvious. We have searched for it in the following ways:

(1) The *spectrum* (Figure 2) of microflares below 10 keV is dominated by a thermal component with a temperature of the order of 10 MK. In 24% of the cases

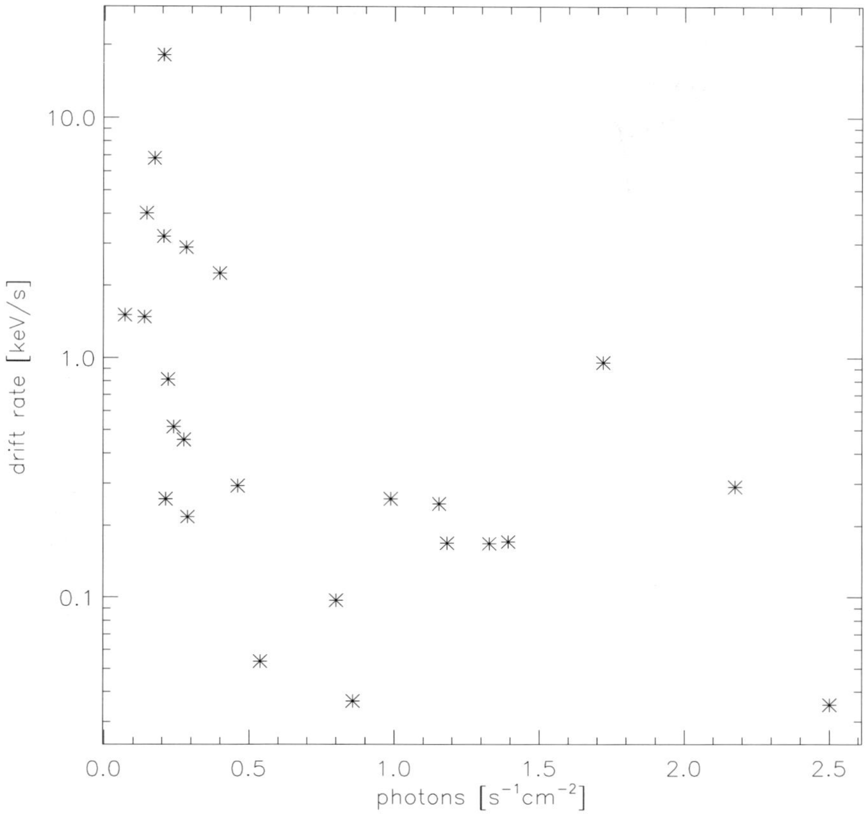


*Figure 6.* Drift rate of the light curves of the 1 keV channels between 3 and 9 keV vs. peak photon flux in the 7–8 keV channel.

an enhanced high-energy tail is observed extending up to 15 keV. It could also be interpreted by a hot thermal kernel.

(2) The *spectrogram* (Figure 4) shows more clearly the difference between the two components: (i) The low-energy thermal component having a triangular shape with a long decay at the lowest photon energies and (ii) a short vertical structure extending to higher energies in the rise phase of the former component.

(3) The drift of the *cross-correlation peak* at low energies is a combination of two effects: (i) The maximum of the light curve is more and more delayed the lower the energy, and (ii) the shape of the microflare becomes asymmetric at lower energies having a longer decay time than rise time (Figure 3). Cross-correlation is thus also a method to identify the two components and to separate them in energy.

There are two major indications for a 'non-thermal' origin of the high-energy component:

– A hot kernel of thermally emitting plasma also fits the 9–14 keV spectrum, but would produce a drift in the correlogram during cooling that is not observed in the second component.

– The second component has shorter duration, occurs in the early phase of the thermal component and vanishes at the peak flux of the thermal component. It resembles the non-thermal component of regular flares in its relation to the first component known as the Neupert effect. The second component thus exhibits the characteristics of thick target bremsstrahlung of an electron beam.

Therefore, we interpret the second component of microflares as non-thermal emission.

An important result is the increase of the emission measure throughout the event (Figure 3). It indicates that the amount of high-temperature material increases (excluding the unrealistic possibility of compression). At the same time the temperature decreases. A surprising characteristic is the flat top of the product $T\sqrt{\mathcal{M}}$ from the time of the temperature peak early in the flare until the late decay phase. Under the assumption of constant volume, the flat top suggests a constant thermal energy content.

Thermal conduction could explain such a behavior if radiative energy loss can be ignored. The thermal component may start with a high temperature and a small mass. Thermal conduction then adds more material to the temperature range observed, but decreases the mean temperature at the same time. It is consistent with the observation of the thermal component, having a later peak and longer duration at low energies.

Conduction can also explain the drift of the cross-correlation (Figure 5). As the temperature decreases, the count rate in the higher energy channels diminishes more rapidly than at low energies where the increasing emission measure even delays the maximum flux. The smaller the flare size, the faster this takes place. Thus the cross-correlation peak drifts the faster, the smaller the flare. Considering the previously reported relation between flare size and peak flux (see Introduction), this may explain the observed inverse relation between drift rate and peak flux (Figure 6).

Conduction models have been proposed to interpret a Neupert-like relationship between higher and lower energy photons (reviewed in Dennis and Schwartz, 1989). Although the second component of microflares at high energies could be fitted by the thermal emission of a hot kernel, it would not contain enough energy to allow for conductive heating of the first component. Therefore, conduction may explain the evolution of the first component, but cannot explain the relation between the first and second component.

We propose here that the flare duration at 3–9 keV (thermal component) is not controlled by radiation loss, but by thermal conduction distributing the energy to a larger mass reservoir at lower temperature.

## 4. Conclusions

The X-ray spectrum from 3–15 keV of solar microflares in active regions has been observed with high-energy resolution by the RHESSI satellite. The thermal component can be identified in all microflares and traced up to some 12 keV in the spectrogram. It can be well separated from a second component at higher energies, not visible in all events, that peaks simultaneously at all energies, has short duration and occurs during the rise phase of the first component. The second component can be traced in energy down to about 8 keV. The existence of two components excludes a previous interpretation (see Introduction) as one non-thermal electron population having a cut-off distribution in energy.

The similarities of microflares to regular flares suggest identical physical processes and similar plasma conditions. The properties of the thermal and non-thermal components are consistent with the standard flare scenario of precipitating electrons heating cold material to flare temperature. The collisional energy loss of beaming electrons causes bremsstrahlung emission observable in the 9–15 keV range and heats cold material to emit thermally at $<12$ keV.

High spectral resolution at the transition between thermal and non-thermal emission has been found to be useful for studying the role of energy deposition by beams and thermal conduction. No-shutter observations by RHESSI are well suited to further quantify these results. This will lead to firm estimates on the various forms of energies involved in microflares and determine their role in heating the active-region corona.

## Acknowledgements

RHESSI is a NASA Small Explorer Mission with an important hardware contribution by the Paul-Scherrer Institute (Villigen, Switzerland). The HESSI Experimental Data Center is supported by ETH Zürich. Many people have contributed to the successful launch and operation of RHESSI. We thank all of them, and in particular Pascal Saint-Hilaire for help and explanations, Hugh Hudson and Säm Krucker for helpful comments. The RHESSI work at ETH Zürich is partially supported by the Swiss National Science Foundation (grant No. 20-67995.02).

## References

Benz, A. O. and Krucker, S.: 2002, *Astrophys. J.* **568**, 413.

Brosius, J. W., Davila, J. M., Thomas, R. J., and Monsignori-Fossi, B. C.: 1996, *Astrophys. J.* **106**, 143.

Dennis, B. R. and Schwartz, R. A.: 1989, *Solar Phys.* **121**, 75.

Feldman, U., Doscheck, G. A., Behrung, W. E., Dennis, B. R., and Phillips, K. J. H.: 1996, *Astrophys. J.* **460**, 1034.

Hurford, G. J. *et al.*: 2002, *Solar Phys.*, this volume.
Krucker, S., Benz, A. O., Bastian, T. S., and Acton, L. W.: 1997, *Astrophys. J.* **488**, 499.
Krucker, S., Christe, S., Lin, R. P., Hurford, G. J., and Schwartz, R. A.: 2002, *Solar Phys.*, this volume.
Lin, R. P., Feffer, P. T., and Schwartz, R. A.: 2001, *Astrophys. J.* **557**, L125.
Lin, R. P., Schwartz, R. A., Kane, S. R., Pelling, R. M., and Hurley, K. C.: 1984, *Astrophys. J.* **283**, 421.
Lin, R. P. *et al.*: 2002, *Solar Phys.*, this volume.
Neupert, W. M.: 1968, *Astrophys. J.* **153**, L59.
Peres, G., Orlando, S., Reale, F., Rosner, R., and Hudson, H. S.: 2000, *Astrophys. J.* **528**, 537.
Reale, F., Peres, G., and Orlando, S.: 2001, *Astrophys. J.* **557**, 906.
Saint-Hilaire, P. and Benz, A. O.: 2002, *Solar Phys.*, this volume.
Shimizu, T.: 1995, *Publ. Astron. Soc. Japan* **47**, 251.
Simnett, G. M., Mouradian, Z., Martres, M.-J., and Soru-Escaut, I.: 1989, *Astron. Astrophys.* **224**, 284.
Smith, D. M. *et al.*: 2002, *Solar Phys.*, this volume.
Watanabe, T. *et al.*: 1995, *Solar Phys.* **157**, 169.

# HARD X-RAY MICROFLARES DOWN TO 3 keV

SÄM KRUCKER[1], STEVEN CHRISTE[1] , R. P. LIN[1,2], GORDON J. HURFORD[1] and RICHARD A. SCHWARTZ[3]

[1]*Space Sciences Lab, University of California, Berkeley, CA 94720-7450, U.S.A.*
[2]*Physics Department, University of California, Berkeley, CA 94720-7450, U.S.A.*
[3]*SSAI & Laboratory for Astronomy and Solar Physics, NASA Goddard Space Flight Center, Greenbelt, MD 20771, U.S.A.*

(Received 7 August 2002; accepted 17 September 2002)

**Abstract.** The excellent sensitivity, spectral and spatial resolution, and energy coverage down to 3 keV provided by the Reuven Ramaty High-Energy Solar Spectroscopic Imager mission (RHESSI) allows for the first time the detailed study of the locations and the spectra of solar microflares down to 3 keV. During a one-hour quiet interval (GOES soft X-ray level around B6) on 2 May, 1:40–2:40 UT, at least 7 microflares occurred with the largest peaking at A6 GOES level. The microflares are found to come from 4 different active regions including one behind the west limb. At $7''$ resolution, some events show elongated sources, while others are unresolved point sources. In the impulsive phase of the microflares, the spectra can generally be fitted better with a thermal model plus power law above $\sim 6-7$ keV than with a thermal only. The decay phase sometimes can be fitted with a thermal only, but in some events, power-law emission is detected late in the event indicating particle acceleration after the thermal peak of the event. The behind-the-limb microflare shows thermal emissions only, suggesting that the non-thermal power law emission originates lower, in footpoints that are occulted. The power-law fits extend to below 7 keV with exponents between $-5$ and $-8$, and imply a total non-thermal electron energy content between $10^{26}-10^{27}$ erg. Except for the fact that the power-law indices are steeper than what is generally found in regular flares, the investigated microflares show characteristics similar to large flares. Since the total energy in non-thermal electrons is very sensitive to the value of the power law and the energy cutoff, these observations will give us better estimates of the total energy input into the corona. (Note that color versions of figures are on the accompanying CD-ROM.)

## 1. Introduction

The Reuven Ramaty High-Energy Solar Spectroscopic Imager (RHESSI) instrument (see Lin *et al.*, 2002) provides uniquely high sensitivity in the energy range from $\sim 3$ to 15 keV, together with $\sim 1$ keV FWHM spectral resolution and simultaneously imaging down to 2.3 arc sec. This energy range covers emissions from hot, $\geq 5$ MK thermal plasmas and from accelerated, non-thermal low-energy electrons. Previous solar hard X-ray instruments, such as HXRBS flown on SMM and HXT on *Yohkoh*, have entrance windows that absorb emission below $\sim 15$–25 keV to avoid saturation from the intense thermal emissions in large flares (Kane and Anderson, 1970). For balloon-borne instruments (Lin *et al.*, 1984), the Earth's atmosphere absorbs solar hard X-rays below $\sim 15$ keV. The RHESSI instrument

*Solar Physics* **210:** 445–456, 2002.

accommodates medium and large flares by automatically inserting shutters in front of detectors to absorb low-energy photons and avoid saturation. Instruments designed to measure solar flare soft X-rays generally have very small effective areas above $\sim 3$ keV. The instrument most comparable to RHESSI in this energy range, the Hard X-ray Imaging Spectrometer (HXIS) on the *Solar Maximum Mission*, provides effective areas of $\sim 0.7$ mm$^2$ at $\sim 3.5$ keV rising to $\sim 7–26$ mm$^2$ at $\sim 10$ $-15$ keV (Van Beek *et al.*, 1980). For comparison RHESSI, with shutters out, has an effective area of $\sim 10$ mm$^2$ at $\sim 3.5$ keV, rising rapidly to $\sim 250$ mm$^2$ at 5 keV and $\sim 3500$ mm$^2$ at 10–15 keV; a factor of $\sim 14$ to $\geq 130$ times larger. Thus, the RHESSI observations from $\sim 3$ to 15 keV with the shutters out can provide new information on low-level energy releases, whether they result in heating of hot thermal plasmas or in the acceleration of low-energy electrons.

It is evident from previous observations that the Sun releases energy in transient outbursts, ranging from major flares ($\sim 10^{32}–10^{33}$ ergs) down to microflares and even nanoflares, with the frequency of the releases increasing as the energy released decreases (see discussion in Aschwanden *et al.*, 2000). For flares, the energy releases often appear to be dominated by accelerated $\sim 10$s of keV electrons and sometimes $\sim 1$ MeV nucl$^{-1}$ ions. Hard ($> 20$ keV) X-ray microflares, tiny bursts with $\sim 10^{27}$ to $10^{28}$ ergs in $> 20$-keV electrons, were discovered with a balloon-borne instrumentation to occur on average once every $\sim 6$ min near solar maximum (Lin *et al.*, 1984), leading to speculation that the energy released in accelerated electrons, summed over HXR bursts of all sizes, might contribute significantly to the heating of the active corona.

Lin, Feffer, and Schwartz (2001) investigated HXR microflares at lower energies using the BATSE SPEC detectors on the Compton Gamma-Ray Observatory, whose thresholds were occasionally set as low as 8 keV. They found that only one third of all non-thermal (hard spectra unlikely to be thermal) events detected above 8 keV are observed above $\sim 25$ keV. Additionally, the generally steep HXR spectra (power-law fits with exponent of 3–7) reveal that most of the flare energy is in the non-thermal electrons at lowest energy. Furthermore, the similarity of frequency vs. energy release distribution of these $> 8$ keV bursts to that of active-region transient brightenings (ARTBs) seen in soft X-rays by the *Yohkoh* SXT instrument (Shimizu, 1995) suggests that these accelerated electrons may provide the energy for ARTBs. Here, the first RHESSI spectra and images of microflares detected above 3 keV are presented and discussed.

## 2. Data Analysis

Standard RHESSI software (Hurford *et al.*, 2002) was used to analyze data. First, GOES and RHESSI data were searched for times with low solar activity when both shutters were out; the shutters (Lin *et al.*, 2002) suppress the response below 10 keV by a factor of 50 or more. Then, the microflares were identified and a

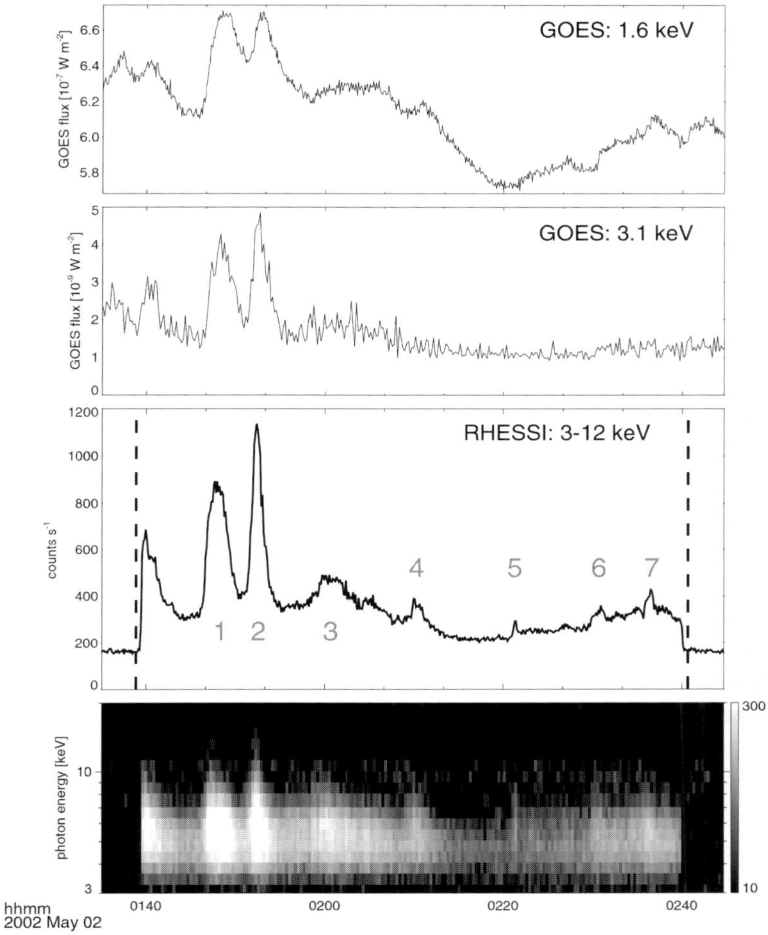


*Figure 1.* RHESSI and GOES X-ray observations during times of low solar activity. Around one hour of observations are shown corresponding to one RHESSI day. The *vertical dashed lines* give the sunrise and sunset, respectively. *From top to bottom*: softer and harder GOES channel plotted in a linear scale, the RHESSI X-ray total uncalibrated counts between 3 and 12 keV, and the RHESSI count spectrogram plot (background subtracted). The microflares studied in detail are marked with numbers (for color version of figures see the accompanying CD-ROM).

full disk image of each microflare was made with a resolution of $\sim 60''$ to obtain the centroid position. In a second step, cleaned images of each microflare around the flare position are made using higher resolution. Since grid 2 (4 arc sec) has a $\sim 20$ keV threshold, these images are reconstructed with $7''$ resolution (grid 3). Initial imaging with the finest resolution (2.3 arc sec) by grid 1 did not improve the image, suggesting no structures at the $2''$ level. However, more analysis is needed.

The spectra shown here are produced using the SPEX software package. Energy bins corresponding to the FWHM resolution of 1 keV are used. The relatively low number of counts and the absence of a wide range in energy where non-background counts are detected make fitting of the microflare spectra challenging. Microflares are typically seen from 3 keV up to 15 keV, giving 12 data points to fit. Additional complications are given by the iron line at 6.7 keV (cf., Tanaka *et al.*, 1984) that are ignored in this preliminary study. Furthermore, the fitted parameters depend on the background subtractions. Generally, the total pre-event background – instrumental, non-solar, and solar background – are subtracted before fitting the spectra. This is appropriate if the pre-event contribution of the flaring source can be neglected, but inappropriate if the flaring loop is already contributing significantly to the pre-event emission. One way to distinguish these two scenarios is to image the pre-event emission. If the pre-event emission is coming from a different location than the flare site, then the total pre-event background should be subtracted. If not, subtracting the background spectra observed during RHESSI night might give better results. The background subtractions used in this paper are still preliminary and more investigations are needed (see also discussion in Benz and Grigis 2002). The spectra shown in Figure 4 are obtained by subtracting the night time background for microflares 1 and 2 and the pre-event background for microflare 3.

Two models were fitted to the derived spectra and compared: (a) a thermal fit only, (b) a thermal fit plus a non-thermal power law. (For a discussion on a double thermal model see Benz and Grigis (2002, this issue.) The thermal fit contains only two parameters, temperature and emission measure; the thermal plus non-thermal fit has three more parameters: the power law exponent, the low energy cutoff below which the non-thermal spectrum is assumed flat, and the intensity. There are clear cases when a thermal alone does not fit the data at all, as often observed in the rise phase of an event; or when the thermal alone already fits the data well enough. The less clear cases are when the thermal fit gives only moderate agreement with the data and an additional non-thermal fit provides smaller residuals; but often the derived power law index is very steep with values around 10.

The derived power-law exponent is reasonably accurate, giving consistent values within $\pm 0.5$, even when using different values for the subtracted background. At lower energies where the thermal is fitted, the uncertainties are introduced mostly by the selection of the background as described above. At energies above $\sim 10$ keV, the component of the pre-event solar background is smaller than the non-solar background and the selection of the solar background is less critical. Therefore, the derived values for the temperature and emission measure from the thermal fit are more uncertain and still preliminary. The accuracy of the derived temperatures is estimated to $\pm 2$ MK from multiple fitting of the same events with different background subtractions and different initial guesses of parameters. Subtracting the solar pre-event background gives generally higher temperatures (up to $\sim 2$ MK) than subtracting the background observed during night. The accuracy of the logarithm of the emission measure is around $\pm 0.5$ and is inversely correlated to the error in temperature. For the largest microflares that produce a clear enhancement in both GOES channels, independent temperature estimates can be derived

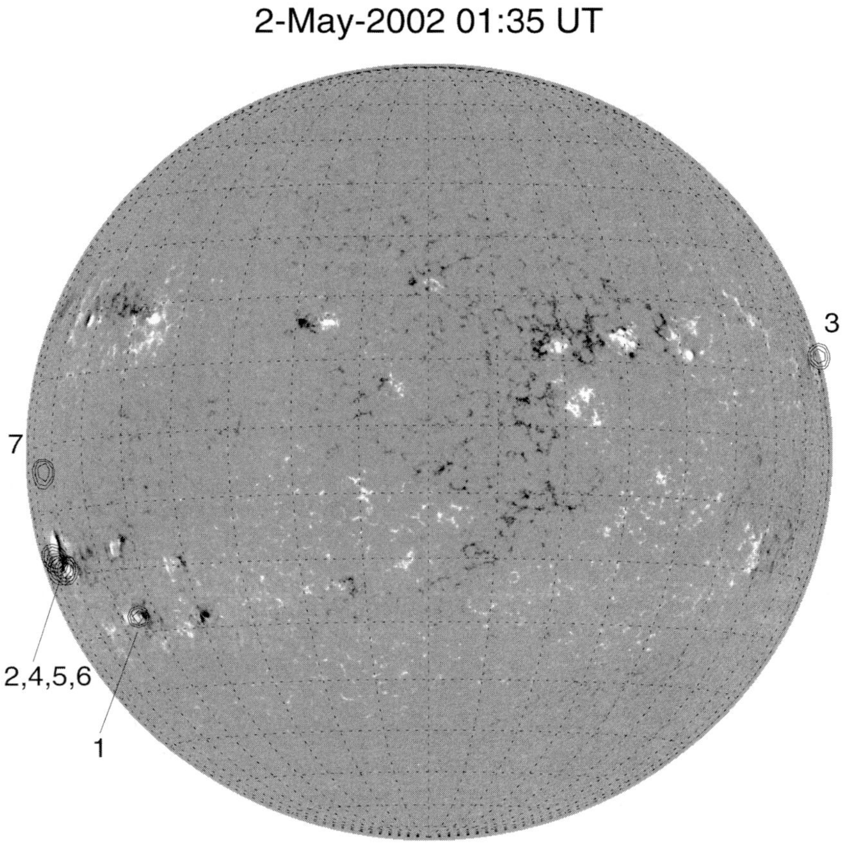


*Figure 2.* The locations of the 7 microflares marked in Figure 1 are shown on an MDI magnetogram.

using the ratio of the two GOES channels. The derived temperatures from GOES are generally smaller but roughly consistent within the error estimates given above.

## 3. Observations

One orbit of RHESSI data is shown here as an example of microflare activity seen down to 3 keV (Figure 1). The time period selected is 2 May 2002, 01:40–02:40 UT, with a GOES soft X-ray background level around B6. The same time interval is also discussed by Benz and Grigis (2002). During this period, at least 7 microflares could be easily identified (marked with numbers in Figure 1). A number of even smaller transient increases – e.g., at 02:04, 02:19, 02:27, 02:32, and 02:36 UT – appear to be significant in the 3–12 keV count rates. After subtracting the GOES background flux, the largest microflare has a peak GOES soft X-ray classification of A6, while the smallest of the 7 microflares is around A0.1,

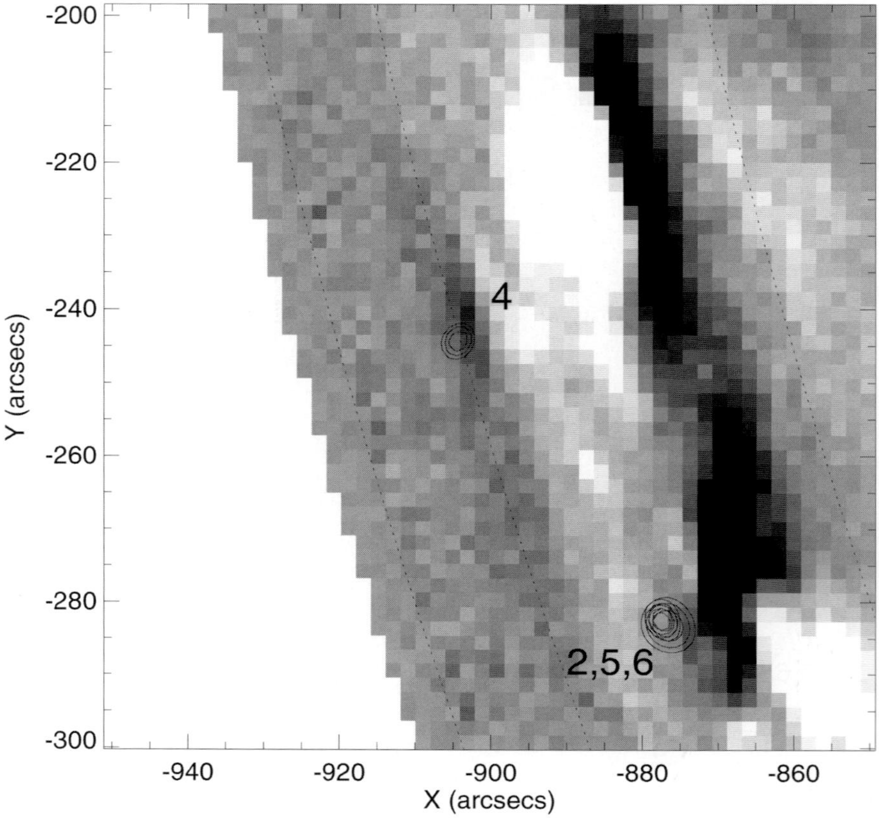


*Figure 3.* Zoom of Figure 2 showing the 4 microflares occurring in AR 9934 with 7″ resolution.

hardly detectable above the GOES background. Compared to the GOES survey of flare sizes reported by Feldman *et al.* (1996), the largest microflare presented here corresponds to the smallest flares presented in their study.

### 3.1. Microflare imaging

Figure 2 shows cleaned images of the microflares numbered in Figure 1 superimposed on a full-disk MDI magnetogram. All the HXR microflares analyzed so far are found to originate from active regions. The 7 microflares outlined in Figure 1 come from 4 different active regions: microflare 1 is from the southeast (NOAA AR 9932), microflares 2,4,5,6 from the nearby active region AR 9934, microflare 3 originates from just behind the limb (AR 9915), and microflare 7 occurs in a newly emerging active region near the east limb (AR 9938). Imaging at 7″ resolution show some microflares with elongated sources, while others are spatially not resolved. Figure 3 is an enlarged image of microflares 2, 4, 5, and 6, showing that microflare 4 occurs in a different part of the active region than 2, 5, and 6.

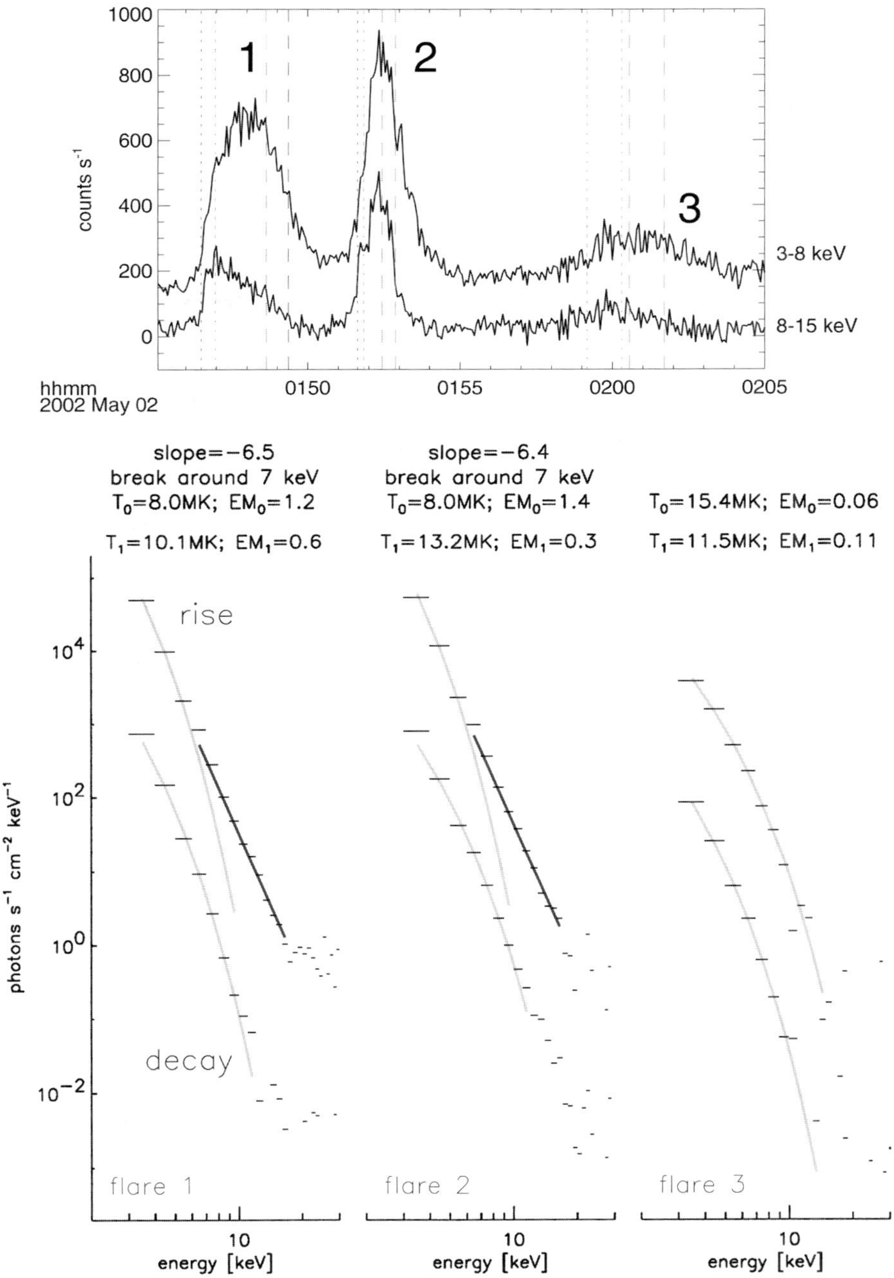


*Figure 4.* (a) *Top plot* shows the time series of the 3–8 keV and the 8–15 keV energy range that roughly correspond to the thermal and non-thermal energy regime. For a clearer representation, the 8–15 keV time profile is multiplied by a factor of 3. The *gray dotted and dashed lines* give the time intervals used for the fits shown below. (b) Spectra during the impulsive phase (shown shifted up by two decades) and the decay phase of the microflares labeled 1 to 3 in Figure 1. The impulsive phase is fitted with both a thermal and a non-thermal component, the decay phase with a thermal component only. For the behind-the-limb microflare (flare 3), a thermal alone fits the data well enough. The shown curves give the range fitted; values above $\sim 15$ keV are dominated by noise.

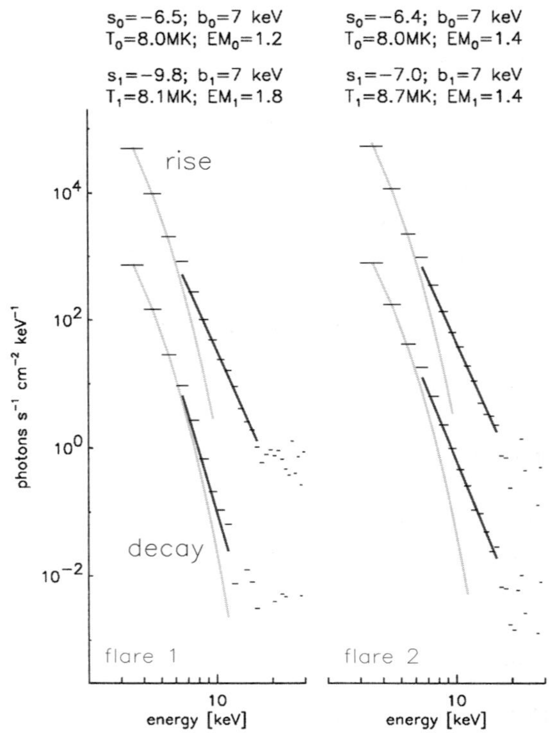


*Figure 5.* The spectra of microflares 1 and 2 are shown again, but the decay phase is fitted with both a thermal and a non-thermal component.

## 3.2. Microflare Energy Spectra

As a preliminary analysis, spectral fits are performed separately to the impulsive phase and the decay phase of the microflares. Examples of spectra of the first three microflares are shown in Figure 3. A thermal fit alone cannot represent the data of microflares 1 and 2 during the impulsive phase, but a thermal plus a non-thermal power-law produces a good fit. The non-thermal power-law fit reveals steep spectra for the 2 May, 01:40–02:40 UT, time interval with exponents ranging from $-5$ to $-8$. These are much steeper values than what is generally observed for normal sized flares where the indices are often between $-3$ and $-4$, and are somewhat steeper than the $> 8$ keV microflares (exponents of $-3$ to $-7$) reported by Lin, Feffer, and Schwartz (2001). The power laws extend to low energies and the best fit gives low energy cutoffs of around 7 keV for the two biggest microflares with best statistics. The smaller microflare 3 that occurs just behind the limb does not show a clear non-thermal component in the spectra and is best fitted with a thermal spectra only, suggesting that the non-thermal emission is occulted. The fitted temperatures of microflare 3 are rather high compared to the other 6 microflares. Since hotter flares are generally larger flares (e.g., Feldman *et al.*, 1996), this suggests that most of the microflare emission is occulted and that this microflare might be much larger than the A0.1 level observed by GOES.

The spectra of the decay phases do not show a picture as clear as the impulsive phase. Some events like microflare 1 are moderately well fitted with a thermal emission only, others like microflare 2 are better represented with a thermal plus a non-thermal fit (Figure 5). The thermal fit of microflare 2 alone does not represent the decay phase at 3 keV and above 10 keV (cf., Figure 4). This difference in the spectrum for these two microflares can also be seen in the uncalibrated time profiles above and below 8 keV (Figure 4, top, for calibrated time profiles see Benz and Grigis, 2002): Microflare 1 shows a typical impulsive onset above 8 keV and a more gradual onset at lower energies, reminiscent of the Neupert effect (Neupert, 1968), but for microflare 2, the time profile looks very similar in both energy ranges. Still, the emission below 8 keV has a slightly longer decay phase than above 8 keV. For the decay of microflare 1, the thermal plus non-thermal fit gives a very steep (and therefore questionable) power-law index close to $-10$, however, it does suggest some non-thermal HXR emission. The thermal fit alone gives a temperature around $\sim 8$ MK for microflare 1 (Figure 5, left), but does not fit the emission at 3 keV or above 10 keV well.

### 3.3. Microflare seen in RHESSI and TRACE

To place these HXR microflares in the context of lower temperature flare emission, a microflare occurring on 6 May 2002, 09:01 UT (Figure 6), which is comparable to the largest flare occurring in the 2 May time interval (GOES A8 level), is presented here. The RHESSI time profiles above and below 8 keV show two peaks about 2 minutes apart with the first peak harder than the second one. No TRACE overlap is available for the 2 May time interval except for the behind limb flare that did not show any enhanced emission in EUV (195 Å).

The 6 May microflare is compared with TRACE images in four different phases: (1) pre-flare phase, (2) rise to first peak, (3) main (second) peak, and (4) late in the decay phase. In EUV (195 Å) this event shows two bright spots flaring that are already visible in the pre-flare image. These 195 Å sources show a very similar time profile as in X-rays; with two peaks as well. The X-ray emission below 8 keV occurs between the two EUV spots, suggesting a hot loop in between. Above 8 keV, the X-ray image looks different with a strong source near the western EUV spot and a possible weaker source possibly related to the eastern EUV spot. The image above 8 keV is reconstructed with relatively low statistics and therefore is of lower quality. Still, the rise phase of this microflare can be interpreted as a single loop brightening with footpoints seen above 8 keV and the loop top seen at lower energies.

During the main peak of the event, the emission below and above 8 keV comes only from the loop. About 8 minutes after the rise phase, a very fine and slightly curved loop is seen with TRACE connecting the earlier flaring footpoints. The X-

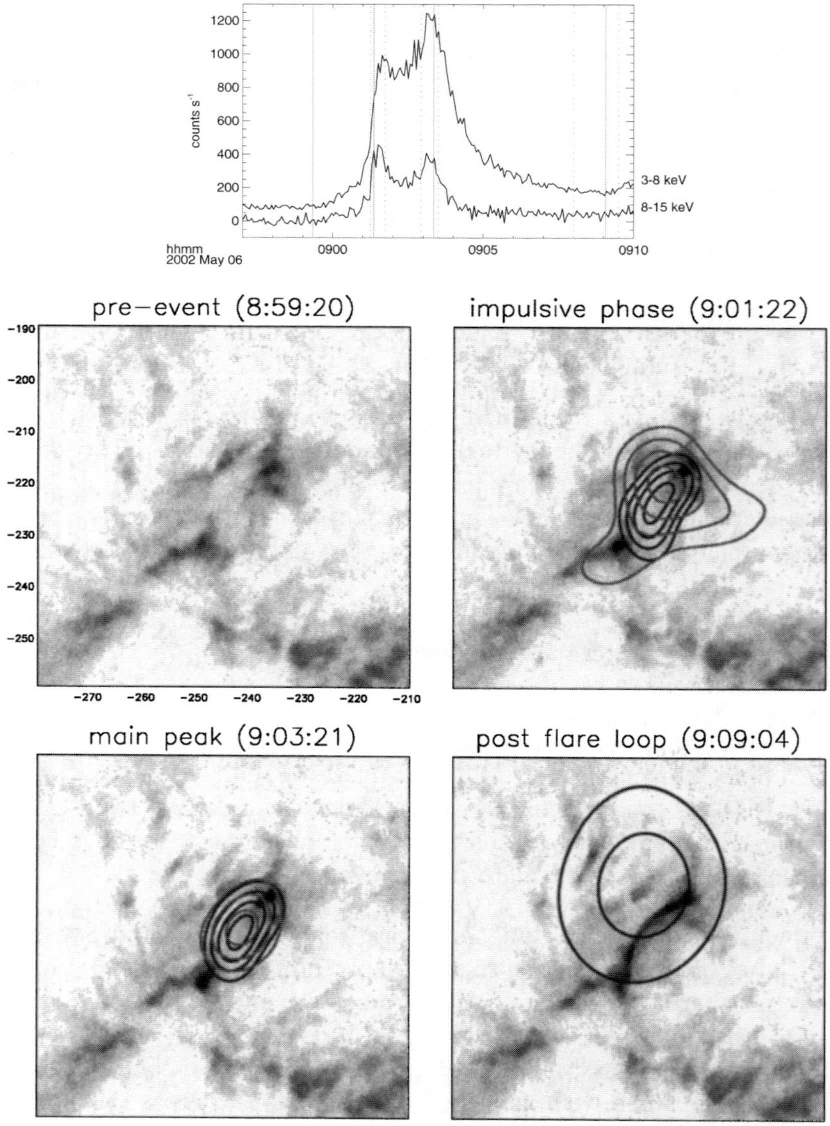


*Figure 6.* (a) *Top panel*: RHESSI time profiles below and above 8 keV are shown, with the integration times of the RHESSI images given by *dashed gray lines*; the *solid lines* give the time of the TRACE images. The time series above 8 keV is multiplied by a factor of 4 for a clearer representation. (b) *Lower panels* show RHESSI and TRACE observations of the microflare (GOES A8) occurring on 6 May, 09:01 UT. The images show TRACE 195 Å observations scaled to the maximum emission during the flare, superimposed are RHESSI X-ray contours (3–8 keV in *black*, 8–15 keV in *gray*; *white and red* in the color version on the CD-ROM). *Top left*: pre-event image. *Top right*: initial impulsive phase, note the different location of the 3–8 keV and 8–15 keV emission. *Bottom left*: image during main peak of the X-ray emission. *Bottom right*: later, a loop is seen with TRACE connecting the earlier flaring footpoints. The contours are 30, 50,70 90% except for the last time interval where only the 70 and 90% levels are shown.

ray emission seen with RHESSI at that time is extremely faint with the total flux in the source more than 100 times below the peak value. Using grids 7 (60 arc sec) and higher, the imaging shows a source slightly above the loop seen with TRACE. This is similar to what is observed with RHESSI in large flares (Gallagher *et al.*, 2002). We note here that the same region flares again just minutes after the microflare discussed here. In addition to the two flaring footpoints seen in EUV, much fainter brightenings occurring in the same active region are also observed, but their connection to the microflare is unclear and probably energetically unimportant.

## 4. Discussion and Conclusions

The microflares presented in this work can be interpreted as small flares: they occur in active regions, show non-thermal energy input as well as thermal heating with temperature and emission measure roughly following the relation found in flares (Feldman *et al.*, 1996). The main difference is that observed slopes of the non-thermal power-law spectra are steep with values between $-5$ to $-8$ whereas the spectra in normal flares are generally harder. These results are robust with respect to background subtraction while the thermal fit is much more sensitive and the given temperatures and emission measures are therefore preliminary. The non-thermal emission is observed to extend down to $\sim 7$ keV, in some microflares possibly below. Recent RHESSI observations of large flares (Gallagher *et al.*, 2002; Krucker and Lin, 2002) report that non-thermal emission is seen down to about 10 keV; below that the thermal emission dominates. Since the energy spectra are steep, the total energy in non-thermal electrons required to produce the observed emission is strongly dependent on the lower energy cutoff $E_0$ of the non-thermal spectrum and the power-law exponent, $\gamma$. The energy loss rate $P$ of electrons above $E_0$ is given by Brown (1971) and Lin (1974; see Lin, Feffer, and Schwartz, 2001, for corrected factors):

$$P(> E_0) \sim E_0^{-(\gamma-1)}. \tag{1}$$

Hence, the knowledge of $E_0$ is crucial. In earlier work with observations above 20–25 keV, the cutoff energy was often set to 25 keV (e.g., Crosby, Aschwanden, and Dennis, 1993). For regular flares, the use of 25 keV instead of 10 keV introduces a factor of $\sim 10$. For the microflares present in this work, the factor is $\sim 500$, since the spectra are steeper and $E_0$ smaller. Hence, the correction for smaller events seems to be larger. This could possibly change the flare frequency distribution published by Crosby, Aschwanden, and Dennis (1993) and lead to a re-evaluation of microflare contribution to coronal heating.

To obtain the total energy in energetic electrons, the non-thermal fit of the spectra was averaged over the rise phase of the event. Then Equation (1) is used to get the energy loss rate. By multiplying by the duration of the rise phase, a rough approximation of the total energy in energetic electrons can be derived. This does not include the non-thermal emission seen in the decay phase of some events. The derived total energies for the microflares discussed are between $10^{26}$ to $10^{27}$ erg.

The (instantaneous) total thermal energy content can be derived from the thermal fit using the emission measure and temperature together with a estimate of the source size from the RHESSI images. As a first approximation, only an upper limit of the source size is used taken from the images with $7''$ resolution. These estimates of the thermal energy content are extremely rough, but give energies of several times $10^{27}$ ergs, comparable to the derived total non-thermal electron energy. Of course, to obtain the total energy input to the thermal plasma requires an estimate of the energy losses as well.

Future RHESSI microflare observations will lead to more confidence in current fitting models and appropriate background subtraction. More accurate flare size estimates (see Schmahl and Hurford, 2002) will yield improved thermal energy estimates and a better understanding of low-count imaging will improve RHESSI's resolution for small flares.

## Acknowledgements

This research is supported by NASA contract NAS 5-98033. We acknowledge the RHESSI team's outstanding effort in making this mission a success under adverse conditions.

## References

Aschwanden, M. J., Tarbell, T. D., Nightingale, R. W., Schrijver, C. J., Title, A., Kankelborg, C. C., Martens, P., and Warren, H. P.: 2000, *Astrophys. J.* **535**, 1047.

Benz, A. O. and Grigis, P. C.: 2002, *Solar Phys.*, this volume.

Crosby, N. B., Aschwanden, M. J., and Dennis, B. R.: 1993, *Solar Phys.* **143**, 275.

Feldman, U., Doschek, G. A., Behring, W. E., and Phillips, K. J. H.: 1996, *Astrophys. J.* **460**, 1034.

Gallagher, P., Dennis, B. R., Krucker, S., Schwartz, R. A., and Tolbert, A. K.: 2002, *Solar Phys.*, this volume.

Hurford, G. J. *et al.*: 2002, *Solar Phys.*, this volume.

Kane, S. R. and Anderson, K. A.: 1970; *Astrophys. J.* **162**, 1003.

Krucker, S. and Lin, R. P.: 2002, *Solar Phys.*, this volume.

Lin, R. P., Schwartz, R. A., Kane, S. R., Pelling, R. M., and Hurley, K. C.: 1984, *Astrophys. J.* **283**, 421.

Lin, R. P., Feffer, P. T., and Schwartz, R. A.: 2001, *Astrophys. J.* **557**, L125.

Lin *et al.*: 2002, *Solar Phys.*, this volume.

Neupert, W. M.: 1968, *Astrophys. J.* **153**, L59.

Schmahl, E. J. and Hurford, G. J.: 2002, *Solar Phys.*, this volume.

Shimizu, T.: 1995, *Publ. Astron. Soc. Jap.* **47**, 251.

Tanaka, K., Watanabe, T., and Nitta, N.: 1984, *Astrophys. J.* **282**, 793.

van Beek, H. F., Hoyng, P., Lafleur, B., and Simnett, G. M.: 1980, *Solar Phys.* **65**, 39.